# Strain and Its Implications in Organic Chemistry
Organic Stress and Reactivity

# NATO ASI Series

## Advanced Science Institutes Series

*A Series presenting the results of activities sponsored by the NATO Science Committee, which aims at the dissemination of advanced scientific and technological knowledge, with a view to strengthening links between scientific communities.*

The Series is published by an international board of publishers in conjunction with the NATO Scientific Affairs Division

| | |
|---|---|
| **A Life Sciences**<br>**B Physics** | Plenum Publishing Corporation<br>London and New York |
| **C Mathematical**<br>  **and Physical Sciences**<br>**D Behavioural and Social Sciences**<br>**E Applied Sciences** | Kluwer Academic Publishers<br>Dordrecht, Boston and London |
| **F Computer and Systems Sciences**<br>**G Ecological Sciences**<br>**H Cell Biology** | Springer-Verlag<br>Berlin, Heidelberg, New York, London,<br>Paris and Tokyo |

Series C: Mathematical and Physical Sciences - Vol. 273

# Strain and Its Implications in Organic Chemistry

Organic Stress and Reactivity

edited by

## Armin de Meijere
Institut für Organische Chemie,
Universität Hamburg, Hamburg, F.R.G.

and

## Siegfried Blechert
Institut für Organische Chemie,
Universität Bonn, Bonn, F.R.G.

**Kluwer Academic Publishers**

Dordrecht / Boston / London

Published in cooperation with NATO Scientific Affairs Division

Proceedings of the NATO Advanced Research Workshop on
Strain and Its Implications in Organic Chemistry
Organic Stress and Reactivity
Ratzeburg, F.R.G.
28 August – 2 September 1988

**Library of Congress Cataloging in Publication Data**

```
Strain and its implications in organic chemistry : organic stress and
   reactivity / edited by Armin de Meijere and Siegfried Blechert.
        p.    cm. -- (NATO ASI series. Series C, Mathematical and
   physical sciences ; vol. 273)
     "Published in cooperation with NATO Scientific Affairs Division."
     Includes index.
     ISBN 0-7923-0176-5
     1. Chemical bonds--Congresses.  2. Chemistry, Organic--Synthesis-
   -Congresses.  3. Organic compounds--Congresses.  I. Meijere, A. de.
   II. Blechert, Siegfried, 1946-    . III. North Atlantic Treaty
   Organization. Scientific Affairs Division. IV. Series: NATO ASI
   series. Series C, Mathematical and physical sciences ; no. 273.
   QD461.S875 1989
   547.1'39--dc20                                            89-31035
```

ISBN 0–7923–0176–5

---

Published by Kluwer Academic Publishers,
P.O. Box 17, 3300 AA Dordrecht, The Netherlands.

Kluwer Academic Publishers incorporates the publishing programmes of
D. Reidel, Martinus Nijhoff, Dr W. Junk and MTP Press.

Sold and distributed in the U.S.A. and Canada
by Kluwer Academic Publishers,
101 Philip Drive, Norwell, MA 02061, U.S.A.

In all other countries, sold and distributed
by Kluwer Academic Publishers Group,
P.O. Box 322, 3300 AH Dordrecht, The Netherlands.

*printed on acid free paper*

---

All Rights Reserved
© 1989 by Kluwer Academic Publishers
No part of the material protected by this copyright notice may be reproduced or
utilized in any form or by any means, electronic or mechanical, including photo-
copying, recording or by any information storage and retrieval system, without written
permission from the copyright owner.

Printed in The Netherlands

In Memoriam Hans Musso

# TABLE OF CONTENTS

**Foreword**    XIII

**List of Participants**    XV

**B. M. Trost**
Some Synthetic Implications of the Concept of Cyclopropyl Units as Pseudo-Functional Groups    1

**K. N. Houk**
Stereoselective Electrocyclizations and Sigmatropic Shifts of Strained Rings : Torquoelectronics !    25

**E. Piers**
The Preparation and Thermal Rearrangement of Functionalized 6-(1-Alkenyl)bicyclo[3.1.0]hex-2-enes. Applications to Synthesis.    39

**H. U. Reissig\*, R. Zschiesche, A. Wienand, M. Buchert**
Synthesis with Donor-Acceptor-Substituted Vinylcyclopropanes    51

**W. Ando\*, N. Tokitoh**
Synthesis of Functionalized Episulfides    59

**L. A. Paquette\*, J. Dressel**
Through-Bond Interaction via Cyclobutane Relay Orbitals as a Means of Extending Conjugation    77

**T. Hudlicky\*, A. Fleming, T. Lovelace, G. Seoane, K. Gadamasetti, G. Sinai-Zingde**
[2+3]-Carbo- and Heterocyclic Annulation in the Design of Complex Molecules    109

**M. S. Baird\*, J. R. Al-Dulayymi, H. H. Hussain**
Synthetic and Mechanistic Aspects of the Cyclopropene to Vinylcarbene Rearrangement    117

**M. Christl\*, M. Braun**
Generation and Interception of 1-Oxa-2,3-cyclohexadiene and 1,2,4-Cyclohexatriene    121

**K. J. Shea**
On the Relationship Between Strain and Chemical Reactivity of Torsionally Distorted Carbon-Carbon Double Bonds    133

**P. G. Gassman**
Photoinduced Single Electron Transfer from Strained Rings    143

**S. Nishida**
Chemical Behaviour of Cation Radicals Derived from Strained Molecules in Solution. Spiro-Activation in the Cleavage of Some Arylcyclopropanes upon Treatment with TCNE or DDQ   169

**C. J. Suckling**
The Cyclopropyl Group in Studies of Enzyme-Catalysed Reactions   177

**B. Ernst*, A. de Mesmaeker, H. Greuter, S. J. Veenstra**
Intramolecular [2+2]-Cycloadditions of Ketenes and Olefins   207

**L. Ghosez*, C. Lian Yong, B. Gobeaux, C. Houge, I. Marko, M. Perry, H. Saimoto**
Strained Iminium Salts in Synthesis   235

**R. D. Miller*, W. Theis, G. Heilig, S. Kirchmeyer**
The Generation and Rearrangement of 2-Diazocarbonyl Cyclobutanones: The Formation of 5-Spirocyclopropyl-2(5H)furanones   255

**E. Osawa*, J. M. Rudzinski, D. A. Barbiric, E. D. Jemmis**
Computational Studies of Prismane, Helvetane/Israelane and Asterane   259

**Z. Goldschmidt*, H. E. Gottlieb**
Cyclobutane-Cyclopentane Interconversion of Coordinated Adducts of Cycloheptatriene and Tetracyanoethylene via the [2,2]-Sigmahaptotropic Rearrangement   263

**G. Mehta**
Quest for Higher Prismanes   269

**R. Keese*, W. Luef, L. Mani, S. Schüttel**
Planarizing Distortions in Polycyclic Carbon Compounds   283

**H. Hopf*, C. Marquard**
Strain Release in Aromatic Molecules: The [2n]-Cyclophanes   297

**A. Krief*, P. Surleraux, W. Dumont, P. Pasau, Ph. Lecomte, Ph. Barberaux**
Trip around the Three-Membered Cycles Syntheses   333

**K. B. Wiberg*, D. R. Artis**
Experimental and Theoretical Studies of Bridged Cyclopropenes   349

**G. Szeimies**
From Bicyclo[1.1.0]butanes to [n.1.1]Propellanes   361

**Z. Yoshida**
New Aspects of Highly Strained Ring Chemistry   383

**B. Zwanenburg*, A. J. H. Klunder**
Strained Cage Systems : Synthetic and Structural Implications   405

L. Fitjer*, A. Kanschik, M. Majewski
Studies on the Synthesis of Cyclopentanoic Sesquiterpenes via Rearrangement
Routes: (±)Modhephene and (±)Isocumene       431

C. J. M. Stirling
Directionality in Formation of Small Rings by Intramolecular Nucleophilic Substitution       439

C. Almansa, M. L. Garcia, C. Jaime, A. Moyano, M. A. Pericás, F. Serratosa*
Synthesis, Chiroptical Properties and Synthetic Applications of Perhydrotriquinacene-1,4,7-trione       447

W. E. Billups
Reactions of Strained Carbon-Carbon Bonds with Metal Atoms       451

Y. Apeloig*, M. Karni, D. Arad
Cyclopropabenzenes and Alkylidenecyclopropabenzenes. A Synergistic Relation between Theory and Experiment       457

J. Michl*, P. Kaszynski, A. C. Friedli, G. S. Murthy, H.-C. Yang, R. E. Robinson, N. D. McMurdie, T. Kim
Harnessing Strain: From [1.1.1]Propellanes to Tinkertoys       463

W. E. Billups*, M. M. Haley, G.-A. Lee
Synthesis of Cyclopropenes       483

R. Boese*, D. Bläser
How to Get Structures of Strained Compounds: Low Temperature Structures and X-X Electron Deformation Densities       485

S. Braverman*, Y. Duar
Novel Intramolecular 1,1-Cycloadditions of Diazoallenes       489

A. Fadel, B. Karkour, J. Ollivier, J. Salaün*
Efficient Route to Optically Active Strained Compounds       493

W. Grimme*, W. Neukam
Anionic (3+2)-Cycloreversion of Strained Cage Nitriles       495

A. Krebs*, A. Jacobsen-Bauer, E. Haupt, M. Veith, V. Huch
Reaction of an Angle Strained Cycloheptyne with a Stabilized Stannylene Synthesis and X-Ray Structure of a Distannacyclobutene System       497

O. M. Nefedov*, S. P. Kolesnikov, M. P. Egorov, A. Krebs, Yu. T. Struchkov
Structure and Reactivity of Small Heterocycles Containing Germanium Atoms       499

H. Bothe, A.-D. Schlüter*
Relief of Strain Energy as Driving Force in Ring-Opening Polymerizations of a [1.1.1]Propellane       503

**T. Thiemann, S. Kohlstruk, G. Schwär, A. de Meijere***
Diels-Alder Reactions of Siloxyallylidenecyclopropanes: Facile Syntheses of Spiro-[2.5]octan-3-ones                                                                                                 507

**N. Krass, L. Wessjohann, D. Yu, A. de Meijere***
Cyclopropyl Group Containing Amino Acids from α-Chlorocyclopropylidenacetates                                                                                                                    509

**S. Wyn Roberts, C. J. M. Stirling***
Eliminative Fission of Strained Cycloalkoxides - The Question of Electrophilic Catalysis                                                                                                           513

**P. A. Krackman, W. H. de Wolf*, F. Bickelhaupt**
Special Reactivity of 8,11-Dihalo[5]metacyclophane                                                                                                                                                  515

**G. B. M. Kostermans, W. H. de Wolf*, F. Bickelhaupt**
Small [n]Cyclophanes. Where is the Limit?                                                                                                                                                           517

**Index**                                                                                                                                                                                           519

# FORMAT OF THE CONFERENCE AND CORRESPONDING CONTRIBUTIONS TO THIS VOLUME

**Invited lectures**

| | |
|---|---|
| B. M. Trost | 1 |
| K. N. Houk | 25 |
| E. Piers | 39 |
| H. U. Reissig*, R. Zschiesche, A. Wienand, M. Buchert | 51 |
| W. Ando*, N. Tokitoh | 59 |
| L. A. Paquette*, J. Dressel | 77 |
| P. G. Gassman | 143 |
| S. Nishida | 169 |
| C. J. Suckling | 177 |
| B. Ernst*, A. de Mesmaeker, H. Greuter, S. J. Veenstra | 207 |
| L. Ghosez*, C. Lian Yong, B. Gobeaux, C. Houge, I. Marko, M. Perry, H. Saimoto | 235 |
| G. Mehta | 269 |
| R. Keese*, W. Luef, L. Mani, S. Schüttel | 283 |
| H. Hopf*, C. Marquard | 297 |
| A. Krief*, P. Surleraux, W. Dumont, P. Pasau, Ph. Lecomte, Ph. Barberaux | 333 |
| K. B. Wiberg*, D. R. Artis | 349 |
| G. Szeimies | 361 |
| Z. Yoshida | 383 |
| B. Zwanenburg*, A. J. H. Klunder | 405 |
| J. Michl*, P. Kaszynski, A. C. Friedli, G. S. Murthy, H.-C. Yang, R. E. Robinson, N. D. McMurdie, T. Kim | 463 |

**Contributed papers**

| | |
|---|---|
| T. Hudlicky*, A. Fleming, T. Lovelace, G. Seoane, K. Gadamasetti, G. Sinai-Zingde | 109 |
| M. S. Baird*, J. R. Al-Dulayymi, H. H. Hussain | 117 |
| M. Christl*, M. Braun | 121 |
| K. J. Shea | 133 |
| R. D. Miller*, W. Theis, G. Heilig, S. Kirchmeyer | 255 |
| E. Osawa*, J. M. Rudzinski, D. A. Barbiric, E. D. Jemmis | 259 |
| Z. Goldschmidt*, H. E. Gottlieb | 263 |
| L. Fitjer*, A. Kanschik, M. Majewski | 431 |
| C. J. M. Stirling | 439 |
| C. Almansa, M. L. Garcia, C. Jaime, A. Moyano, M. A. Pericás, F. Serratosa* | 447 |
| W. E. Billups | 451 |
| Y. Apeloig*, M. Karni, D. Arad | 457 |

**Posters**

| | |
|---|---|
| W. E. Billups*, M. M. Haley, G.-A. Lee | 483 |
| R. Boese*, D. Bläser | 485 |
| S. Braverman*, Y. Duar | 489 |
| A. Fadel, B. Karkour, J. Ollivier, J. Salaün* | 493 |
| W. Grimme*, W. Neukam | 495 |
| A. Krebs*, A. Jacobsen-Bauer, E. Haupt, M. Veith, V. Huch | 497 |
| O. M. Nefedov*, S. P. Kolesnikov, M. P. Egorov, A. Krebs, Yu. T. Struchkov | 499 |
| H. Bothe, A.-D. Schlüter* | 503 |
| T. Thiemann, S. Kohlstruk, G. Schwär, A. de Meijere* | 507 |

N. Krass, L. Wessjohann, D. Yu, A. de Meijere*  509
S. Wyn Roberts, C. J. M. Stirling*  513
P. A. Krackman, W. H. de Wolf*, F. Bickelhaupt  515
G. B. M. Kostermans, W. H. de Wolf*, F. Bickelhaupt  517

# FOREWORD

The topic "Stress and Strain" of this conference was ideally constrasted by the remoteness and quiet atmosphere of the meeting place Hotel Seehof in Ratzeburg, a small medieval town situated on a peninsula in lake "Küchensee" east of Hamburg in northern Germany. With the participation of 53 leading experts from all over the world, the workshop covered the widest possible range from the advancement of bonding theory, new mechanistic insights into chemical transformations and physical properties of highly strained compounds to their use as building blocks in organic synthesis and even as probes into the detection of enzyme mechanisms. Because of their specific reactivities small ring units can uniquely play their role in the construction of composite functionalities. Such functionalities can increase the elegance in natural and non-natural products syntheses, since they help to develop more convergent synthetic routes and improve the necessary chemo-, regio- and stereo-selectivity.

This book presents all of the 20 invited lectures and is complemented with short versions of 12 contributed papers and 13 poster presentations. I am convinced that it will stimulate further rapid development of this field of organic chemistry, which recently has seen extensions into the bioorganic area as well as towards new materials. In fact, several "supra-natural" - at first sight exotic - compounds are already available in useful quantities and are being exploited to create vastly new molecular devices, i.e. compounds with unprecedented molecular functions and polymers with unconventional properties. May this Proceedings volume have the ultimate impact on the chemical community around the world and stimulate more fascinating ideas in this and other advancing areas of chemistry.

This meeting was only made possible through a grant from the NATO Scientific Affairs Division and generous financial contributions by the following companies:

> AKZO (Arnheim, NL), BASF (Ludwigshafen), BAT (Hamburg), BAYER (Leverkusen), CIBA-GEIGY (Basel, Switzerland), DEGUSSA (Frankfurt), TH. GOLDSCHMIDT (Essen), HENKEL (Düsseldorf), HOECHST (Frankfurt), HÜLS (Marl), KALI-CHEMIE (Hannover), E. MERCK (Darmstadt), SCHERING (Berlin), THOMAE (Biberach).

I gratefully acknowledge the help of the Organizing Committee and International Advisory Board, Professors Siegfried Blechert, Paul Gassmann, Hans Musso, Jaques Salaün, Lars Skatteböl. I am indebted to a large number of my co-workers at the University of Hamburg, especially Holger Gebauer, Rolf Lackmann, Christian Militzer, Oliver Reiser, Heide von Rekowski, Ludger Wessjohann and Dahai Yu, who have actually done all the time-consuming little things.

Without the initiation of the late Professor Hans Musso, however, this meeting would not have taken place at all. Consequently special credit goes to him and it is only appropriate to dedicate this book *"In Memoriam Hans Musso"*.

Armin de Meijere

**List of Participants**

Prof. W. Ando
Dept. of Chemistry
University of Tsukuba
Sakura-mura
Ibaraki 305
Japan

Prof. Y. Apeloig
Dept. of Chemistry
Technion-Israel Institute of Technology
Haifa 32000
Israel

Dr. M. Baird
Dept. of Chemistry
The University
Newcastle upon Tyne
Great Britain

Prof. M. Balci
Dept. of Chemistry
Atatürk Üniversitesi
Fen-edeb.Fak., Kiya Bölümü
Erzurum
Turkey

Prof. M.T. Barros da Silva
Universidade Nova de Lisboa
Faculdade de Ciencias e Technologia
Quinta da Torre
2825 Monte da Caparica
Portugal

Dr. D. Bellus
CIBA-GEIGY AG
Division AGRO
Pflanzenschutz R-1004.6.36
CH-4002 Basel
Switzerland

Prof. E. Billups
Department of Chemistry
Rice University
P.O.Box 1892
Houston, Texas 77251
USA

Prof. S. Blechert
Inst. f. Org. Chemie
Universität Bonn
Gerhard-Domagk-Str. 1
5300 Bonn 1
Federal Republic of Germany

Dr. R. Boese
Inst. f. Org. Chemie
Universität Essen
Universitätsstr. 5
4300 Essen
Federal Republic of Germany

Dr. K. Bott
BASF Aktiengesellschaft
ZHR / Organische Reaktionen B9
6700 Ludwigshafen
Federal Republic of Germany

Prof. S. Braverman
Department of Chemistry
Bar-Ilan University
Ramat-Gan, 52100
Israel

Prof. M. Christl
Inst. f. Org. Chemie
Universität Würzburg
Am Hubland
8700 Würzburg
Federal Republic of Germany

Dr. B. Ernst
CIBA-GEIGY AG
Zentrale Forschungslaboratorien
CH-4002 Basel
Switzerland

Prof. L. Fitjer
Inst. f. Org. Chemie
Universität Göttingen
Tammannstr. 2
3400 Göttingen
Federal Republic of Germany

Prof. P. Gassman
Department of Chemistry
University of Minnesota
207 Pleasant Street S.E.
Minneapolis, Minnesota 55455
USA

Prof. L. Ghosez
Laboratoire de Chimie
Université Cath. de Louvain
Place Louis Pasteur 1
B-1348 Louvain-La-Neuve
Belgium

Prof. Z. Goldschmidt
Department of Chemistry
Bar-Ilan University
Ramat-Gan, 52100
Israel

Prof. W. Grimme
Inst. f. Org. Chemie
Universität Köln
Greinstr. 4
5000 Köln 41
Federal Republic of Germany

Prof. H.M.R. Hoffmann
Inst. f. Org. Chemie
Universität Hannover
Schneiderberg 1b
3000 Hannover 1
Federal Republic of Germany

Prof. H. Hopf
Inst. f. Org. Chemie
Techn. Univ. Braunschweig
Hagenring 30
3300 Braunschweig
Federal Republic of Germany

Prof. K. Houk
Department of Chemistry
Univ. of Calif. at Los Angeles
Los Angeles, CA 90024
USA

Prof. T. Hudlicky
Chemistry Department
Virginia Polytechnic
Institute and State University
Blacksburg, Virginia 24061
USA

Dr. H. Kaulen
p/A Bayer AG
Zentralbereich Forschung und Entwicklung
5090 Leverkusen, Bayerwerk
Federal Republic of Germany

Prof. R. Keese
Inst. f. Org. Chemie
Universität Bern
Freie Str. 3
CH-3012 Bern
Switzerland

Prof. A. Krebs
Inst. f. Org. Chemie
Universität Hamburg
Martin-Luther-King-Platz 6
2000 Hamburg 13
Federal Republic of Germany

Prof. A. Krief
Lab. de Chimie Organique
Univ. Notre-Dame de la Paix
Rue de Bruxelles 61
B-5000 Namur
Belgium

Prof. H. Mayr
Inst. f. Org. Chemie
Mediz. Universität zu Lübeck
Ratzeburger Allee 160
2400 Lübeck
Federal Republic of Germany

Prof. G. Mehta
School of Chemistry
University of Hyderabad
Hyderabad 500 134
India

Prof. A. de Meijere
Inst. f. Org. Chemie
Universität Hamburg
Martin-Luther-King-Platz 6
2000 Hamburg 13
Federal Republic of Germany

Prof. J. Michl
Department of Chemistry
University of Texas at Austin
Austin, Texas 78712
USA

Prof. R.D. Miller
IBM Research K91/282
5600 Cottle Road
San Jose, CA 95193
USA

Prof. O. Nefedov
N.D. Zelinskii Institute
Acad. of Sciences of the USSR
Moscow
UdSSR

Prof. S. Nishida
Department of Chemistry
Hokkaido University
Sapporo 060
Japan

Prof. E. Osawa
Dept. of Chemistry
Hokkaido University
Faculty of Science
Sapporo 060
Japan

Prof. L.A. Paquette
Department of Chemistry
Ohio State University
140 West 18th Avenue
Columbus, Ohio 43210
USA

Prof. E. Piers
Department of Chemistry
Univ. of British Columbia
2036 Main Mall
Vancouver, BC  V6T 1Y6
Canada

Dr. M. Psiorz
Fa.Thomae
Birkendorfer Straße 65
7959 Biberach
Federal Republic of Germany

Prof. H.U. Reissig
Inst. f. Org. Chem.
Techn. Hochschule Darmstadt
6100 Darmstadt
Federal Republic of Germany

S. Roberts
School of Molecular Sc.
University Coll. of N. Wales
Bangor
Gwynedd LL57 2UW
Great Britain

Prof. J. Salaün
Lab. des Carbocycles
Université de Paris-Sud
Batiment 420
F-91405 Orsay Cédex
France

Dr. A.D. Schlüter
Max-Planck-Institut für Polymerforschung
Postfach 3148
6500 Mainz
Federal Republic of Germany

Prof. F. Serratosa
Dep. de Quimica Organica
Universitat de Barcelona
Barcelona -28
Spain

Prof. K. Shea
Dept. of Chemistry
UC Irvine
Irvine, CA 92717
USA

Prof.L. Skattebøl
Department of Chemistry
University of Oslo
P.O.Box 1033 Blindern
0315 Oslo 3
Norway

Prof. C.J.M. Stirling
School of Molecular Sc.
University Coll. of N. Wales
Bangor
Gwynedd LL57 2UW
Great Britain

Prof. C.J. Suckling
Department of Chemistry
University of Strathclyde
295, Cathedral Street
Glasgow G1 1XL
Great Britain

Prof. G. Szeimies
Inst. f. Org. Chemie
Universität München
Karlstr. 23
8000 München
Federal Republic of Germany

Prof. B. Trost
Department of Chemistry
Stanford University
Stanford, CA 94305
USA

Prof. K.B. Wiberg
Department of Chemistry
Yale University
P.O.Box 6660
New Haven, Connecticut 06511
USA

Prof. E. Winterfeldt
Inst. f. Org. Chemie
Universität Hannover
Schneiderberg 16
3000 Hannover
Federal Republic of Germany

Prof. W.H. de Wolf
Vakgroep Org. Chemie
Vrije Universiteit
De Boelelaan 1083
081 HV Amsterdam
The Netherlands

Prof. Z. Yoshida
Department of Chemistry
Kyoto University
Yoshida
Kyoto 606
Japan

Prof. B. Zwanenburg
Dept. of Chemistry
University of Nijmegen
Toenooiveld
6525 ED Nijmegen
The Netherlands

Barry M. Trost
Department of Chemistry
Stanford University
Stanford, CA 94305
USA

INTRODUCTION

Cyclopropanes constitute a unique ring system because they combine thermodynamic instability with kinetic reactivity. Release of the 27 kcal/mol of ring strain can provide a thermodynamic driving force. However, as the comparison of cyclopropanes with cyclobutanes (which have similar strain energies) illustrates, release of such strain energy may not be sufficient for high reactivity. The $\pi$ character of the ring bonds of a cyclopropane provides the kinetic opportunity to initiate the unleashing of the strain. As such, the cyclopropane should be considered as a novel functional group. In this lecture, I wish to consider the implications of this functional group concept in terms of generating new reagents for organic synthesis.[1]

1-Phenylthiocyclopropyl Carbinols in Halogenative Rearrangements

We have previously established the utility of lithiated cyclopropylphenyl sulfide as a conjunctive reagent in the synthesis of cyclobutanones and cyclopentanones.[2-5] This work stimulated the development of other metalated heteroatom substituted cyclopropanes for synthesis.[6,7] In exploring further directions, we were attracted to the concept of a cyclobutene synthesis because of the utility of the latter as a precursor of dienes according to eq 1. The notion was to generate the initial cyclopropylcarbinyl cation under conditions in

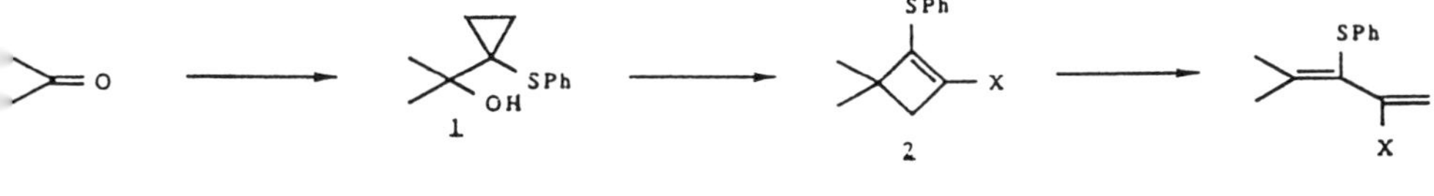

(1)

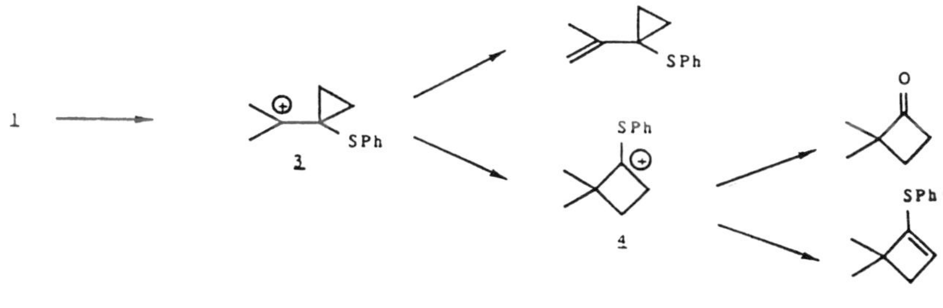

(2)

A. de Meijere and S. Blechert (eds.), Strain and Its Implications in Organic Chemistry, 1–23.
© 1989 by Kluwer Academic Publishers.

which it would live long enough to allow rearrangement to the sulfur stabilized cyclobutyl cation that would eliminate a proton and generate a cyclobutene. Problems that arose included premature loss of a proton to form the vinylcyclopropane and capture of the cyclobutyl cation by water to form a cyclobutanone (eq 2),[3] as well as fragmentation leading to open chain products.[5]

Satisfactory results utilized conditions such as thionyl chloride in pyridine, p-toluenesulfonic acid in refluxing benzene, and (carbomethoxysulfamoyl) triethylammonium hydroxide inner salt (Burgess' reagent[8]) in benzene at reflux.[3,9] As shown in the table, the adducts of aldehydes and hindered ketones rearranged smoothly, but the adducts of unhindered ketones rearranged in lower yields and/or gave the types of by-products outlined in eq 2. The adduct of acetone only gave 2,2-dimethylcyclobutanone using p-toluenesulfonic acid in benzene regardless of all attempts to remove the water as effectively as possible. In this case, Burgess' reagent gave a 1:1 mixture of the cyclobutene and the vinylcyclopropane.

The utility of this process could be enhanced if the cyclobutene could be further derivatized to permit formation of more highly substituted cyclobutenes and, subsequently, dienes. From our previous work on the bromination of enol thioethers[10] and the utility of vinyl bromides as substrates for cross-coupling reactions[11] and as precursors for organometallics, we examined the rearrangement under conditions that might simultaneously introduce bromine (eq 1, **1** to **2**, X = Br). Indeed, treatment of the acetone adduct **5** with phosphorus pentabromide in methylene chloride at -78° led directly to the

brominated vinyl sulfide **6**. This reaction proved to be more general (eq 4-6) than we experienced with the non-halogenative version. Its success stands in contradiction to the reactions of the carbonyl adducts **5** with Lewis acids, which lead to ring opening products.[5] The vinyl bromides underwent brominative hydrolysis to the dibromocyclobutanones upon exposure to NBS in moist acetonitrile, as

Table 1. Rearrangement of 1-Phenylthiocyclopropylcarbinols to Cyclobutenes

| Entry | Cyclopropylcarbinol | Reagent | Cyclobutene | Isolated Yield |
|---|---|---|---|---|
| | 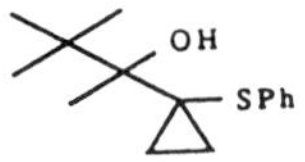 (OH, SPh substituted cyclopropane) | TsOH[a] | 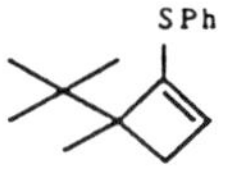 (SPh-cyclobutene) | 95% |
| | 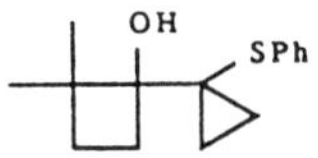 (OH, SPh cyclobutyl-cyclopropane) | CH$_3$O$_2$CNSO$_2$N$^-$(C$_2$H$_5$)$_3^+$ [a] | 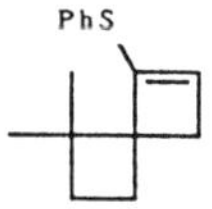 (PhS-cyclobutene) | 92 |
| | 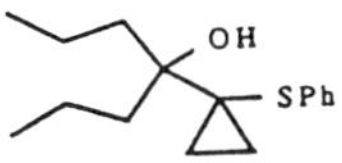 (OH, SPh with propyl chains) | TsOH[a] | 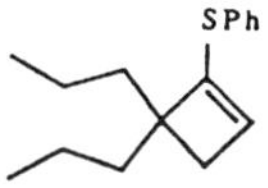 (SPh-cyclobutene with propyl chains) | 43% |
| | | CH$_3$O$_2$CNSO$_2$N$^-$(C$_2$H$_5$)$_3^+$ [a] | | 22%[b,c] |
| | 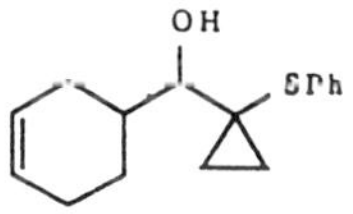 (OH, SPh cyclohexenyl) | SOCl$_2$[b] | 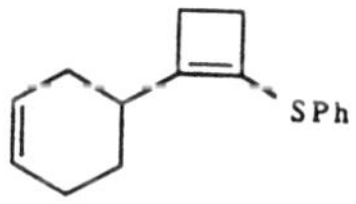 (SPh-cyclobutene with cyclohexenyl) | 94% |
| | 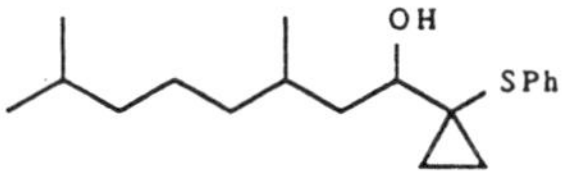 (OH, SPh with long alkyl chain) | CH$_3$O$_2$CNSO$_2$N$^-$(C$_2$H$_5$)$_3^+$ [a] | 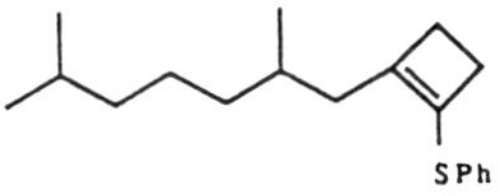 (SPh-cyclobutene with alkyl chain) | 93% |
| | 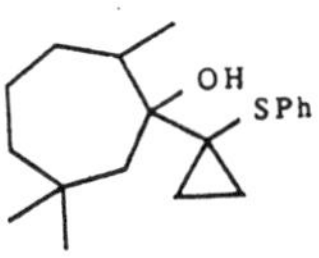 (OH, SPh with gem-dimethylcycloheptyl) | TsOH[a] | 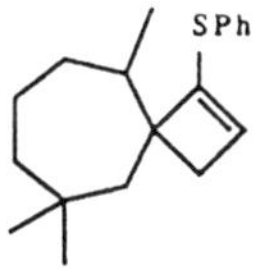 (SPh-cyclobutene spiro) | 40% |
| | | CH$_3$O$_2$CNSO$_2$N$^-$(C$_2$H$_5$)$_3^+$ [a] | | 50%[d] |

a) Reaction performed in benzene at reflux. b) Reaction performed in pyridine at reflux. c) Vinylcyclopropane also isolated in 26% yield. d) Vinylcyclopropane also isolated in 27% yield.

upon exposure to NBS in moist acetonitrile, as shown in eq 4. Since the dibromocyclobutanones undergo nucleophilically triggered ring cleavage,[12] the overall process becomes a useful geminal alkylation of carbonyl compounds as summarized in eq 7.

Of most interest, was the viability of a lithiated cyclobutene exemplified by __11__, whose stability towards $\beta$-elimination is assured by the unfavorability of a cyclobutyne. Metal halogen exchange with t-butyllithium proceeds uneventfully to give __11__, whose presence was demonstrated by protonolysis to the cyclobutene __12__. While such cyclobutenes may arise directly from the carbinols, the greater generality of the brominative rearrangement makes this process a useful alternative. It should be recalled that thermolysis of __12__ in xylene produces the desired diene and Raney nickel desulfurization in ethanol completely saturates the ring (eq 9).

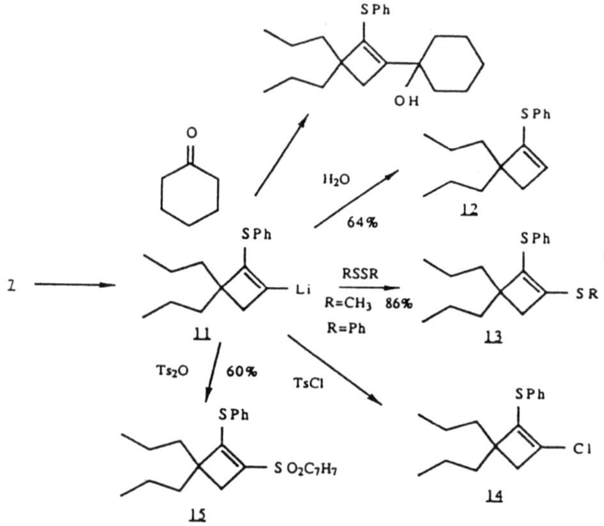

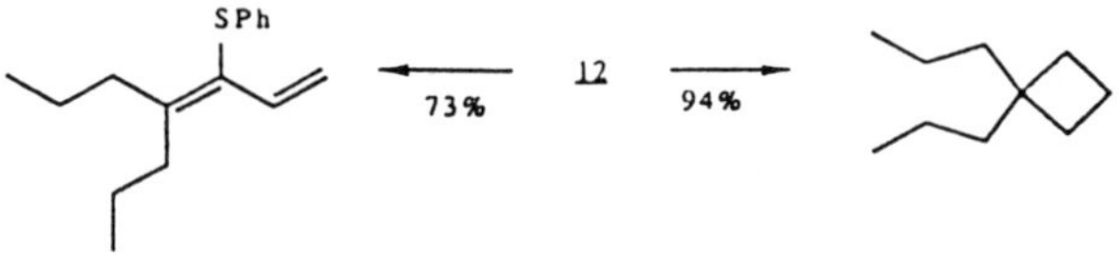

Other electrophiles including ketones and disulfides may be employed. The bis-sulfenylated cyclobutenes 13 (R = Ph, CH$_3$) were thermolyzed to their corresponding dienes, which provides the equivalent of a carbonyl olefination that could not be done in a direct

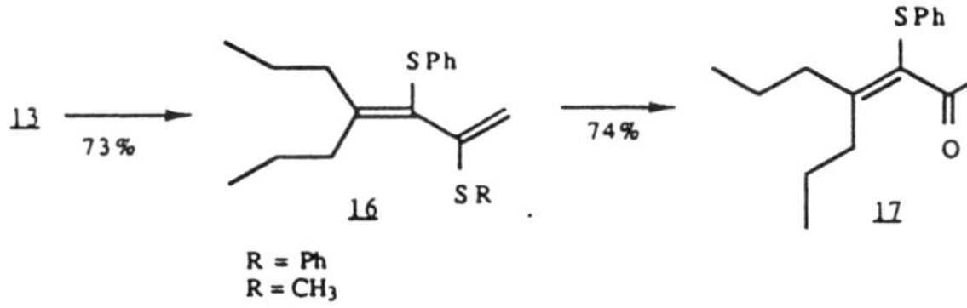

$$13 \xrightarrow{73\%} 16 \xrightarrow{74\%} 17 \quad (10)$$

R = Ph
R = CH$_3$

manner. The methylthio compound 16 (R = CH$_3$) may be readily hydrolyzed to the enone 17 with mercuric chloride in moist acetonitrile. Since a cross aldol condensation between a β-ketosulfide[13] and a ketone is not possible, this alternative provides a reasonable access. A virtue of these sulfenylated enones is the ease with which they undergo conjugate addition, as illustrated in eq 11, in contrast to their unsulfenylated analogues.

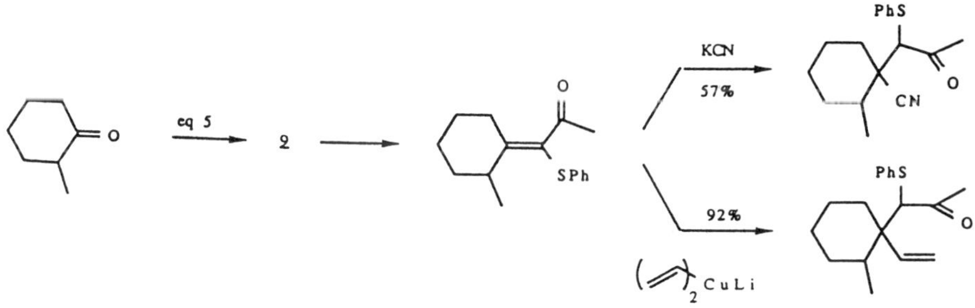

$$\xrightarrow{eq\ 5} 2 \longrightarrow \quad (11)$$

The monosulfones were of special interest. Direct arylsulfonation of organolithium compounds has not been recorded because sulfonyl chlorides serve as chlorinating agents to give 14 (eq 8) rather than sulfonylating agents. Nevertheless, use of p-toluenesulfonic anhydride[14] does give the desired sulfone 15 in what appears to be a general reaction. Table 2 summarizes some of the examples, as well as illustrates the formation of the corresponding diene monosulfide monosulfone upon thermolysis in xylene.

Table 2. Preparation of Cyclobutene and Diene Monosulfides-Monosulfone

| Entry | Bromide | Cyclobutene | Isolated Yield | Diene | Isolated Yield |
|---|---|---|---|---|---|
| 1 | R=R'=CH$_3$ | | 40 | | 85 |
| 2 | R=R'=nC$_3$H$_7$ | | 60 | | 85 |
| 3 | R=CH$_3$, R'=nC$_6$H$_{13}$ | | 77 | | 47 |
| 4 | n = 1 | | 30 | | N.D.[a] |
| 5 | n = 2 | | 50 | | 80 |
| 6 | | | 64 | | 76 |

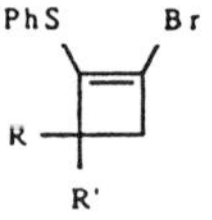

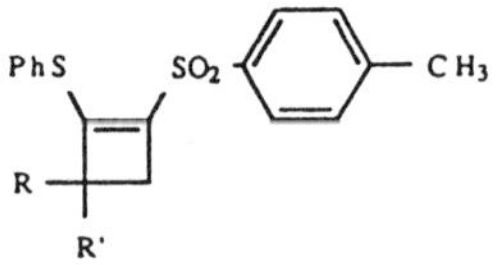

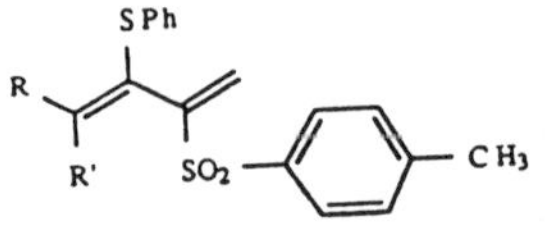

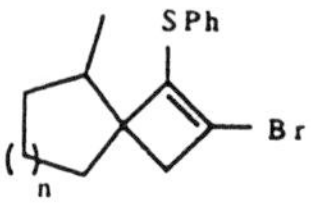

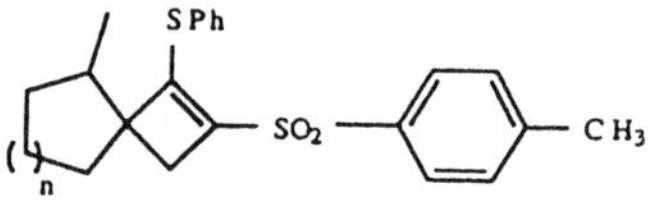

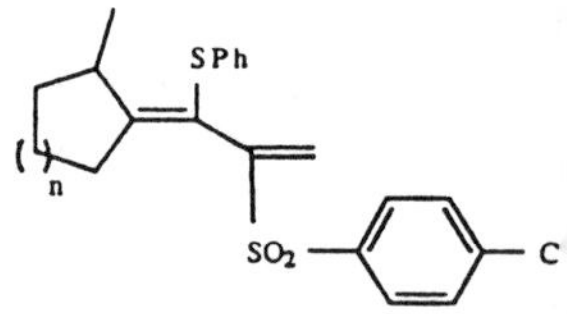

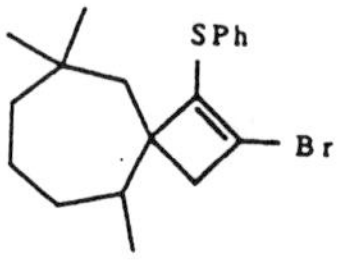

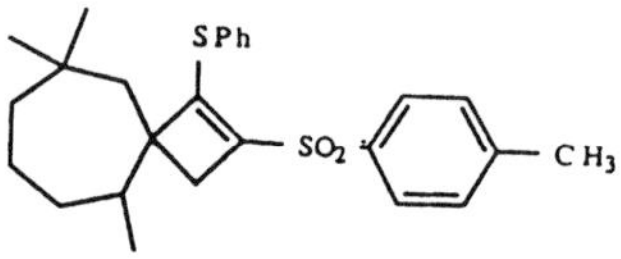

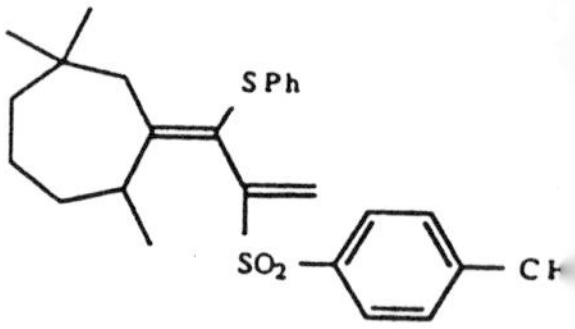

a) N.D. = not determined

## 2-Phenylthio-3-p-toluenesulfonylbutadienes as 1,3-Dication Synthons

The ready availability of diene sulfide-sulfones led to an investigation of their reactivity towards organocuprates. Treatment with 1 eq of lithium di-n-butylcuprate generated from n-butyllithium and tri-n-butylphosphine-cuprous iodide in ether at -78° led to elimination accompanying addition to form allene sulfones (eq 12). The allenyl sulfone 18 may undergo a second conjugate addition under

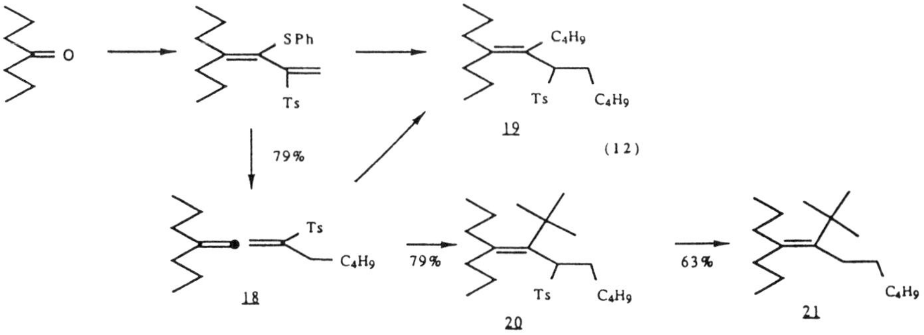

similar conditions but at 0° to the allyl sulfone 20. Reductive desulfonylation with 6% sodium amalgum in buffered methanol gives a tetrasubstituted double bond with a fair degree of flexibility with respect to the alkyl chains. If the original 1,3-diene is treated with excess cuprate and the temperature allowed to rise to 0° before quenching, diaddition to give allyl sulfone 19 may occur directly. Eq 13 illustrates some of the flexibility offered by this methodology.

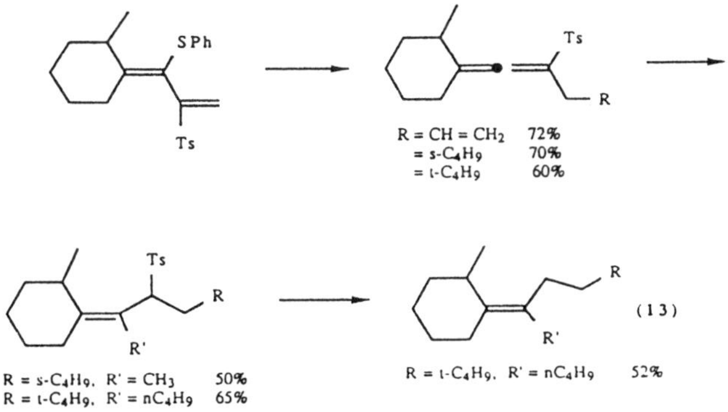

In this sequence, we are considering the diene sulfide-sulfone as a synthon for the dication 22. Since carbonyl compounds bearing $\alpha$, $\alpha'$-hydrogens may be equated with the dianion 23, a six membered ring annulation may be envisaged as outlined in eq 14. Indeed, this annulation proceeds remarkably well upon heating a DME solution of a ketone and the dienes at reflux in the presence of 1.1 eq of

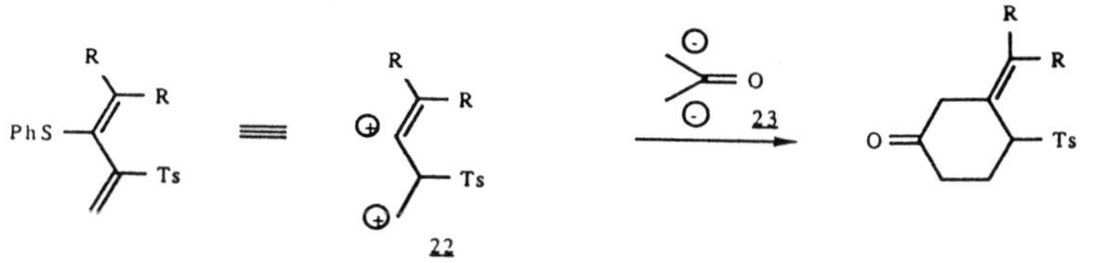

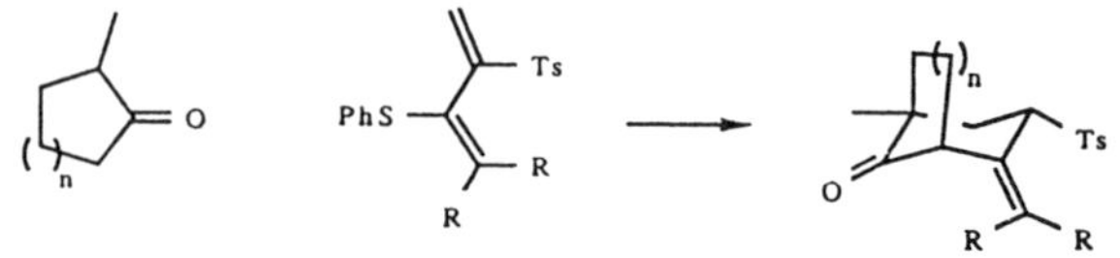

| n | R | |
|---|---|---|
| 1 | $CH_3$ | 55% |
| 1 | $nC_3H_7$ | 42% |
| 2 | $CH_3$ | 51% |
| 2 | $nC_3H_7$ | 47% |

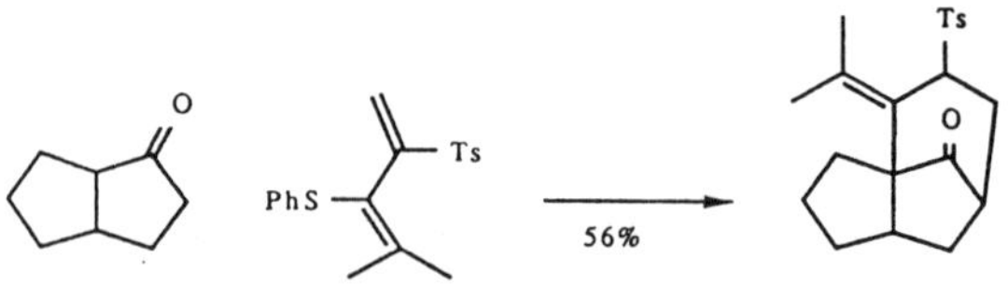

potassium t-butoxide, as illustrated by the examples in eq 15. The tricyclic nucleus of several natural products may be simply constructed as shown in eq 16.

Vinylcyclopropanols as Composite Pseudofunctional Groups

In the foregoing, we used the ability to control the reactivity of the cyclopropyl cation to generate novel cyclobutene derivatives. The role of the sulfur in stabilizing such intermediates was not discussed. For example, we might consider that cyclopropyl carbinyl cations such as 24a may better be represented by strongly delocalized structures

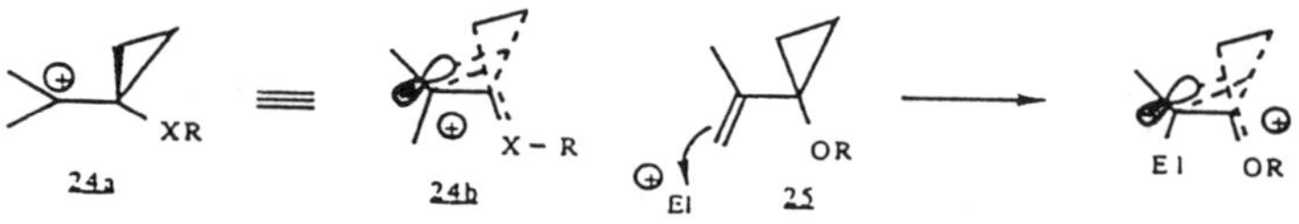

such as 24b. The latter suggests that the presence of a heteroatom such as sulfur, or even better, oxygen, may stabilize such a species. A manifestation of such a concept might be enhanced reactivity of vinylcyclopropanol derivatives such as 25 toward electrophilic attack.

An alternative view of a vinylcyclopropanol like 25 is as a composite of an olefin, an alkoxy group, and a cyclopropane. Consider the composite of an alkoxy group and an olefin to make an enol ether (eq 17). Such oxygenated olefins have unique reactivity that derives

$$\diagup\!\!\!\diagup \ + \ OR \ \Longrightarrow \ \diagup\!\!\!\diagup^{OR} \ +\triangle \ \rightleftharpoons \ \diagup\!\!\!\diagup\!\!\!\diagdown_{OR} \quad (17)$$

from electron delocalization from oxygen to the π system, and creates the most important composite functional group in organic chemistry. What might be the consequences of splicing a cyclopropane ring between the olefin and oxygen of an enol ether? The π-nature of the cyclopropane ring may still transmit electron density from oxygen to the olefin albeit attenuated.

To explore this concept qualitatively, we examined the reactions of the vinylcyclopropane 26 to the vinylcyclopropanols 27 and 28 in their reactions with alkoxycarbonium ions (eq 18 and 19).[15]

$$\text{26} \quad \xrightarrow[-40°\text{ to rt}]{\text{TMSOSO}_2\text{CF}_3 \atop \text{PhCH(OCH}_3)_2} \quad \text{CH}_3\text{O}\diagup\!\!\!\diagdown \text{OCH}_3 \atop \text{Ph} \quad (18)$$

$$\text{27} \quad R = TMS \quad -78°$$
$$\text{28} \quad R = Ac \quad -40° \text{ to } -10°$$

$$\xrightarrow[\text{PhCH(OCH}_3)_2]{\text{TMSOSO}_2\text{CF}_3} \quad (19)$$

Vinylcyclopropane 26 required room temperature to go to completion and produced many products - that depicted being isolated in about 26% yield. In stark contrast, vinylcyclopropanol 27 reacted almost instantly at -78° to give the cyclobutanone in good yield. Further support that oxygen is involved in the rate determining step derives from the effect of changing the oxygen substituent from trimethylsilyl to acetyl, which substantially slows reaction.

We explored the reactivity of a vinylcyclopropanol towards several electrophiles in order to determine if the cyclopropane ring participated. For example, bromination of 29 led only to the bromocyclobutenones 30 and 31 with high chemoselectivity, but with low diastereoselectivity.[16] Epoxidation with t-butylhydroperoxide in the presence of a catalytic amount of vanadyl acetoacetonate in toluene led to rearrangement to hydroxycyclobutanone 32 with complete diastereoselectivity which corresponds to that expected for a trans addition to the double bond (eq 20).

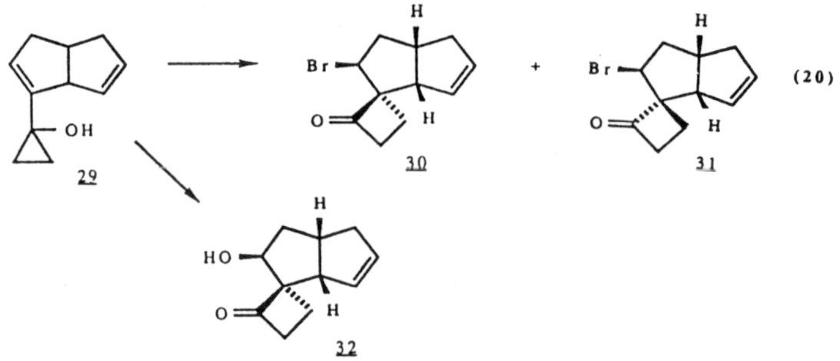

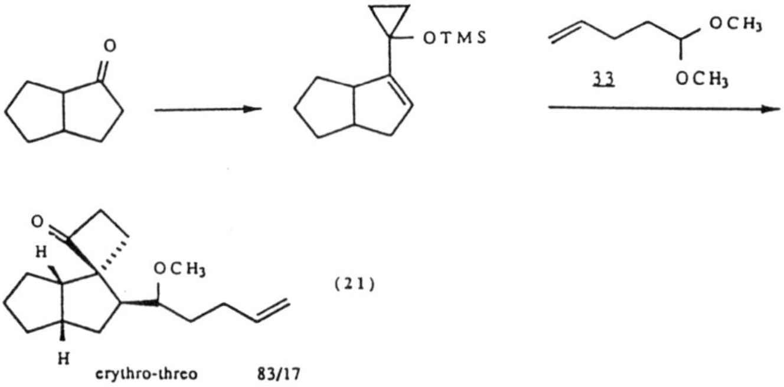

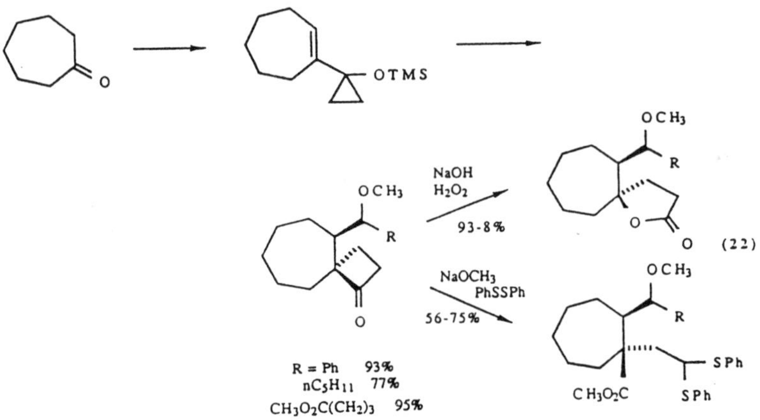

The use of carbon electrophiles was of special interest since the overall process from a starting carbonyl partner provides attachment of three carbon-carbon bonds.[15] As shown in eq 21, chemoselective condensation with acetals in the presence of trimethylsilyl trifluoromethanesulfonate proceeds smoothly.[17] The higher reactivity of the vinylcyclopropanol double bond compared to a simple double bond is highlighted by this example in which the electrophilic partner 33 could have reacted intramolecularly with an olefin. The utility of this tris-alkylation methodology is suggested by the transformations of eq 22, in which the cyclobutanone units in the products become a focal point for further skeletal developments as a result of their strain.[18,19]

The indication that vinylcyclopropanols possess exceptional nucleophilic properties led us to examine their role as cyclization terminators. As summarized in eq 23 and 24, the intramolecular versions to generate six[19] and seven[20] membered rings proceeds well with modest diastereoselectivity. In each case, the major diastereomer of the product, which is the only one depicted, may be rationalized via the syn conformations of the alkoxycarbonium ion intermediates as the most reactive ones.

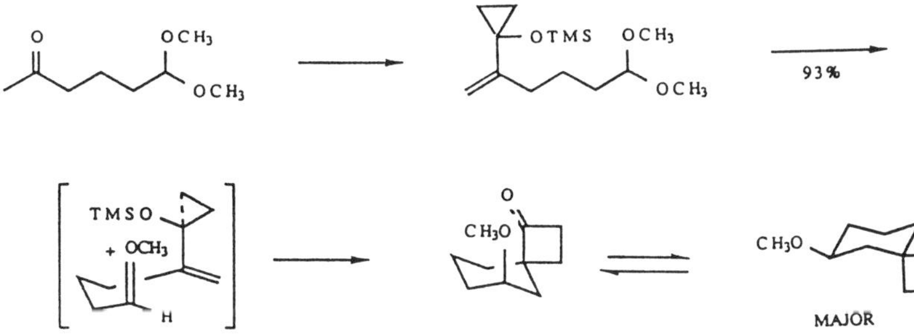

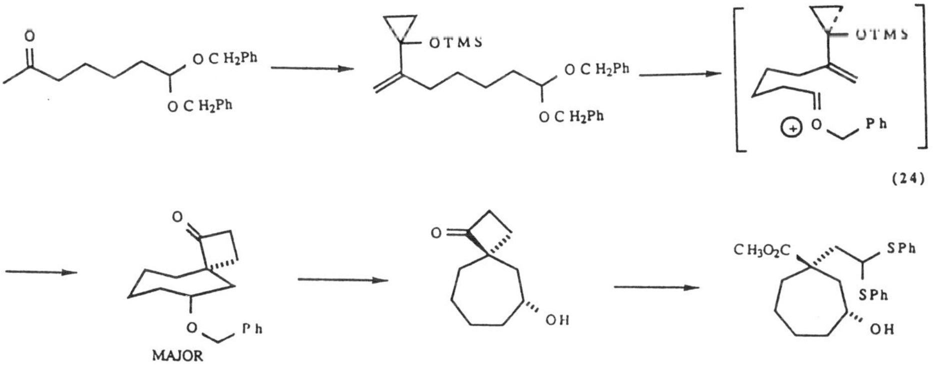

Eq 25 and 26 illustrate (1) the diastereoselectivity when an additional stereogenic center is present in the tether connecting the acetal to the trimethylsilyl ether of the vinylcyclopropanol, and (2) the ready availability of the cyclization precursors. For example, ozonolysis and acetal formation of the keto-olefin provides the ketoacetal precursor of substrate 35 (eq 25). The high diastereoselectivity by having a methyl group in a 1,3-orientation, with respect to the cyclopropanol, arises because of a very severe 1,3-

eclipsing interaction between these two substituents for the conformation leading to the alternative diastereomer. On the other hand, moving the substituent down the tether by one, now allows access to both conformations with no overriding steric effect, and, thus,

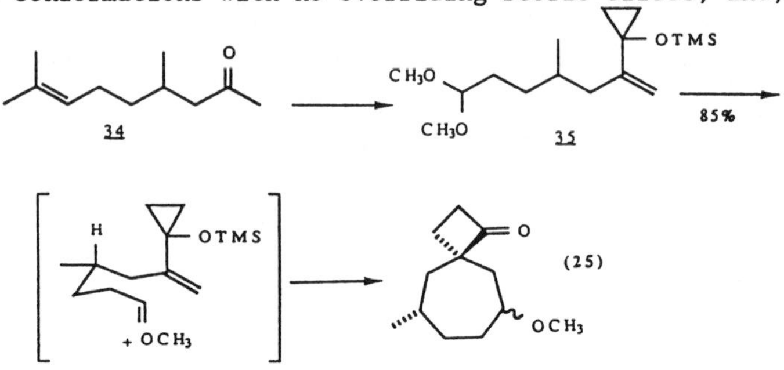

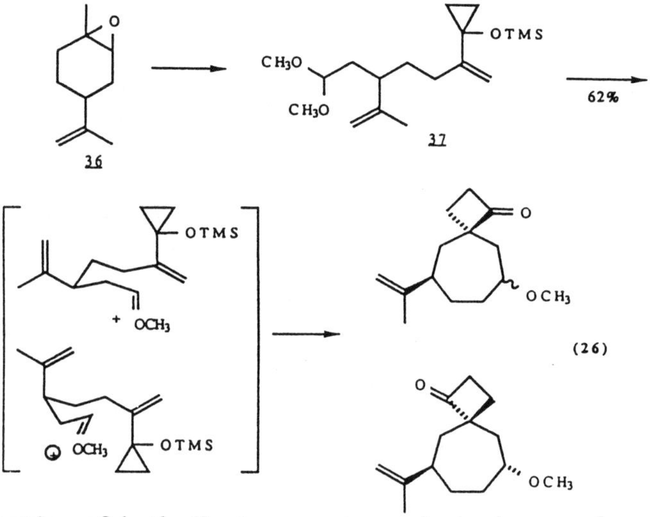

formation of both diastereomeric products in nearly equal amounts, as shown in eq 26. In this case, the ketoacetal precursor of the cyclization substrate 37 was easily available by periodic acid cleavage of limonene oxide (36).

Many surprises arose during a study of the effect of ring size on the efficiency of this cyclization.[20] Five membered rings could not be formed. Apparently, the geometric restrictions imposed upon a 5-endo trig cyclization allow alternative pathways to dominate. On the other hand, eight membered rings formed with remarkable ease even at 0.01 M concentrations (eq 27). Similarly, a thirteen membered ring also formed at moderate concentrations (eq 28).

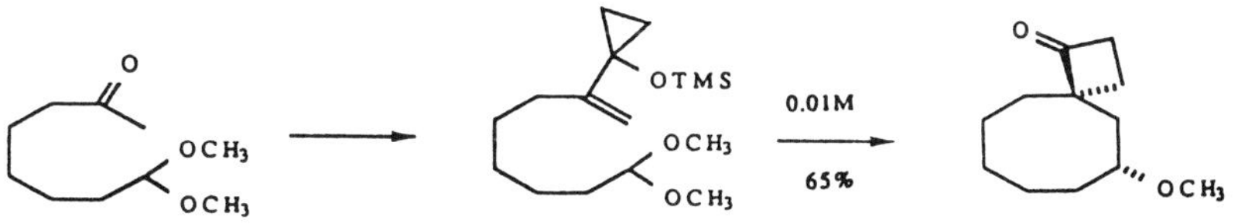

(27)

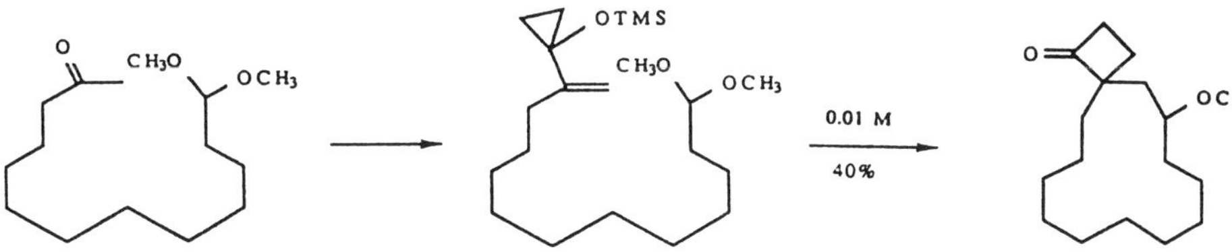

(28)

The evolution of this cyclization into an annulation depends upon the availability of the requisite ketoacetal. An olefin may be a very convenient precursor to an aldehyde either by oxidative cleavage or by hydroboration (or hydroalumination)-oxidation, in which case an acetylcycloalkene becomes the basic building block. For example, reaction of allyltrimethylsilane with 1-acetylcyclohexene in the presence of titanium tetrachloride (Sakurai reaction[21]) gave the ketoolefin 38. Ozonolysis and acetal formation gave the ketoacetal precursor of the cyclization substrate 39. Cyclization proceeded smoothly with 10 mol% trimethylsilyl trifluoromethanesulfonate to give the annulated product 40 with good (7:1) diastereoselectivity (eq 29).

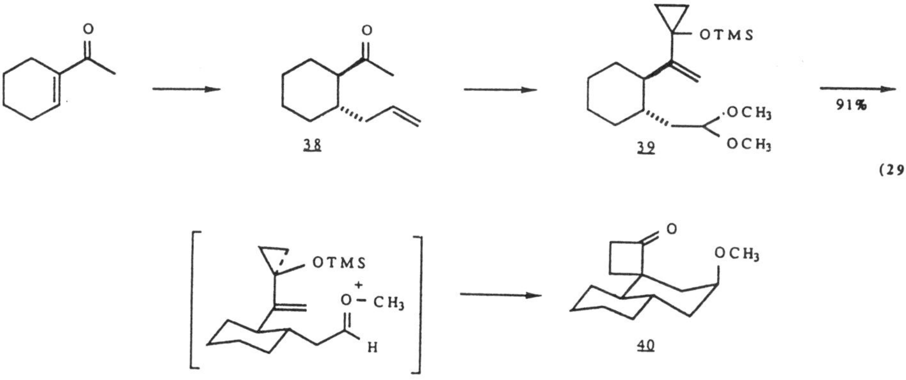

(29)

Alternatively, the olefin obtained by the Sakurai reaction may be hydroborated and oxidized to form the ketoacetal precursor. This sequence is illustrated by eq 30 for formation of the perhydroazulene skeleton from 1-acetylcyclopentene. Thus, from the same ketoalkene, either a six or seven membered ring may result depending upon the mode of elaboration into the aldehyde.

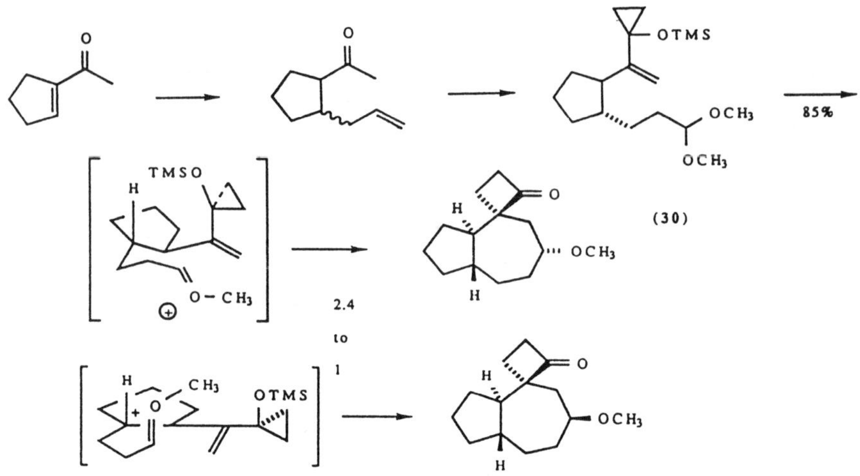

Alternatively, copper iodide catalyzed addition of 3-butenylmagnesium bromide also provides the requisite ketoalkene as shown in eq 31. While oxidative cleavage can lead to a perhydroazulene, the ability to annulate an eight membered ring onto a five to create a bicyclo[6.3.0]undecane, a ring system common to a growing number of natural products, would be available via the hydroboration-oxidation sequence. As shown in eq 31, this sequence was particularly effective in elaborating this system, giving virtually a single diastereomeric product.

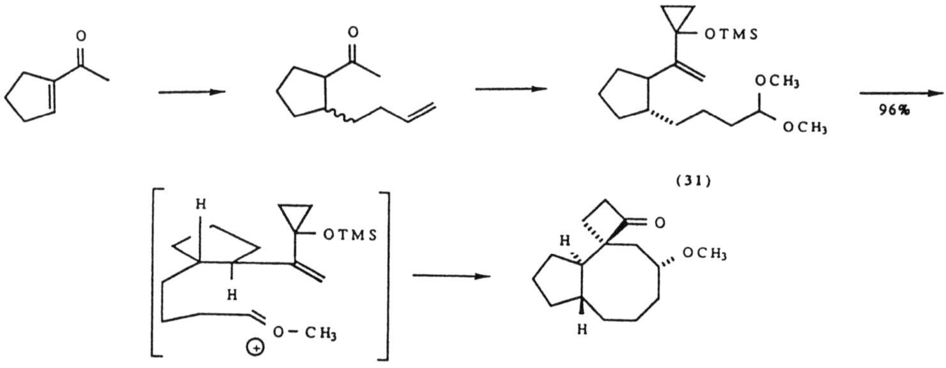

The success of the vinylcyclopropanol as a cyclization terminator suggested the acetylenic version may be even better. An alternative strategy to the requisite alkynyl cyclopropanols according to eq 32 was employed.[22] The reaction of the lithium acetylide with

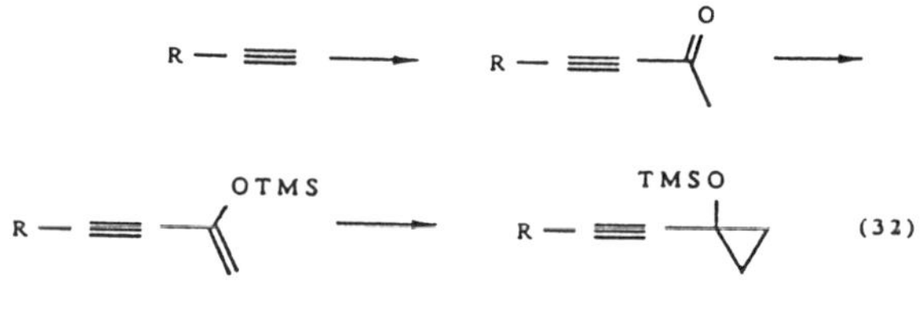

dimethylacetamide gave the ketoacetylene, which was converted in standard fashion to its enol silyl ether. Furakawa's modification of the Simmons-Smith reaction using diethylzinc and methylene iodide[23] provided the requisite substrates.[24] As predicted, the acetylenic version proceeded smoothly to generate six membered rings. Eq 33 illustrates formation of a carbocycle; whereas, eq 34 illustrates formation of a heterocycle. In contrast to the vinyl case, a <u>trans</u> addition is not necessarily preferred, as shown in eq 33. Whether this stereochemistry arises because of an equilibration of the product or of the electronic symmetry of an acetylene cannot be ascertained at present.

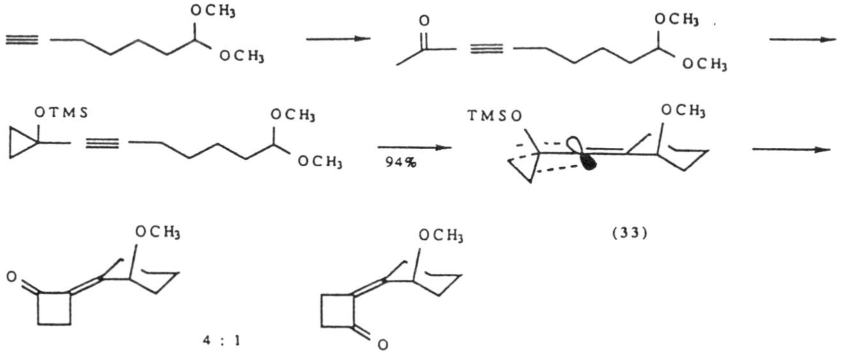

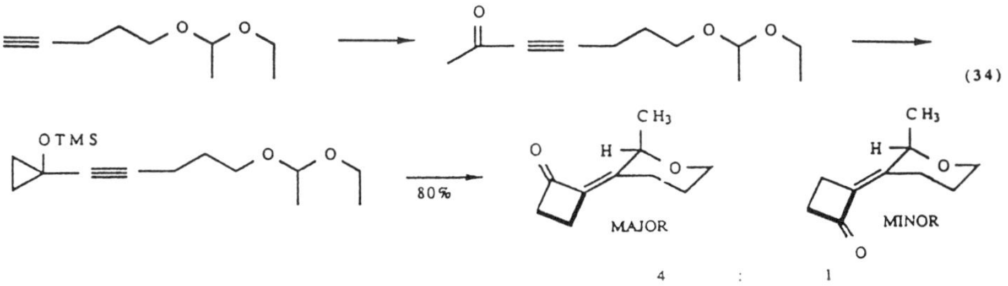

The ring size effects are strikingly different. Five membered ring formation, which fails in the vinyl case, occurs smoothly (eq 35). The fact that the acetylene is exocyclic to the forming ring relieves

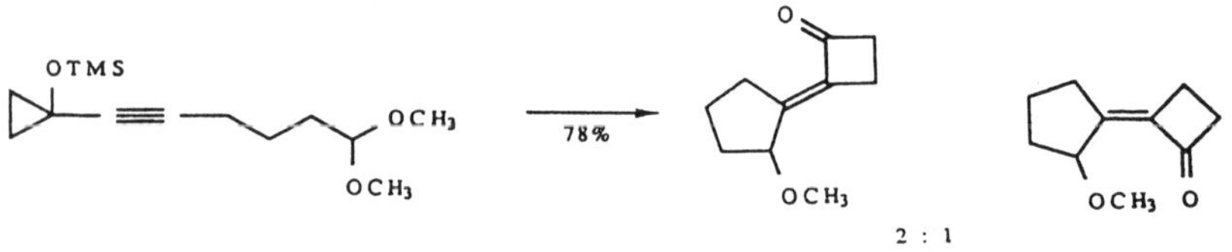

$$\text{(35)}$$

the cyclization from geometric constraints that inhibit such a process. On the other hand, eight membered ring formation does not occur. By placing the sp hybridized carbons exocyclic to the forming ring, the transannular interactions that disfavor cyclization to medium sized rings are no longer minimized.

In evaluating the reasons for the efficacy of the alkenyl-and alkynylcyclopropanols as terminators in cyclizations, it is difficult to separate the importance of the special bonding characteristics from the strain factors. Since a cyclobutyl system is so close in energy to a cyclopropyl system, the differential reactivity of cyclopropyl and cyclobutyl systems suggests that the special bonding characteristics would appear to be the most important. On the other hand, a tremendous amount of strain energy is relieved in going from a four to a five membered ring. To evaluate whether relief of strain can be important in such processes, we briefly examined the use of an alkynylcyclobutanol as a terminator. As shown in eq 36, cyclization does proceed to generate a mixture of the four possible ring expansion products. Obviously, this is an area where further work is needed.

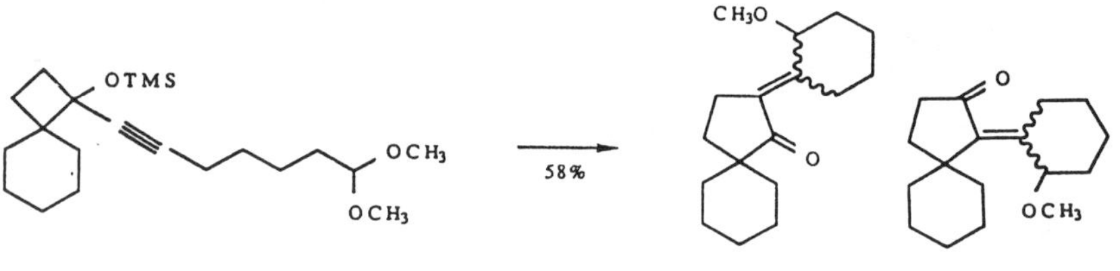

$$\text{(36)}$$

An Application

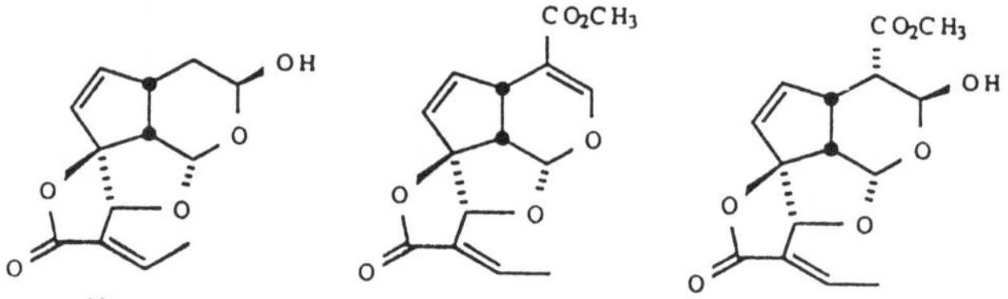

The unique properties of a 1-vinylcyclopropanol as a composite pseudofunctional group may lead to new concepts in synthetic strategy directed toward natural products. The plumera and allamanda iridoids, represented by allamcin (41), plumericin (42), and allamandin (43), are attractive targets since (1) members of this class exhibit cytotoxic, antileukemic, antimicrobial, and antifungal activities; (2) their densely functionalized skeletons offer a major challenge; and (3) no total syntheses of this family, except for the simplest member fulvoplumierin, had previously been recorded.[25] Scheme 1 offers a retrosynthetic analysis focussing on two key strategies: (1) the elaboration of the butenolide 44 from the butanolide 45 using

Scheme 1. Retrosynthetic Analysis of Allamandin, Plumericin, and Allamcin

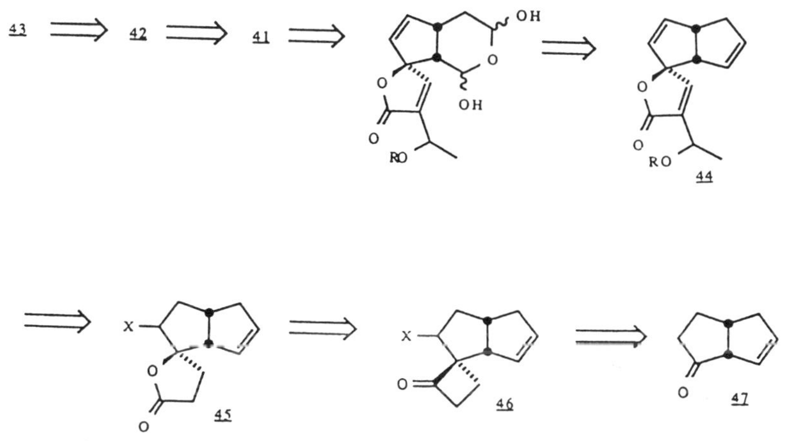

organosulfur mediated aldol[26] and olefin formation chemistry;[13,27] and (2) the elaboration of the butanolide 45 from a simple ketone 47. This latter structural modification requires the introduction of the carbon chain, as well as a substituent X that would permit easy introduction of a cyclopentene double bond. The methodology proceeding through a vinylcyclopropanol, which is easily derived from a saturated ketone, is ideally suited for such a task.

Scheme 2 outlines the synthesis. Condensation of the bicyclo[3.3.0] oct-7-en-2-one with cyclopropyldiphenylsulfonium fluoroborate in DMSO in the presence of powdered potassium hydroxide generates the oxaspiropentane, which is directly extracted into pentane. The pentane extracts are directly treated with lithium diethylamide. By this simple operation, the vinylcyclopropanol 48, which is remarkably stable even to flash chromatography, is available in pure form. Exposure of the vinylcyclopropanol to benzeneselenenyl bromide at -78° to -45° effects electrophilically initiated ring expansion to give the selenyl substituted cyclobutanone spiroannulation product 49 as a single diastereomer (>100:1). The chemo- and diastereoselectivity of this process illustrates the unusual nature of the vinylcyclopropanol.

Scheme 2. Biomimetic Synthesis of Allamcin, Plumericin, and Allamandin

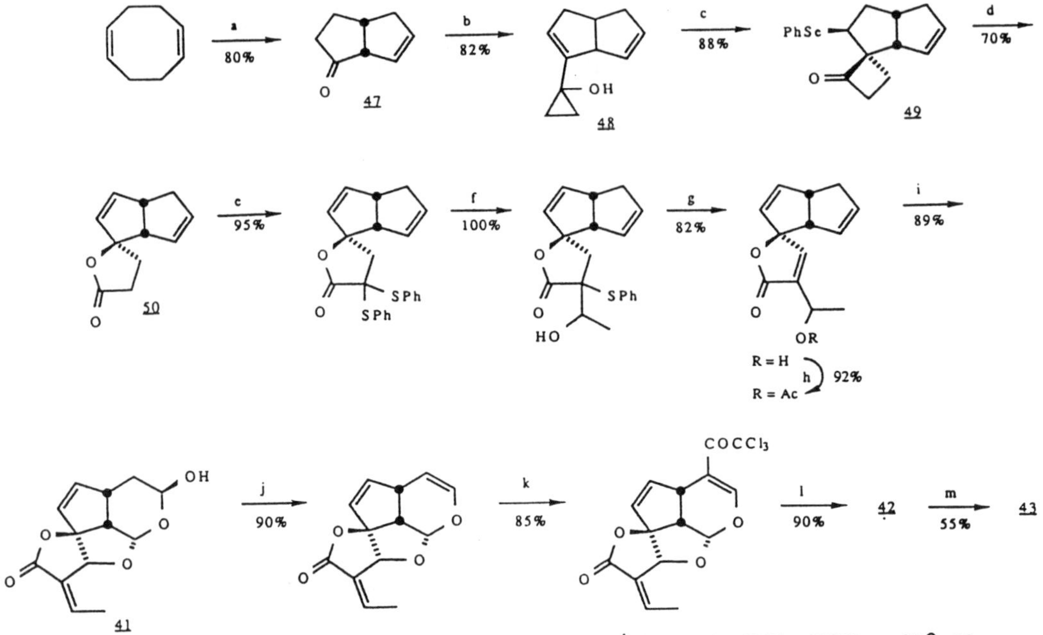

(a) 3 steps, see ref 16; (b) i. c-C$_3$H$_5$S$^+$Ph$_2$BF$_4^-$, KOH, DMSO, 25° ii. LiN(C$_2$H$_5$)$_2$, pentane; (c) PhSeBr, CH$_2$Cl$_2$, (C$_2$H$_5$)$_3$N, -45°; (d) MCPBA, CH$_2$Cl$_2$, -78° then warm to 0°, C$_2$H$_5$OCH=CH$_2$, rt; (e) LDA, THF, PhSSO$_2$Ph, -78°; (f) C$_2$H$_5$MgBr, THF, 0° then CH$_3$CHO; (g) MCPBA, CH$_2$Cl$_2$, -78° then CaCO$_3$, CCl$_4$, reflux; (h) Ac$_2$O, C$_5$H$_5$N, DMAP, CH$_2$Cl$_2$, 0°; (i) i. cat. OsO$_4$, NMO, THF, H$_2$O, 0°; ii. NaIO$_4$, ether, H$_2$O then NaOAc; (j) i. Ac$_2$O, (iC$_3$H$_7$)$_2$NC$_2$H$_5$, DMAP, CH$_2$Cl$_2$; ii. FVT, 500° @ 0.005 mm Hg; (k) CCl$_3$COCl, 2,6-di-t-butylpyridine, CH$_2$Cl$_2$, rt; (l) Mg(OCH$_3$)$_3$, CH$_3$OH, -45°; (m) 0.01 N HClO$_4$, H$_2$O, 95-100°.

The choice of selenium as the electrophilic partner to initiate the ring expansion derives from (1) the desire to have a cis - syn elimination process for introduction of the double bond in the five membered ring because of the geometrical limitations imposed upon vicinal eliminations in five membered rings, and (2) the ability to coalesce the olefin-forming and Baeyer-Villiger reactions into a single operation. Indeed, both objectives were met since 49 undergoes Baeyer-Villiger oxidation, oxidation to the selenoxide, and elimination of the selenoxide to give butanolide 50 simply by being exposed to two equivalents of MCPBA initially at -78° and then at 0°. It is interesting to note that use of hydrogen peroxide produces an isomeric unsaturated lactone 51 (eq 37).[28] The involvement of a cyclic peroxide such as 52 that restricts which cyclobutyl bond can adopt the required trans periplanar relationship to the peroxide bond accounts for the abnormal migration observed. MCPBA precludes the formation of the cyclic peroxide and thus leads to the normal product. The remaining stages of the synthesis are outlined in Scheme 2. Suffice it to say that by combining the above route with methodologies developed in our laboratories for elaborating structures via $\beta$-ketosulfides and for carbomethoxylating enol ethers, a sequence evolved which totally avoids the use of any protecting groups. Thus, the synthesis of allamcin, plumericin, and allamandin requires only 9, 13, and 14 steps respectively, from the previously known bicyclic ketone 47 which is available in only 3 steps from cycloocta-1,5-diene. It is interesting to note that synthetic intermediate 41 was prepared in our laboratories in racemic form before it was isolated as a natural product.

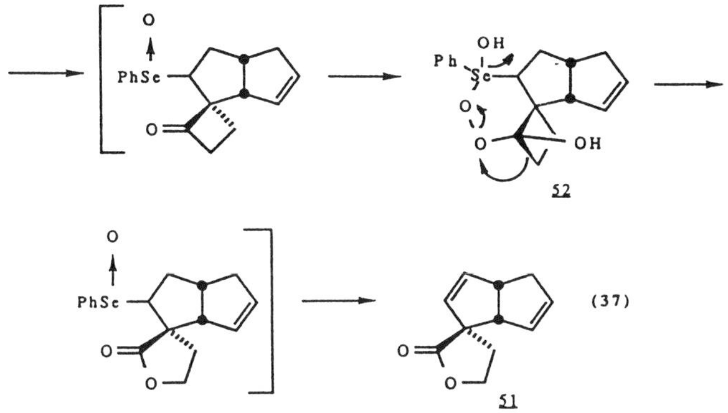

Conclusion

Small ring conjunctive reagents continue to represent a fertile area for searching out unusual reactivity. The presence of the vast amount of potential energy in the form of strain energy provides an unusual opportunity to manipulate structures in a highly selective fashion.

Scheme 3 summarizes some of the structural entities that can be generated utilizing the halogenative rearrangement of $\beta$-phenylthiocyclopropyl carbinols.

Similarly, the vinylcyclopropanols take on a new dimension. Previously, we noted their value as intermediates for cyclopentanone construction.[2,3] The present work demonstrates their sensitivity to electrophilically initiated ring expansions as shown in eq 38. Since

Scheme 3. Structural Entities via Halogenative Rearrangement

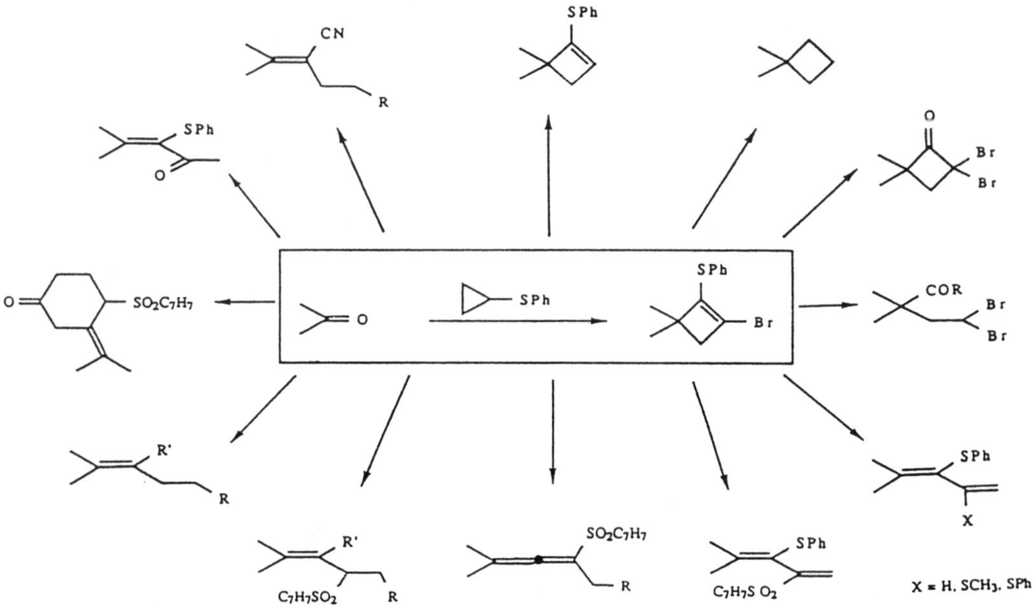

the product retains a small strained ring, a cyclobutanone, further structural changes that rely on release of strain energy, some of which are shown in eq 38, may be performed with high selectivity. Since the vinylcyclopropanols are so easily available via our diphenylsulfonium cyclopropylide chemistry, the overall transformations shown in eq 38

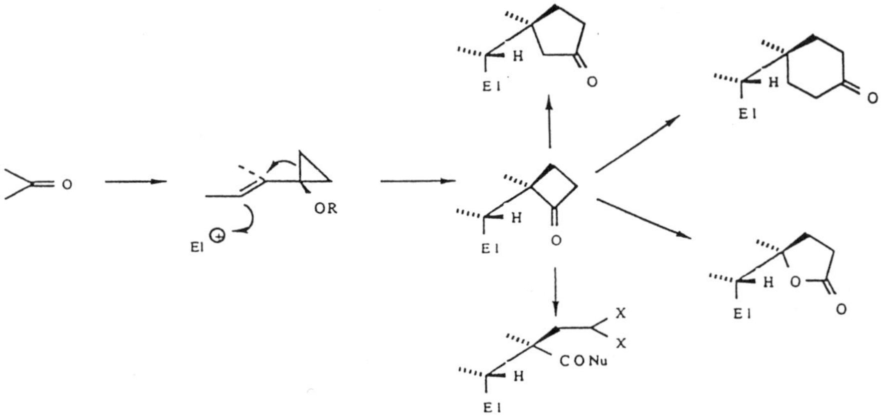

(38)

may be performed starting from the ubiquitous carbonyl compounds.[1] Of special note is the high reactivity of the vinylcyclopropanol as a terminator in cationic initiated cyclizations. The fact that six, seven, and, remarkably, eight membered rings are readily accessible does suggest that these vinylcyclopropanols are reacting as a unit. In that sense, they justifiably may be referred to as composite pseudofunctional groups. Eq 39-41 summarize the cyclizations and annulations that are now available. Once again, the cyclobutanone moiety may be easily further modified as shown in eq 38 to impart great versatility upon this cyclization methodology.

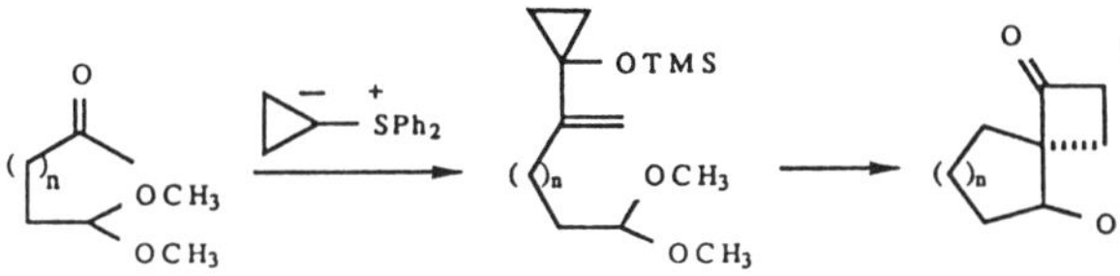

(39)

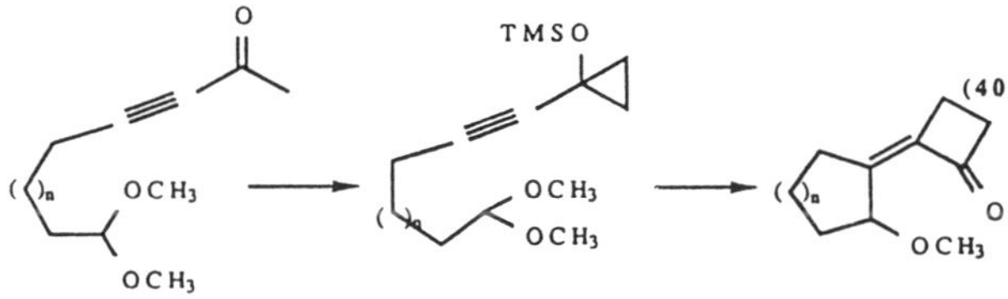

(40)

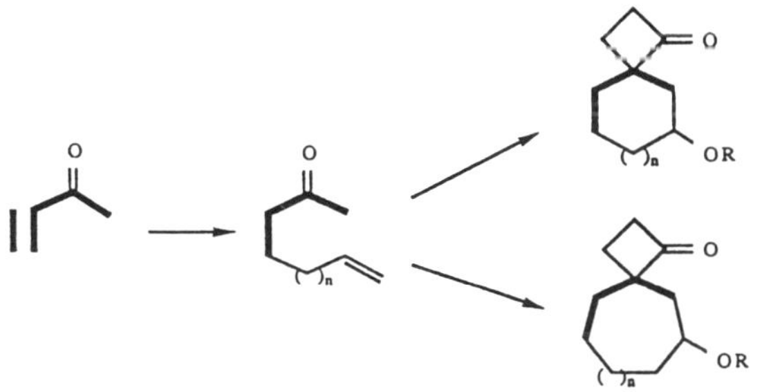

(41)

Acknowledgement

I thank the National Science Foundation and the National Institutes of Health for their generous support of our programs. My students involved in these programs are identified in the references. Special acknowledgment goes to Henry C. Arndt, who did all of the work on the halogenative rearrangement of cyclopropyl carbinols, Donna C. Lee, who did almost all the work on the use of vinylcyclopropanols as cyclization terminators, and James M. Balkavoc and Michael K.-T. Mao, who carried out the total synthesis of allamcin, plumericin, and allamandin. The contributions of this talented group of co-workers to both concept and practice made this chemistry possible.

## References

1. For reviews of earlier work, see Trost, B.M. *Accounts Chem. Res.*, **1974**, *7*, 85; Trost, B.M. *Gazz. Chim. Ital.*, **1984**, *114*, 139; Trost, B.M. *Topics Curr. Chem.*, **1986**, *133*, 3.

2. Trost, B.M.; Keeley, D.; Bogdanowicz, M.J. *J. Am. Chem. Soc.*, **1973**, *95*, 3068; **1974**, *96*, 1252; **1976**, *98*, 248.

3. Trost, B.M.; Keeley, D.E.; Arndt, H.C.; Rigby, J.H.; Bogdanowicz, M.J. *J. Am. Chem. Soc.*, **1977**, *99*, 3080; Trost, B.M.; Keeley, D.E.; Arndt, H.C.; Bogdanowicz, M.J. *J. Am. Chem. Soc.*, **1977**, *99*, 3088.

4. Also see Cohen, T.; Matz, J.R. *Syn. Commun.*, **1980**, *10*, 311; Cohen, T.; Daniewski, W.M.; Weisenfeld, R.B. *Tetrahedron Lett.*, **1978**, 4665; Tanaka, K.; Unema, H.; Matsui, S.; Koji, A. *Bull. Chem. Soc. Japan*, **1982**, *55*, 2965; Gadwood, R.C.; Lett, R.M. *J. Org. Chem.*, **1982**, *47*, 2268.

5. For other applications see Miller, R.D.; McKean, D.R. *Tetrahedron Lett.*, **1979**, 583; Miller, R.D.; McKean, D.R.; Kaufmann, D. *Tetrahedron Lett.*, **1979**, 587; Miller, R.D.; McKean, D.R. *Tetrahedron Lett.*, **1980**, 2639; Bumgardner, C.L.; Lever, J.R.; Purrington, S.T. *Tetrahedron Lett.*, **1982**, *23*, 2379.

6. Oxygen analogue: Cohen, T.; Matz, J.R. *J. Am. Chem. Soc.*, **1980**, *102*, 6900; Gadwood, R.C. *Tetrahedron Lett.*, **1984**, *25*, 5851.

7. Selenium analogue: Halazy, S.; Lucchetti, J.; Krief, A. *Tetrahedron Lett.*, **1978**, 3971.

8. Burgess, E.M.; Penton, H.R.; Taylor, E.A. *J. Org. Chem.*, **1973**, *38*, 26.

9. Arndt, H.C., unpublished work in our laboratories.

10. Trost, B.M.; Lavoie, A. *J. Am. Chem. Soc.*, **1983**, *105*, 5075.

11. Negishi, E-I. *Accounts Chem. Res.*, **1982**, *15*, 340; Heck, R.F. *Org. Reactions*, **1982**, *27*, 345; Stille, J.K. *Angew. Chem. Int. Ed. Eng.*, **1986**, *25*, 508.

12. Trost, B.M.; Bogdanowicz, M.J. *J. Am. Chem. Soc.*, **1973**, *95*, 2038; Trost, B.M.; Bogdanowicz, M.J.; Kern, J. *J. Am. Chem. Soc.*, **1975**, *97*, 2218.

13. Trost, B.M. *Accounts Chem. Res.*, **1978**, *11*, 453; *Chem. Rev.*, **1978**, *78*, 363.

14. Field, L.; McFarland, J.W. *Org. Syn. Coll. Vol. 4*, **1963**, 940.

15. For early investigations of vinylcyclopropanol with electrophiles, see Wasserman, H.H.; Hearn, M.J.; Cochoy, R.E. *J. Org. Chem.*, 1980, *45*, 2874.

16. Trost, B.M.; Mao, M.K-T *J. Am. Chem. Soc.*, 1983, *105*, 6753; Trost, B.M.; Mao, M.K-T; Balkovic, J.M.; Buhlmayer, P. *J. Am. Chem. Soc.*, 1986, *108*, 4965.

17. Trost, B.M.; Brandi, A. *J. Am. Chem. Soc.*, 1984, *106*, 5041.

18. Trost, B.M.; Bogdanowicz, M.J. *J. Am. Chem. Soc.*, 1973, *95*, 5321.

19. Trost, B.M.; Rigby, J. *J. Org. Chem.*, 1976, *41*, 3217; 1978, *43*, 2938.

20. Lee, D.C., unpublished work in our laboratories.

21. Sakurai, H. *Pure Appl. Chem.*, 1982, *54*, 1.

22. In our hands, the reaction of alkynyllithiums with cyclopropanone hemiacetals proved less convenient than cyclopropanation of enol silyl ethers. For the former method, see Salaun, J. *Chem. Rev.*, 1983, *83*, 619.

23. Furukawa, J.; Kawabata, N.; Nishimura, J. *Tetrahedron*, 1968, *24*, 53; 1969, *25*, 2647; Miyano, S.; Hashimoto, H. *Bull. Chem. Soc. Japan*, 1973, *46*, 892.

24. Cf Girard, C.; Amice, P.; Barnier, J.P.; Conia, J.M. *Tetrahedron Lett.*, 1974, 3329; Conia, J.M. *Pure Appl. Chem.*, 1975, *43*, 317; Girard, C.; Conia, J.M. *J. Chem. Res., Synop.*, 1978, 182.

25. Trost, B.M.; Balkovec, J.M.; Mao, M.K.-T. *J. Am. Chem. Soc.*, 1986, *108*, 4974.

26. Trost, B.M.; Mao, M.K.-T. *Tetrahedron Lett.*, 1980, *21*, 3523.

27. Trost, B.M.; Salzmann, T.N.; Hiroi, K. *J. Am. Chem. Soc.*, 1976, *98*, 4887.

28. Trost, B.M.; Buhlmayer, P.; Mao, M.K-T *Tetrahedron Lett.*, 1982, *23*, 1443.

STEREOSELECTIVE ELECTROCYCLIZATIONS AND SIGMATROPIC SHIFTS OF STRAINED RINGS:
*TORQUOELECTRONICS*!


Kendall N. Houk
Department of Chemistry and Biochemistry
University of California, Los Angeles
Los Angeles, California 90024 USA


ABSTRACT. The stereoselectivities of cyclobutene electrocyclizations have been investigated computationally with *ab initio* molecular orbital calculations, and unusual predictions have been tested experimentally. Outward rotations are favored by donors and inward rotations by powerful acceptors. The electronic effects which cause this stereoselectivity are found to influence the ground-state geometries of cyclobutenes and other substituted small-ring compounds. The study of the electronic effects which promote twisting in a predictable fashion is a new aspect of stereoelectronics which we have named (only partly in jest) "torquoelectronics"!


## 1. INTRODUCTION

### 1.1. Introductory remarks

It is a pleasure for me to participate in this conference, and I would like to take this opportunity to thank Professor Armin de Meijere for the invitation. It gave me the chance to organize my thoughts about several subjects we have studied which are related to the topic of this conference, having to do with structures and reactions of molecules containing three and four membered rings. The subjects which I will talk about can be loosely classified together under two terms, torquoselectivity and torquoelectronics, which began as a joke, but I hope do not end as a joke! These words now serve a purpose, to describe a newly-recognized type of stereoelectronic effect.

### 1.2 Organization of lecture

I will begin with a short history of our entry into small-ring chemistry, with some interesting and puzzling observations made by Wolfgang Kirmse's group in Bochum. This collaboration led us to study cyclobutene electrocyclizations theoretically in great detail, and eventually to develop a new theory to explain rates and stereochemical effects. We have also studied the influence of these substituents on the structures of small-rings.
  I will also describe how the same ideas can be applied to predict what is likely to happen in other electrocyclizations, and in ground states of molecules such as oxiranes, aziridines, and methylenecyclopropanes. Torquoelectronics serves to relate ground state distortions to transition state energetic effects.

Next, I will describe our studies of retro-ene reactions in small rings. The calculations show quite clearly dramatic stereoelectronic effects, which are relevant to a variety of related reactions in which cyclopropane or cyclobutane single bonds are broken.

Finally, I will cite some additional data from Kirmse's laboratories, and suggest that torquoelectronics could be a modifier of substituent effects on sigmatropic shifts and stepwise reactions. Torquoelectronic effects serve as a bridge between the understanding of concerted and diradical processes.

## 2. STEREOSELECTIVITY OF CYCLOBUTENE ELECTROCYCLIZATIONS.

Our initial inspirations involved the data from Kirmse's and Dolbier's groups. Kirmse measured the rates of electrocyclization of various substituted cyclobutenes, and deduced the following substituent effects on the activation energy for ring opening (1).

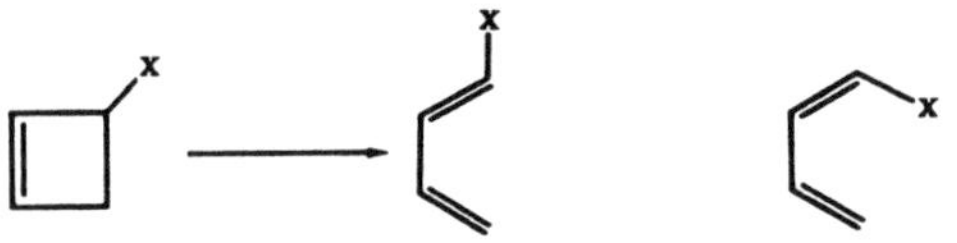

| X | $\Delta E_a$ (out) | $\Delta E_a$ (in) | $\Delta\Delta E_a$ |
|---|---|---|---|
| H | $\equiv 0^a$ | $\equiv 0^a$ | 0 |
| Me | -1 | +3 | 4 |
| Cl | -3 | +6 | 9 |
| OR | -5 | +10 | 15 |

[a]The activation energy for cyclobutene opening is 32 kcal/mol.

The order of substituent effects indicates that electronic effects, not just steric effects, are influencing rates and stereoselectivities. About the same time, Dolbier reported the interesting result shown below (2). The perfluorodimethylcyclobutene opens up to give the product arising from a highly crowded transition state.

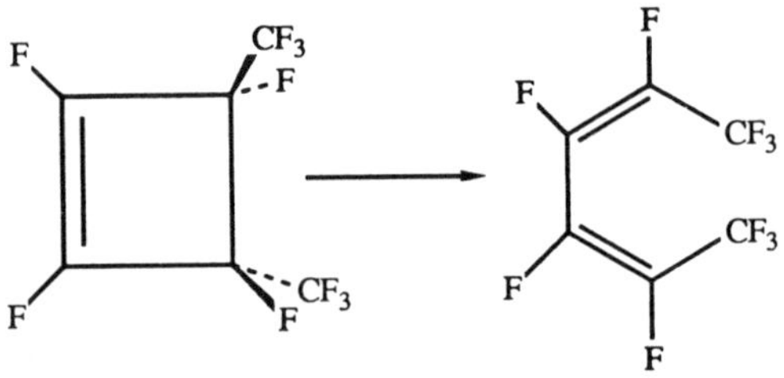

Formation of the product in which both trifluoromethyls rotate outward is disfavored by 19 kcal/mol, a huge difference indeed. In response to these data, Nelson Rondan, in my group at Pittsburgh, undertook calculations which led to a theoretical explanation, as well as interesting predictions which were ultimately verified and which are still leading to quite a variety of predictions about structural, reactivity, and stereoselectivity effects, particularly in small ring systems that this Symposium is all about.

The explanation can be made with the aid of the orbital diagram shown in Figure 1.

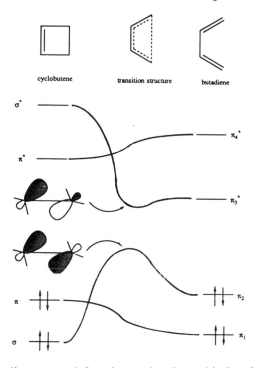

Figure 1: Correlation diagram and frontier molecular orbitals of transition state for cyclobutene conrotatory electrocyclic opening.

The calculations revealed something that we had not anticipated from the qualitative correlation diagram for this reaction. The HOMO is primarily a stretched and twisted σ bond orbital, while the LUMO is primarily a stretched and twisted σ* orbital. Electron-donor substituents at a saturated carbon of cyclobutene have very little effect on the stability of the reactant but stabilize the transition state to a large extent due to overlap and mixing with the σ* orbital. Electron-withdrawing groups also influence the transition state, but not the reactant energies.

This provided what ultimately was a simple explanation of the Kirmse and Dolbier results. Shown below are the σ and σ* orbital of the cyclobutene-opening transition state, and a p orbital of a substituent. When this p orbital is the doubly-occupied orbital of an electron-donor, then upon outward rotation, the substituent orbital interaction with the transition structure σ orbital lends to destabilization, since both of these orbitals are occupied. This is more than compensated for by the stabilizing interaction with σ* of the transition state. This two-electron interaction dominates, and electron donors therefore lower the activation energy.

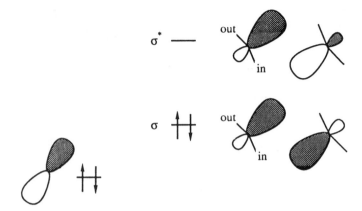

Upon inward rotation, the four-electron interaction is larger because of the increased overlap. This is destabilizing, while the two-electron interaction with σ* is diminished because of lesser overlap. Put another way, the inward rotation introduces an anti-aromatic four-electron interaction, while outward rotation results in a stabilizing interaction due to linear conjugation (1). These orbital interactions intensify with better donors, so the difference in activation energy for outward versus inward rotation is extremely large with good donors, and can even become favorable for inward rotation if the substituent is a sufficiently good acceptor.

Our subsequent studies led to predictions about the magnitude of other substituent effects and to the more exciting prediction that some substituents would rotate inward preferentially (4-9). Table 1 summarizes our calculations and the available experimental data, substituents are listed in the table in the order of increasing $\sigma_R°$ values. Note the switchover to preferred inward rotation, between apparently similar groups of electron-withdrawing groups.

Table 1. Substituent Effects on the Stereochemistry of Cyclobutene Electrocyclizations. R is at C-3.

| R- | $\sigma_R°$ | $E_a$ (in-out)[a] (6-31G*//3-21G) | $E_a$ (in-out)[b] Experimental |
|---|---|---|---|
| $NH_2$ | -0.48 | 17.5 | — |
| OH | -0.43 | 17.2 | 14[c] |
| F | -0.34 | 16.9 | — |
| $CH_3$ | -0.11 | 6.8 | 4 |
| H | 0.0 | 0.0 | 0 |
| CN | 0.13 | 4.3 | — |
| $NO_2$ | 0.15 | 7.4 | — |
| CHO | 0.24 | -4.6 | -3[d] |
| NO | 0.32 | -2.6 | — |
| $BH_2$ | — | -18.2 | — |

[a] kcal/mol
[b] Kirmse; Rondan; and Houk, ref. 1.
[c] experimental value for alkoxy
[d] Rudolf; Spellmeyer; and Houk, ref. 5.

In experimental work in our laboratories and at least one other, predictions have been tested and confirmed (5-10). Several of the more interesting results are described below.

David Spellmeyer performed calculations that indicated a 4.6 kcal/mol preference for inward rotation of a formyl group. In the experiment, only the product of inward rotation, the (Z)-aldehyde was detected (5).

The measured activation energy, 26 kcal/mol, is very close to that predicted, 25 kcal/mol, which is 7 kcal/mol below the activation energy for cyclobutene opening (5).

The large sensitivity of the inward or outward rotational preference to the nature of the substituent is demonstrated by calculations on a series of electron-withdrawing substituents which are summarized in Table 2.

Table 2. Predicted Change in Activation Energy of Cyclobutene Opening ($E_a$ = 32 kcal/mol) by Electron-withdrawing Substitutents. Calculations are 6-31G*//3-21G.

| Substituent | $\Delta E_a$(out) | $\Delta E_a$(in) | $\Delta \Delta E_a$ |
|---|---|---|---|
| $-CO_2^-$ | -6.4 | +0.9 | +7.3 |
| $-CN$ | -2.3 | +2.3 | +4.6 |
| $-CO_2Me$[a] | -4.2 | -1.7 | +2.5 |
| $-CO_2H$ | -4.3 | -2.0 | +2.3 |
| $-CHO$ | -2.3 | -6.9 | -4.6 |
| $-CO_2H_2^+$ | -10.2 | -15.8 | -5.6 |

a) Only 3-21G results were obtained, with a standard methyl group.

Of particular interest is the sensitivity of activation energies to minor changes in the nature of electron-withdrawing substituents. All of the groups in Table 2 lower the activation energy of reaction upon outward rotation. Upon inward rotation, the activation energy is only lower than that for outward rotation for the formyl group and the protonated carboxylic acid.

Some of these predictions have been tested. Piers found that the $CO_2Me$ group has a small preference for outward rotation.[10] We have studied several compounds with cyano groups or acid groups and α-methyl groups at C-3.

For X=CN, we find only the product arising from Me rotation out and cyano in. This implies that cyano has less than a 2.5 kcal/mol preference for outward rotation. Calculations predict the correct product, but only because they overestimate the preference for outward rotation of both Me (7 kcal/mol calculated; 4 kcal/mol experimental) and CN (5 kcal/mol calculated; <3 kcal/mol experimental). The corresponding acid gives a 60:40 mixture at 120° C, but when a Lewis-acid catalyst ($BF_3$ or $CF_3CO_2H$) is added, only the compound with the acid rotating in is observed, and the reaction proceeds at room temperature. This change of stereochemistry by acid or base catalysis is a new aspect of stereochemistry of potential general interest. Calculations indicate that the small preference for outward rotation by a carboxylic acid should become larger in the conjugate base. On the other hand, protonation of the acid produces a powerful electron-withdrawing group, and inward rotation is expected. The latter has already been verified experimentally.

These ideas about the influence of donor and acceptor substituents on adjacent bonds of cyclobutenes can be applied to a wide variety of σ-bond-breaking processes. A few of those being studied by our group or elsewhere are shown below.

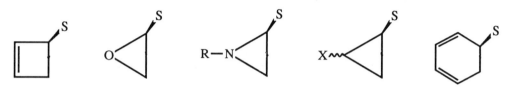

The conrotatory opening of other substituted cyclobutenes is being studied in our labs and others. Analogous effects should occur in other conrotatory openings such as those of oxiranes and aziridines. We have undertaken a study of the thermal electrocyclic ring-opening of oxiranes and aziridines. Since the carbonyl ylides and azomethine ylides that are formed can be observed directly in favorable cases or trapped in others, it is possible that the direction of electrocyclic opening can be determined.

Disrotatory reactions should show similar effects to conrotatory, except the magnitude of the effects may be diminished. This is shown in the drawings below.

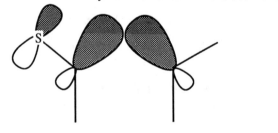 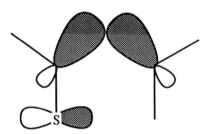

                Outward                               Inward

The difference in overlap between inward versus outward rotation of the substituent is smaller in disrotatory ring opening. Two cases under study are noted above. Dolbier and Phanstiel have already found that a fluorine prefers to rotate outward in a cyclopropyl solvolysis.[12] The leaving group determines the direction of rotation, so the preference had to be determined by rate measurements. The cyclohexadiene-hexatriene interconversion is an additional target, although the triene is often less stable than the diene, so rate measurements on the closure reactions of isomeric reactants will be necessary.

The transition state effect which we have noted should also be important in ground states of a variety of molecules. That is, donors should repel adjacent σ bonds, while good conjugative acceptors should attract them. This effect is quite large in transition structures, because breaking or forming σ-bonds have a high-energy HOMO and low-energy LUMO. Nevertheless, in ground states there must be similar, if smaller, effects which will be manifested in ground state distortions. Indeed, David Spellmeyer discovered these distortions in calculated structures of substituted cyclobutene ground states.[7] The 3-21G optimized structures of several of these are shown below:

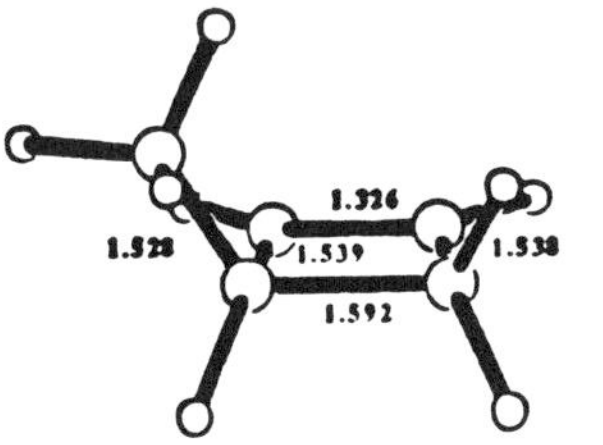

3-methylcyclobutene

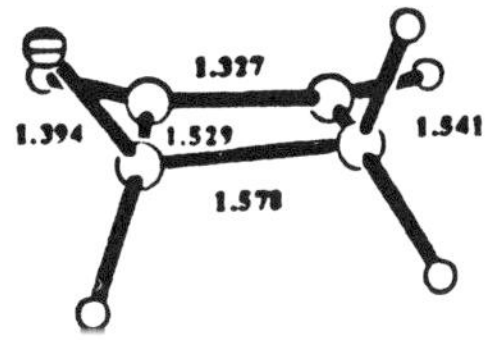

3-fluorocyclobutene

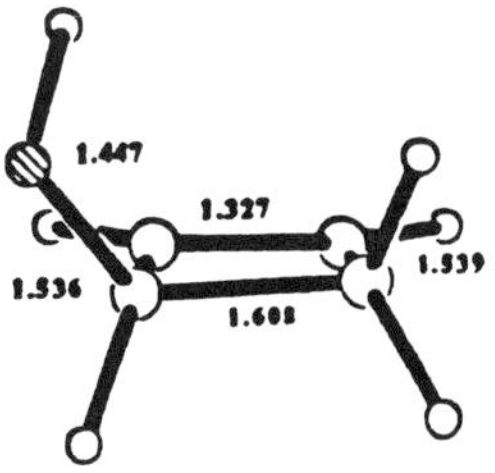

3-hydroxycyclobutene

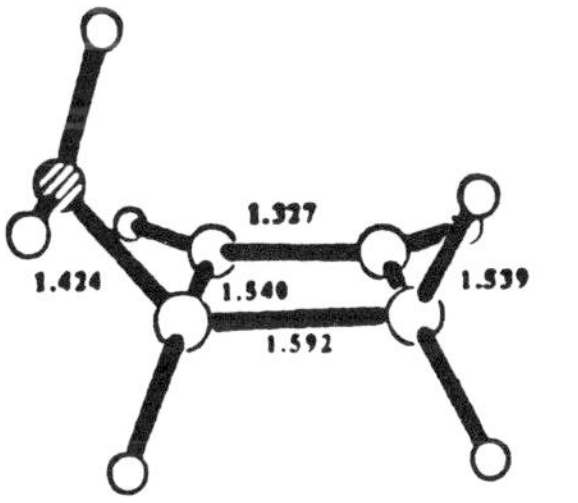

3-aminocyclobutene

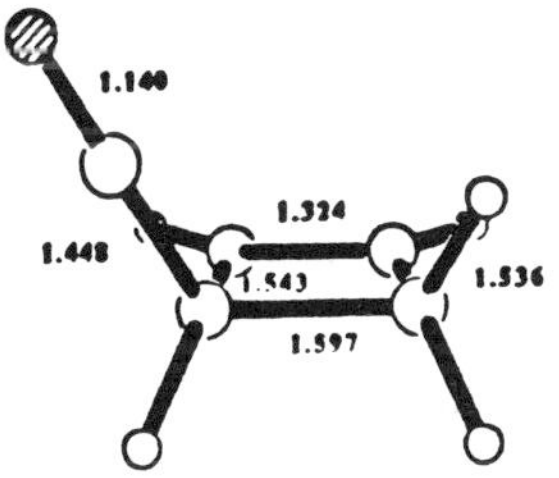

3-cyanocyclobutene

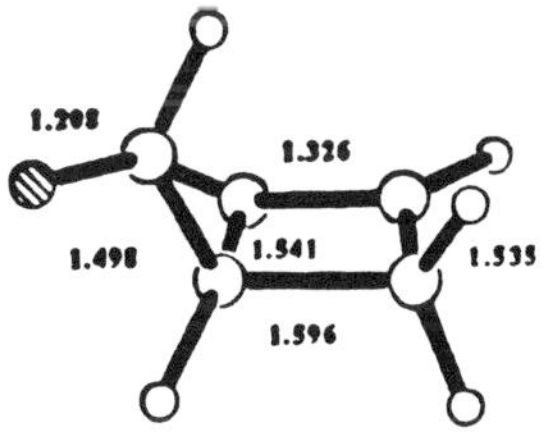

3-formylcyclobutene

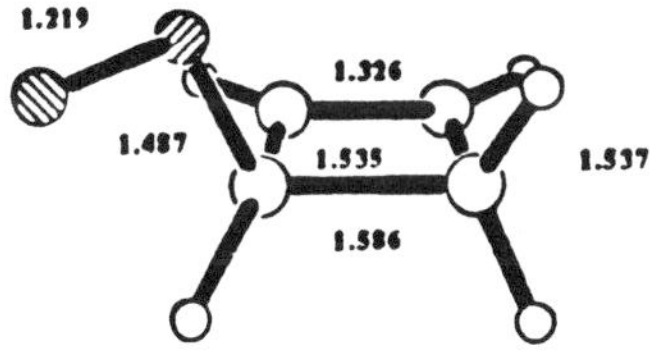

3-nitrosocyclobutene

In fact, there is a remarkable correlation of the ground-state molecular distortion and the $\sigma_R^\circ$ constant of the substituents, just as there was for transition state energetic effects. The graph below summarizes the data. The dihedral deviation is the difference between the dihedral angle $XC_3C_2C_1$ of a substituted cyclobutene and that of the H of cyclobutene, $HC_3C_2C_1$.

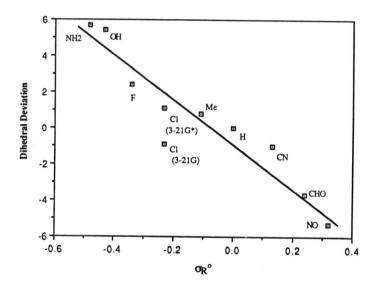

These effects are also present in oxiranes as found in calculations by Hiroshi Tezuka and Jeffrey Evenseck[13], and summarized below.

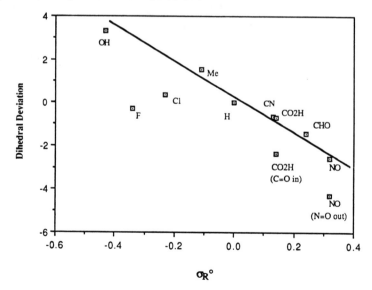

The same type of effect is observed in ring-substituted methylenecyclopropanes as well.

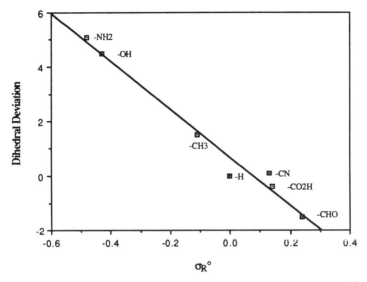

We expect that these ground state effects will be reflected in larger transition state effects, and consequently that the stereoselectivity of conrotatory ring-opening of these molecules will be controlled in much the same way as the cyclobutene electrocyclizations.

Such a connection between ground-state distortions and stereoselectivity is increasingly common. We noticed it some time ago for norbornenes[14], and John Baldwin recently noted the connection for ground-state distortions and 1,3-sigmatropic shift stereoselectivities of methylenecyclopropanes.[15] Baldwin noted that the X-ray and neutron diffraction structures of a substituted methylenecyclopropane revealed small distortions of the carbon skeleton in the direction corresponding to the major pathway of thermal isomerization. The distortion is shown below, along with a drawing which shows the major direction of isomerization of several derivatives.

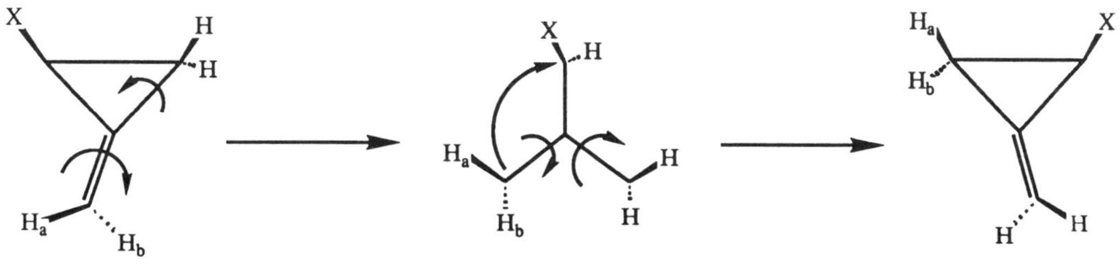

While our calculations have focussed upon a different distortion, they do suggest a reason that the preferred isomerization mode is that shown above: it occurs so as to cause outward rotation of the donor substituents. Only a powerful acceptor like a formyl group is likely to rotate inward.

Now we will turn to another aspect of strained-ring reactions, that of the stereochemistry of the retro-ene reactions of vinylcyclopropanes and vinylcyclobutanes. We have studied the reactions shown below[17]:

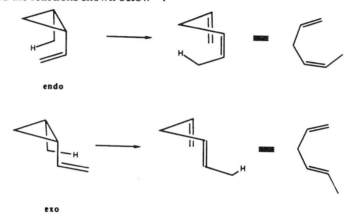

Daub and Berson[18] estimated that the endo transition state is at least 12 kcal/mol lower in energy than the exo transition state. This free energy difference was estimated from experiments on substituted derivatives. Berson, following an earlier suggestion by Winstein[19], proposed that this rate difference "is plausibly associated with orbital overlap factors".[18]

The homodienyl hydrogen shift of cis-1-methyl-2-vinylcyclopropane was studied with the 3-21G basis set. Our calculated activation energy is too high, but the activation energy difference of 17 kcal/mol between the endo and exo transition structures should be relatively reliable. This difference is in accord with Berson's experimental estimate of ≥12 kcal/mol.

The transition structures for the parent ene reaction of propene with ethylene, the endo transition structure, TS-endo, and exo transition structure, TS-exo of the homodienyl hydrogen shift are shown below.

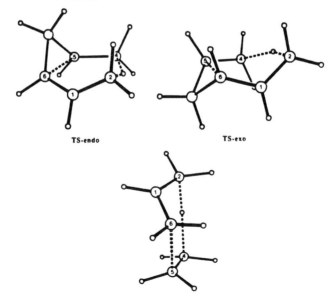

The endo transition structure is a bridged seven-membered ring with a chairlike geometry. The exo transition structure, TS exo, can also be viewed as a seven-membered ring with chairlike geometry, but differs from TS-endo in that the atoms of the six-membered ring are in a boatlike conformation, whereas those of the *endo* are in a half-chair conformation.[17]

The differences between the endo and exo transition structure geometries occur because they maximize overlap of the breaking CH bond with the breaking CC bond of the cyclopropane ring. As shown below in a comparison of the propene plus ethylene transition structures (A and B), the *endo* (C and D) and *exo* (E and F) transition structures, the *endo* requires little distortion from reactant in order to orient the molecule so that the hybridized orbital of $C_4$ aligns with the hybridized orbital on $C_5$. The $\pi$-orbital at $C_1$ is oriented as to maximize overlap with the developing double bond orbital at $C_6$, while still maintaining $\pi$ bonding with the hybrid orbital on $C_2$.

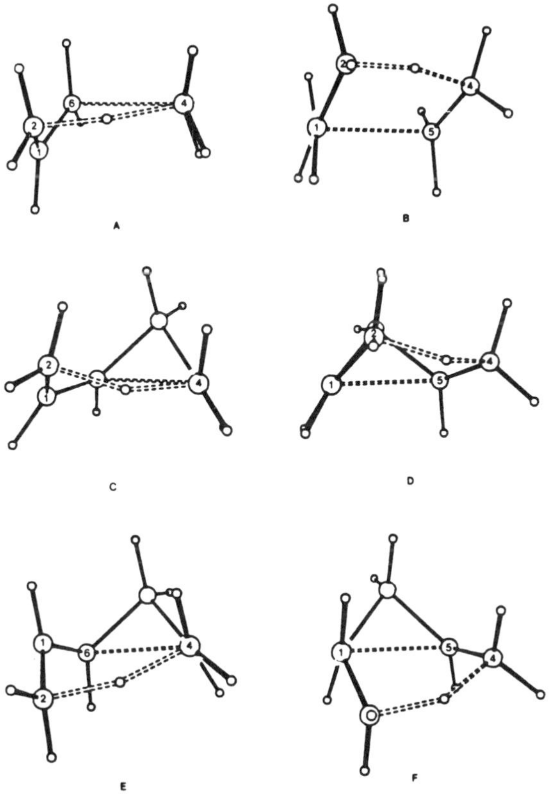

In the *exo* transition structure, the overlap of the hybrid orbitals on $C_4$ and $C_5$ is diminished somewhat. In order to have overlap of the hydrogen 1s orbital with a p orbital at $C_2$, and still have overlap of the hybrid orbitals on $C_4$ and $C_5$, the CH bond breaking is much more advanced (1.446 Å). The nature of this distortion, which is brought about by the need to complete the hydrogen transfer, is easily seen by the staggered position of the methyl group and is reflected by the similar forming CH bond lengths of 1.480 and 1.453 Å for TS-endo and TS-exo, respectively. As seen in E and F, the $\pi$-orbital of $C_1$ is again oriented to maximize overlap of the developing double bond while maintaining overlap with the hybrid orbital of the former double bond. The enormous preference of *endo* reaction arises primarily from distortions of the alkyl fragment in the transition state.

We have carried out studies of the cyclobutane case as well, and estimate a less than 5 kcal/mol preference in that case for the *endo* transition state. The same general features apply, but there is a less profound preference because there is less bending of the breaking CC bond in cyclobutane.

These reactions prompt us to return to general torquoelectronic considerations. As shown below, when a σ bond breaks, the "torque" about this bond will influence the magnitude of donor and acceptor substituent interactions. If the twist is enforced by the type of reaction, as in the cyclobutene opening, the influence on substituent effects can be very dramatic.

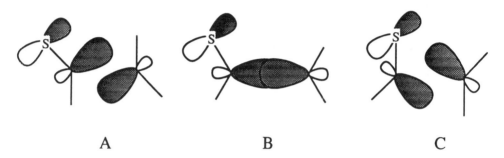

A          B          C

For example, an "inside" rotating acceptor has a larger stabilizing effect upon inward rotation than it has upon outward. The opposite is true for donors, which actually destabilize the breaking σ bond upon inward rotation. There will be discovered cases which are like one or the other of these situations, and substituent effects will follow different patterns as a result. Future experimental and theoretical work is designed to uncover such reactions.

ACKNOWLEDGEMENT. I acknowledge the brilliance and diligence of my coworkers, many of whom are named in the references. Professor Hiroshi Tezuka and Jeffrey Evanseck have also done many of the calculations described here. Our work has been supported generously by the University of California, Los Angeles, the National Science Foundation, and the National Institutes of Health.

REFERENCES AND NOTES

(1) Kirmse, W.; Rondan, N. G.; Houk, K. N. *J. Am. Chem. Soc.*, **1984**, *106*, 7989.
(2) Dolbier, W. R.; Koroniak, H.; Burton, D. J.; Bailey, A. R.; Shaw, G. S.; Hansen, S. W. *J. Am. Chem. Soc.*, **1984**, *106*, 1871.
(3) Rondan, N.; Houk, K. N. *J. Am. Chem. Soc.*, **1985**, *107*, 2099.
(4) Spellmeyer, D. C.; Houk, K. N. *J. Am. Chem. Soc.*, **1988**, *110*, 3412.
(5) Rudolf, K.; Spellmeyer, D. C.; Houk, K. N. *J. Org. Chem.*, **1987**, *52*, 3708.
(6) Jefford, C. W.; Rimbault, C. G.; Wang, Y.; Miller, R. D.; Spellmeyer, D. C.; Houk, K. N. *J. Org. Chem.*, **1988**, *53*, 2125.
(7) Spellmeyer, D. C. Ph. D. *Dissertation*, U.C.L.A., 1988.
(8) Kallel, A.; Houk, K. N., unpublished results.

(9) Wang, Y.; Duh, H.-Y.; Buda, A.; Houk, K. N., unpublished results.
(10) Piers, E., unpublished results.
(11) Buda, A.; Houk, K. N., unpublished results.
(12) Dolbier, W.; Phanstiel, O. *Tetrahedron Lett.*, **1987**, *29*, 53.
(13) Tezuka, H.; Evanseck, J., unpublished results.
(14) Houk, K. N.; Rondan, N. G.; Brown, F. K.; Jorgensen, W. L.; Madura, J. D.; Spellmeyer, D. C. *J. Am. Chem. Soc.*, **1983**, *105*, 5980.
(15) Baldwin, J. *J. Org. Chem.*, **1987**, *52*, 1173.
(16) Thomas, B.; Houk, K. N., unpublished results.
(17) Loncharich, R. J.; Houk, K. N. *J. Am. Chem. Soc.*, **1988**, *110*, 2089.
(18) Daub, J. P.; Berson, J. A. *Tetrahedron Lett.*, **1984**, *25*, 4463.
(19) Glass, D. S.; Boikess, R. S.; Winstein, S. *Tetrahedron Lett.*, **1966**, 999.

THE PREPARATION AND THERMAL REARRANGEMENT OF FUNCTIONALIZED
6-(1-ALKENYL)BICYCLO[3.1.0]HEX-2-ENES. APPLICATIONS TO SYNTHESIS


Edward Piers
Department of Chemistry
University of British Columbia
Vancouver, B.C., Canada V6T 1Y6


ABSTRACT. In connection with developing methods that would be applicable to the total synthesis of natural products such as sinularene (6), prezizaene (7), and quadrone (8), the thermolytic rearrangements of a number of substituted 6-(1-alkenyl)bicyclo[3.1.0]hex-2-enes were investigated. Upon thermolysis, compounds 9, 12, 36, 38, 42, 44, and 58 underwent clean [3,3]-sigmatropic (Cope) rearrangement to provide excellent yields of the functionalized bicyclo[3.2.1]octa-2,6-dienes 10, 15, 37, 30, 43, 45, and 59, respectively. In contrast, substances 29, 46, and 55 proved to be very poor substrates for Cope rearrangement. Thus, heating of these materials gave low yields or none of the corresponding Cope rearrangement products 30, 47, and 56, respectively.

1. INTRODUCTION

The thermal $\sigma 2_s + \pi 2_s + \pi 2_s$ (Cope) rearrangement of 6-endo-vinylbicyclo[3.1.0]hex-2-ene (2) to bicyclo[3.2.1]octa-2,6-diene (3) (eq. [1]) was first reported in 1964.[1] In this work, the cis "divinylcyclopropane" 2 was not isolated since, under the conditions of its formation (reaction of the aldehyde 1 with methylenetriphenylphosphorane in refluxing tetrahydrofuran-diethyl ether), it rearranged rapidly to 3. In 1965, Brown[2] isolated 2 and determined that the thermal conversion of 2 into 3 occurs with a half-life of approximately one day at 25 °C ($E_a$ = 22.9 kcal/mol). It appears to be well accepted that 2 is transformed into 3 via a concerted process involving a transition state that can be represented by A (eq. [1]).[3-5]

More than a decade after the seminal studies of Schleyer[1] and Brown,[2] Baldwin and Gilbert[6] showed that thermolysis of 6-exo-vinylbicyclo[3.1.0]hex-2-ene (4) at elevated temperatures (e.g. 195 °C) also provides the bridged bicyclic diene 3 (eq. [2]). Furthermore, using optically active substrates, these researchers found that the overall conversion of 4 into 3 takes place with one-center epimerization at C-6. Presumably,[3-6] transformation of 4 into 3 involves initial formation of the diradical 5, which undergoes appropriate bond rotation and subsequent ring closure to the <u>endo</u> isomer 2. The latter substance then undergoes facile bond reorganization to provide 3.

During the period 1964-1980, the thermal rearrangement of 6-(1-alkenyl)bicyclo[3.1.0]hex-2-enes received relatively little attention from synthetic organic chemists. However, the current decade has seen appreciable activity in this area and it has become clear that this type of transformation constitutes a versatile and effective method for the preparation of functionalized bicyclo[3.2.1]octa-2,6-dienes. Since many natural products, particularly in the terpenoid family, incorporate into their structures the bicyclo[3.2.1]octane carbon skeleton, the possibility of employing this rearrangement for the total synthesis of naturally occurring substances is particularly attractive. Thus, our recent work in this area has focussed on the preparation and thermal rearrangement of highly functionalized 6-(1-alkenyl)bicyclo[3.1.0]hex-2-enes, with the aim of producing substances that would serve as suitable intermediates for the total synthesis of the sesquiterpenoids sinularene (6), prezizaene (7), and quadrone (8). The following summary and discussion of the results of this work will concentrate on the preparation and thermal behavior of a number of bicyclic and tricyclic "divinylcyclopropane" systems.

## 2. RESULTS AND DISCUSSION

### 2.1 Studies Related to the Synthesis of (±)-Sinularene (6)

Sinularene (6) is a structurally unusual marine natural product isolated from the soft coral <u>Sinularia mayi</u>.[7] In model studies, it was shown[8] that the <u>trans</u> "divinylcyclopropane" 9 rearranged smoothly to the bicyclic diene 11 (eq. [3]). Presumably, this transformation involves the intermediacy of the <u>endo</u> isomer 10, formed from 9 by way of a diradical (<u>vide supra</u>). Bond reorganization of 10 then proceeds via the transition state **B** to afford the product 11. Interestingly, the Cope rearrangement of 10 is a facile process even though the presumed transition state is destabilized by a notable steric interaction between the isopropyl group and the ring methylene group (see **B**). In any case, the success of the conversion 9 → 11 showed the viability of using this rearrangement methodology to prepare a bicyclo[3.2.1]octa-2,6-diene possessing an <u>exo</u>-oriented C-4 isopropyl group.

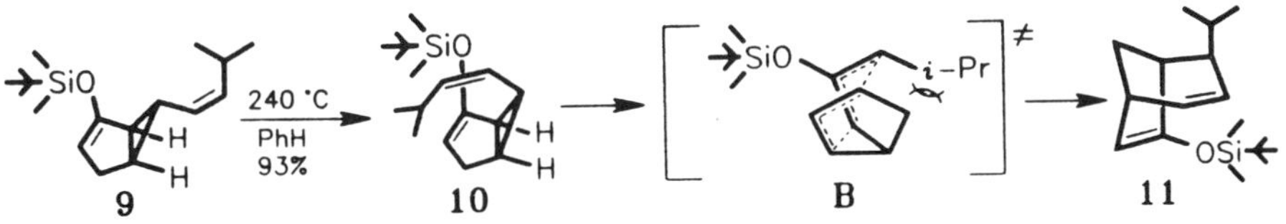

[3]

The key intermediate for the synthesis of (±)-sinularene (6) was envisaged to be the highly substituted bicyclo[3.1.0]hexene 12 (Scheme 1).[9] However, it is necessary to note at the outset that thermolysis of 12 could, in theory, result in (at least) two different modes of bond reorganization. Thus, in addition to (or instead of) the desired Cope rearrangement (12 → 15), it was possible that a homo-[1,5]-sigmatropic hydrogen migration (12 → 16) could occur. In the model study (eq. [3]), the latter type of reaction was precluded, since substrate 9 did not possess an angular methyl group.

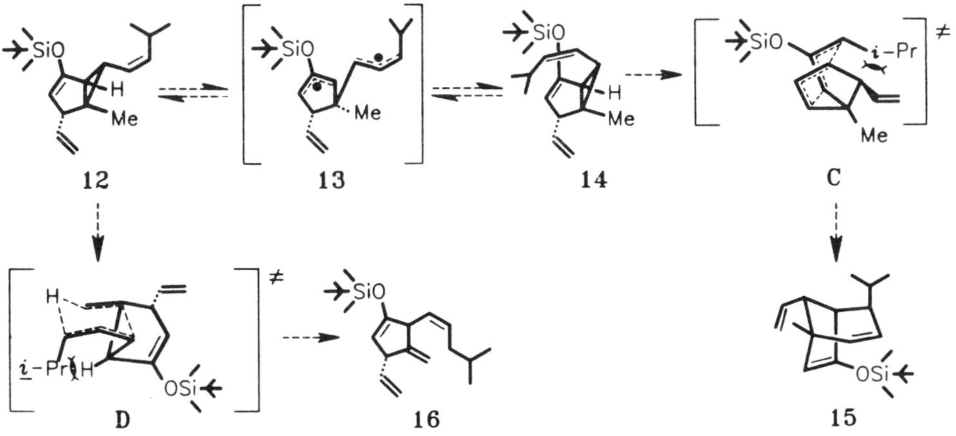

*Scheme 1*

Which of the two pathways outlined in Scheme 1 would be favored? Although a clear-cut prediction is not straightforward, the chemical literature does contain pertinent, helpful data.

For the "parent" homo-[1,5]-sigmatropic hydrogen migration (17 → 18, eq. [4]), which, presumably, proceeds via a transition state that can be represented by E,[4,10] the energy of activation $E_a$ is 31.2 kcal/mol.[11] The presumed transition state for the transformation of 12 into 16 suffers from a serious steric interaction as shown in D (Scheme 1). Therefore, $E_a$ for this process would be expected to be significantly larger than 31.2 kcal/mol.

$$17 \xrightarrow[\text{kcal/mol}]{\text{heat} \atop E_a=31.2} [E]^{\ddagger} \longrightarrow 18 \quad [4]$$

It is highly likely that the rate-limiting step of the overall conversion of 12 into 15 would be the isomerization of 12 into 14, via the diradical 13. For example, the values of $E_a$ for the transformations 19 → 20 and 21 → 20 (eq. [5]) are 32.8 and 20.5 kcal/mol, respectively.[12] Furthermore, as noted previously, $E_a$ for the Cope rearrangement of 6-endo-bicyclo[3.1.0]hex-2-ene (2) (eq. [1]) is 22.9 kcal/mol and, even though the transition state for the Cope rearrangement of 14 is destabilized by a notable steric repulsion (see C, Scheme 1), it is unlikely that this interaction would raise the value of $E_a$ for the latter process to the region 32-35 kcal/mol.

$$19 \xrightarrow[\text{kcal/mol}]{\text{heat} \atop E_a=32.8} 20 \xleftarrow[\text{kcal/mol}]{\text{heat} \atop E_a=20.5} 21 \quad [5]$$

On balance, if one assumes that $E_a$ for 12 → 14 would be similar to that for 19 → 20 (32.8 kcal/mol) and that $E_a$ for 12 → 16 would be considerably higher than that for 17 → 18 (31.2 kcal/mol), it appears reasonable to surmise that thermolysis of 12 would proceed in the desired manner to give the bicyclic compound 15.

The synthesis of the key "divinylcyclopropane" 12 was achieved as outlined in Scheme 2.[9] Orthoester-based Claisen rearrangement of the allylic alcohol 22 afforded the ester 23, which was converted via standard transformations into the diazo ketone 24. Carbenoid cyclization of 24, followed by partial hydrogenation of the resultant bicyclic keto alkyne 25, provided the keto alkene 26. Conjugate addition of lithium divinylcuprate to the enone 27 (derived from 26 by the method of Saegusa and coworkers[13]) was expected to take place preferentially from the sterically more accessible convex face of the substrate. In the event, the reaction produced the desired (expected) product 28 and the corresponding C-4 epimer in a ratio of approximately 9:1, respectively.

Routine transformation of 28 into the enol silyl ether 12 set the stage for investigation of the crucial thermolysis reaction. It was gratifying to find that this process led to the formation of a single product (86 %), which was readily identified as the bicyclic triene 15. Thus, thermal bond reorganization of 12 occurred exclusively in

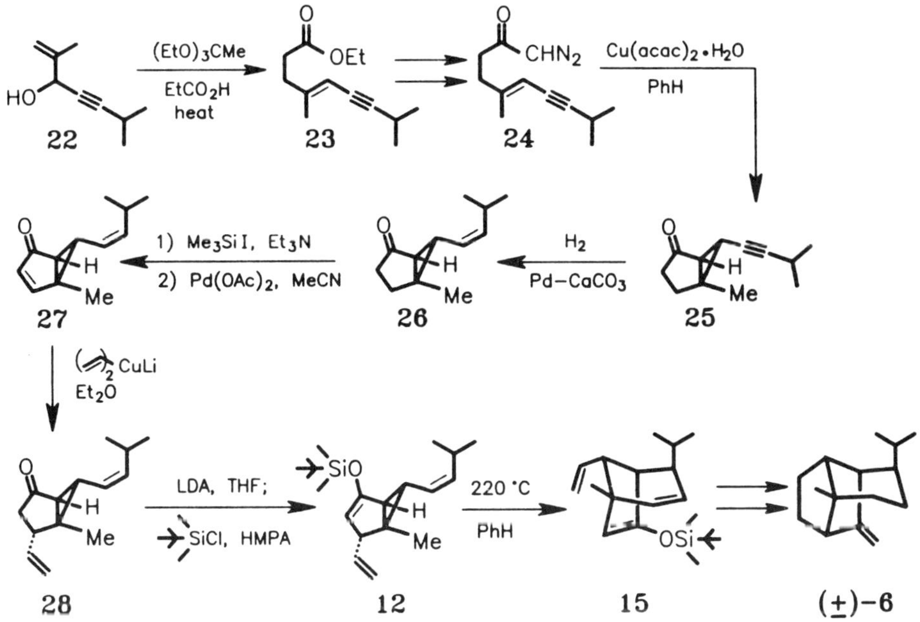

*Scheme 2*

the desired (Cope) sense. The product 15 was transformed, via an efficient four-step sequence of reactions, into (±)-sinularene.[9]

## 2.2 Studies Related to the Synthesis of (±)-Prezizaene (7)

The hydrocarbon (-)-prezizaene (7) has been isolated from *Eremophila georgei*.[14] On the basis of retrosynthetic analyses, it appeared that the synthesis of this natural product might be accomplished via a route in which the thermal rearrangement of a substituted 6-(1-alkenyl)bicyclo[3.1.0]hex-2-ene would play a key role. For example, preparation and successful Cope rearrangement of the substrate 29 would give 30, which, presumably, could be transformed into (±)-prezizaene (7).

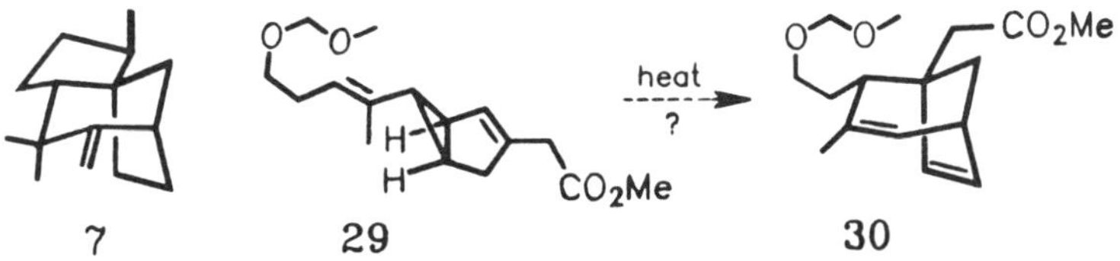

The synthetic route employed to prepare 29 is outlined in Scheme 3.[15] Reaction of the dibromocyclopropane 31 with n-butyllithium under conditions that allow equilibration of the resultant carbenoid species[16] gave, stereoselectively, the exo bromide 32. This material was converted into the corresponding cyclopropylzinc halide, which, upon treatment with the vinyl iodide 33 in the presence of a catalytic amount of tetrakis(triphenylphosphine)palladium(0), gave the bicyclic alkene 34. This overall coupling process (32 + 33 → 34) illustrates a newly developed method for preparing vinyl- and divinylcyclopropanes. A number of other palladium(0)-catalyzed couplings of cyclopropylzinc halides with alkenyl iodides have been carried out[17] and this type of transformation appears to be quite general.

Transformation of 34 into the ketone 35 was accomplished in routine fashion. Subjection of 35 to a Wittig-Horner reaction, followed by deconjugation of the resultant α,β-unsaturated ester, provided the required "divinylcyclopropane" 29.

Unfortunately, thermolysis of 29 at various temperatures (155-220 °C) in benzene (sealed tube) invariably gave a mixture of many products from which the desired substance 30 could be isolated in only poor yields (~ 25 % at best).

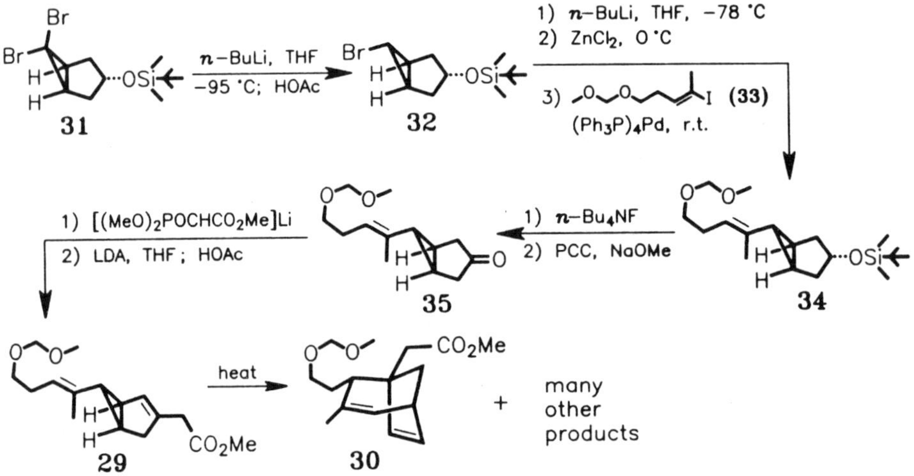

*Scheme 3*

It appears, perhaps not surprisingly, that the presence of the β,γ-unsaturated ester function in 29 is primarily responsible for the poor results obtained from the (high temperature) thermolysis of this material. Thus, for example, thermal rearrangement of the enol ether 36 (readily derived from the ketone 35) provided cleanly and efficiently (87 %) the bicyclic diene (37) (eq. [6]).[15]

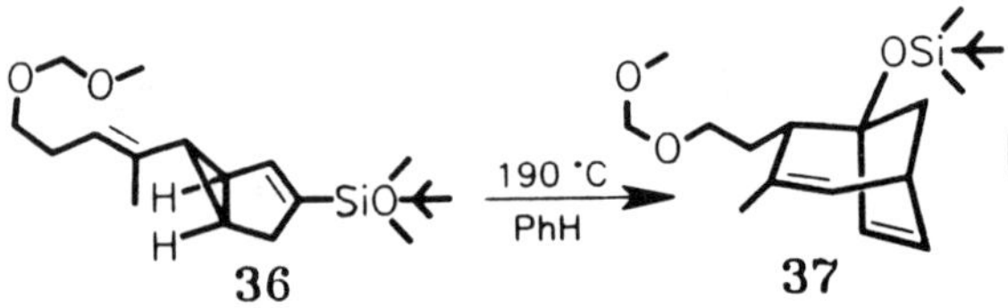

One way to circumvent the difficulties associated with the thermolysis of 29 would be to prepare the C-6 (endo) epimer 38. In the latter material, the two "vinyl" groups of the "divinylcyclopropane" system are *cis* to one another and, consequently, one would expect the Cope rearrangement of 38 to occur at relatively low temperatures. Presumably, the β,γ-unsaturated ester function present in 38 would be stable under these mild conditions.

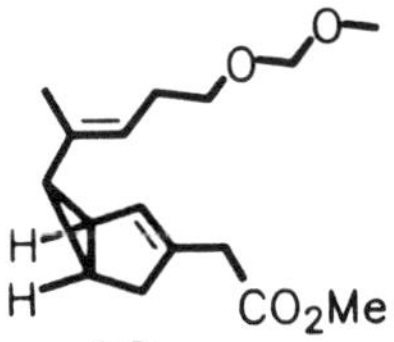

**38**

Scheme 4 outlines the route used to prepare 38.[17] Reaction of the dichlorocyclopropane 39 with excess *tert*-butyllithium under conditions in which the resultant carbenoid species do not equilibrate gave, after protonation of the intermediates, a mixture of 40 and the corresponding C-6 epimer in ratio of 4:1, respectively. Reaction of 40 with lithium 4,4'-di-*tert*-butylbiphenylide,[18] conversion of the resultant lithio species into the corresponding cyclopropylzinc chloride, and subsequent palladium(0)-catalyzed coupling with the vinyl iodide 33 afforded 41 in 62 % yield. Conversion of 41 into the required diene 38 was effected via a route very similar to that outlined previously (34 → 35 → 29, Scheme 3).

Substrate 38 undergoes Cope rearrangement slowly at room temperature and, upon distillation under reduced pressure, is converted into the bicyclo[3.2.1]octa-2,6-diene 30 in high yield (98 %). Compound 30 has been converted into (±)-prezizaene (7).[17]

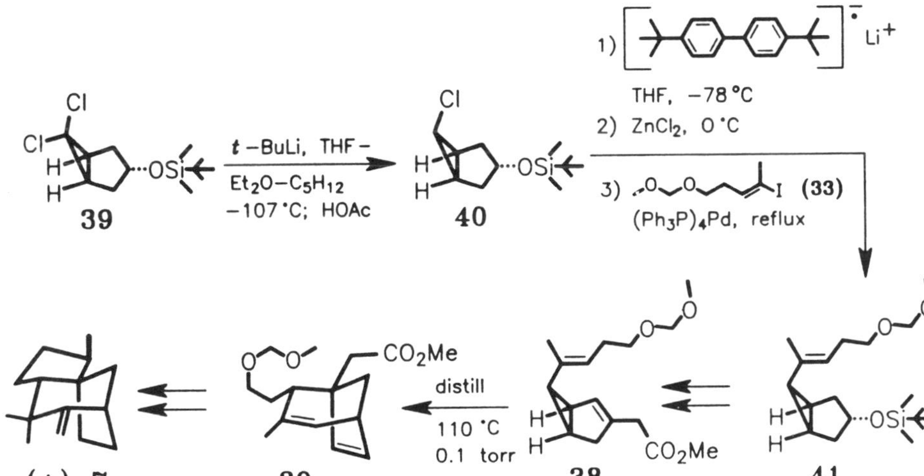

*Scheme 4*

As mentioned previously, both endo- and exo-6-vinylbicyclo-[3.1.0]hex-2-ene undergo efficient thermal rearrangement to bicyclo[3.2.1]octa-2,6-diene (eqs. [1] and [2]). Furthermore, it was found that the highly substituted 6-exo-(1-alkenyl)bicyclo[3.1.0]-hex-2-ene 12 serves as an excellent substrate for the preparation of the Cope rearrangement product 15 (Scheme 2). However, on the basis of the work just described, it is clear that there are occasions when it is necessary to take note of the fact that cis divinylcyclopropanes generally undergo Cope bond reorganization at much lower temperatures than the corresponding trans isomers. Thus, while the endo isomer 38 is an excellent synthetic precursor of the bicyclic diene 30 (Scheme 4), the corresponding exo isomer 29 is not (Scheme 3).

2.3 Studies Related to the Synthesis of (±)-Quadrone (8)

The cytotoxic sesquiterpenoid (-)-quadrone, isolated from the fungus Aspergillus terreus, has been shown to possess the constitution and absolute configuration shown in 8.[19] In model studies, it had been shown[20,21] that the trans "divinylcyclopropanes" 42 and 44 undergo smooth and efficient (> 90 %) Cope rearrangement to the tricyclic dienes 43 and 45, respectively (eqs. [7] and [8]). Thus, it appeared possible that thermolysis of appropriately substituted versions of 42 or 44 could provide substances that would serve as suitable intermediates for the synthesis of quadrone.

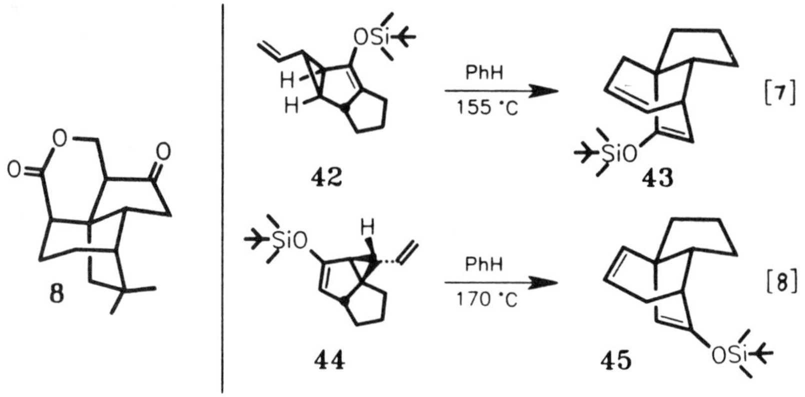

It was decided to investigate initially the preparation and thermal behavior of compound 46. Successful Cope rearrangement of this material would provide 47. Presumably, the latter substance could easily be transformed into the tricyclic keto aldehyde 48, which had been converted previously into (±)-quadrone (8).[22]

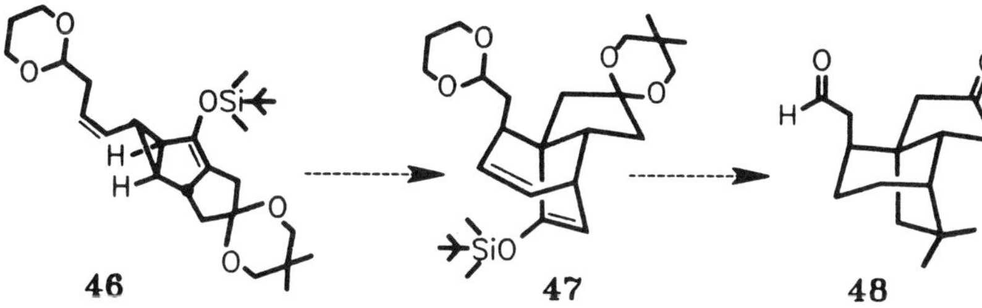

Rhodium(II) acetate-catalyzed addition of ethyl diazoacetate to the alkene 49 (Scheme 5)[23] afforded a mixture of esters 50, which was readily converted into the stereochemically homogeneous aldehyde 51. Wittig reaction of 51 with the phosphorane 52, followed by cleavage of the tert-butyldimethylsilyl ether and oxidation of the resultant alcohol, provided the cis alkene 53, accompanied by minor amounts of the corresponding trans isomer. The nicely crystalline cis compound could be obtained by fractional crystallization of the mixture.[21] Transformation of 53 into the enol silyl ether 46 set the stage for investigating the key thermolysis reaction.

Unfortunately, thermolysis of 46 in benzene produced a mixture of at least three products. Furthermore, careful analysis of the crude product mixture by proton nuclear magnetic resonance spectroscopy failed to detect the presence of the desired material 47.[21]

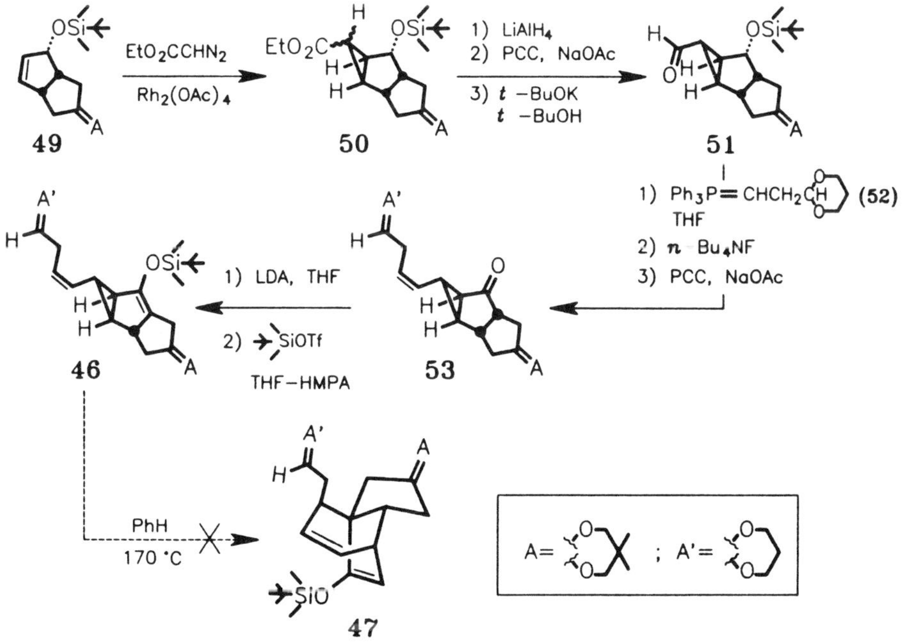

Scheme 5

In an effort to probe the reason(s) underlying the failure of 46 to undergo the Cope rearrangement, the structurally simpler substrate 55 was prepared from the aldehyde 54 (Scheme 6).[21] Thermolysis of 55 in benzene produced a mixture of three major products, accompanied by a number of minor products. Furthermore, partial separation of the mixture by column chromatography, along with careful spectral analysis of the various fractions, showed clearly that none of the major compounds was the desired Cope rearrangement product 56.[21] Thus, both substrates 46 (Scheme 5) and 55 (Scheme 6) failed to rearrange in the required (Cope) sense.

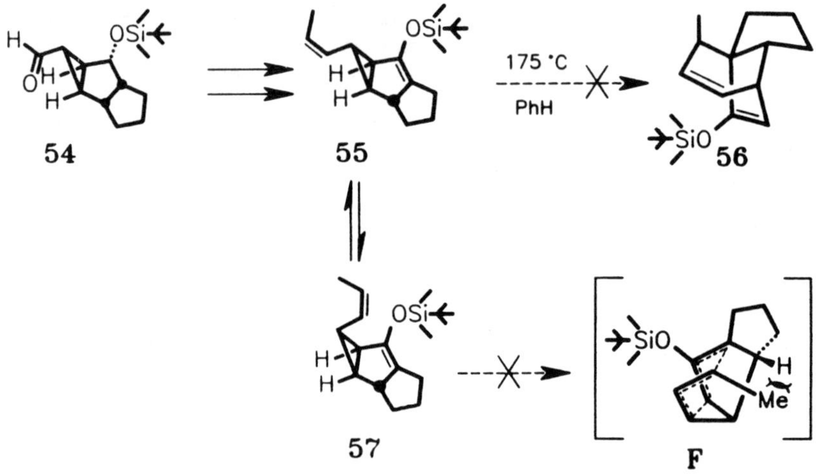

*Scheme 6*

As noted earlier, substrate 42 undergoes clean, efficient Cope rearrangement (eq. [7]). Structurally, the only difference between 42 and 55 is that the former substance possesses at C-4 a simple vinyl group while the latter compound contains at C-4 a cis-1-propenyl function. Why does this rather subtle structural change result in such a profound difference in the thermal behavior of 42 and 55? It is highly likely that the answer to this question is to be found by considering the Cope rearrangement step. That is, on the basis of many reported examples of Cope rearrangement of trans "divinylcyclopropane" systems, there is little doubt that the isomerization of 55 into 57 should proceed smoothly. However, molecular models show that, due to the highly rigid ring system in 57, the (presumed) transition state for the Cope rearrangement of this material would be destabilized by a severe steric interaction between the methyl group of the cis-1-propenyl function and the angular proton, as shown in F (Scheme 6). Apparently, this destabilizing repulsion precludes the possibility of a Cope rearrangement and other modes of bond reorganization become operative. A similar line of reasoning can be used to rationalize the fact that thermolysis of 46 does not produce 47 (Scheme 5).

In view of the failures described above, it was necessary to modify the synthetic strategy associated with the projected synthesis of (±)-quadrone (8). Thus, employing chemistry similar to that outlined in Scheme 5, the aldehyde 51 was converted into the trans "divinylcyclopropane" 58 (Scheme 7).[23] Not unexpectedly, the latter substance, upon thermolysis in benzene, was transformed smoothly and efficiently into the tricyclic Cope product 59. Cleavage of the enol silyl ether function in 59 provided the ketone 60, which was converted via a fairly standard sequence of reactions into the keto aldehyde 48.[23] As mentioned earlier, compound 48 had been converted into (±)-quadrone (8) by Burke et al.[22] Thus, eventually, a formal total synthesis of (±)-quadrone was achieved via a route in which a Cope rearrangement of a "divinylcyclopropane" system played a key role. However, it should be noted that, due to the lack of success in the originally planned route and the consequent necessity to use a modified approach, the sequence that was finally used was quite lengthy.

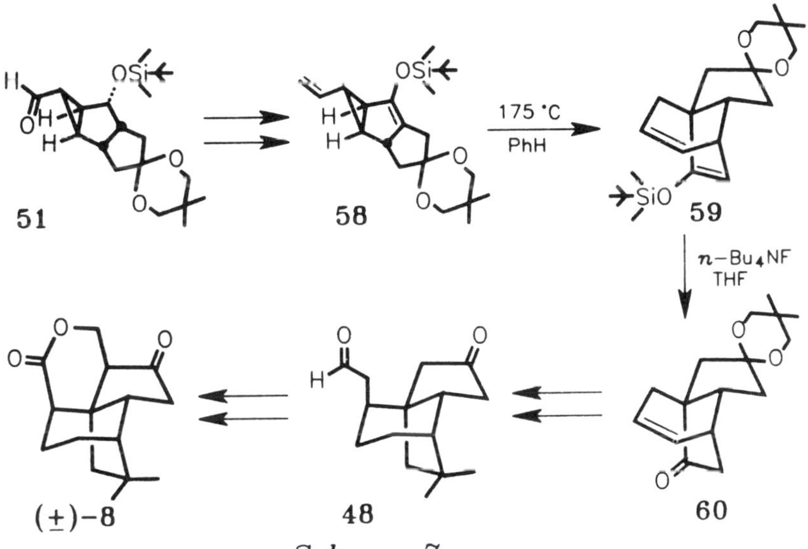

Scheme 7

## 3. CONCLUSION

The studies summarized above show that the thermal rearrangement of functionalized 6-(1-alkenyl)bicyclo[3.1.0]hex-2-ene systems can serve as an excellent method for the synthesis of substances that contain as part of their structures the bicyclo[3.2.1]octane carbon skeleton. For example, thermolysis of the substrates 9, 12, 36, 38, 42, 44, and 58 provides, in each case, an excellent yield of the corresponding

[3,3]-sigmatropic rearrangement product. On the other hand, these investigations also revealed (some) limitations of this type of process. Thus, heating of the "divinylcyclopropanes" **29**, **46** and **57** produces little or none of the corresponding Cope rearrangement products.

## 4. REFERENCES

1. C. Cupas, W.E. Watts, and P. von R. Schleyer, Tetrahedron Lett. 2503 (1964).
2. J.M. Brown, Chem. Commun. 226 (1965).
3. S.J. Rhoads and N.R. Raulins, Org. React. 22, 1 (1975).
4. E.M. Mil'vitskaya, A.V. Tarakanova, and A.V. Plate, Russ. Chem. Rev. 45, 469 (1976).
5. J.J. Gajewski, Hydrocarbon Thermal Isomerizations, Academic Press, Inc., New York, NY, 1981, pp. 215-216, 258-260.
6. J.E. Baldwin and K.E. Gilbert, J. Am. Chem. Soc. 98, 8283 (1976).
7. C.M. Beechan, C. Djerassi, J.S. Finer, and J. Clardy, Tetrahedron Lett. 2395 (1977).
8. E. Piers, G.L. Jung, and E.H. Ruediger, Can. J. Chem. 65, 670 (1987).
9. E. Piers and G.L. Jung, Can. J. Chem. 65, 1668 (1987).
10. C.W. Spangler, Chem. Rev. 76, 187 (1976).
11. R.J. Ellis and H.M. Frey, J. Chem. Soc. 5578 (1964).
12. W. Pickenhagen, F. Naf, G. Ohloff, P. Muller, and J.-C. Perlberger, Helv. Chim. Acta, 56, 1868 (1973).
13. Y. Ito, T. Hirao, and T. Saegusa, J. Org. Chem. 43, 1011 (1978).
14. P.J. Carrol, E.L. Ghisalberti, and D.E. Ralph, Phytochemistry, 15, 777 (1976).
15. E. Piers and P.S. Marrs, Unpublished work.
16. D. Seyferth, R.L. Lambert, Jr., and M. Massol, J. Organomet. Chem. 88, 255 (1975).
17. E. Piers, M. Jean, and P.S. Marrs, Tetrahedron Lett. 28, 5075 (1987).
18. P.K. Freeman and L.L. Hutchinson, J. Org. Chem. 48, 4705 (1983).
19. R.L. Ranieri and G.J. Calton, Tetrahedron Lett. 499 (1978); G.J. Calton, R.L. Ranieri, and M.A. Espenshade, J. Antibiot. 31, 38 (1978); K. Kon, K. Ito, and S. Isoe, Tetrahedron Lett. 25, 3739 (1984).
20. E. Piers, G.L. Jung, and N. Moss, Tetrahedron Lett. 25, 3959 (1984).
21. E. Piers and N. Moss, Unpublished work.
22. S. Burke, C.W. Murtiashaw, J.O. Saunders, J.A. Oplinger, and M.S. Dike, J. Am. Chem. Soc. 106, 4558 (1984).
23. E. Piers and N. Moss, Tetrahedron Lett. 26, 2735 (1985).

# SYNTHESIS WITH DONOR-ACCEPTOR-SUBSTITUTED VINYLCYCLOPROPANES

H.-U. Reissig*, R. Zschiesche, A. Wienand, M. Buchert
Institut für Organische Chemie
Technische Hochschule Darmstadt
Petersenstr. 22, D-6100 Darmstadt, FRG

ABSTRACT. The concept of using 2-alkenylsubstituted 2-siloxycyclopropanecarboxylates as protected enones is introduced. Efficient synthesis of these compounds as well as simple ring cleavage variants are described. In-situ reaction of the liberated enones with hetero nucleophiles or carbon nucleophiles allows preparation of a variety of functionalized γ-oxocarboxylate derivatives. Of special interest are adducts obtained from nitro alkanes in high yields. These are precursors for natural products as the pheromone chalcogran or the macrolides pyrenophorin and vermiculine, respectively. Intramolecular 1,3-dipolar cycloaddition of a nitrone - gained from one of these nitro alkane adducts - as well as intramolecular Diels-Alder-reactions provide stereoselective entries to polycyclic systems. Implications of these strategies for natural product synthesis are discussed.
 An alternative approach to donor-acceptor-substituted vinyl-cyclopropanes by use of Fischer carbene complexes is also represented. Thus, vinyl-substituted Cr(0) carbene complexes transfer this ligand cleanly to electron-deficient alkenes to afford vinylcyclo-propanes. The second route to these compounds involves methyl dienoates and their highly stereoselective reaction with simple Cr(0) carbene complexes.

## INTRODUCTION

Donor-acceptor-substituted cyclopropanes of general structure **1** are interesting and versatile building blocks for organic synthesis[1,2]. To this class of strained bifunctional synthons also belong methyl 2-siloxycyclopropanecarboxylates **2**, which are easily available in reasonable quantity from silyl enol ethers and methyl diazoacetate. Cyclopropanes **2** can further be substituted or functionalized at C-1 by generation of the ester enolate using LDA and reaction with apt electrophiles[2]. Smooth regioselective ring cleavage of **2** is achieved by fluoride reagents to afford γ-oxocarboxylate derivatives **3**.

Combined with certain other transformations this concept allows efficient preparations a variety of heterocyclic compounds **4**[2,3].

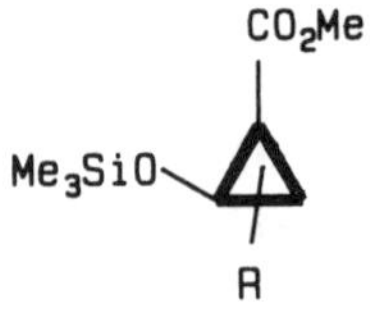

## 2-ALKENYL 2-SILOXYCYCLOPROPANECARBOXYLATES

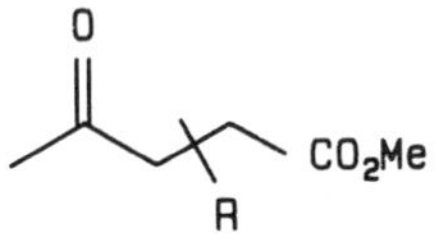

As special derivatives of 2-siloxycyclopropane carboxylates those having an additional 2-alkenyl group, e.g. **6**, are of high preparative interest, because of their <u>equivalence to functionalized enones</u>. At the cyclopropane stage this enone moiety is protected, and therefore many transformations are possible - for instance, elaboration of the substitution pattern required in later steps of a synthesis - without interference of this rather labile functional group.

The most simple compound in this series - cyclopropane **6** - can be obtained in good yield[4] from siloxybutadiene **5**, which is available from methylvinylketone[5].

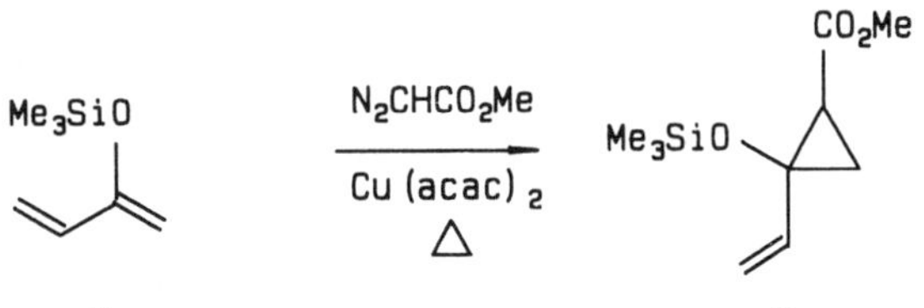

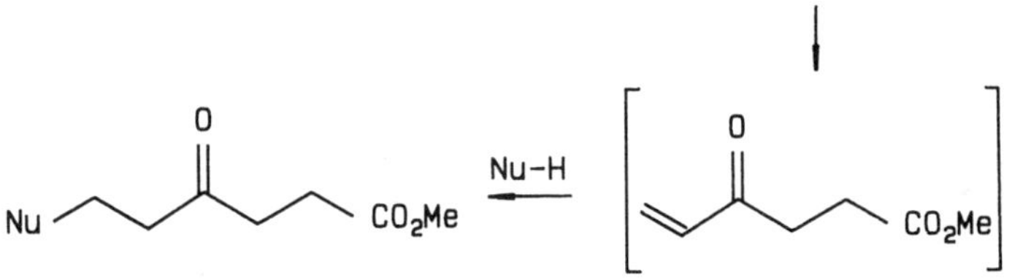

Ring cleavage of **6** provides the isolable, yet, rather sensitive enone **7**. Therefore, we have developed one-pot-procedures to trap **7** in-situ by suitable nucleophiles. This protocol brings about efficient syntheses of adducts **8** and includes oxygen, nitrogen, sulfur, or carbon nucleophiles[4]. For instance, nitrite addition provides nitro compound **9**, whereas methyl acetoacetate gives tetrafunctional adduct **10** in good yield from **6**. Combination of slightly more complex components in this Michael-addition makes available an adduct convertable to a steroid derivative within a few steps[6].

9         10

11         12

Nitroalkanes as CH-acids react with **6** in the presence of Triton B via **7** to afford compounds **11** in almost quantitative yield[7]. Considering the possibility of a Nef-reaction – which can be executed with good efficiency, indeed – these nitro compounds are equivalents of systems with three differentiated carbonyl groups in a 1,4,7-distance.

This pattern of functional groups is of high synthetic value, since its generation by carbon-carbon bond forming reactions usually requires two steps of <u>Umpolung</u>: during synthesis of **11** we have used the "<u>cyclopropane trick</u>" and the <u>nitronate anion</u> $d^1$-<u>synthon equivalence</u>[8].

Starting from 1-nitropropane and **6** the adduct **11** (R = Et) is an excellent precursor for a short-cut synthesis of the bark beetle pheromone chalcogran **12**. Compound **11** (R = Me) could be converted to a known precursor of the macrolide pyrenophorin, whereas the more complex nitro compound **13** – easily prepared from **6**, too – should be transformable into the macrolid vermiculine[9].

Twofold addition of enone **7** to nitromethane gives the symmetrical pentafunctional compound **14**. Following these simple procedures, a variety of polyfunctional compounds having synthetic potential are available[7].

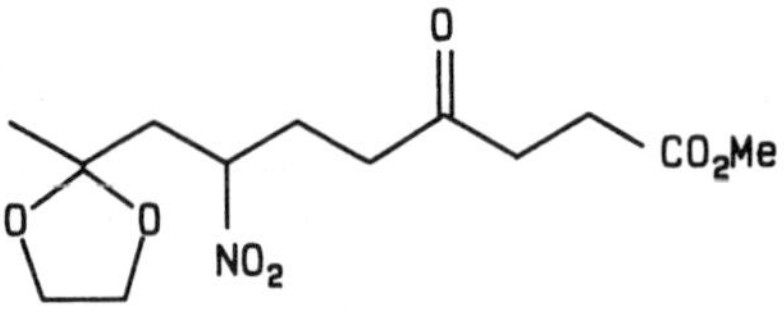

**13**

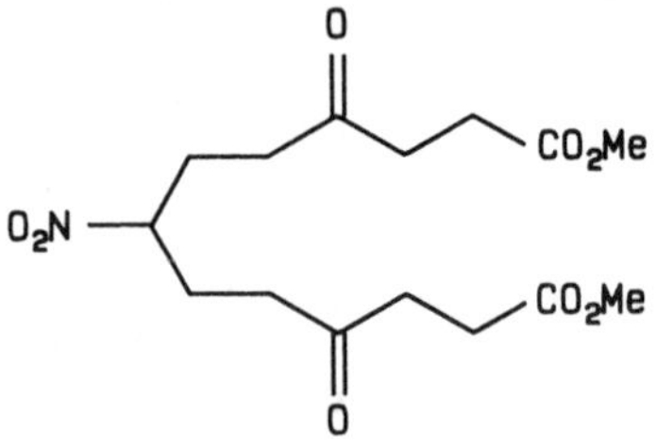

**14**

The reduction of simple γ-nitroketones and of more complex compounds as **11** or **13**, respectively, using ammonium formiate as hydrogen donor and palladium on carbon as catalyst, allows smooth and practical preparation of fivemembered cyclic nitrones **15**[10]. Unfortunately, this method is not applicable for substrates carrying alkenyl side chains as **16** - gained from **6** in two steps[7]. Therefore the classical Zn / NH₄Cl procedure has to be used, which affords nitrone **17**, albeit only in low yield. Subsequent intramolecular 1,3-dipolar cycloaddition provides tricyclic compound **18** in good yield[6]. This approach should have potential for preparation of functionalized pyrrolidine derivatives.

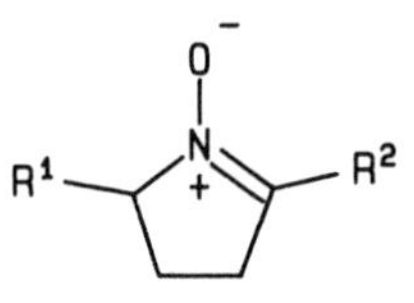

**15**

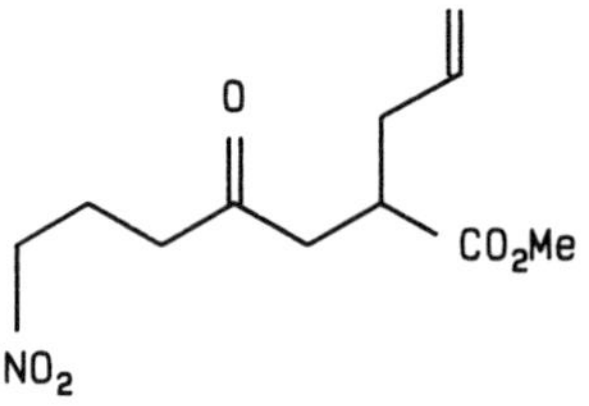

**16**

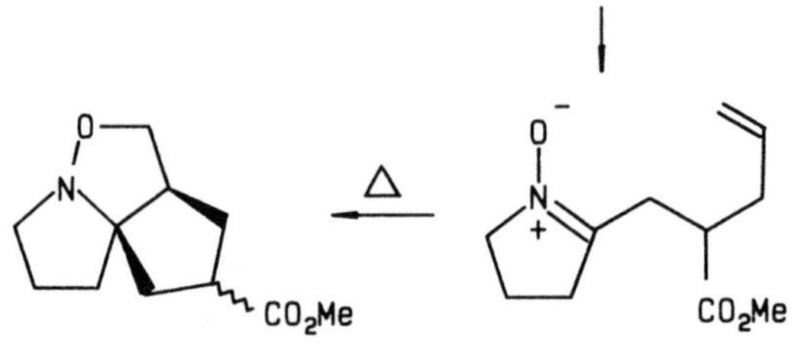

**18**

**17**

On the other hand, intramolecular Diels-Alder reaction is easily achievable by use of trienone **20**. This crucial intermediate is readily prepared by alkylation of **6** with 5-bromo-1,3-pentadiene and subsequent ring opening of **19** with fluoride. Trienone **20** is perfectly suited for the [4+2]-process for electronic and steric reasons. Thus, cis-**21a** and cis-**21b** are obtained after reaction at room temperature as a 6:1 mixture[11].

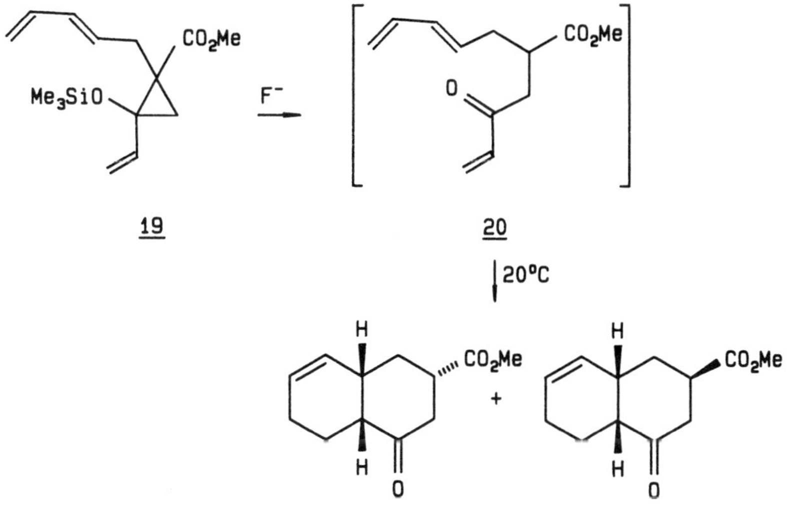

[Structure **19** with Me₃SiO, CO₂Me, vinyl and butadienyl substituents on cyclopropane] →F⁻→ [Structure **20**: open-chain enone with CO₂Me] →|20°C|→ cis-**21a** + cis-**21b** (84 : 16)

[Two decalinone structures shown with H stereochemistry indicators and CO₂Me groups]

The cis-stereochemistry with regard to the ring junction requires an endo-transition state, whereas the location of the CO₂Me group in the predominant cis-**21a** is interpreted by a favoured boatlike conformation of the linkage between diene and dienophile as compared with the chair conformation in the transition states **TS**.

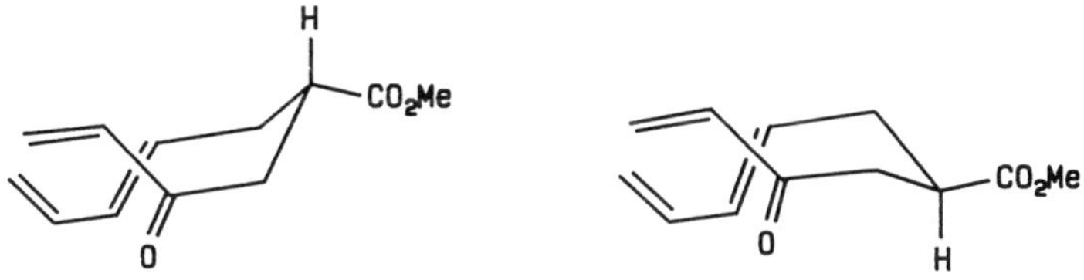

[Transition state structures **TSa** and **TSb**]

Diels-Alder-reaction of a dimethyl substituted analogue of **20** provides a cycloadduct which is a suitable precursor for the terpene α-eudesmol[6].

Summing up the chemistry of vinylcyclopropane **6**, one recognizes that even start with this very simple compound allows approach to several classes of natural products. However, the most notable feature of our concept should be _flexibility_ that is guaranteed by easily available and inexpensive starting materials and that would allow straightforeward synthesis of analogues with good _efficiency_.

Therefore, many future applications of the concept disclosed here are to be expected using vinylcyclopropanes of type 6 as

* strained, but nevertheless stable intermediates,
* protected version of highly reactive, functionalized enones.

## CARBEN COMPLEX ROUTE TO DONOR-ACCEPTOR CYCLOPROPANES

Preparation of donor-acceptor-substituted cyclopropanes as 2 involves combination of an acceptor-substituted carbenoid with an electron-rich olefin. From a synthetic point of view it would be very useful if this polarity pattern could be reversed, that means, addition of a donor-substituted carben(oid)e to an electron deficient alkene. Actually, the cyclopropanation reaction discovered by E.O. Fischer and K.H. Dötz in the beginings of carbene complex chemistry fulfils this requirement[12]. We therefore reinvestigated this reaction of Cr(0) carbene complexes with acceptor olefins to explore scope and limitations of this potentially useful process[13,14].

For synthesis of vinylcyclopropanes two possibilities could be realized. Styryl-substituted carbene complex 22 transfers the carbene ligand to acceptor olefins 23 to provide vinylcyclopropanes 24 in good yield[13].

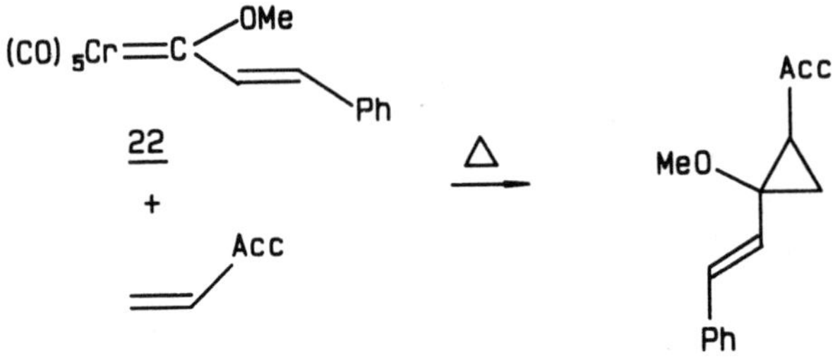

Alternatively, the simple carbene complex 25 adds to methyl dienoates 26 and affords compounds 27 with moderate to good efficiency, however, with very good stereoselectivity[14].

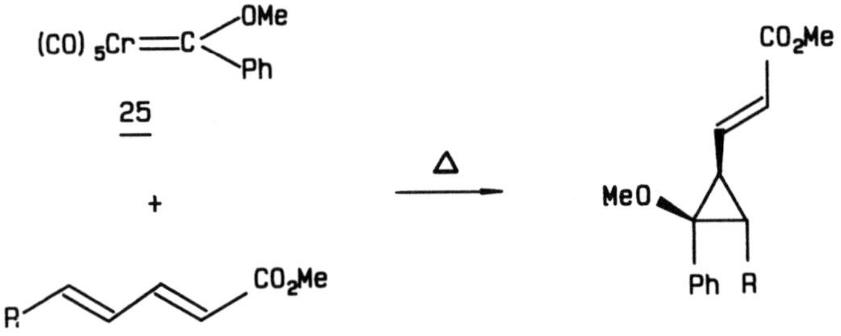

Besides the regioselective formation of **27**, it is interesting to note that both reactions providing **24** or **27**, respectively, follow only the [2+1]-path. There is no evidence for formation of cyclopentene derivatives as result of a [4+1]-cycloaddition. However, these fivemembered carbocycles should be accessible by thermolysis of the vinylcyclopropanes.

Future investigations have to prove whether trifunctional compounds of type **24** and **27** obtained by this carbene complex route are suitable starting materials for other synthetically useful transformations.

ACKNOWLEDGEMENT. We are very grateful to the Deutsche Forschungsgemeinschaft, the Stiftung Volkswagenwerk, the Fonds der Chemischen Industrie, the Vereinigung von Freunden der Technischen Hochschule zu Darmstadt, and the Karl-Winnacker-Stiftung (Hoechst AG, Frankfurt) for generous support of these investigations. Some preliminary experimental contributions of Dipl.Chem. E.L. Grimm and supplementary studies of Dipl.Ing. Th. Hafner are thankfully appreciated.

## REFERENCES

[1] Review: H.-U. Reissig, Organic Synthesis via Cyclopropanes: Principles and Applications in The Chemistry of the Cyclopropyl Group (Z. Rappoport, Ed.), p. 375, J. Wiley & Sons, Chichester 1987.

[2] Review: H.-U. Reissig, Donor-Acceptor-Substituted Cyclopropanes: Versatile Building Blocks in Organic Synthesis, Top. Curr. Chem. **144** (1988) 78.

[3] For recent publications see: a) C. Brückner, H.-U. Reissig, Liebigs Ann. Chem. **1988**, 465. - b) C. Brückner, B. Suchland, H.-U. Reissig, Liebigs Ann. Chem. **1988**, 471. - c) C. Brückner, H.-U. Reissig, J. Org. Chem. **53** (1988) 2440. - d) C. Brückner, H. Holzinger, H.-U. Reissig, J. Org. Chem. **53** (1988) 2450. - e) H.-U. Reissig, H. Holzinger, G. Glomsda, Tetrahedron **44** (1988) in press.

[4] E.L. Grimm, R. Zschiesche, H.-U. Reissig, J. Org. Chem. **50** (1985) 5543.

[5] P. Cazeau, F. Duboudin, F. Moulines, O. Babot, J. Dunogues, Tetrahedron **43** (1987) 2089.

[6] R. Zschiesche, Dissertation, Technische Hochschule Darmstadt 1988.

[7] R. Zschiesche, H.-U. Reissig, Liebigs Ann. Chem. **1988**, in press.

[8] Review: D. Seebach, Angew. Chem. **91** (1979) 259; Angew. Chem. Int. Ed. Engl. **18** (1979) 239.

9) R. Zschiesche, T. Hafner, H.-U. Reissig, Liebigs Ann. Chem. **1988**, in press.

10) R. Zschiesche, H.-U. Reissig, Tetrahedron Lett. **29** (1988) 1685.

11) R. Zschiesche, E.L. Grimm, H.-U. Reissig, Angew. Chem. **98** (1986) 1104; Angew. Chem. Int. Ed. Engl. **25** (1986) 1086.

12) E.O. Fischer, K.H. Dötz, Chem. Ber. **103** (1970) 1273. - K.H. Dötz, E.O. Fischer Chem. Ber. **105** (1972) 1356.

13) A. Wienand, H.-U. Reissig, Tetrahedron Lett. **29** (1988) 2315.

14) M. Buchert, H.-U. Reissig, Tetrahedron Lett. **29** (1988) 2319.

# SYNTHESIS OF FUNCTIONALIZED EPISULFIDES

Wataru Ando* and Norihiro Tokitoh
Department of Chemistry, University of Tsukuba,
1-1-1 Tennohdai, Tsukuba, Ibaraki 305,
Japan

**ABSTRACT.** The results of our recent studies on the synthesis and reactions of the strained episulfides bearing exocyclic double bonds are described. Various types of substituted allene episulfides and 1,2,3-butatriene episulfide derivatives are readily obtained in the reactions of diazo compounds with thioketenes, the rearrangement of sulfur ylide formed by an intramolecular carbene reaction, the thionation of methylenecyclopropanones, and the alkenylidene carbene addition with thioketones. Some properties of the unique episulfides thus obtained are also discussed.

## 1. INTRODUCTION

Much interest has long been focused on the synthesis of strained three-membered ring systems bearing exocyclic double bonds such as methylenecyclopropanes,[1] allene oxides,[2] and allene episulfides[3] in view of their unique reactivities especially in the interconversion 1⇌2⇌3 as has been theoretically demonstrated.[4] As for the parent system of allene

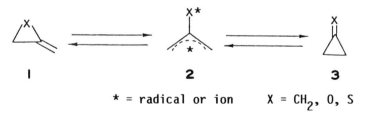

* = radical or ion    X = $CH_2$, O, S

episulfide, in 1978 Block suggested the thermal rearrangement of the initially formed cyclopropanethione into allene episulfide in the reaction of the trithiane **4**.[5] In this reaction the intermediacy of the thioxyallyl ion was postulated in connection with the analogous oxygen system. The equilibrium of allene episulfides **6** and **7** through the thioxyallyl ion **8** was also shown by Block in a labeling experiment.[5] Pyrolysis of the spirothiirane **5** labelled with $^{13}C$ or $^2H$ on the thiirane ring afforded the allene episulfide with the label equally distributed between the ring and exocyclic methylene group.

4

5 6 8 7

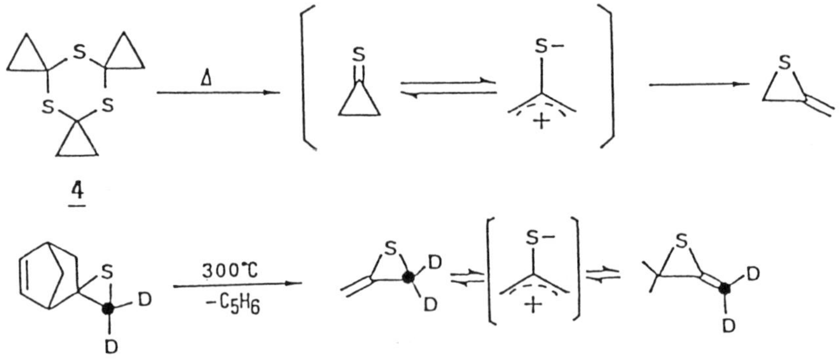

However, in contrast to the other systems the chemistry of the sulfur-containing tautomeric system has not been fully investigated due to the limited examples of allene episulfide synthesis and the lack of stable cyclopropanethione,[5,6] and the nature of the intermediary thioxyallyl species is still a current topics.

Furthermore, very little is known about the synthesis of the 1,2,3-butatriene episulfides, which attract much attention as the methylene homologues of allene episulfides, in contrast to the wide chemistry of the alkenylidenecyclopropanes and bisalkylidenecyclopropanes.[7]

From this point of view, we present here the synthesis of a number of allene episulfides and 1,2,3-butatriene episulfides, and the studies on their thermal and acid-catalyzed isomerization reactions, the substituent effect of which enlightened us on the intrinsic character of thioxyallyl intermediate.

## 2. RESULTS AND DISCUSSION

### 2. 1. Synthesis of a Variety of Substituted Allene Episulfides

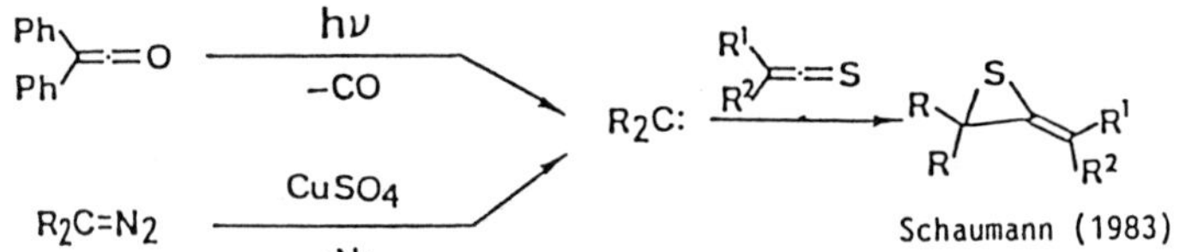

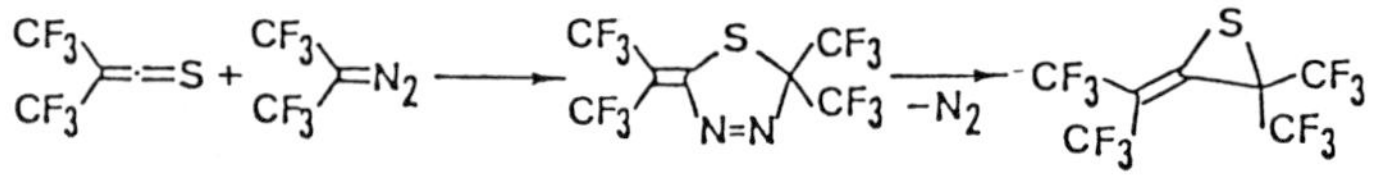

While several kinds of allene episulfides bearing bulky substituents have been readily synthesized by the direct carbene addition reaction to sterically protected stable thioketenes by Schaumann et al.,[8] only one

Schaumann (1983)

R=Ph, fl
$R^1=R^2={}^tBu, -CMe_2(CH_2)_3CMe_2-$

Middleton (1969)

example has been known for the alternative method of allene episulfide formation in the case of trifluoromethyl substituted system by the thermal decomposition of the corresponding 2-alkylidene-1,3,4-thiadiazoline derivative.[9] With a view to designing a variety of substituted allene episulfides even from the rather unstable thioketenes, we have investigated here the cycloaddition of diazo compounds with several thioketenes and their subsequent thermal denitrogenation.

Reaction of t-butylthioketene, prepared by the pyrolysis of 4-t-butyl-1,2,3-thiadiazole, with diphenyl- and dimethyldiazomethane at low temperature(-30 °C and -78 °C, respectively) gave the corresponding 2-alkylidene-1,3,4-thiadiazolines **9** and **10** in high yields. Thermal denitrogenation of **9** and **10** readily proceeded to afford the expected allene episulfides **11** and **12** in 26 and 36 % yields, respectively. However, under these reaction conditions isomeric allene episulfide **11'** and mercaptohexadiene **13** were obtained as the main product, respectively.[10]

On the other hand, treatment of t-butylthioketene with phenyldiazomethane at -78 °C followed by the stirring at room temperature gave no thiadiazoline derivatives but the allene episulfide **14** was isolated directly in 75% yield as a single product without any isomerization. Similarly, cycloaddition of bis(trimethylsilyl)thioketene with diphenyldiazomethane in chloroform followed by the heating of the reaction mixture at 50 °C produced the allene episulfide **15** in 58% yield. The isolation of the intermediary thiadiazoline derivative unsuccessful.[11]

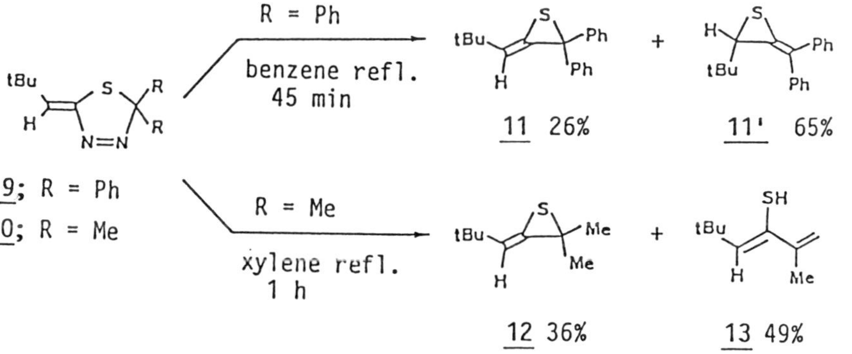

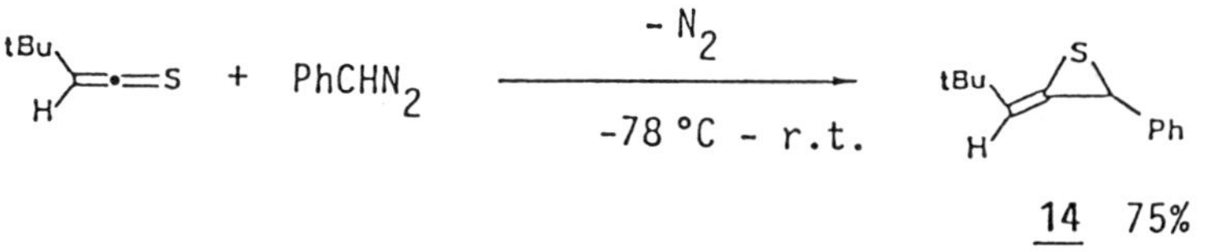

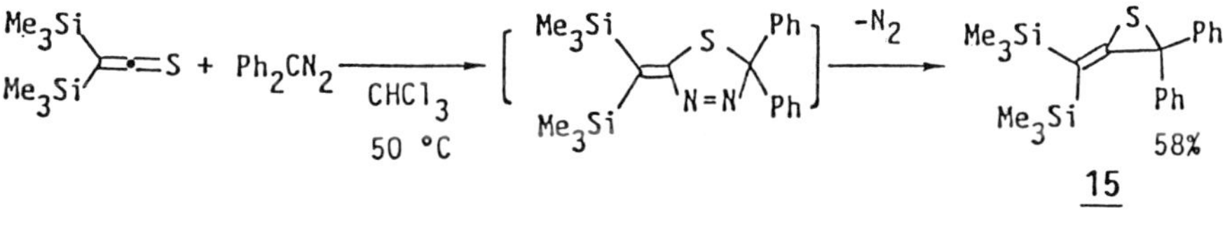

Thermolysis of the tetraaryl substituted 2-alkylidene-1,3,4-thiadiazolines such as **16** and **17** gave rather complicated results. **16** was decomposed in refluxing xylene to give the allene episulfide **18** along with the benzothiolene derivative **19** in 10 and 75% yields, respectively. Evidently **19** was formed by the combination of a vinyl radical with a phenyl group through a 1,5-radical **20**. The decomposition of **17** under

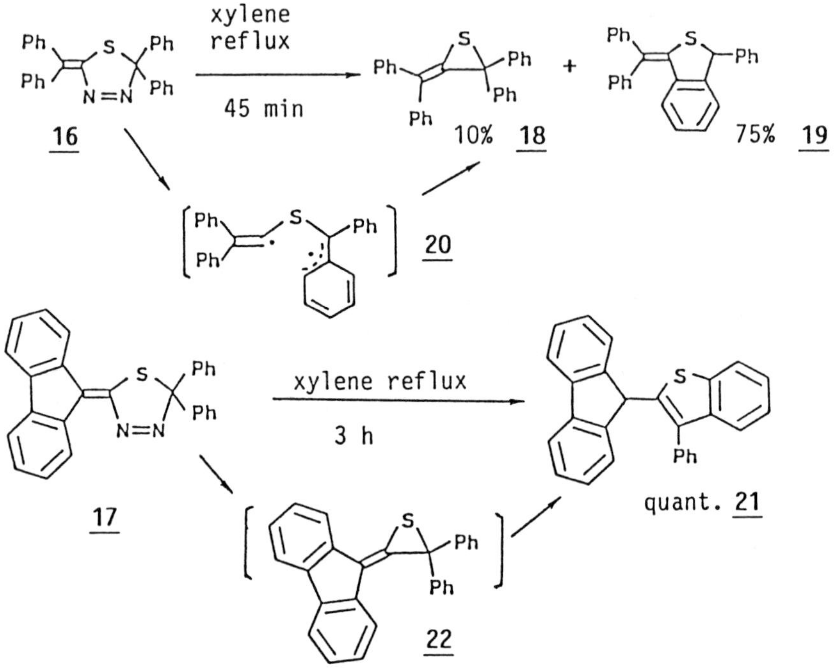

similar reaction conditions resulted in a quantitative formation of benzothiophene derivative **21**, which is well interpreted by an initial generation of an allene episulfide **22** followed by the thermal C-S bond cleavage leading to a thioxyallyl diradical or by an acid-catalyzed isomerization via thioxyallyl cation during the silica gel column chromatography.[12]

## 2. 2. Thermal and Acid-catalyzed Isomerization of Allene Episulfide

2. 2. 1. <u>Kinetic studies on the isomerization of allene episulfide.</u> In tautomerism of allene episulfide with cyclopropanethione, a novel strained small ring system, thioxyallyl intermediate has played a well-fitting and very important role, and has been attracting much current interest. Recently, we have described the kinetic studies on the thermal C-S bond cleavage of tetramethylallene episulfide **23** followed by the 1,4-hydrogen shift via thioxyallyl intermediate **24** giving the corre-

sponding vinyl sulfide derivative **25**.[13] As for the structure of the thioxyallyl intermediate we have already reported its MCSCF calculations to show that $^1B_1$ state is more stable than $^1A_1$ state and it has some dipole moment. However, there has been little experimental arguments for its charge distribution.

**23** → Δ, Diglyme → [**24**] → 1,4-H shift → **25** 69%

$E_a = 26.5$ kcal mol$^{-1}$, $\Delta H^{\ddagger} = 26.2$ kcal mol$^{-1}$, $\Delta S^{\ddagger} = -12.1$ e.u.

Here, we delineate a kinetic investigation of thermal valence isomerization of 1,1-di-t-butyl-3,3-diphenylallene 2-episulfide **26a** and 1,1-bis(trimethylsilyl)-3,3-diphenylallene 2-episulfide **26b**, the substituent effect of which reveals the partial ionic character of thioxyallyl intermediate.[11,14]

When **26a** was heated in o-dichlorobenzene at 150 °C for 2 h, 69% of isomerized allene episulfide **28a** was obtained along with 31% of allene **29a** at 10% conversion. Similarly, **26b** afforded its valence isomer **28b** in 62% yield together with 33% of allene **29b** at 72% conversion by heating in o-dichlorobenzene at 120 °C for 2 h. Under the reaction conditions **28a** and **28b** gave only the corresponding allenes very slowly without any isomerization backward into the original allene episulfides.

| | | | |
|---|---|---|---|
| **26a**; R=$^t$Bu | 150 °C, 2h | **28a** 69% | **29a** 31% |
| **26b**; R=Me$_3$Si | 120 °C, 2h | **28b** 62% | **29b** 33% |

Then, in order to examine the substituent effect on the isomerization of allene episulfide and a nature of thioxyallyl intermediate, kinetic studies on the thermolysis of **26a** and **26b** were carried out in o-dichlorobenzene and diglyme by measuring the decreasing rate of the substrate using $^1$H-NMR spectroscopy. The first-order rate constants and the activation parameters thus obtained are listed in Table I. Comparison of the rate constants and parameters for the isomerization of **26a** and **26b** shows that the replacing of the two t-butyl groups by the more electropositive trimethylsilyl groups resulted in a noticeable increase of the rate of isomerization. When thermolyzed in a polar

Table I. First-order Rate Constants and the Activation Parameters on the Thermolysis of Allene Episulfides 26a and 26b.

| Substrate | Temp./K | Solvent[a] | Rate Constant $k/s^{-1}$ | Activation Parameters |
|---|---|---|---|---|
| 26a | 453 | A | $46.2 \times 10^{-5}$ | $E_a$ = 24.3 kcal mol$^{-1}$ |
|  | 443 | A | $23.3 \times 10^{-5}$ |  |
|  | 433 | A | $12.5 \times 10^{-5}$ | $\Delta H^{\ddagger}$ = 23.5 kcal mol$^{-1}$ |
|  | 423 | A | $6.80 \times 10^{-5}$ | $\Delta S^{\ddagger}$ = -14.8 e.u. |
|  | 423 | B | $11.1 \times 10^{-5}$ |  |
| 26b | 411 | A | $62.5 \times 10^{-5}$ | $E_a$ = 19.2 kcal mol$^{-1}$ |
|  | 398 | A | $29.0 \times 10^{-5}$ |  |
|  | 383 | A | $12.8 \times 10^{-5}$ | $\Delta H^{\ddagger}$ = 18.4 kcal mol$^{-1}$ |
|  | 373 | A | $5.85 \times 10^{-5}$ | $\Delta S^{\ddagger}$ = -29.0 e.u. |
|  | 363 | A | $2.74 \times 10^{-5}$ |  |
| 26b | 397 | B | $34.8 \times 10^{-5}$ | $E_a$ = 14.5 kcal mol$^{-1}$ |
|  | 387 | B | $24.0 \times 10^{-5}$ | $\Delta H^{\ddagger}$ = 13.8 kcal mol$^{-1}$ |
|  | 363 | B | $6.35 \times 10^{-5}$ | $\Delta S^{\ddagger}$ = -40.1 e.u. |

a) A; o-Dichlorobenzene, B; Diglyme.

solvent such as diglyme, both **26a** and **26b** showed a slight accerelation indicating the partial ionic nature of the possible transition states and/or the intermediate due to the dipole moment of the C-S bond. These results imply that the mechanism of the valence isomerization of allene episulfide involves the thioxyallyl intermediate with a positive charge on sulfur atom and a negative charge on the allylic part as shown in the scheme, which is in good agreement with the preliminarily calculated charge distributed structure of thioxyallyl intermediate.[15]

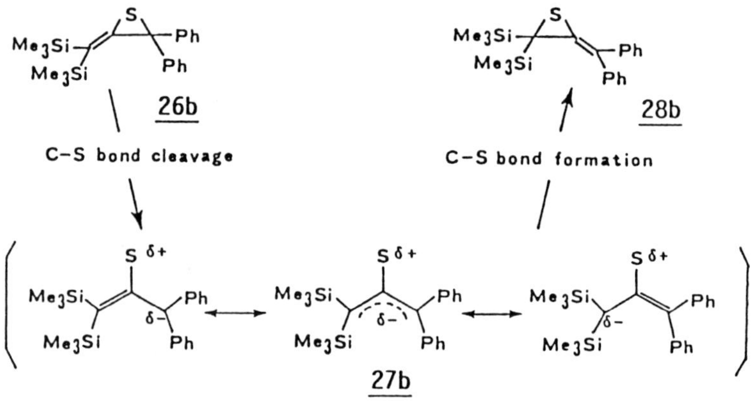

It seems to us that the alpha-effect of the neighboring trimethylsilyl groups contributed in cooperation with the resonance stabilization by the phenyl groups to the stabilization of the resonated structure **27b** which intervenes in the C-S bond forming step, though one cannot totally abandon the possible stabilizing effect of silyl substituents to alpha-radical center.

2. 2. 2. <u>Acid-catalyzed isomerization of allene episulfide.</u> An analogous valence isomerization of allene episulfides does take place readily with acid catalysts.[16] When **26a** was treated with a catalytic amount of trifluoroacetic acid or $BF_3 \cdot Et_2O$ in chloroform at room temperature, the $^1$H-NMR spectrum monitored immediately after the addition showed a quantitative isomerization into **28a**. Under the reaction conditions once produced **28a** underwent further intramolecular cyclization very slowly to give the benzothiophene **31** via the thioxyallyl cation **30**. Similarly, silyl substituted allene episulfide **26b** was directly converted into the benzothiophene derivative **32** by $BF_3 \cdot Et_2O$ in chloroform and no isomeric allene episulfide **28b** was obtained. Allene episulfide **11** also isomerized into **11'** completely even on silica gel column, however, the treatment of **11** and/or **11'** with trifluoroacetic acid resulted in a exclusive formation of the indene derivative **33** in an excellent yield.

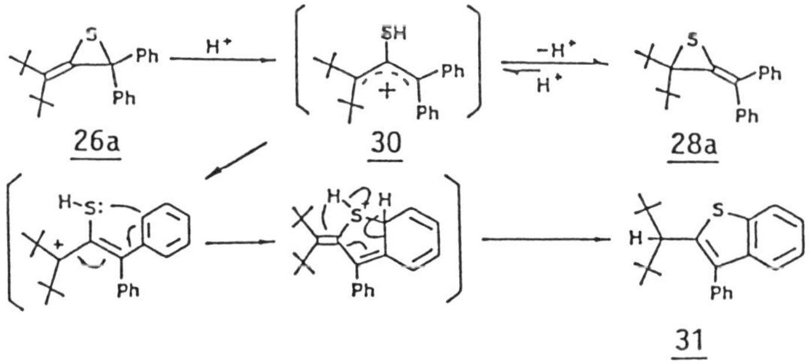

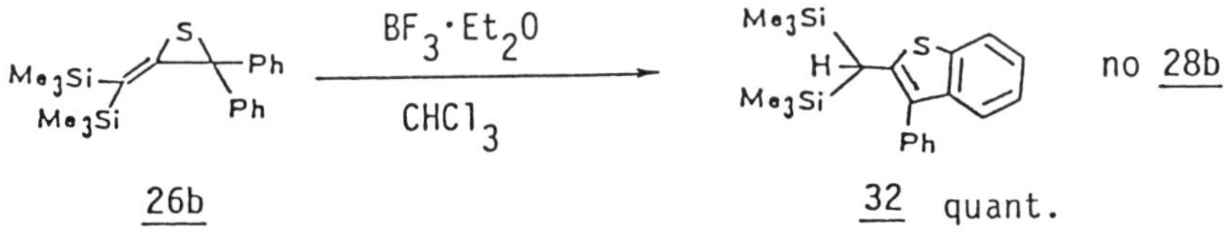

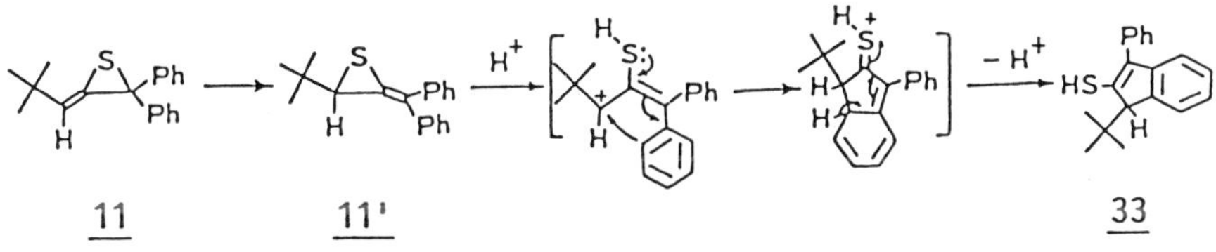

Acid-catalyzed isomerization of allene episulfides described here might be interpreted with an intermediacy of the thioxyallyl ion as illustrated in the scheme, and the direction of the subsequent intramolecualr cyclization leading to the benzothiophene or indene skeleton seems to be controlled by the steric hindrance around the cationic reaction center.

## 2. 3. Synthesis of 1,2,3-Butatriene Episulfides

The direct sulfurization of 1,2,3-butatrienes did not give the expected episulfides but resulted in a novel formation of 6,7-bisalkylidene-1,2,3,4,5-pentathiepanes,[17] we have established the following new methods for the construction of 1,2,3-butatriene episulfide skeleton and investigated their molecular structure and the relative stability between the 1-episulfide and 2-episulfide derivatives.

2. 3. 1. <u>Rearrangement of sulfur ylide via intramolecular carbene reaction: Formation of 1,1,4,4-tetramethyl-1,2,3-butatriene 2-episulfide.</u>
The thietanone 34 was treated with excess tosylhydrazine in the presence of catalytic amount of $BF_3 \cdot Et_2O$ in ethanol at 50-70 °C, yielding the corresponding hydrazone 35 along with the unreacted 34 by chromatographic separation. The purified hydrazone 35 was treated with an equimolar amount of BuLi at -70 °C in THF and the resulting lithium salt 36 was subjected to the vacuum pyrolysis (130-150 °C/$10^{-4}$ Torr.) to afford 1,1,4,4-tetramethyl-1,2,3-butatriene 2-episulfide 37, also called the thiiranoradialene, in 70% yield (mp. 42.5-43.5 °C).[18]

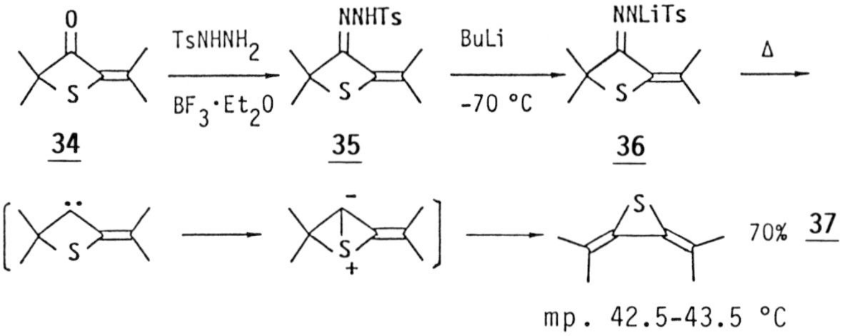

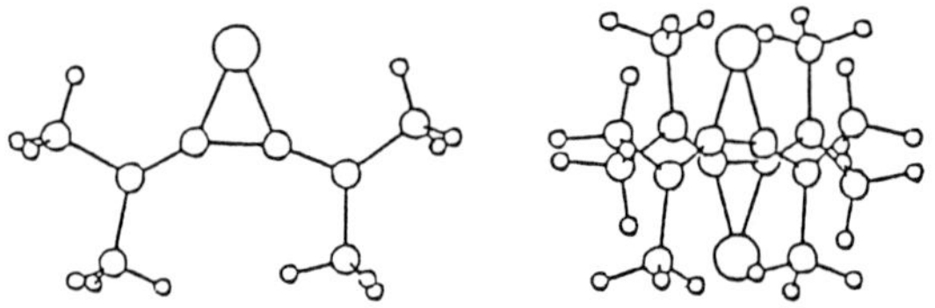

Figure 1. X-Ray Structure Analysis of <u>37</u>

The structure of 37 was confirmed by $^1$H- and $^{13}$C-NMR, IR, and MS spectra, and finally determined by X-ray crystallographical structure analysis as shown in Fig.1. The molecule is almost planar. The C(1)-C(2)-C(3) and C(2)-C(3)-C(4) angles (147.7 and 159.6°) are similar to those in methylenecyclopropane (150°). This indicates that the termini of the diene chromophore are separated by a greater distance than in normal cisoid dienes in a solid states. In this rearrangement of the intermediary sulfur ylide, no 1,2,3-butatriene 1-episulfide was obtained suggesting the higher thermodynamic stability of the 2-episulfide skeleton than the 1-episulfide one.

2. 3. 2. Formation of 1,2,3-butatriene episulfides by thionation of methylenecyclopropanones. Much interest has been focused on cyclopropanethione from a standpoint of tautomerism with allene episulfide via thioxyallyl intermediate, but few examples of intermediary cyclopropanethione have been reported. Recently, we have described a facile formation of methylenecyclopropanones 39 by the peracid oxidation of sterically hindered 1,2,3-butatrienes 38.[19] One might expect that the thionation of 39 would be a facile route to produce 1,2,3-butatriene episulfides via methylenecyclopropanethione. The reaction of methylenecyclopropanones 39a and 39b with an equimolar amount of phosphorous pentasulfide in pyridine at 80 °C gave the 1,2,3-butatriene 1-episulfides 40a and 40b as stable white crystals along with the corresponding butatrienes and allenes, respectively. Similarly, the thionation of methylenecyclopropanone 39c with phosphorous pentasulfide in pyridine at 60 °C under the irradiation of ultrasound afforded the 1,2,3-butatriene 2-episulfide 41 as white crystals along with the butatriene 38c, and no 1-episulfide derivative 40c was isolated. When this reaction was carried out at 80 °C without the irradiation of ultrasound, only the complex tarry mixture was obtained.[20]

The 1,2,3-butatriene episulfides 40a, 40b, and 41 were almost certainly formed by the tautomerization of the initially formed methylenecyclopropanethione **42** via the thioxyallyl-type intermediate **43** as shown in the scheme. The direction of the thiirane ring formation of **40** and **41** is attributed to the preferential coupling of the radical reaction centers to give the less sterically hindered 1,2,3-butatriene episulfide skeleton. The direct desulfurization via **43** can be written for the formation of 1,2,3-butatrienes, and such processes have been postulated to occur in the thermolysis of allene episulfides as depicted in 2. 2. 1. The allenes definitely seem to be the products of the direct thermal decarbonylation of the starting methylenecyclopropanones **39**.

The structures of these 1,2,3-butatriene episulfides **40a, 40b,** and **41** were confirmed by NMR, IR, UV, and MS spectra, and elemental analysis. Of particular note among these spectral data are the unsymmetricity and the observation of allenic carbon absorptions in $^1$H- and $^{13}$C-NMR spectra of **40a** and **40b**, which convinced us of their 1,2,3-butatriene 1-episulfide skeletons, and the molecular symmetricity and the absence of the allenic unit in the NMR and IR spectra of **41** indicating its thiiranoradialene structure.

2. 3. 3. <u>Molecular structure of 1,2,3-butatriene episulfides.</u> Most of the known allene episulfides were characterized in a gas phase or isolated as an oil and few examples have been synthesized in a crystalline form so far as we know, and only a few reports have dealt with its molecular structure which should be closely related to the ring strain and reactivity. From a viewpoint of making a comparative study with the theoretically optimized molecular structure of the parent allene episulfide, we performed the X-ray structure analysis of the newly obatined 1,2,3-butatriene episulfides, which are the methylene homologues of allene episulfide.

The X-ray diffraction of **40b** gave the following crystal data and the refined molecular structure as shown Fig. 2 together with the selected bond lengths and angles; crystal data of **40b**: $C_{22}H_{36}S$, MW 332.25, monoclinic, space group $P2_1/c$, a = 10.837(1) Å, b = 12.027(1)Å, c = 16.398(1) Å, β = 103.65(2)°, V = 2076.9(2) Å$^3$, $D_c$ = 1.06 g cm$^{-3}$, μ(Mo-K$_α$) = 1.55 cm$^{-1}$, z = 4, R = 0.055.

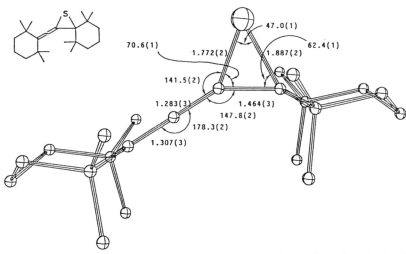

**Figure 2.** ORTEP drawing of 1,2,3-butatriene 1-episulfide **40b** and selected bond lengths (Å) and angles (degree).

The analysis of **40** is worthy of attention from the viewpoint of not only confirming the alkenylidene thiirane ring structure whose allene unit was well preserved despite the ring strain and steric congestion but also the first example of the crystallographical structure analysis of allene episulfide derivative. Of particular note in the unsymmetrically substituted thiirane ring structure of **40b** are the marked elongation of C(1)-S(1) bond (1.887(2) Å) and the expansion of C(1)-C(2)-S(1) bond angle (70.6(1)°) evidently due to the strained allene episulfide structure, which are in good agreement with the optimized geometries of the parent allene episulfide skeleton determined by RHF Closed-Shell SCF calculation with STO-3G basis set as shown in Fig 3.[21] Furthermore, the X-ray structure of **40b** is consistent with that observed in the case of parent allene episulfide using microwave spectroscopy by E. Block.[5]

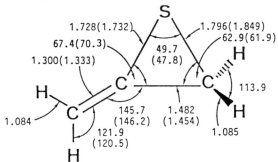

(Values in parentheses are the observed ones with microwave.)

**Figure 3.** Optimized geometries of parent allene episulfide.

Meanwhile, **41** was proved to have the nearly symmetrical thiirano-radialene structure with a characteristic shortening of the C(2)-C(3) bond (1.427(7) Å), the value of which is slightly shorter than that of common episulfide derivatives (ca. 1.49 Å) or 1,3-butadiene derivatives (ca. 1.46 Å), as well as in the case of previously described 1,1,4,4-tetramethyl-1,2,3-butatriene 2-episulfide in 2. 3. 1.  The bond angles of C(1)-C(2)-C(3) and C(2)-C(3)-C(4) in **41** were found to be almost equal to each other (158.1(6) and 158.7(6)°).  Crystal data for **41** are as follows; $C_{20}H_{32}S$, MW 304.22, monoclinic, space group $P2_1$, a = 8.136(3) Å, b = 10.247 Å, c = 12.128 Å, β = 112.71(4), V = 932.7(5) Å$^3$, $D_c$ = 1.08 g cm$^{-3}$, μ(Mo-Kα) = 1.54 cm$^{-1}$, z = 2, R = 0.087.

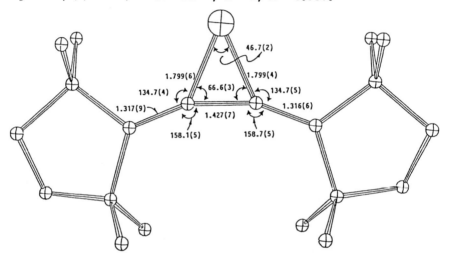

**Figure 4.** ORTEP drawing of 1,2,3-butatriene 2-episulfide **41** and selected bond lengths (Å) and angles (degree).

These results suggest that the bulky substituents on the terminal carbons of the 1,2,3-butatriene episulfides **40** and **41** did not affect the intrinsic nature of these novel skeletons but effectively protected the reactive and strained molecules, since the symmetricity of the thiirano-radialene structure is more perfectly reflected in **41** than in the aforementioned tetramethyl derivatives.

**2. 3. 4.** <u>Facile formation of 1,2,3-butatriene 1-episulfides by alkenylidene carbene addition to thioketone.</u> Although in the formation of 1,2,3-butatriene episulfides by thionation of sterically protected methylenecyclopropanones it was found that the direction of thiirane ring formation from the thioxyallyl-type intermediate **43** was influenced by the nature of terminal substituents, we could not ascertain the intrinsic stability of the two types of 1,2,3-butatriene episulfides **40** and **41**.

In order to elucidate their relative stability, we have established here a new and systematic synthetic route for 1,2,3-butatriene 1-episulfide by alkenylidene carbene addition to thioketone using several kinds of chloroallenes and/or chloroalkyne as the carbene source under the phase-transfer catalytic reaction conditions.[22]

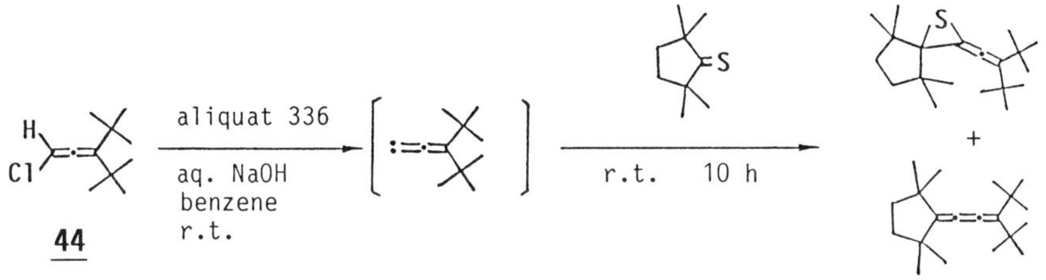

When the benzene solution of 2,2,5,5-tetramethylcyclopentanethione and a slight excess amount of 1,1-di-t-butyl-3-chloroallene **44** was added to the suspension of methyltrioctylammonium chloride (ailquat 336) and 50% aq. NaOH solution at room temperature, the expected 1-episulfide **45** was obtained as white crystals in 38% yield along with 13% of 1,2,3-butatriene **46** and recovered chloroallene (**44**, 23%). The results using the other carbene sources are listed in Table II.

Table II. Alkenylidene Carbene Addition to Thioketone.

| Entry | Carbene Sources | Products and Yields(%) | | |
|---|---|---|---|---|
| 1 | 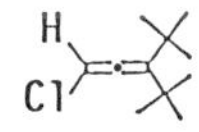 **44** | 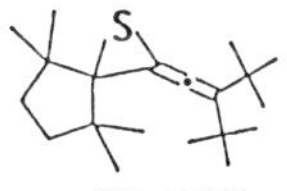 **45** 38% | — | 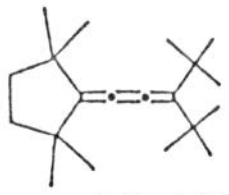 **46** 13% |
| 2 | 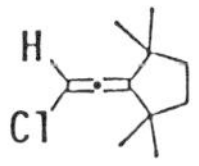 **47** | 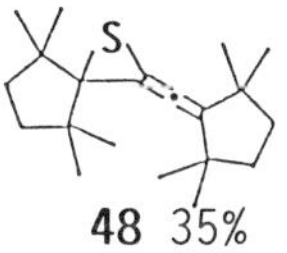 **48** 35% | 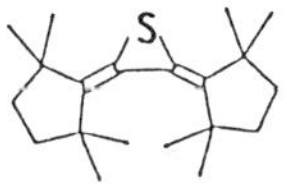 **41** 8% | 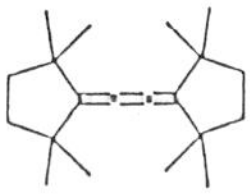 **49** 3% |
| 3 | 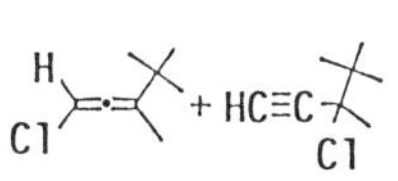 **50** + **50'** | 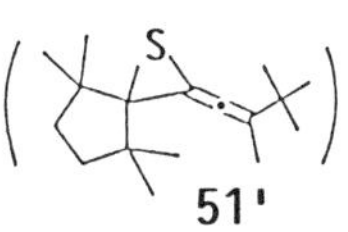 (**51'**) | 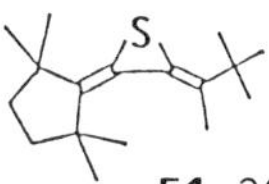 **51** 39% | 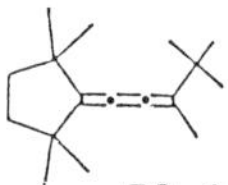 **52** 15% |
| 4 | 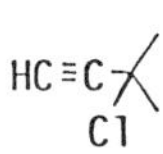 **53** | 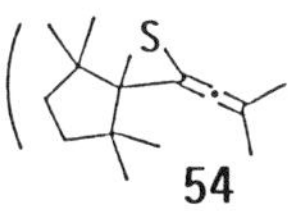 (**54** and/or **55**) | 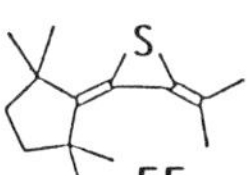 | 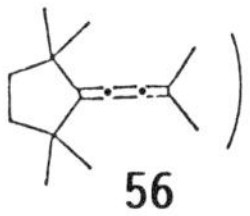 + **56** |

In the case of chloroallene **47**, besides the desired 1-episulfide(**48**, 35%) the corresponding 2-episulfide(**41**, 8%) was isolated together with the butatriene(**49**, 3%) as by-products. Furthermore, the reaction with the sterically less hindered carbene source (**50** and/or **50'**) afforded mainly the 2-episulfide **51** in 39% yield along with the butatriene(**52**, 15%) after silica gel chromatography and the reaction with the alkenylidene carbene derived from 3-chloro-3-methyl-1-butyne **53** resulted in a formation of an unstable mixture of the 1,2,3-butatriene episulfides(**54** and/or **55**) and the butatriene **56**. The structure of the episulfides and butatrienes obtained here were confirmed by $^1$H- and $^{13}$C-NMR, IR, and MS spectra and elemental analysis.

The 1,2,3-butatrienes **46** and **49** are probably formed by desulfurization of once produced 1,2,3-butatriene episulfides **45, 48,** and **41** with the alkenylidene carbene, since the isolated episulfides **45, 48,** and **41** are very stable at room temperature. The formation of 2-episulfides **41** and **51** and the lack of 1-episulfide derivative in the case of **51** might be interpreted with the acid-catalyzed isomerization of initially generated 1-episulfides on silica gel suggesting the higher thermodynamic stability of 2-episulfide skeleton over the 1-episulfide one. By HPLC purification no 2-episulfide **51** but an unseparable mixture of the 1-episulfide **51'** and **51** was obtained from the crude mixture of the reaction using **50** and/or **50'**, where the steric repulsion is avoidable between the substituents facing each other in the 2-episulfide structure. **51'** was rather unstable and readily isomerized into **51** by TLC separation on silica gel.

2. 3. 5. <u>Thermal tautomerization of 1,2,3-butatriene episulfides.</u> Photochemically both the 1,2,3-butatriene 1-episulfide and 2-episulfide were easily desulfurized into the corresponding 1,2,3-butatrienes and no isomerization was observed even in the early stage of the photolysis.

On the other hand, we have found the thermal interconversion between the two types of 1,2,3-butatriene episulfides in the case of **41** and **48**.[23] Heating of the o-dichlorobenzene solution of 2-episulfide **41** at 120 °C for 40 min resulted in a formation of equilibrated mixture with the 1-episulfide **48** which was ascertained by the $^1$H-NMR and HPLC monitoring and the characteristic IR absorption of the allenic carbon-carbon stretching at 1980 cm$^{-1}$. As for **48**, the similar thermal equilibration with **41** was observed within the temperature range from 90 to 120°C, and the equilibrium ratio of **41** to **48** became smaller at the higher reaction temperature. The thermal instability of **41** conflicting with the essential stability of the 2-episulfide skeleton relative to the 1-episulfide one might be attributable to the increasing steric repulsion of the inner methyl groups on the cisoid butadiene unit of **41** caused by the augmentative molecular vibration at higher temperature. The steric congestion between the inner methyl groups of **41** has been certainly confirmed by the X-ray molecular structure analysis as described in 2. 3. 3.

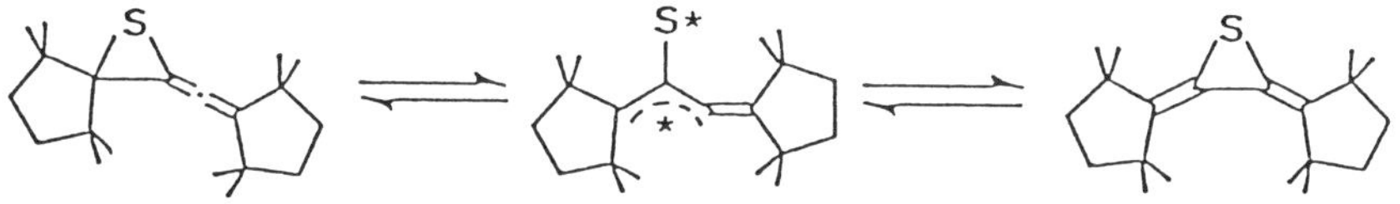

**48** ⇌ ⇌ **41**

[48]/[41]; 1.52(90 °C) and 1.16(120 °C)

Among the results of our successful synthesis and reactions of 1,2,3-butatriene episulfides we were lucky enough to receive both type of episulfides **41** and **48** as stable crystals in the case of tetramethylcyclopentane ring substituted system, and the thermal interconversion between **41** and **48** is worthy of attention as the first example of the sulfur-analogous tautomeric system of alkenylidenecyclopropane and bisalkylidenecyclopropane.

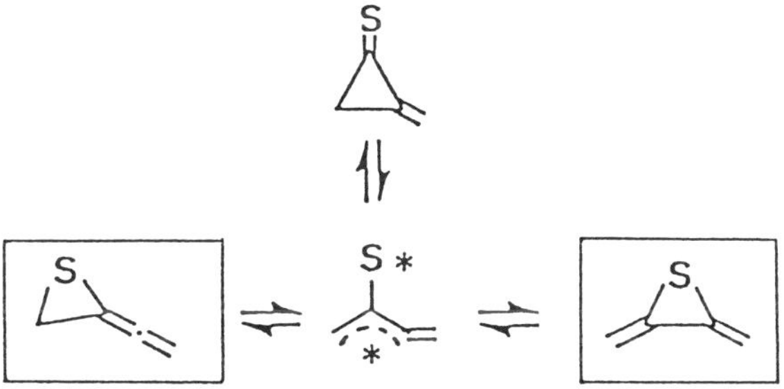

## 3. SUMMARY

Utilizing the steric protection by the bulky substituents and the characteristic properties of the thiocarbonyl compounds we have established the synthesis of a variety of substituted allene episulfides and 1,2,3-butatriene episulfides, novel and unique strained episulfides bearing exocyclic double bonds. Studies on their thermal and acid-catalyzed isomerization reaction and an interesting substituent effect on which provide us with the new aspects in the sulfur-analogous tautomeric systems of alkylidenecyclopropane and bisalkylidenecyclopropane, especially in the nature of intermediary thioxyallyl species.

## ACKNOWLEDGEMENT

We gratefully acknowledge the invaluable help of our colleagues named in the references, together with Professor Keiji Morokuma and Dr. Katsuhisa Ohta of Institute of Molecular Science for theoretical studies and Dr. Midori Goto of National Chemical Laboratory for Industry and Dr. Katsuhiko Ueno of Research Institute for Polymers and Textile for the X-ray analysis.

## REFERENCES

1.  a) J. P. Chesick, J. Am. Chem. Soc., **1963**, 85, 2720; b) J. J. Gajewski, J. Am. Chem. Soc., **1968**, 90, 7178; c) J. C. Gilbert and J. R. Butler, J. Am. Chem. Soc., **1970**, 92, 2168; d) W. von E. Doering and H. D. Roth, Tetrahedron, **1970**, 26, 2825; e) M. J. Dewar, J. Am. Chem. Soc., **1974**, 93, 3081; f) W. T. Borden, J. Am. Chem. Soc., **1974**, 96, 3754; g) J. H. Davis and W. A. Goddard, III., J. Am. Chem. Soc., **1976**, 98, 303.
2.  a) N. J. Turro, Acc. Chem. Res., **1969**, 2, 25; b) T. H. Chan and B. S. Ong, J. Org. Chem., **1978**, 43, 2994.
3.  A. G. Hortmann and A. Bhattacharijya, J. Am. Chem. Soc., **1976**, 98, 7081.
4.  a) Y. Osamura, W. T. Borden, and K. Morokuma, J. Am. Chem. Soc., **1984**, 106, 5112; b) H. Bock, S. Mohmand, T. Hirabayashi, and A. Semkow, Chem. Ber., **1982**, 115, 1339; c) O. Kikuchi, H. Nagata, and K. Morihashi, J. Mol. Struct., **1985**, 124, 261.
5.  E. Block, R. E. Penn, M. D. Ennis, T. A. Owens, and S. -L. Yu, J. Am. Chem. Soc., **1978**, 100, 7436.
6.  E. Longejan, Th. S. v. Buys, H. Steinberg, and Th. J. de Boer, Recl. Trav. Chim. Pays-Bas, **1978**, 97, 214.
7.  a) J. K. Crandall and D. R. Paulson, J. Am. Chem. Soc., **1966**, 88, 4320; b) D. R. Paulson, J. K. Crandall, and C. A. Bunnel, J. Org. Chem., **1970**, 35, 3708; c) R. Bloch, P. Le. Perchec, and J. -M. Conia, Angew. Chem., Int. Ed. Engl., **1970**, 9, 798; d) M. E. Hendrick, J. A. Haride, and M. Jones, Jr.,J. Org. Chem., **1971**, 36, 3061; e) G. Kobrick and B. Bosner, Tetrahedron Lett., **1973**, 2031; f) D. H. Aue and M. J. Meshishnek, J. Am. Chem. Soc., **1977**, 99, 223; g) J. Belzner and G. Szeimies, Tetrahedron Lett., **1986**, 27, 5839.
8.  a) E. Schaumann, H. Behr, G. Adiwidjaja, A. Tangerman, B. H. M. Lammerink, and B. Zwannenburg, Tetrahedron, **1981**, 37, 219; b) H. Behr, Thesis, University of Humburg, FRG, **1981**.
9.  W. J. Middleton, J. Org. Chem., **1969**, 34, 3201.
10. Unpublished results; T. Furuhata, Thesis, University of Tsukuba, **1987**.
11. N. Tokitoh, N. Choi, and W. Ando, Chemistry Lett., **1987**, 2177.
12. T. Furuhata and W. Ando, Tetrahedron Lett., **1987**, 28, 1179.
13. T. Furuhata and W. Ando, Tetrahedron, **1986**, 42, 5301.
14. W. Ando, A. Itami, T. Furuhata, and N. Tokitoh, Tetrahedron Lett., **1987**, 28, 1787.
15. W. Ando and T. Furuhata, Nippon Kagaku Kaishi, **1987** 1293.

16. The direct observation of thioxyallyl cation has been achieved by the low temperature NMR spectroscopy; W. Ando, Y. Hanyu, T. Furuhata, and T. Takata, J. Am. Chem. Soc., **1983**, 105, 6151.
17. N. Tokitoh, H. Hayakawa, M. Goto, and W. Ando, Tetrahedron Lett., **1988**, 28, 1935.
18. a) W. Ando, Y. Hanyu, and T. Takata, Tetrahedron Lett., **1981**, 22, 4815; b) W. Ando, Y. Hanyu, Y. Kumamoto, and T. Takata, Tetrahedron, **1986**, 42, 1989.
19. W. Ando, H. Hayakawa, and N. Tokitoh, Tetrahedron Lett., **1986**, 27, 6357.
20. W. Ando, H. Hayakawa, and N. Tokitoh, Tetrahedron Lett., **1987**, 28, 1803.
21. N. Tokitoh, H. Hayakawa, M. Goto, and W. Ando, Chemistry Lett., **1988**, 961.
22. P. J. Stang, Chem. Rev., **1978**, 78, 383 and references cited therein.
23. N. Tokitoh, H. Hayakawa, and W. Ando, Tetrahedron Lett., **1988**, 29, 5161.

THROUGH-BOND INTERACTION VIA CYCLOBUTANE RELAY ORBITALS AS A MEANS OF
EXTENDING CONJUGATION


LEO A. PAQUETTE* AND JÜRGEN DRESSEL
Evans Chemical Laboratories
The Ohio State University
120 West 18th Avenue
Columbus, Ohio 43210


ABSTRACT. The hydrocarbon series represented by tricyclo[5.5.0.0$^{2,8}$]-
dodecatetraene (11), tricyclo[5.3.0.0$^{2,8}$]deca-3,5,9-triene (12), and
9,10-dimethylenetricyclo[5.3.0.0$^{2,8}$]deca-3,5-diene (13) has been
synthesized from dimethyl ε-truxillate. Stepwise belting of a preformed
all-trans 1,2,3,4-tetrasubstituted cyclobutane with proper differentiation of the 1,3- and 2,4-positions is crucial to successful twofold
annulation. The response of these structurally unusual hydrocarbons to
thermal and photochemical activation is described. Especially revealing
are their photoelectron spectra. For 11, a substantial interaction
exists between the orthogonal 1,3-diene systems via the central four-
membered ring. The level of "ring conjugation" far exceeds that found
in any spiroconjugated molecule. The contrasting behavior of 12 and 13
is reported as well. Finally, the experimental data are compared with
theoretical predictions.


I. Introduction

1.1. THE SPIROCONJUGATION PHENOMENON

As the result of pioneering work by several research groups,[1-6] the
ability of perpendicularly oriented π networks linked together by a
common tetrahedral carbon atom to exhibit through-space interaction is
now well recognized.[7] Since spiroconjugation of this type is con-
trolled by the interaction matrix term ($\beta^{\mu\nu}_{spiro} = <\pi^{\mu}_{\alpha}|H|\pi^{\mu}_{A'}>$), maximum
effects should be seen when $\beta^{\mu\nu}_{spiro}$ is large and the π fragments within
a given molecule are identical or nearly so.
However, within the Mulliken approximation,[9] the interaction matrix
elements are seen to remain small throughout ($\beta^{\mu\nu}_{spiro} = kS^{\mu\nu}$). This
restriction arises because the overlap integrals $S^{\mu\nu}$ between the
coupling atomic orbitals are small in magnitude, typically about 0.025.[1]
This value is on the order of 1/20th that of a normal σ bond.[10] Despite
these constraints, spiro[4.4]nonatetraene (1, Scheme 1) exhibits a
splitting of its highest filled (and highest unfilled) diene π (π*)

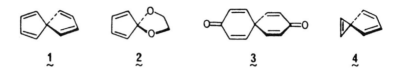

Scheme 1. Spiroconjugated (1, 2) and non-spiroconjugated molecules (3, 4).

orbitals amounting to 1.23 eV, as revealed by its UV[3d] and PE spectra.[4c] Surprisingly, the higher HOMO energy of 1 does not lead to a rate enhancement in its Diels-Alder reaction with dimethyl acetylenedicarboxylate. This absence of anticipated kinetic acceleration may find its origin in other considerations such as steric factors, absence of HOMO-LUMO control, and the like.[3d]

A low-energy red shift of 30 nm in the electronic spectrum of cyclopentadienone ethylene ketal (2)[11] indicates that its lone-pair oxygen orbitals are interacting effectively with the 1,3-diene orbitals.[2a,11] Although the rates of Diels-Alder reaction of the cyclopentadienone ketals with electron-deficient dienophiles have not been reported in detail, they seem unexceptional[12] and appear to pose identical theoretical questions. On the other hand, the dimerization of 2 is strongly accelerated compared to cyclopentadiene, a result which has been attributed to spiroconjugative enhancement of the (diene) HOMO energy.[11] The same ease of dimerization or polymerization[12] is apparent in 1.[3d] Attempts have been made to reconcile these apparently incongruous experimental observations.[2d]

The ESR spectrum of the radical anion of 3 did not provide evidence of negative charge delocalization over both rings, fully as expected on theoretical grounds.[13] Also, spiroconjugation is simply too weak in derivatives of 4 to override the destabilization brought on by charge transfer.[3c,14-17]

## 1.2. EARLY EVIDENCE FOR EXISTENCE OF THE CYCLOBUTANE RELAY EFFECT

The complication encountered with spiroconjugated systems stems from the mandatory carbon spacer that stringently limits the proximity of the relevant termini and consequently the extent of their overlap. Should this insulating structural element be replaced by a suitable orbital relay network through which electronic communication would flow more efficiently, extended conjugation of mutually perpendicular $\pi$-ribbons could in principle be broadly realized. In a seminal paper, Gleiter noted that the cyclobutane ring has the intrinsic ability to fulfill all of the criteria necessary for efficient through-bond orbital interaction:[19] (i) the Walsh orbitals of four-membered carbocycles lie relatively close in energy to the molecular orbitals of cyclic alkenes and 1,3-alkadienes;[6b,20] (ii) the puckered cyclobutane ring has the same $D_{2d}$ symmetry as the two mutually perpendicular $\pi$-ribbons; and (iii) the cyclobutane relay orbitals are only one C-C single bond removed from the $\pi$-ribbon termini, thereby allowing for more pronounced overlap.

These claims were soundly based on the results of photoelectron spectroscopic (PE) analysis of several vinylcyclobutanes.[21,22] Formulas 5,[21a] 6[22,23] and 7 (Scheme 2) represent three ways in which a four-

**5**   **6**   **7**

Scheme 2. Possible ways of linking a $\pi$-ribbon to a cyclobutane ring.

membered ring can be linked to a $\pi$ ribbon. Of these, the latter hydrocarbon deserves special consideration.[24] The parent diene is known, having been synthesized first in these laboratories more than a decade ago.[26,27] The 7,7-dimethyl[28,29] and endo,endo-7,8-diphenyl derivatives[30] have also been prepared and subjected to PE and X-ray analysis. Of particular relevance, the resonance integral ($\beta$) resulting from interaction between the cyclobutane Walsh and diene $\pi$ orbitals in 7 is -1.9 eV,[26] a value that compares very favorably to that found for a double bond and cyclopropane ring.[31]

Although the properties of 7 augur well for the possible existence of a substantial "relay effect" under the right circumstances, experimental evaluation of the tantalizing predictions awaited the successful synthesis of doubly bridged systems. Tricyclo[3.3.0.0$^{2,6}$]octa-3,7-diene (8, Scheme 3), earlier synthesized by Meinwald[32] and by Zimmerman,[33] exhibits in its UV spectrum a bathochromic shift to 300 nm (contrast the $\lambda_{max}$ for 10 at 250 nm), that has invited theoretical analysis.[18,34] Unfortunately, the thermal lability of 8 (estimated $t_{\frac{1}{2}}$ of 10 min at room temp) and its ready isomerization to semibullvalene (9) precluded PE analysis.

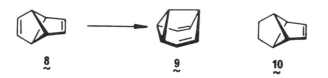

**8**   **9**   **10**

Scheme 3. Examples of small, doubly-bridged cyclobutane systems.

For these reasons, we set out to prepare the hydrocarbon series represented by tricyclo[5.5.0.0$^{2,8}$]dodecatetraene (11), tricyclo-[5.3.0.0$^{2,8}$]deca 3,5,9 triene (12), and 9,10-dimethylenetricyclo-[5.3.0.0$^{2,8}$]deca-3,5-diene (13, Scheme 4). In the sequel, we detail our synthetic approaches to these molecules,[35,36] discuss their response to thermal activation,[37] and document by means of photoelectron spectroscopy the presence of strong relay conjugation in 11.[38]

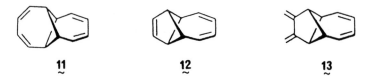

**Scheme 4.** Doubly π-annulated cyclobutane targets.

## II. Synthetic Considerations

### 2.1. RETROSYNTHETIC ANALYSIS

Any synthetic route to 11-13 must ultimately deal with the timing of cyclobutane ring installation. Those disconnections that we and other have considered are shown in Scheme 5 for tetraene 11. Although the

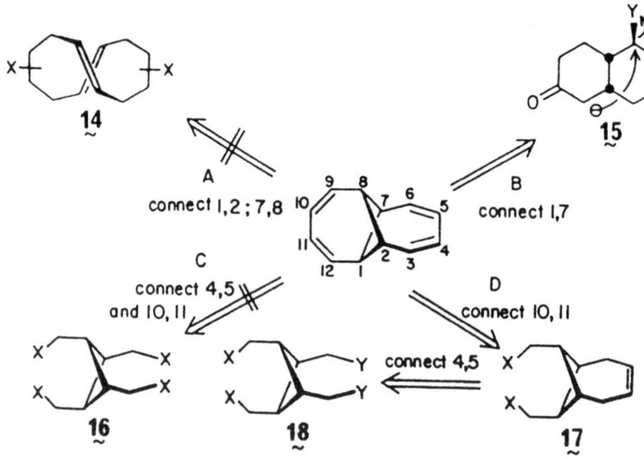

**Scheme 5.** Retrosynthetic disconnections associated with the preparation of 11.

light-induced closure of a monocyclic diene (path A) has been utilized to gain access to the tricyclo[3.3.0.0$^{2,6}$]octane framework,[39] the tricyclo[4.4.0.0$^{2,8}$]decane skeleton,[40] and the tricyclo[5.3.0.0$^{2,8}$]-decane framework,[41] photocyclization of 14 was fully anticipated to lead exclusively to a linear 6-4-6 tricyclic product[42] and was therefore not examined. The particular pathway followed is well recognized to depend heavily on ring size[44,45] and a $C_{12}$ system is simply not conducive to cross bonding.

The intramolecular $S_N2$ displacement illustrated within 15 was developed by Heathcock for the synthesis of copaene and ylangene.[46] More recently, the technique has been applied by Gleiter to transform suberone and 7-methoxysuberone to functionalized tricyclo[5.4.0.0$^{2,8}$]-undecanones.[47] In the present context, use of this protocol would require the subsequent enlargement of both bridges. Furthermore, the process inherently lacks the flexibility we desired to arrive directly at all three target molecules. For these reasons, this pathway was also not pursued.

In their attempts to synthesize truncated tetrahedrane, Scott[48a] and Brousseau,[48b] working in Woodward's laboratory, made recourse to a strategem involving proper cyclization of an all-trans 1,2,3,4-tetrasubstituted cyclobutane of the type 16. However, attempted Dieckmann or acyloin cyclization of 19 failed completely, and exposure of tetrabromide 20 to sodium sulfide in HMPA led only to 21 (Scheme 6). These

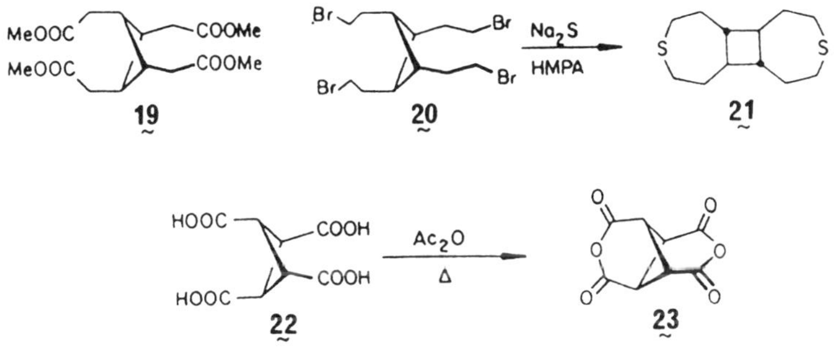

Scheme 6. Ring closures around a cyclobutane ring core.

findings illustrate an important feature of cyclobutane chemistry. The unsubstituted four-membered ring is not flat ($D_{4h}$) but puckered ($D_{2d}$) by 35°,[49-51] with a barrier to planarization of about 1.5 kcal/mol as determined by IR[52] and Raman spectroscopy.[53] This puckering destroys the equivalency of geminal substituents and differentiates axial and equatorial positions. An all-trans 1,2,3,4-tetrasubstituted cyclobutane is therefore strongly expected to prefer the all-equatorial conformation in order to avoid the intense repulsive 1,3-transannular interactions experienced in the all-axial conformation.[54] Thus, whenever 1,2-bridging is geometrically feasible, it is virtually certain to dominate kinetically over 1,3-bridging, and by a substantial margin. Only when the bridge size becomes too short for 1,2-closure will 1,3-annulation perhaps operate. The conversion of tetraacid 22 to dianhydride 23[55] nicely illustrates this last point.

In our view, the route holding the greatest promise for leading successfully to 11-13 was path D in Scheme 5. The tactic of dividing the substituents into two discrete sets (see 18) so as to differentiate the 1,3- and 2,4-positions was expected to enable stepwise belting of two chains to alternate carbon atoms of the four-membered ring. Competing 1,2-cyclization is conveniently skirted and the subsequent second stage belting of 17 was expected to be geometrically feasible.

## 2.2. SYNTHESIS OF TRICYCLO[5.5.0.0$^{2,8}$]DODECA-3,5,9,11-TETRAENE (11).

A cyclobutane precursor that aptly fulfills the conditions demanded by path D is dimethyl ε-truxillate (24).[30,56] Lithium aluminum hydride reduction of this readily available diester and conversion to 26 with triphenylphosphine dibromide set the stage for oxidative degradation of the phenyl groups (Scheme 7). It was necessary to implement degradation

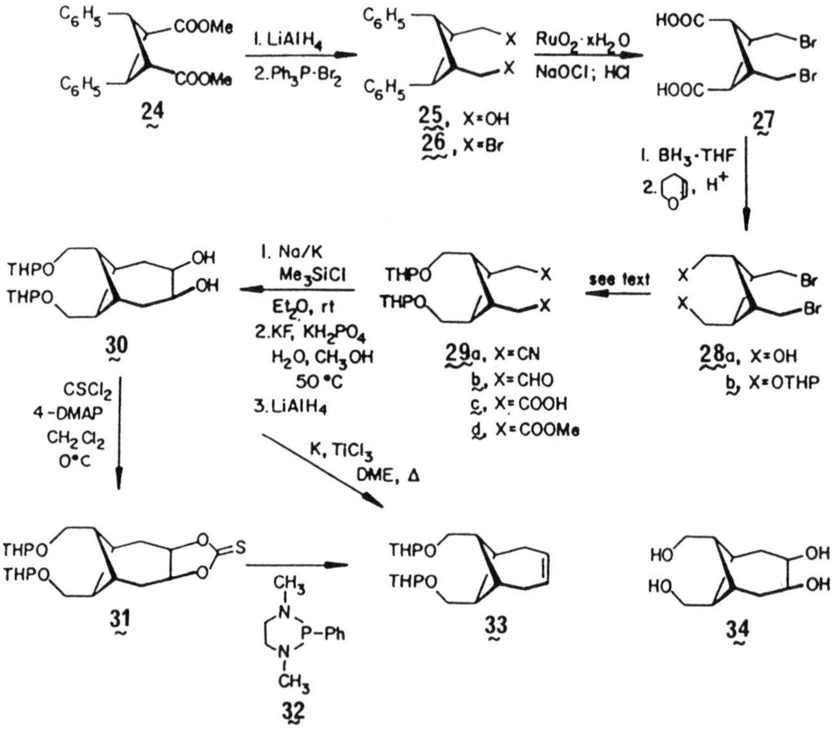

Scheme 7. First stage of the synthesis of 11.

of the aromatic rings at a stage in the sequence when chemoselectivity would operate to the maximum. Following preliminary experiments which established that primary bromides are unreactive to ruthenium tetroxide,[57] 26 was subjected to analogous conditions and found to undergo smooth convesion to diacid 27. Direct reduction of 27 with the borane-tetrahydrofuran complex afforded 28a in 70% overall yield on 0.1 mol scale. Curiously, the rate enhancement that Carlson, et al[58] and Chakraborti and Ghatak[59] observed upon adding acetonitrile as cosolvent for such oxidations could not be reproduced in our system. Also, substitution of sodium metaperiodate for sodium hypochlorite as cooxidant for the ruthenium tetroxide (or ruthenium trichloride) resulted in incomplete oxidation.

To ensure noninterference by the hydroxyl groups in 28a, conversion to the bis(tetrahydropyranyl) ether preceded homologation to diester 29d. Saponification of bisnitrile 29a proceeded most satisfactorily when temperatures of 140-150 °C were employed. In this manner, the overall yield for the four-step conversion of 28a to 29d was 71%.

At this juncture, the first of two planned acyloin condensations was implemented. The initial expectation was that the all-equatorial disposition of the four pendant groups in 29d would be reflected in a reduced rate of cyclization. Certainly, the Schrapler-Ruhlmann modification[60] was preferred. To our amazement, however, the cyclization was much more efficiently accomplished with 1:1 sodium-potassium alloy and chlorotrimethylsilane in ether *at room temperature* than with sodium and Me$_3$SiCl in refluxing toluene.[61] In fact, intramolecular coupling of the diester occurred with sufficient rapidity that high dilution conditions were not necessary. Additionally, 86 g (0.2 mol) of 29d could be efficiently cyclized in only 350 mL of solvent! Thus, little difficulty was encountered with compression of the four extraannular groups into close spatial proximity.

Since direct reduction of the bissilyl enol ether with sodium borohydride in refluxing ethanol proved problematic, the hydrolysis and reduction steps were separated. Unmasking the α-hydroxy ketone was best realized with potassium fluoride in a phosphate buffer at pH 5. When exposed to lithium aluminum hydride in anhydrous tetrahydrofuran, this intermediate was converted exclusively into cis diol 30. Because the asymmetry introduced by the 2-tetrahydropyranyloxy groups made differentiation between the cis- and trans-diols difficult, 30 was deprotected with p-toluenesulfonic acid in ethyl acetate[62] or with acid-washed Dowex 50 resin in methanol.[63] The resulting tetraol displayed 7 lines in its broadband-decoupled $^{13}$C NMR spectrum, as expected uniquely for the $C_s$ symmetry of 34. The trans isomer possesses $C_2$ symmetry and should therefore exhibit only 5 carbon signals.

The hydride quite probably deprotonates the hydroxyl group of the acyloin prior to reduction of the carbonyl function. The resulting aluminate could theoretically serve two purposes. It could deliver a hydride atom intramolecularly to the carbonyl group, thereby giving rise to the trans diol. This reaction pathway has been proposed earlier in other contexts.[64,65] Alternatively, the bulky aluminate could serve as a blocking group, sterically shielding the proximal face of the molecule against intermolecular attack by a second hydride. Force field calculations (MMP2) and semiempirical quantum mechanical calculations (MINDO/3)[26] on bicyclo[4.1.1]octane suggest that the four-carbon bracket

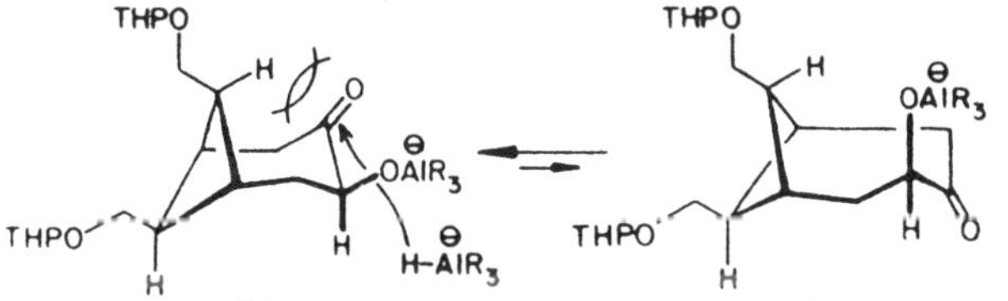

**Scheme 8.** Possible course of hydride reduction of acyloin intermediate.

can adopt a more or less ideal chair conformation together with the cyclobutane ring. The aluminate should prefer the equatorial position (35e) and the 7-endo hydrogen will serve to hinder β-approach of an incoming hydride (Scheme 8). α-Attack then affords the cis diol. Akhtar and Marsh have used the steric bulk of an aluminate to rationalize the stereoselective reduction of cholestan-5α-ol-3-one.[66]

Knowledge of the stereochemistry of the vicinal hydroxyl groups in 30 was crucial to the ensuing olefination. More specifically, the Corey-Winter olefination[67] and its improved variant[68] proceed stereospecifically and in smaller cyclic systems require the diol to be cis. On the other hand, the Vedejs modification[69] is tolerant of mixtures of cis and trans diols. Since 30 had alone been produced, the longer Vedejs methodology could be avoided. Following conversion to thionocarbonate 31, reaction with 2,5-dimethyl-1-phenyl-2,5-diazaphospholidine (32) resulted in the formation of 33 (86%). Neat phospholidine was used as the reaction medium because dilution with tetrahydrofuran caused noticeable rate retardation. Advantageously, the tetrahydropyranyl ether groups in 33 were not at all affected by this reagent.[70] However, the capricious nature of the conversion of 30 to 31 (yields ranged from 43 to 89%) prompted continued search for an improved means for arrival at 33. Unfortunately, dialdehyde 29b could not be induced to undergo deoxygenative coupling directly to 33 in the presence of low valent titanium.[71,72] McMurry had earlier uncovered in his synthesis of civetone[73] that ethylenedioxy groups do not survive the coupling conditions. However, the lability of the tetrahydropyranyloxy groups is seemingly not at issue, since cis diol 30, the presumed intermediate in the McMurry process,[72] could be converted to 33 in 70% yield under standard conditions. This route is to be preferred. The colorless crystalline 33 exhibits $^1$H and $^{13}$C NMR spectra fully consonant with its symmetry.

The bicyclo[4.1.1]octene part structure of 33 was left at the monounsaturated stage during the second chain homologation sequence to obtain 36d (Scheme 9) in order to avoid labilizing the cyclobutane bonds.[30] The overall yield of the diester after Kugelrohr distillation was 93%. In this instance, the acyloin condensation leading ultimately to 37 proved more efficacious when carried out with an excess of lower melting sodium-potassium alloy (1:4, mp -11 °C). Hydrolysis and reduction as before likewise delivered the cis diol, as evidenced by its 9-line $^{13}$C NMR spectrum. Although reductive elimination of the hydroxyl groups in 37 by the Corey-Hopkins protocol[68] provided pivotal diene 40 in 70% yield, two interesting side reactions were noted along this two-step sequence. Thus, adsorption of thiocarbonate 35a on basic alumina (activity III) as a prelude to purification led to partial hydrolysis and formation of carbonate 38b.[74] Chromatography instead on silica gel delivered ketone 39, the end result of a pinacol-like rearrangement.

Again as in the olefination of the first bridge, direct deoxygenation of diol 37 with the titanium trichloride-tetrahydrofuran complex and potassium in refluxing dimethoxyethane avoided the complications associated with thiocarbonate formation. In a very clean reaction, diene 40 can be obtained directly from 37 in 77% yield.

Diene 40 proved to be a colorless crystalline solid. For the first

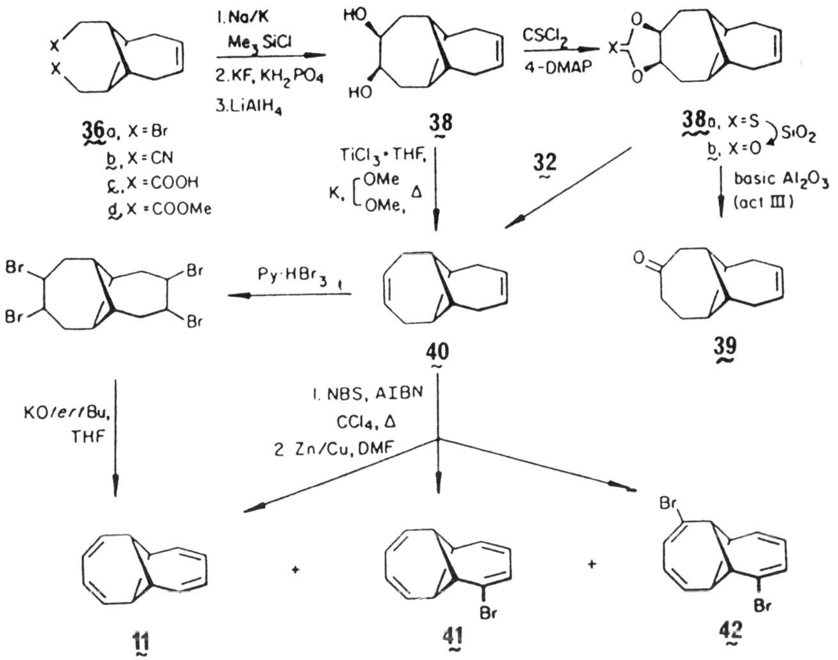

Scheme 9. Completion of the synthesis of 11.

time, the high $D_{2d}$ symmetry of the target tetraene was made evident from the simplified $^1$H (three widely spaced broadened singlets) and $^{13}$C NMR spectra (three lines) of 40.

Allylic bromination of 40 with 4.0-4.7 equiv of N-bromosuccinimide and a catalytic quantity of azobisisobutyronitrile under sunlamp irradiation and heat afforded an extensive mixture of bromides. Direct treatment of this mixture with freshly prepared zinc-copper couple in anhydrous dimethylformamide at room temperature for 12 h delivered tetraene 11 (12%), monobromide 41 (34%), and the dibromo derivative 42 (4%). Coaddition of potassium iodide and iodine, as recommended by others,[26,75] had no major effect on the outcome of this reaction. However, when the reduction was allowed to proceed for almost 4 days, only tetraene 11 could be isolated in 57% yield. It was of crucial importance to exclude oxygen at all times. Also, all manipulations at elevated temperatures (50 °C and above) and exposure to laboratory light had to be avoided, for reasons that will be discussed below.

Tricyclo[5.5.0.0$^{2,8}$]dodecatetraene (11), a soft white solid with an intense musty odor, was easily identified by the simplicity of its $^{13}$C NMR spectrum (3 lines) and by the strong resemblance of the olefinic portion of its $^1$H NMR spectrum to that of bicyclo[4.1.1]octa-2,4-diene (7).[26,27b] The locus of the bromine atom in 41 was determined from the coupling pattern of its olefinic protons and by comparison with data reported for 3-bromobicyclo[4.1.1]octadiene.[27b] The identity of dibromide 42, which was not obtained completely pure, was again inferred from its $^1$H NMR spectrum.

Attention is called to the fact that 11 has the same $D_{2d}$ symmetry as allene. The $C_2$-symmetric dibromide 42 is therefore chiral and in principle resolvable. As expected, 41 and 42 were most efficiently metallated and protonated under Seebach's conditions (2 equiv of tert-butyllithium, THF, -78 °C; CH$_3$OH).[76] In actuality, direct treatment of the allylic bromide mixture with tert-butyllithium proved to be the simplest and most rapid means for obtaining 11 (53%).

2.3. THE ROUTE TO TRICYCLO[5.3.0.0$^{2,8}$]DECA-3,5,9-TRIENE (12).

As in spiroconjugated systems, the level and consequences of through-bond interaction between two π-ribbons linked orthogonally across a cyclobutane ring is intimately related to the number of electrons involed.[18] Whereas the basis orbital energies calculated for 11 predict that this tetraene should be destabilized, those present in 12 were anticipated to be marginally stabilizing. In order to make possible a direct comparison of these systems, 12 was prepared as follows.[35,36]

Dibromide 36a was heated with a solution of anhydrous sodium sulfide in HMPA,[77] and 43 was obtained quantitatively. Moisture had to be precluded during the subsequent α-chlorination in order to avoid adventitious hydrolysis of this reactive intermediate. Notwithstanding, oxidation of the α-chloro sulfide proceeded chemoselectively in the presence of two equivalents of MCPBA to give 44, exposure of which to potassium tert-butoxide in tetrahydrofuran[78] resulted in very rapid conversion to 45 (51% overall, Scheme 10).[79] In the $^1$H NMR spectrum of 45, the vinyl and bridgehead protons appear as two well-spaced triplets, a feature seemingly characteristic of exo-5, anti-6 disubstituted bicyclo[2.1.1]hex-2-enes. This is a result of identical vicinal and allylic W-plan coupling in the isolated AA'XX' spin system.[80]

When attempts to achieve allylic bromination in 45 failed,[81] the pair of bromine atoms were introduced instead at the α-chloro sulfone stage.[82] However, exposure of 46 to potassium tert-butoxide in tetrahydrofuran resulted in unexpectedly rapid monodehydrobromination. Careful monitoring of the progress of this reaction enabled the isolation of 47a (as a mixture of epimers) in 45% yield, thereby demonstrating that desulfonylative ring contraction was, in fact, the slower of the two processes.

These complications were bypassed entirely by initial electrophilic bromination of the double bond in 43. Preliminary studies on 48 made clear the fact that base-promoted twofold dehydrobromination could be performed satisfactorily.[83] Because the conversion of diene sulfide 49 into α-chloro sulfone 47b could be effected only inefficiently (45%), it proved more expedient first to prepare 50 and *to accomplish the three chemical steps necessary to arrive at 12 in a single maneuver*. When 50 was treated with potassium tert-butoxide in THF-d$_8$ at -78 to -30 °C in an NMR tube, desulfonylative ring contraction was seen to occur to the virtual exclusion of dehydrobromination. The latter chemistry took place only in the -30 to 0 °C temperature range. At the preparative level, 12 was obtained efficiently at 0 °C, although admixed with

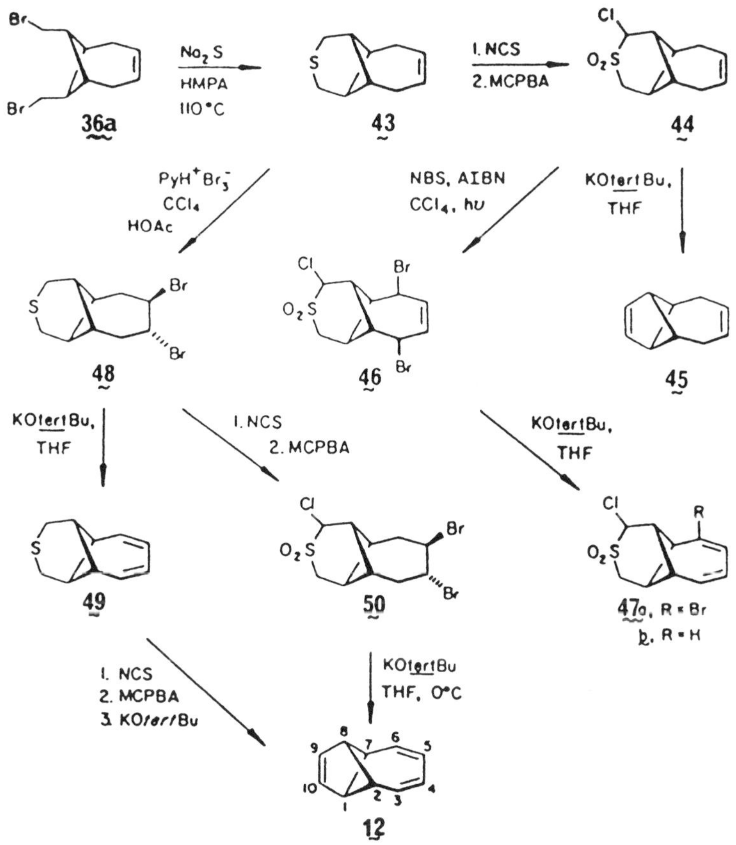

Scheme 10. Synthetic approaches to 12.

approximately 10% of 45. Pure 12 that was not contaminated with 45 could be obtained by installing the diene bridge first and then forming the monoene bridge, i.e., by Ramberg-Bäcklund ring contraction of 47b.

The spectral properties of 12 are in complete agreement with the structural assignment. The characteristic pair of mutually coupled triplets ($J$ = 2.3 Hz) again make their appearance, but in this instance vinyl protons H-9 and H-10 experience a record deshielding (to $\delta$ 7.28) while bridgehead protons H-1,8 are upfield shifted to $\delta$ 1.17.[84] The same effects are seen in the associated carbon atoms. The vertical electronic transition measured for 12 (274 nm) resembles, but is not identical to, the $\lambda_{max}$ values recorded for 7 (277 nm)[26] and the 9,10-dihydro derivative of 12 (283 nm).[42a]

From the strategy standpoint, the pathway to 12 that ultimately proved successful confirmed that it is advisable to delay installation of the bicyclo[2.1.1]hexene double bond until as late as possible. In so doing, the strain and reactivity in this structural segment has no opportunity to provide complications that eventuate in destruction of the ring system.

## 2.4. PREPARATION OF 9,10-DIMETHYLENETRICYCLO[5.3.0.0$^{2,8}$]DECA-3,5-DIENE (13).

Not only the termini of a π-ribbon, but also its central carbons, can be connected to the cyclobutane ring. Tetraene 13 is isomeric with 11. Its two-carbon bridge can be approximated to be ca 1.53 Å in length, somewhat longer than the ethylene bridge in 2 (ca 1.36 Å).[18,85] The ring strain in 13 should, therefore, be intermediate between that residing in 11 and 12. On the other hand, the energy levels of the MO's in 13 resemble more closely those of 11 and the symmetries of the FMO's in both 11 and 13 are opposite to those present in 12.[18] Consequently, 13 offers a unique opportunity to gain information on the relative importance of ring strain and electronic character in these systems.

In order to assess the reactivity of 13 relative to 11 and 12, we have also synthesized this third member of the series,[35,36] the first cyclobutane to have hybrid connectivity to two butadiene units. We chose to adopt the Capozzi-Hogeveen method[86] for introducing the vicinal exo-dimethylene moiety and it was therefore necessary to gain access to dimethyl diene 54. Scheme 11 outlines the crafting of this tricyclic

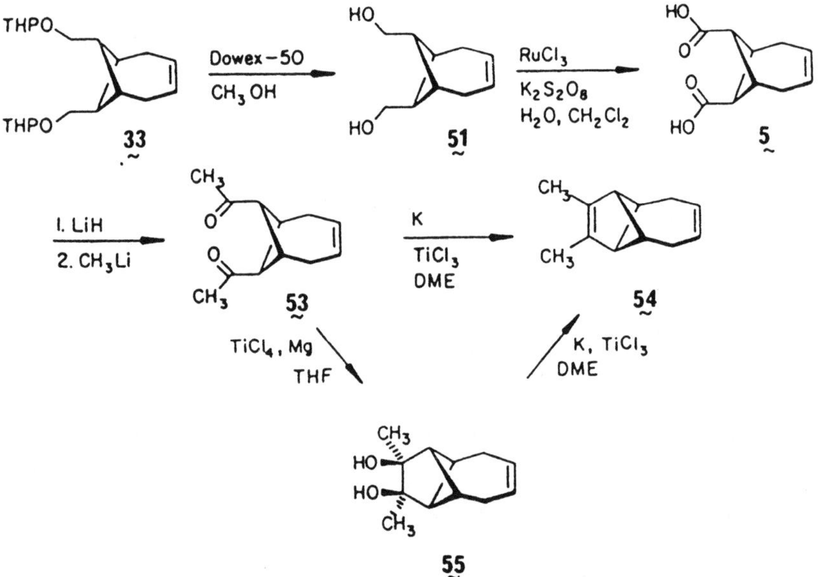

Scheme 11. Arrival at the key diene intermediate 54.

diene intermediate. The tetrahydropyranyl groups in 33 were removed by stirring with acid-washed Dowex-50 resin in methanol.[87] In order to bypass lactone formation[88] during the ensuing oxidation of 51, recourse was made to potassium ruthenate,[89] as generated from alkaline potassium persulfate and ruthenium trichloride hydrate. Addition of methyllithium to the dilithium salt of 52[90] gave rise to 53.

Pinacolic coupling within 53 was best promoted by titanium tetrachloride and magnesium in tetrahydrofuran.[91] Although application of the Capozzi-Hogeveen bromination/dehydrobromination protocol to 54 did lead to 57, no further use was made of 54 as a precursor to 13. Instead, dibromide 57 was prepared much more efficiently by reaction of 55 with pyridinium perbromide and heating of the resulting polyfunctional compound 56 in benzene with the Burgess reagent[92] (Scheme 12). Ultimate

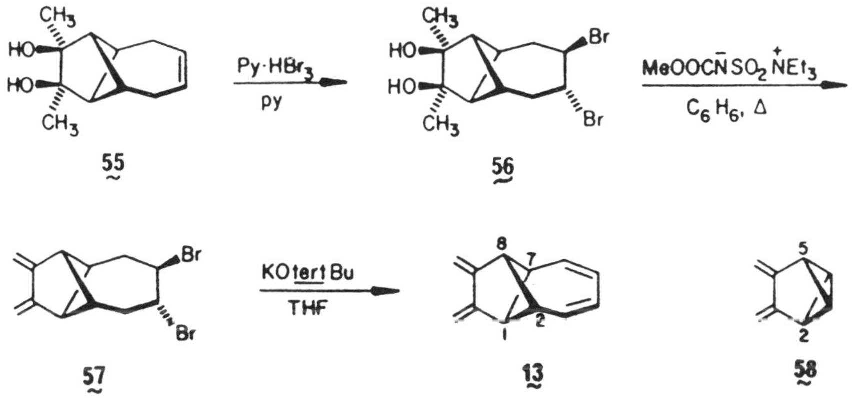

Scheme 12. Acquisition of 13.

dehydrobromination of 57 with potassium *tert*-butoxide in tetrahydrofuran was also achieved without complication.

Tetraene 13, which has proven to be a fairly stable substance, was easily identified on the basis of its $^1$H NMR spectrum. The typical AA'BB'XX' pattern of the cyclic 1,3-diene unit is accompanied by sharp singlets for the exo methylene and bridgehead protons, in agreement with the usual absence of spin-spin interaction there. Contraction of the flap angle of the cyclobutane ring by installation of a short bridge as in 58[93] or widening this same angle by the interposition of a 1,4-butadienyl ribbon as in 13 has little consequence on the chemical shift of the exo methylene protons. However, H-1,8 in 13 experience a notable upfield shift to $\delta$ 1.68 in a manner paralleling the shielding encountered in 12 (compare 58: $\delta$ H$_{2,5}$, 2.72).

III. Response of 11-13 to Thermal and Photochemical Activation.

3.1. THE BEHAVIOR OF TETRAENE 11.

Tricyclo[5.5.0.0$^{2,8}$]dodecatetraene has been predicted to be a benzene dimer, i.e., a (CH)$_{12}$ hydrocarbon,[94] of reasonable stability.[18] Nevertheless, when heated in C$_6$D$_6$ at 102 °C, 11 isomerized cleanly and quantitatively to 59[95] ($k = 1.8 \times 10^{-4} s^{-1}$; $t_{1/2} = 63$ min).[37] For the purpose of isotopic labeling, 42 was subjected to lithium-halogen exchange and subsequent $D_2O$ quench. The $D_{2d}$ symmetry of 11-$d_2$ caused C-3, 6, 9, and 12 to be equivalent as in 60 (Scheme 13). Consequently,

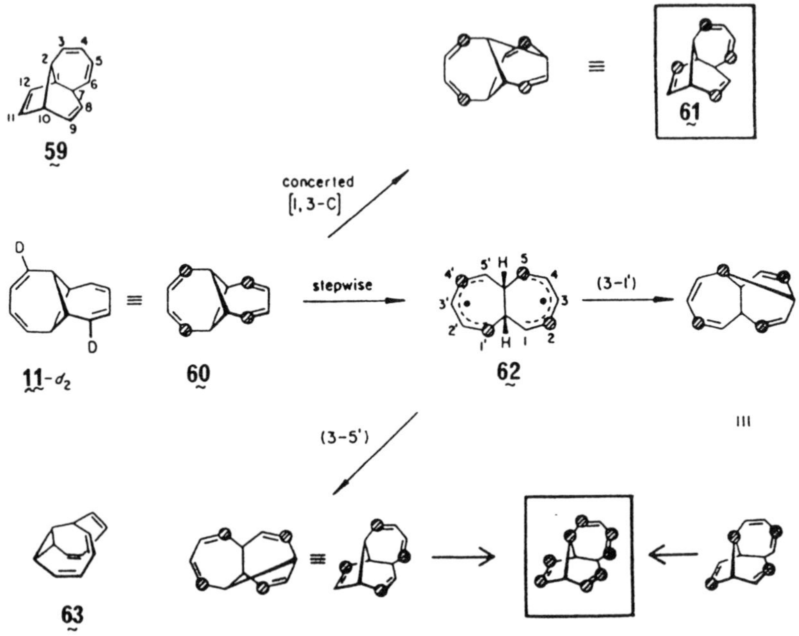

Scheme 13. Course of the thermal isomerization of 11-$d_2$.

11-$d_2$ can be considered to be quadruply labeled as shown in 60. When 11-$d_2$ in carbon tetrachloride solution was thermolyzed for 5 h at 110-115 °C in a sealed NMR tube, only four of the twelve possible positions showed deuterium incorporation as in 61, all with approximately the same intensity ($^2$H NMR analysis at 77 MHz).

Of the various mechanistic alternatives available to 11, only a formally concerted [1,3]-C migration[96] is consistent with these findings. Full symmetrization via 62, which would have distributed the isotopic label over eight sites, clearly does not obtain. Also, any passage through isomer 63 can also be discounted, since the heat of formation of this semibullvalene-like structure (102.35 kcal mol$^{-1}$) reveals it to be of considerably higher energy than 11 (90.50 kcal mol$^{-1}$) and consequently not likely attainable by thermal activation of this magnitude.

When irradiated at 366 nm, 11 again gave rise to 59. Thus, 11 finds [1,3] sigmatropic migration to be most accessible from its ground and excited states.

## 3.2. THERMAL ISOMERIZATION OF 12.

At 20 °C in CDCl$_3$ solution, 12 isomerizes relatively rapidly to the known isobullvalene (64).[97,98] After a short time, $^1$H NMR signals due to lumibullvalene (65)[95b,99,100] developed and ultimately replaced those due to 64. The experimental first-order rate data can be found in Scheme 14.

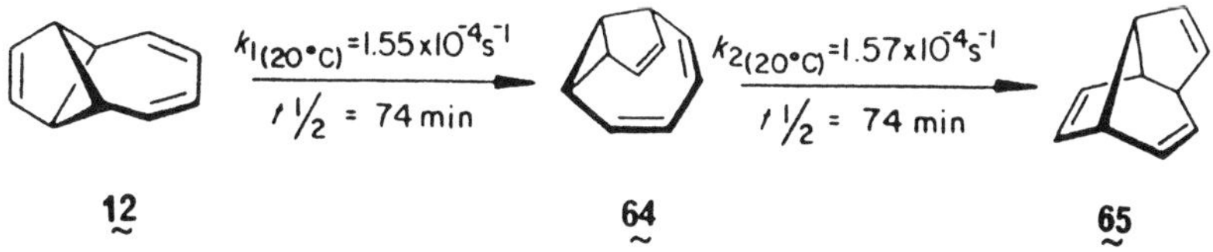

$$12 \xrightarrow[t\frac{1}{2} = 74 \text{ min}]{k_{1(20°C)} = 1.55 \times 10^{-4} \text{s}^{-1}} 64 \xrightarrow[t\frac{1}{2} = 74 \text{ min}]{k_{2(20°C)} = 1.57 \times 10^{-4} \text{s}^{-1}} 65$$

*Scheme 14.* Thermal isomerization of 12.

The thermal isomerization of 12 to 64 can occur by three different mechanisms (Scheme 15): (1) a concerted, thermally allowed [1,5] carbon shift across the diene bridge (12 → 66); (2) a concerted, thermally forbidden [1,3] carbon shift across the ethylene bridge (12 → 67);[96] and (3) a stepwise diradical process involving homolysis of any of the four symmetry-equivalent cyclobutane σ bonds (12 → 68).[101] Suitable isotopic

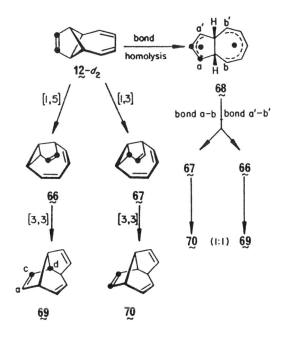

*Scheme 15.* Mechanistic options available to 12-$d_2$.

labeling of 12 can differentiate between these hypothetical pathways, particularly since the conversion of 64 to 65 has previously been demonstrated to be the exclusive result of pure [3,3] sigmatropy.[98b] Thus, reliable accounting can be made of the fate of the individual carbons in the second stage of the isomerization, i.e., 66 → 69 and 67 → 70.

As the result of the ability of α-chloro sulfones[102] and episulfones[103] to undergo rapid hydrogen-deuterium exchange alpha to sulfonyl, 12-$d_2$ was readily prepared by exposure to an excess of potassium tert-butoxide in cold (-70 °C) tetrahydrofuran containing deuterium oxide, and gradual warming of this mixture to 0 °C. A total of 0.46 D was incorporated into the ethylenic bridge. Although the triene was not likely equivalently deuterated at both sites, this issue is unimportant since the $C_{2v}$ symmetry of 12 does not allow independent distinction of these positions.

The labeling patterns within lumibullvalene expected from the three candidate processes are (see Scheme 15): (1) $H_a:H_c$, 1:1; (2) $H_c:H_d$, 1:1; (3) $H_a:H_c:H_d$, 1:2:1. At the experimental level, an intensity ratio of 1:2.1:1 was observed. This finding is uniquely compatible with intervention of biradical 68 since this intermediate alone possesses a mirror plane.[37] The divergency in mechanistic response of 11 and 12 is quite striking.

3.3. THERMALLY-INDUCED REARRANGEMENT OF 13.

When heated in benzene at 80 °C, 13 underwent smooth first-order rearrangement ($t_{\frac{1}{2}}$ = 65 min) to give 72 (Scheme 16). The formation of 72 can result in direct concerted [1,5] sigmatropic rearrangement or via a diradical process mediated by 71 and no distinction has been made between these options.[37]

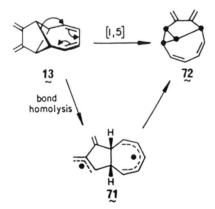

Scheme 16. Pathways open to 13 when heated.

In summary, one sees that within the tricyclic hydrocarbons that have the latent potential for interaction of two mutually perpendicular π-ribbons through a cyclobutane ring, thermal lability is seen to decrease dramatically from 8 through 12 and 13 to 11 (Scheme 17). This order conforms to the progressive reduction in ring strain and bears no necessary relationship to the prevailing electronic state of affairs.

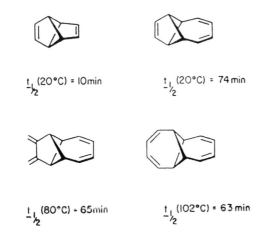

*Scheme 17.* Summary of kinetic data.

Additionally, the general trend is characteristically one where the divinylcyclopropane isomer is formed initially, except in the case of 11. For this system only, this product type is of higher energy than that of the starting cyclobutane and is not thereby accessible.

## IV. Nuclear Magnetic Resonance Effects

Christl and Herbert[104] have advanced the hypothesis that the unusually high deshielding of C-5 and C-6 in bicyclo[2.1.1]hex-2-ene (**A**, 68.0 ppm) relative to those in bicyclo[2.1.1]hexane (**B**, 39.3 ppm) stems from interaction between filled cyclobutane orbitals and the $\pi$ LUMO in the olefin. The resultant reduction of charge density would explain the downfield shift (Scheme 18).

Sander and Gleiter[105] have pointed out the same deshielding trend in the chemical shifts of *protons* 1 and 6 in tricyclo[3.3.0.0$^{2,6}$]oct-3-ene (**C**, $\delta$ 3.45) as compared to tricyclo[3.3.0.0$^{2,6}$]octane (**D**, $\delta$ 1.82). The presence of a cyclohepta-1,3-diene subunit in place of a cyclopentene part structure has precisely the opposite effect. Thus, H-1,8 in tricyclo[5.3.0.0$^{2,8}$]deca-3,5-diene (**E**, $\delta$ 1.27) are shielded in reference to the same protons in tricyclo[5.3.0.0$^{2,8}$]decane (**F**, $\delta$ 2.23). The same is true for C-1 and C-8 (28.31 vs 44.55 ppm). However, MINDO/3, MNDO, and EHT calculations seemingly disagree with the preceding correlation of charge density at the carbon atoms of interest and their chemical shift. As an alternative, Sander and Gleiter implicated an anisotropic magnetic field effect that was deshielding above monoenes and shielding above dienes.

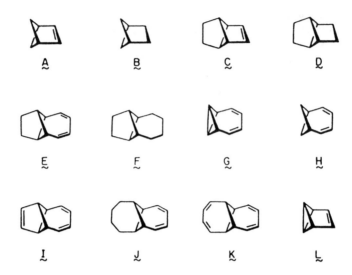

*Scheme 18.* Candidate molecules showing $^1$H NMR effects.

Subsequently, Christl and Herzog[106] noted that the anisotropic field hypothesis would predict increased shielding for an homoallylic carbon the more it is compressed into the region occupied by the diene moiety. A progression of this type is found in going from octavalene (G) to E to bicyclo[4.1.1]octa-2,4-diene (H). However, the relevant spectral data clearly show the reverse to be true. The experimental $\Delta\delta$ values (Table I) happen to be 23.92, 16.24, and 8.48 ppm, corresponding to a progressive shielding relative to the appropriate fully saturated reference molecule. Furthermore, the corresponding proton shift pattern ($\Delta\delta$ = 0.04, -0.96, and -0.20) is completely erratic.

In contrast to the above, the $^{13}$C sequencing in the 1,3-cycloheptadiene series correlates well with decreasing strain in the cyclobutane ring, i.e., a lowering of its $\sigma$ MO's. This observation is empirical, however, and does not constitute a satisfying theoretical explanation. Nonetheless, we see that tricyclo[5.3.0.0$^{2,8}$]deca-3,5,9-triene (I, which contains a bicyclo[2.1.1]hexene moiety) assumes an intermediate position between G (a bicyclobutane) and E (a bicyclo[2.1.1]hexane). The proximity of the $\Delta\delta$ values for J and E are less readily appreciated. However, the $^{13}$C assignments to J have only been tentatively assigned and require more detailed analysis. Whether the strikingly low $\Delta\delta$ value for K (Table I) is a reflection of through bond interaction that has the opposite sign of that in J remains to be determined unequivocally.

Qualitatively at least, the absolute value of $\Delta\delta$ for the cyclopentene-derived systems follow the same trend (Table II). A decrease is apparent in progressing from benzvalene (L) to I. However, the insufficiency of $^{13}$C data and the close similarity of the last three $\Delta\delta$ values in Table II render any interpretation highly speculative.

TABLE I. NMR Analysis of the 1,3-Cycloheptadiene Effect

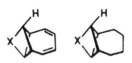

| | δ/ppm | 13C / 1H | 13C / 1H | 1H | Δδ=δ_sat−δ_unsat/ppm 13C |
|---|---|---|---|---|---|
| X = | ∐ | 10.51[a] / 1.23[b] | −13.41[a] / 1.27[b] | −0.04 | 23.92 |
| | ⌐⌐ | 48.9[c] / 2.48[c,e] | 29.61[d] / 1.17[d] | 1.31 | 19.3 |
| | ⌐⌐ | 44.55 / 2.23 | 28.31 / 1.27 | 0.96 | 16.24 |
| | ⌐⌐ | 40.08[f] / 2.01 | 24.97[f] / 2.11 | −0.10 | 15.11 |
| | H−∕ H−∕ | 30.09[a] / 1.44[g] | 21.61[a] / 1.24[g] | 0.20 | 8.48 |
| | ⌐⌐ | 40.91[f] / 2.48 | 32.93 / 2.55 | −0.07 | 7.98 |

[a]Reference 106.  [b]Christl, M.; Herzog, C.; Kenner, P. Chem. Ber. 1986, *119*, 3045.  [c]Reference 42b.  [d]This work.  [e]This assignment is deduced by analogy to tricyclo[5.3.0.- 0^{2,8}]deca-4,9-diene (δ H-1, 2.58; δ H-2, 2.91).  [f]Assignment requires confirmation by 2-D NMR techniques.  [g]Reference 27b.

There remains the need to call specific attention to the striking relationship between the chemical shift of the vinyl protons in those hydrocarbons endowed with bicyclo[2.1.1]hex-2-ene part structures and the overall strain induced by the second bridge within the tricyclic carbon framework (Table III).

TABLE II. NMR Analysis of the Cyclopentene Effect.

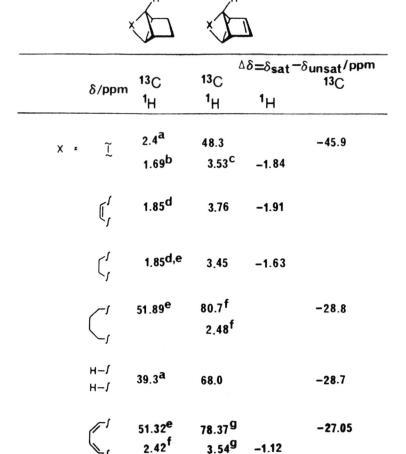

| X = | δ/ppm | $^{13}C$ / $^1H$ | $^{13}C$ / $^1H$ | $^1H$ | $\Delta\delta = \delta_{sat} - \delta_{unsat}$ /ppm $^{13}C$ |
|---|---|---|---|---|---|
| $\tilde{I}$ | | 2.4[a] | 48.3 | | −45.9 |
| | | 1.69[b] | 3.53[c] | −1.84 | |
| (CH₂) | | 1.85[d] | 3.76 | −1.91 | |
| (CH₂) | | 1.85[d,e] | 3.45 | −1.63 | |
| (C=O) | | 51.89[e] | 80.7[f] | | −28.8 |
| | | | 2.48[f] | | |
| H–/H–/ | | 39.3[a] | 68.0 | | −28.7 |
| (C=) | | 51.32[e] | 78.37[g] | | −27.05 |
| | | 2.42[f] | 3.54[g] | −1.12 | |

[a]Reference 104. [b]Christl, M.; Bruntrup, G. Chem. Ber. 1974, 107, 3908. [c]Wilzbach, K.; E.; Ritscher, J. S.; Kaplan, L. J. Am. Chem. Soc. 1967, 89, 1032. [d]Meinwald, J.; Kaplan, B. E. Ibid. 1967, 89, 2611. [e]Reference 105. [f]Reference 42b. [g]This work.

TABLE III.   Correlation of Vinyl Proton Chemical Shift.

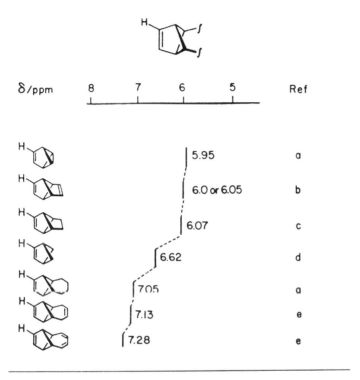

| δ/ppm | | Ref |
|---|---|---|
| | 5.95 | a |
| | 6.0 or 6.05 | b |
| | 6.07 | c |
| | 6.62 | d |
| | 7.05 | a |
| | 7.13 | e |
| | 7.28 | e |

[a]Reference 42b.  [b]References 32 and 33.  [c]Meinwald, J.; Kaplan, B. E. *J. Am. Chem. Soc.* **1967**, *89*, 2611.  [d]Meinwald, J.; Uno, F. *Ibid.* **1968**, *90*, 800.  [e]This work.

## V. Photoelectron Spectroscopic Analysis of Through-bond Interaction via Cyclobutane Relay Orbitals.

### 5.1. SPECTRAL DATA AND SUPPORTING CALCULATIONS

In collaboration with Professor Rolf Gleiter, the PE spectra of 11 and its more saturated congeners 73 and 74, synthesized as outlined in Scheme 19, were recorded. They are illustrated in Figure 1. While several strongly overlapping bands starting around 9.6 eV are seen for 74, one band at 8.4 and a shoulder at 10.3 eV appear in the spectrum of 73 followed by strongly overlapping bands near 11 eV. For 11, two bands with a steep onset at 7.56 and 9.00 eV are encountered, well separated from a series of strongly overlapping bands at 10.0, 10.4 and 10.9 eV. Using other reference data, we can confidently assign the first two bands in 73 to ionizations stemming from $a_2(\pi)$ and $b_1(\pi)$. For 11, its

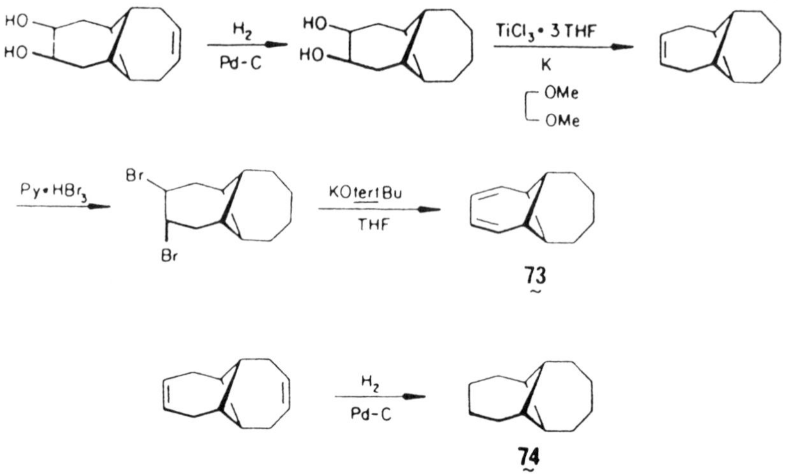

Scheme 19. Synthesis of 73 and 74.

bands 1-4 result from π ionization events; while the first two bands are split by 1,4 eV, bands 3 and 4 overlap strongly because of a Jahn-Teller interaction. The output data of various calculations are summarized in Table IV.[38] Finally, while the electronic absorption spectrum of 11 shows two bands at 315 and 224 nm, that of 73 shows only one band at 284 nm with vibrational spacings.

The PE spectra of 12 and 13 are collected in Figure 2. For reasons of symmetry, interactions in 12 only between the π* orbitals of one π unit and the π-orbitals of the other are expected. Although the cyclobutane unit is a powerful relay, the large energy difference between such bonding and antibonding niveaus were expected to give rise only to small, if not negligible, effects. Indeed, one broad peak due to two close-lying bands is seen for 12 around 8.5 eV (bands 1 and 2) well separated from a second Gaussian band at 9.9 eV. These features reflect a simple superpositioning of the spectra independently derived from 75 and 76 (Scheme 20). This comparison reveals the absence of interaction between the two olefinic fragments in 12.

75
76

Scheme 20. More highly saturated analogues of 12.

For 13, the difference in connectivity lowers the symmetry from $D_{2v}$ (as in 11) to $C_{2v}$. Its PE spectrum has an appearance similar to

TABLE IV. Comparison between the first vertical ionization $I(_{v,j})$ of 11, 73, and 74 with calculated energies ($\epsilon_j$) based on FMO, MINDO/3, and *ab initio* (STO-3G) procedures. All values are in eV.

| Compound | Band | $I_{v,j}$ | Assignment | $-\epsilon_j$(FMO) | $-\epsilon_j$(MINDO/3) | $-\epsilon_j$(STO-3G) |
|---|---|---|---|---|---|---|
| 11 | 1 | 7.56 | $2b_1$ | 7.60 | 8.03 | 5.79 |
|    | 2 | 9.00 | $1a_2$ | 8.50 | 9.13 | 7.21 |
|    | 3 | 10.0 | } 7e | 10.03 | 9.77 | 9.40 |
|    | 4 | 10.4 |       |       |       |       |
|    | 5 | 10.9 | $7a_1$ |       | 10.11 | 10.02 |
|    | 6 | 11.1 | 6e    | 11.46 | 10.80 | 11.08 |
| 73 | 1 | 8.38 | $4a_2$ | 8.08 | 8.23 | 6.37 |
|    | 2 | 10.2 | $8b_1$ | 9.90 | 9.58 | 9.40 |
|    |   |      | { $7b_1$ | 10.40 | 9.81 | 9.65 |
|    | 3 | 10.9 | { $13a_1$ | 10.40 | 9.88 | 10.36 |
| 74 | 1 | 9.6  | $7a_1$ |       | 10.07 | 9.41 |
|    | 2 | 10.0 | 8e    |       | 9.75 | 9.43 |

that of 11 and the interaction scheme present in 11 holds in principle also for 13. The obvious differences can be traced back to the different connectivity between the central ring and the butadiene moieties.

5.2. CONCLUDING ANALYSIS

This work was originated on the premise that the interaction between two perpendicular $\pi$-systems found in spiro compounds should be larger when their central carbon atom is replaced by a four-membered ring system as shown in Figure 3. This concept involves the replacement of the through-space interaction element in the spiro compound by a through-bond interaction feature in the resulting tricyclic system.

The most interesting observation in our examination of 11 is the close similarity of the energy gaps [$\Delta I(1,2)$ = 1.44 eV and $\Delta E(1,2)$ = 1.60 eV)] between the corresponding bands in its PE (I) and electron absorption (E) spectra. A similar equivalence has been observed for 1. For 11, the two $\sigma$ orbitals of its cyclobutane ring are ideally suited to interact with the corresponding linear combinations of the two highest occupied $\pi$-MO's (Figure 4). As a result, *the "relay conjugation" in 11 is more efficient than the spiroconjugation in 1.*

A strong interaction between the two perpendicular $\pi$-systems and the central four-membered ring similar to that in 11 is seen in 13. The origin of the energy difference of the first two bands in the electronic absorption spectrum of 13 is totally different from that of 11. In 13,

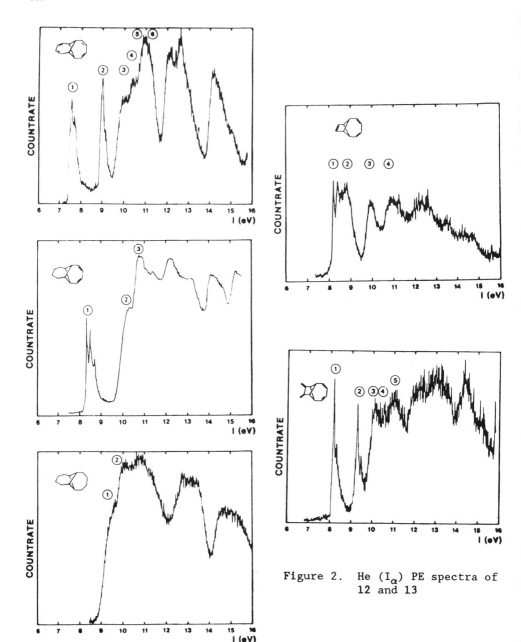

Figure 1. He ($I_\alpha$) PE spectra of 11, 73, and 74

Figure 2. He ($I_\alpha$) PE spectra of 12 and 13

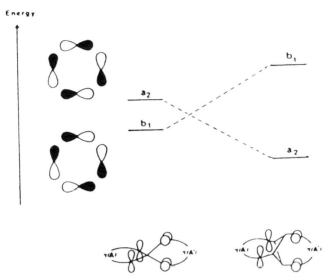

Figure 3. Orbital sequence for $b_1$ and $a_2$ in spiro compounds (left) and the corresponding tricyclic compounds (right).

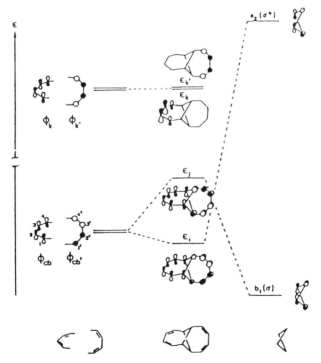

Figure 4. Definition of valence orbitals of the two perpendicular $\pi$-MO's and the $\sigma$-MO's of the four-membered ring in 11 $(D_{2d})$.

the energy difference is mainly due to the splitting of the lowest unoccupied MO's. This can be traced back to the lower symmetry of 13 as compared to 11.

In relevant contrast, the PE spectrum of 12 clearly reveals the absence of interaction between its two olefinic fragments. At least in this instance, theory has been upheld.

**Acknowledgments.** The authors acknowledge with thanks the financial support of the National Science Foundation and permission granted by the American Chemical Society to reproduce Figures 1-4 in the present context.

## VI. References and Notes

(1) Hoffmann, R.; Imamura, A.; Zeiss, G. D. *J. Am. Chem. Soc.* 1967, 89, 5215.
(2) (a) Simmons, H. E.; Fukunaga, T. *J. Am. Chem. Soc.* 1967, 89, 5208. (b) Gordon, M. D.; Fukunaga, T.; Simmons, H. E. *Ibid.* 1976, 98, 8401.
(3) (a) Semmelhack, M. F.; Foos, J. S.; Katz, S. *J. Am. Chem. Soc.* 1972, 94, 8637. (b) Semmelhack, M. F.; DeFranco, R. J. *Ibid.* 1972, 94, 8838. (c) Semmelhack, M. F.; DeFranco, R. J.; Margolin, J.; Stock, J. *Ibid.* 1973, 95, 426. (d) Semmelhack, M. F.; Foos, J. S.; Katz, S. *Ibid.* 1973, 95, 7325.
(4) (a) Boschi, R.; Dreiding, A. S.; Heilbronner, E. *J. Am. Chem. Soc.* 1970, 92, 123. (b) Batisch, C.; Heilbronner, E.; Semmelhack, M. F. *Helv. Chim. Acta* 1973, 56, 2110. (c) Batisch, C.; Heilbronner, E.; Rommel, E.; Semmelhack, M. F.; Foos, J. S. *J. Am. Chem. Soc.* 1974, 96, 7662.
(5) Tajiri, A.; Nakajima, T. *Tetrahedron* 1971, 27, 6089.
(6) (a) Dürr, H.; Gleiter, R. *Angew. Chem., Int. Ed. Engl.* 1978, 17, 559. (b) Gleiter, R. *Top. Curr. Chem.* 1979, 86, 197.
(7) Model calculations have also been performed on spiro compounds with silicon and phosphorus as central atoms [Böhm, M. C.; Gleiter, R. *J. Chem. Soc. Perkin II* 1979, 443]. The influence of 3d participation on the ground state is found to be small, in line with experimental work performed so far in this area.[8]
(8) (a) Märkl, G.; Merz, A. *Tetrahedron Lett.* 1969, 1231. (b) Schweig, A.; Weidner, U.; Hellwinkel, D.; Krapp, W. *Angew Chem., Int. Ed. Engl.* 1973, 12, 310. (c) Shain, A. L.; Ackerman, J. P.; Teague, M. W. *Chem. Phys. Lett.* 1969, 3, 550.
(9) Mulliken, R. S. *J. Chem. Phys.* 1949, 46, 497.
(10) Paquette, L. A.; Wallis, T. G.; Kempe, T.; Christoph, G. G.; Springer, J. P.; Clardy, J. *J. Am. Chem. Soc.* 1977, 99, 6949.
(11) Garbisch, E. W., Jr.; Sprecher, R. F. *J. Am. Chem. Soc.* 1966, 88, 3433, 3434.
(12) Only the rate of disappearance of 1 was measured without product identification.[4d]
(13) Gerson, F.; Gleiter, R.; Moshuk, G.; Dreiding A. S. *J. Am. Chem. Soc.* 1972, 94, 2919.

(14) Dürr, H.; Ruge, B.; Schmitz, H. *Angew Chem. Soc., Int. Ed. Engl.* **1973**, *12*, 577.
(15) Chiang, J. F.; Wilcox, C. F., Jr. *J. Am. Chem. Soc.* **1973**, *95*, 2885.
(16) Bischof, P.; Gleiter, R.; Dürr, H.; Ruge, B.; Herbest, R. *Chem. Ber.* **1976**, *109*, 1412.
(17) Clark, R. A.; Fiato, R. A. *J. Am. Chem. Soc.* **1970**, *92*, 4736.
(18) Bischof, P.; Gleiter, R.; Haider, R. *Angew. Chem., Int. Ed. Engl.* **1977**, *16*, 110; *J. Am. Chem. Soc.* **1978**, *100*, 1036.
(19) Hoffmann, R. *Acc. Chem. Res.* **1971**, *4*, 1.
(20) Hoffmann, R.; Davidson, R. B. *J. Am. Chem. Soc.* **1971**, *93*, 5699.
(21) (a) Bischof, P.; Gleiter, R.; de Meijere, A.; Meyer, L. U. *Helv. Chim. Acta* **1974**, *57*, 1519. (b) Bischof, P.; Gleiter, R.; Kukla, M. J.; Paquette, L. A. *J. Electron Spectrosc. Relat. Phenom.* **1974**, *4* 177. (c) Bruckmann, P.; Klessinger, M. *Chem. Ber.* **1978**, *111*, 944. (d) Bischof, P.; Gleiter, R.; Gubernator, R.; Haider, R.; Musso, H.; Schwarz, W.; Trautmann, W.; Höpf, H. *Ibid.* **1981**, *114*, 994. (e) Gleiter, R.; Haider, R.; Spanget-Larsen, J.; Bischof, P. *Tetrahedron Lett.* **1983**, 1149. (f) Gleiter, R.; Haider, R.; Gubernator, K.; Bischof, P. *Chem. Ber.* **1983**, *116*, 2983. (g) Gleiter, R.; Haider, R.; Bischof, P.; Lindner, H.-J. *Ibid.* **1983**, *116*, 3736.
(22) Bruckmann, P.; Klessinger, M. *Chem. Ber.* **1978**, *111*, 944.
(23) Gleiter, R.; Gubernator, K.; Grimme, W. *J. Org. Chem.* **1981**, *46*, 1247.
(24) The shortcomings associated with systems such as 5 and 6 are briefly discussed in reference 26. Furthermore, it is known that the radical anion of a exhibits an eight times larger $\beta$-hyperfine splitting ($a_H$) with $H_\delta$ than does semidione b. Symmetry-allowed interaction between the naphthalene SOMO and the $a_2$ Walsh orbital (as shown) generates high spin density in the cyclobutane C-C bonds, thereby giving rise to substantive coupling with $H_\delta$. In

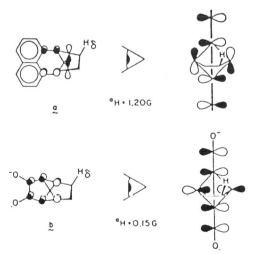

contrast, the semidione SOMO can interact only with the occupied $b_2$ Walsh orbital that contains little spin density in the cyclobutane C-C bonds.

(25) (a) Nelsen, S. F.; Gillespie, J. P. J. Am. Chem. Soc. 1973, 95, 2940. (b) Russell, G. A.; Whittle, P. R.; Keske, R. G. Ibid. 1971, 93, 1467.
(26) Gleiter, R.; Bischof, P.; Volz, W. E.; Paquette, L. A. J. Am. Chem. Soc. 1977, 99, 8.
(27) Alternative syntheses of 7 has been reported more recently: (a) Yin, T.-K.; Lee, J. G.; Borden, W. T. J. Org. Chem. 1985, 50, 531. (b) Christl, M.; Herzog, C.; Nusser, R. Chem. Ber. 1986, 119, 3059.
(28) Young, S. D.; Borden, W. T. Tetrahedron Lett. 1976, 4019.
(29) Young, S. D.; Borden, W. T. J. Org. Chem. 1980, 45, 724. (b) Borden, W. T.; Lee, J. G.; Young, S. D. J. Am. Chem. Soc. 1980, 102, 4841. (c) Gajewski, J. J.; Gortva, A. M.; Borden, W. T. Ibid. 1986, 108, 1083.
(30) Chasey, K. L.; Paquette, L. A.; Blount, J. F. J. Org. Chem. 1982, 47, 5262.
(31) (a) Gleiter, R.; Heilbronner, E.; de Meijere, A. Helv. Chim. Acta 1971, 54, 783. (b) Askani, R.; Gleiter, R.; Heilbronner, E.; Hornung, V.; Musso, H. Tetrahedron Lett. 1971, 4461. (c) Bischof, P.; Gleiter, R.; Heilbronner, E.; Hornung, V.; Schroder, G. Helv. Chim. Acta 1970, 53, 1645.
(32) Meinwald, J.; Tsuruta, H. J. Am. Chem. Soc. 1969, 91, 5877; 1970, 92, 2579.
(33) Zimmerman, H. E.; Robbins, J. D.; Schantl, J. J. Am. Chem. Soc. 1969, 91, 5878.
(34) (a) Gleiter, R.; Kobayashi, T. Helv. Chim. Acta 1971, 54, 1081. (b) Kanda, K.; Koremoto, T.; Imamura, A. Tetrahedron 1986, 42, 4169.
(35) (a) Paquette, L. A.; Dressel, J.; Chasey, K. L. J. Am. Chem. Soc. 1986, 108, 512. (b) Dressel, J.; Paquette, L. A. Ibid. 1987, 109, 2857. (c) Paquette, L. A.; Dressel, J.; Pansegrau, P. D. Tetrahedron Lett. 1987, 4965.
(36) Dressel, J.; Chasey, K. L.; Paquette, L. A. J. Am. Chem. Soc. 1988, 110, in press.
(37) Dressel J.; Pansegrau, P. D.; Paquette, L. A. J. Org. Chem. 1988, 53, in press.
(38) Gleiter, R.; Toyota, A.; Bischof, P.; Krennrich, G.; Dressel, J.; Pansegrau, P. D.; Paquette, L. A. J. Am. Chem. Soc. 1988, 110, in press.
(39) (a) Srinivasan, R. J. Am. Chem. Soc. 1963, 85, 819, 3048. (b) Whitesides, G. M.; Goe, G. L.; Cope, A. C. Ibid. 1969, 91, 2608. (c) Meinwald, J.; Kaplan, E. Ibid. 1967, 89, 2611. (d) Borden, W. T.; Gold, A. Ibid. 1971, 93, 3830. (e) Salomon, R. G.; Kochi, J. K. Tetrahedron Lett. 1973, 2529.
(40) Heathcock, C. H.; Badger, R. A. J. Org. Chem. 1972, 37, 234.
(41) Miyashita, M.; Yoshikoshi, A. J. Am. Chem. Soc. 1974, 96, 1917.
(42) (a) Gleiter, R.; Sander, W.; Butler-Ransohoff, I. Helv. Chim. Acta 1986, 69, 1872. (b) Gleiter, R.; Zimmerman, H.; Sander, W.;

Hauck, M. *J. Org. Chem.* **1987**, *52*, 2644 and references cited therein.
(43) (a) Shani, A.; Tashma, Z. *Helv. Chim. Acta* **1977**, *60*, 1903. (b) Shani, A. *Tetrahedron Lett.* **1972**, 569. (c) Scheffer, J. R.; Wostradowski, R. A. *J. Org. Chem.* **1972**, *37*, 4317. (d) Scheffer, J. R.; Boise, B. A. *J. Am. Chem. Soc.* **1971**, *93*, 5490.
(44) Srinivasan, R.; Carlough, K. H. *J. Am. Chem. Soc.* **1967**, *89*, 4932.
(45) Gleiter, R.; Sander W. *Angew. Chem., Int. Ed. Engl.* **1985**, *24*, 566.
(46) Heathcock, C. H.; Badger, R. A.; Patterson, J. R., Jr. *J. Am. Chem. soc.* **1967**, *89*, 4133.
(47) (a) Gleiter, R.; Muller, G.; Huber-Patz, U.; Rodewald, H.; Irngartinger, H. *Tetrahedron Lett.* **1987**, 1985. (b) Gleiter, R.; Muller, G. *J. Org. Chem.* submitted for publication.
(48) (a) Brousseau, R. J. Ph.D. Thesis, Harvard University, Cambridge, MA, 1977. (b) Scott, L. T. Ph.D. Thesis, Harvard University, Cambridge, MA, 1980.
(49) Moriarty, R. M. *Top. Stereochem.* **1974**, *8*, 271.
(50) Cotton, F. A.; Frenz, B. A. *Tetrahedron* **1974**, *30*, 1587,
(51) Allen, F. H. *Acta Cryst.* **1984**, *B40*, 64.
(52) Stone, J. M. R.; Mills, T. M. *Mol. Phys.* **1970**, *18*, 631.
(53) Miller, F. A.; Campbell, R. J. *Spectrochim. Acta* **1971**, *27A*, 947.
(54) (a) Bauld, N. L.; Cessac, J.; Holloway, R. L. *J. Am. Chem. Soc.* **1977**, *99*, 8140. (b) Adman, E.; Margulis, T. N. *J. Chem. Soc., Chem. Commun.* **1967**, 641.
(55) (a) Griffin, G. W.; Hager, R. B. *Rev. Chim. Acad. Rep. Populaire Roumaine* **1962**, *7*, 901. (b) Vellturo, A. F.; Griffin, G. W. *J. Org. Chem.* **1966**, *31*, 2241.
(56) Farnum, D. G.; Mostahari, A. *J. Org. Photochem. Synth.* **1971**, *1*, 103.
(57) (a) Caputo, J. A.; Fuchs, R. *J. Org. Chem.* **1968**, *33*, 1959. (b) Wolfe, S.; Hasan, S. K.; Campbell, J. R. *J. Chem. Soc. D.* **1970**, 1420.
(58) Carlsen, P. H. J.; Katsuki, T.; Martin, V. S.; Sharpless, K. B. *J. Org. Chem.* **1981**, *46*, 3936.
(59) Chakraborti, A. K.; Ghatak, U. K. *Synthesis* **1983**, 746.
(60) Bloomfield, J. J.; Owsley, D. C.; Nelke, J. M. *Org. React.* **1976**, *23*, 259.
(61) Chasey, K. L. Ph.D. Thesis, The Ohio State University, Columbus, OH 1982.
(62) Miyashita, N.; Yoshikoshi, A.; Grieco, P. A. *J. Org. Chem.* **1977**, *42*, 3772.
(63) Beier, R.; Mundy, B. P. *Synth. Commun.* **1979**, 271.
(64) House, H. O. *"Modern Synthetic Reactions, Second Edition"*, Benjamin/Cummings: Menlo Park, CA, 1972; p 57ff.
(65) Lansbury, P. T.; Bieron, J. F.; Klein, M. *J. Am. Chem. Soc.* **1966**, *88*, 1477.
(66) Akhtar, M.; Marsh, S. *J. Chem. Soc. C.* **1966**, 937.
(67) Corey, E. J.; Winter, R. A. E. *J. Am. Chem. Soc.* **1963**, *85*, 2677.
(68) Corey, E. J.; Hopkins, P. B. *Tetrahedron Lett.* **1982**, 1979.
(69) Vedejs, E.; Wu, E. S. C. *J. Org. Chem.* **1974**, *39*, 3641.
(70) This finding is in contrast to the effect of the Vedejs modifica-

tion on 30. See reference 61.
(71) McMurry, J. E. *Acc. Chem. Res.* **1983**, *16*, 405.
(72) Dams, R.; Malinovski, M.; Westdarp, J.; Geise, H. Y. *J. Org. Chem.* **1978**, *47*, 248.
(73) McMurry, J. E.; Fleming, M. P.; Kees, K. L.; Krepski, L. R. *J. Org. Chem.* **1978**, *43*, 3255.
(74) Compare: Muratake, H.; Takahashi, T.; Natsume, M. *Heterocycles* **1983**, *20*, 1963.
(75) (a) Elix, J. A.; Sargent, M. V.; Sondheimer, F. *J. Chem. Soc., Chem. Commun.* **1966**, 508. (b) Vogel, E.; Bell, W. A.; Lohmer, E. *Angew. Chem., Int. Ed. Engl.* **1971**, *10*, 399.
(76) Seebach, D.; Neumann, H. *Chem. Ber.* **1974**, *107*, 847.
(77) Paquette, L. A.; Wingard, R. E., Jr.; Philips, J. C.; Thompson, G. L.; Read, L. K.; Clardy, J. *J. Am. Chem. Soc.* **1971**, *93*, 4508.
(78) Paquette, L. A. *Org. React.* **1977**, *25*, 1.
(79) Bicyclo[2.1.1]hexenes have previously been prepared by this methodology: Carlson, R. G.; May, K. D. *Tetrahedron Lett.* **1975**, 947.
(80) See: (a) Pomerantz, M. *J. Am. Chem. Soc.* **1966**, *88*, 5350. (b) Wiberg, K. B.; Ubersax, R. W. *J. Org. Chem.* **1972**, *37*, 3827.
(81) Generation of a radical site α to the bicyclo[2.1.1]hexene unit in 45 provides a splendid opportunity for relief of the considerable ring strain.
(82) Paquette, L. A.; Houser, R. W. *J. Am. Chem. Soc.* **1971**, *93*, 4522.
(83) Dehydrobromination so as to form vinyl bromide was not viewed as a potential problem here so long as the bromine atoms find it possible to assume an axial disposition without steric interference.
(84) Compare Sander, W.; Gleiter, R. *Chem. Ber.* **1985**, *118*, 2548.
(85) Jorgensen, W. L.; Borden, W. T. *J. Am. Chem. Soc.* **1973**, *95*, 6649.
(86) Capozzi, G.; Hogeveen, H. *J. Am. Chem. Soc.* **1975**, *97*, 1479.
(87) Beier, R.; Mundy, P. B. *Synth. Commun.* **1979**, *9*, 271.
(88) See, for example: Trost, B. M.; Balkovec, J. M.; Angle, S. R. *Tetrahedron Lett.* **1986**, 1445.
(89) Schroder, M.; Griffith, W. P. *J. Chem. Soc., Chem. Commun.* **1979**, 58.
(90) Paquette, L. A.; Snow, R. A.; Muthard, J. L.; Cynkowski, T. *J. Am. Chem. Soc.* **1978**, *100*, 1600.
(91) Corey, E. J.; Danheiser, R. L.; Chandrasekara, S. *J. Org. Chem.* **1976**, *41*, 260.
(92) Burgess, E. M.; Penton, H. R., Jr.; Taylor, E. A. *J. Org. Chem.* **1973**, *38*, 26.
(93) Lanzendorfer, F.; Christl, M. *Angew. Chem., Int. Ed. Engl.* **1983**, *22*, 871.
(94) Banciu, M.; Popa, C.; Balaban, A. T. *Chem. Scr.* **1984**, *24*, 28.
(95) (a) Paquette, L. A.; Kukla, M. J.; Ley, S. V.; Traynor, S. G. *J. Am. Chem. Soc.* **1977**, *99*, 4756. (b) Kukla, M. J. Ph.D. Thesis, The Ohio State University, Columbus, OH, 1974.
(96) Berson, J. A. *Acc. Chem. Res.* **1972**, *5*, 406.
(97) Hojo, K.; Seidner, R. T.; Masamune, S. *J. Am. Chem. Soc.* **1970**, *92*, 6641.
(98) (a) Katz, T. J.; Cheung, J. J.; Acton, N. *J. Am. Chem. Soc.* **1970**,

92, 6643. (b) Katz, T. J.; Cheung, J. J. *Ibid.* **1969**, *91*, 7772.
(99)  Jones M., Jr. *J. Am. Chem. Soc.* **1967**, *89*, 4236.
(100) Paquette, L. A.; Kukla, M. J. *J. Am. Chem. Soc.* **1972**, *94*, 6874.
(101) Diradical **68** is recognized to be involved in the high-temperature degenerate thermal rearrangement of lumibullvalene.[95b,100]
(102) See, for example: (a) Paquette, L. A. *J. Am. Chem. Soc.* **1964**, *86*, 4085. (b) Paquette, L. A.; Philips, J. C.; Wingard, R. E., Jr. *Ibid.* **1971**, *93*, 4516.
(103) (a) Neurieter, N. P.; Bordwell, F. G. *J. Am. Chem. Soc.* **1963**, *85*, 1209. (b) Neurieter, N. P. *Ibid.* **1966**, *88*, 558.
(104) Christl, M.; Herbert, R. *Org. Magn. Res.* **1979**, *12*, 150.
(105) Sander, W.; Gleiter, R. *Chem. Ber.* **1985**, *118*, 2548.
(106) Christl, M.; Herzog, C. *Chem. Ber.* **1986**, *119*, 3067.

## [2+3] CARBO- AND HETERO-CYCLIC ANNULATION IN THE DESIGN OF COMPLEX MOLECULES.[1]

T. Hudlicky,* A. Fleming, T. Lovelace, G. Seoane, K. Gadamasetti, G. Sinai-Zingde
Chemistry Department, Virginia Tech
Blacksburg, VA 24061
USA

**Abstract**: Vinylcyclopropanes, vinyloxiranes, and vinylaziridines were prepared by either a [2+3] protocol involving the additions of the dienolate of ethyl 2-bromocrotonate to enones, aldehydes, and some imines, or by a [4+1] protocol that utilized the intramolecular cycloadditions of diazoketones or azides to 1,3-dienes. The resulting strained 3-membered rings provided, through their energy contents, the necessary driving force for the corresponding rearrangments to annulated cyclopentenes, dihydrofurans, and pyrrolines respectively. The applications of these processes are highlighted in the syntheses of natural products containing five-membered rings. Retigeranic acid, ipomeamarone, and pyrrolizidine diols are the targets of these applications. An approach to the taxane skeleton is also advanced and is based on the ease with which endo vinylcyclopropanes undergo the divinylcyclopropane Cope rearrangement.

**Introduction.** The efficient formation of five-membered rings has been near the top of the list of priorities within the synthetic community for some time. Because of the inherent functional dissonance[1] of any assembly of an odd number of atoms the accessibility of such systems is limited. Many annulation technologies for the synthesis of five-membered rings have been reported[2], yet the field is far from saturation or provision of a fully reliable method to accomplish such task within the modern guidelines of the trade, i.e. regio-, stereo-, and enantio-selective preparation of these compounds. With the knowledge of the defined constraints of this field we set out some time ago to accomplish the task of developing a general method of synthesis for cyclopentanoids and their heterocyclic counterparts.
**Methodology.** The use of strained intermediates serves well in the formulation of a energy-release based approach to annulated cyclopentenes and their heterocyclic derivatives. For the simplest model study in such a methodology, the preparation of a bicyclo[3.3.0]octane system such as **1**, we eventually arrived at two basic approaches depicted in Fig.1. The first of these involves the intramolecular formation of vinylcyclopropane **2** from a ketocarbenoid **3**, while the second relies on the intermolecular union of an enone with anion **4**. These two approaches involve identical topological arguments. For example, the overall cyclization of **3** to **1** requires the participation of six π-electrons as does the 1,4-addition/halide displacement of **4**. The so-called "retrosynthetic transition

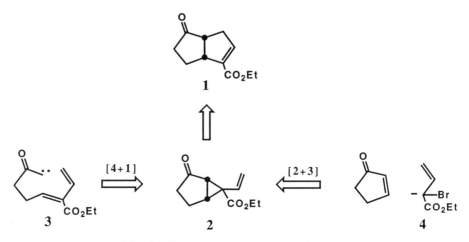

Fig.1. Cyclopentene annulation

state" utilized during the planning stages of this methodology is therefore identical for both approaches and can be depicted as the structure **5,** which is a mixture of hybrids **6** and **7**, Fig.2. Since only the movement of electrons and not of atoms is required to interchange **6** and **7** these systems can be related through the term "retrosynthetic resonance" or "system resonance". It will be seen that both of these processes found a

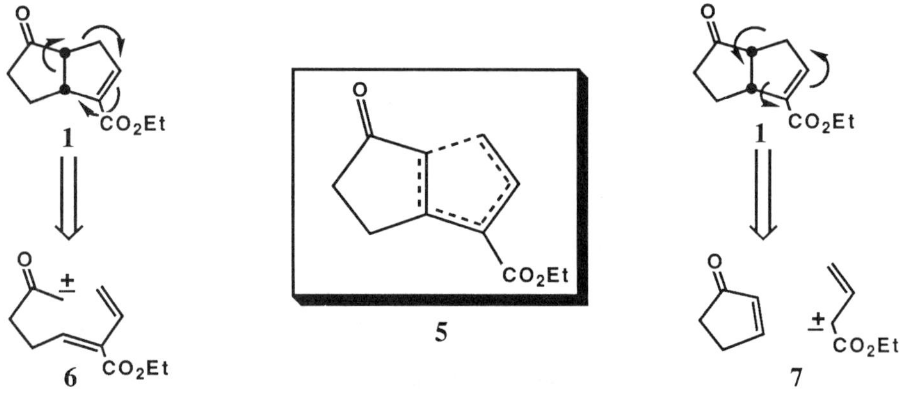

Fig.2.

widespread use in the design of cyclopentanoids and that both are readily adaptable toward the construction of oxygen or nitrogen containing systems.[4]

The exploitation of the intramolecular cyclopropanation / rearrangement in the synthesis of triquinanes has been amply documented[5] as has been the detailed study of stereochemical consequences of these pathways[6]. The logical extrapolation of this process to

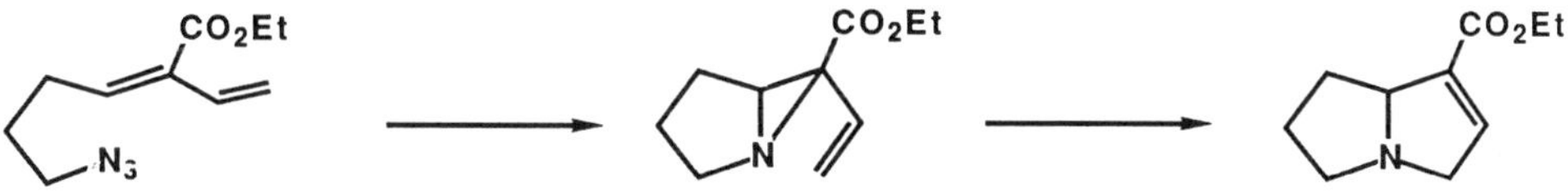

**Fig.3. Pyrroline annulation**

the cycloadditions of azides to dienes has brought forth the intramolecular pyrroline annulation depicted in Fig 3.[7] The [2+3] version of this annulation using **4** is limited by the uneasy receptiveness of imines to dienolate anions - better success will be realized when an equivalent of **4** suitable for use under acid catalyzed conditions is unearthed.

The additions of reagent **4** to aldehydes proved facile and, surprisingly, stereoselective as well, in contrast to the vinylcyclopropanation scheme of enones. Thus an efficient dihydrofuran annulation methodology had been formulated and rendered accessible as shown in Fig.4.[8]

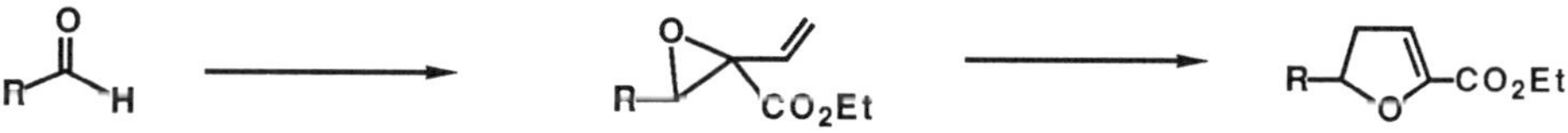

**Fig.4. Dihydrofuran annulation**

Following the necessary investigations that established the limiting features of the above annulations we applied these techniques to the total synthesis of natural products. A brief review of the most recent applications is presented in this manuscript.

**Retigeranic acid.** This unique sesterterpene eluded the chemical synthesis for over a decade until Corey[9] and Paquette[10] succeeded in its preparation. Our approach to this molecule involved the disconnection of the central cyclopentene ring and combining the two halves using the protocol of the [2+3] annulation.[11] This strategy resulted in an extremely short and enantioselective synthesis of retigeranic acid **14**, Fig.5, while

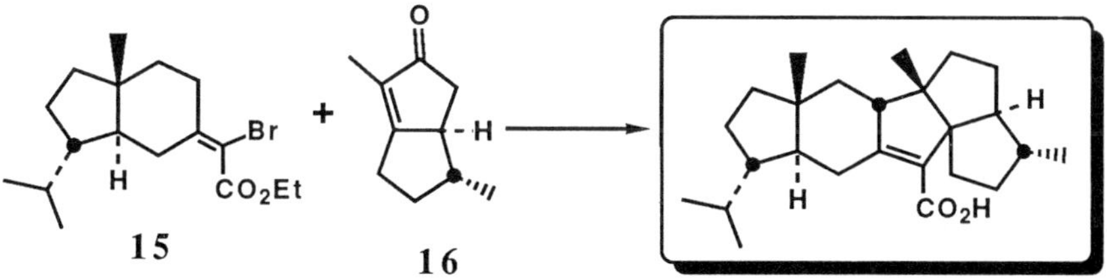

**Fig.5.**

providing a valuable insight into the reactivity of more highly functionalized bromocrotonates such as **15**. The precursor to **15** was previously reported in the literature by Fallis[12] while enone **16** was prepared by Paquette.[13] The chirality of **14** was established by using limonene (for **15**) and pulegone(for **16**) as starting materials.[14]

**Ipomeamarone.** The detailed use of reagent **4** in vinyloxiranation has recently been published.[8] The stereoselectivity of this process has been related to the enolate anion geometry, which has been determined to be E-, as shown in the model for the transition state in the addition of **4** to aldehydes to produce syn vinyloxiranes.**19**, in analogy to the well known aldol process, Fig.6. The synthesis of ipomeamarone was realized by the addition of **4** to furan 3-carboxaldehyde which gave the syn epoxide **21** in excellent yield. This compound yielded dihydrofuran **22** upon flash vacuum pyrolysis in an overall [2+3] annulation. Reduction of the acrylate was accomplished with magnesium in

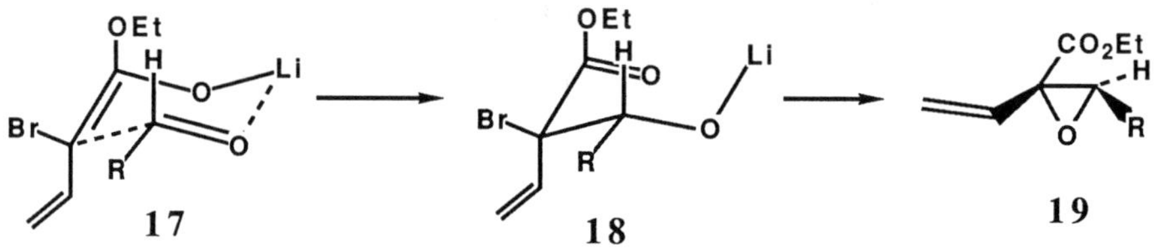

**Fig.6.**

methanol and the resulting ester alkylated with LDA / MeI to produce a separable mixture of methyl esters **23** with the desired stereoisomer predominating in a ratio of 3:2. Further study of the diastereoselectivity of this process is in progress. Following the initial attempts at homologization of acid **24** through Arndt-Eistert type process we turned to the preparation of nitrile **27** via alcohol **25** and its tosylate **26**. The Grignard reaction of isobutyl bromide with **27** than leads to the natural product, Fig.7.[15]

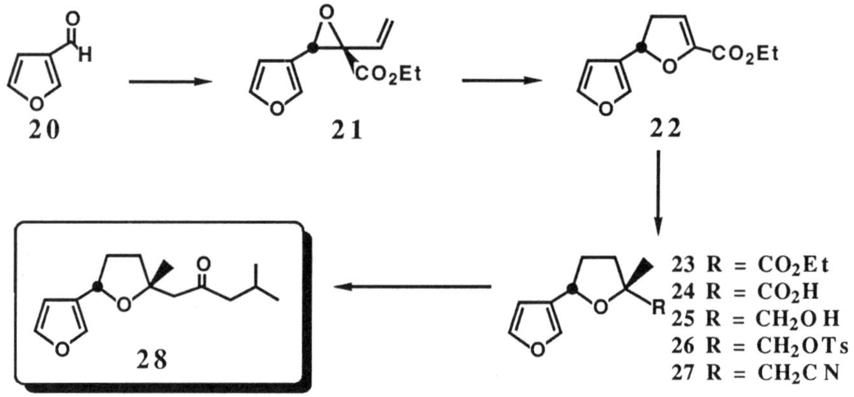

23 R = $CO_2Et$
24 R = $CO_2H$
25 R = $CH_2OH$
26 R = $CH_2OTs$
27 R = $CH_2CN$

**Fig.7.**

The total synthesis of pyrrolizidine diols of the heliotridane type has been previously accomplished by us using the azide-diene cycloaddition in an overall [4+1] pyrroline annulation.[16,17] Further improvements were realized by rendering this process enantioselective through the application of microbial processes. The known alcohol 29[17] was converted to ketoester 30 and this substance subjected to the documented yeast reduction.[18] To our delight a kinetic resolution took place at the secondary chiral site on the ester portion of the molecule. To our knowledge this represents the first example of the resolution of a second chiral center in β-ketoesters that contain prochirality on the ester

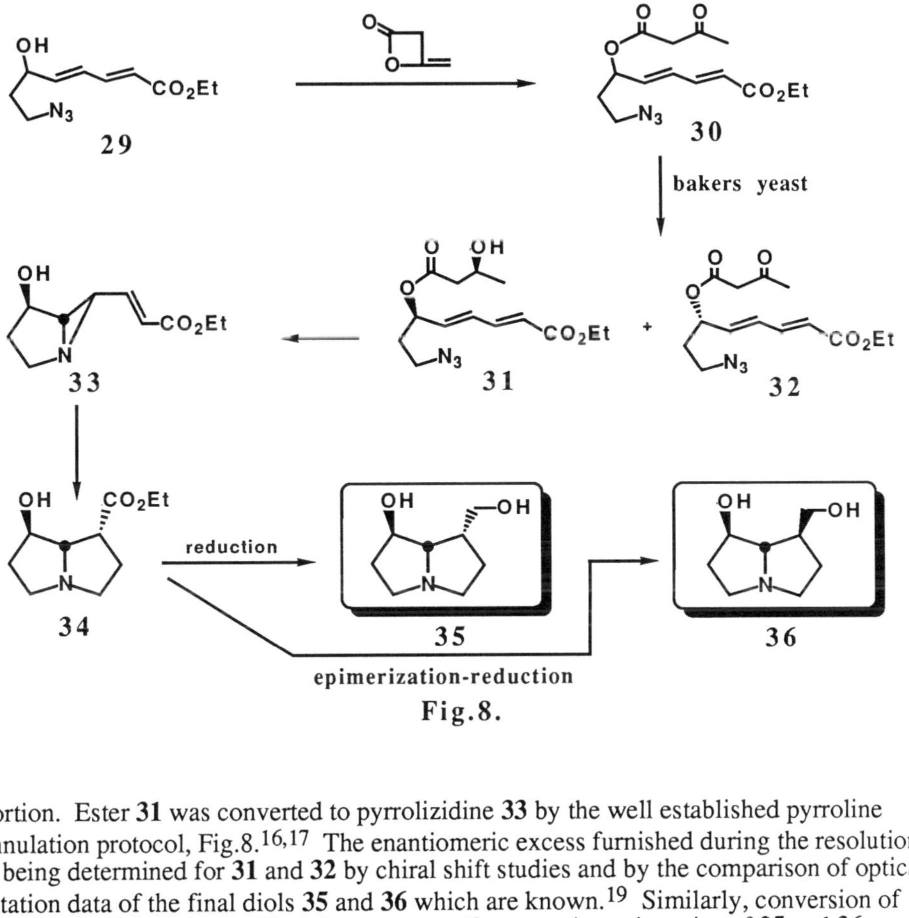

Fig.8.

portion. Ester 31 was converted to pyrrolizidine 33 by the well established pyrroline annulation protocol, Fig.8.[16,17] The enantiomeric excess furnished during the resolution is being determined for 31 and 32 by chiral shift studies and by the comparison of optical rotation data of the final diols 35 and 36 which are known.[19] Similarly, conversion of the unreacted enantiomer 32 to the corresponding enantiomeric series of 35 and 36 proceeds along the identical lines.

The foregoing examples illustrate the versatility of these annulation protocols in the design of complex natural products containing five-membered rings. Additional application presented itself upon the discovery that the endo isomers of vinylcyclopropanes such as 2 rearrange at low temperature to give the product of the divinylcyclopropane rearrangement, bicyclo[3.2.1.]octane **37,** Fig.9.[20] This rearrangement lent itself to the

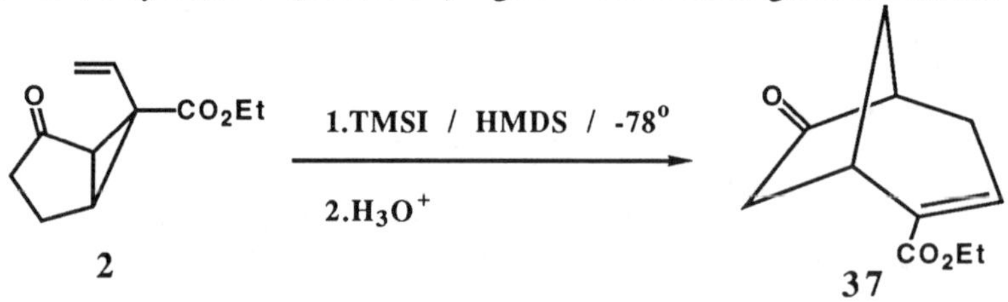

**Fig.9.**

design of an approach to taxane skeleton **38** as shown in Fig.10. The key thought in this approach relies on the cleavage of the known taxinine K skeleton **39**[21] to the simple functionalized taxane ring system **38**. The bicyclo[3.2.1.]octane **40** is generated by the Cope rearrangement of the endo vinylcyclopropane **41**, itself available by the intramolecular addition of the dienolate anion of **42** to the enone moiety. That this should be possible is supported by our evidence that bromocrotonates form E-dienolate species exclusively.[8] It is the E-dienolate that will form the requisite bromolactone appropriately aligned for the internal alkylation to endo isomer **41**. We have prepared the unsaturated ester **42** and are at the moment testing the viability of the internal cyclopropanation-Cope sequence to produce the required taxinine ring system **39**. The precursory enone alcohol **43** was prepared in 14 steps and in chiral form from 2-allyl-2-methylcyclopentanedione and the chirality set by Brooks` yeast reduction.[22]

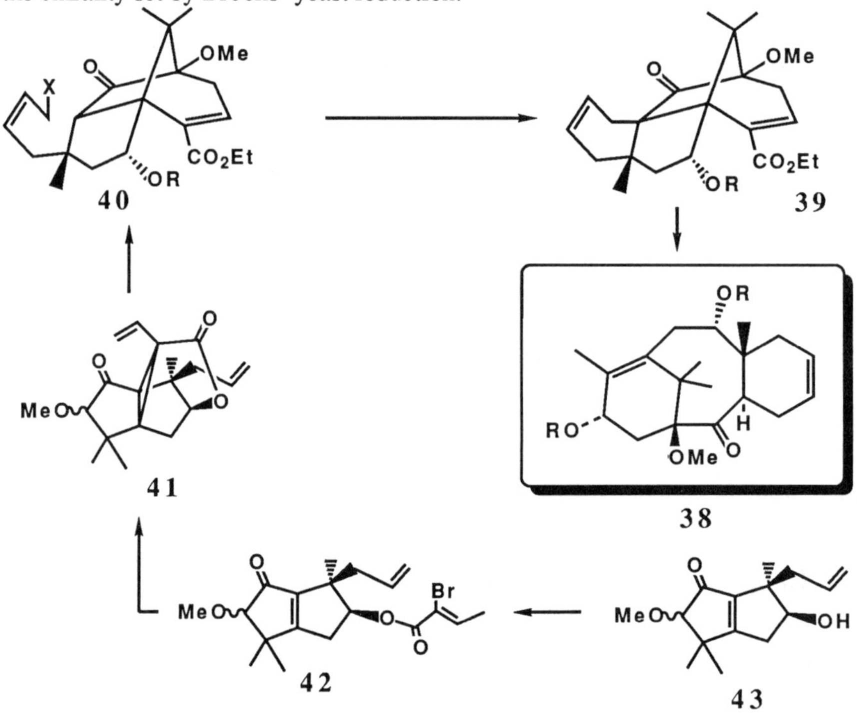

**Fig.10.**

**Summary.** The methodology presented above relies on the energetics of rearrangement-prone 3-membered ring systems to furnish, via diradical or electrocyclic processes, the annulated five-membered ring compounds. The full potential of these techniques is far from being realized as new applications present themselves almost continuously. The detailed studies of the processes that lead to the strained intermediates will no doubt produce even more applications as well as understanding of the factors that govern the stereoselective and efficient preparation of the small ring intermediates, in the context of a system-oriented design of complex natural products.

**Acknowledgments.** The authors wish to thank the following sponsors for the financial support of this work: The National Institutes of Health (AI-19749, AI-00564), The petroleum research Fund (AC-16617, AC-20748), The Jeffress Memorial Fund, T.D.C. research, inc., and the Chemistry Department, Virginia Tech.

**References.**
1. Parts of this contribution have been or will be published in more detail elsewhere.
2. Evans, D.A. Consonant and Dissonant Relationships: An Organizational Model for Organic Synthesis, unpublished manuscript, **1973**.
3. For recent reviews see: Vandewalle, M.; DeClercq, P. *Tetrahedron*, **1985** 1767; Paquette, L.A. *Topics in Current Chemistry*,**1984** 119, 1.
4. For a discussion of the topological approach to the synthesis of five-membered ring compounds see: Hudlicky, T.; Rulin, F.; Lovelace, T.C.; Reed, J.W. Studies in Natural Products, Rahman, A.-U.,ed., Elsevier, **1988**.
5. See for example: Hudlicky, T.; Kutchan, T.M.; Naqvi, S. *Organic React.*, **1985** 33, 247; citations in ref. 3 and 4 above.
6. Hudlicky, T.; Koszyk, F.J.; Dochwat, D.M.; Cantrell, G.C. *J.Org. Chem.*, **1981** 46, 2911.
7. Hudlicky, T.; Frazier, J.O.; Kwart, L.D. *Tetrahedron Lett.*, **1985** 3523; Hudlicky, T.; Sinai-Zingde, G.;Seoane, G. *Syn. Commun.* **1987** 17, 1155; see also ref 16 and 17 below.
8. Hudlicky, T.; Fleming, A.; Lovelace, T.C. *Tetrahedron,* **1988** 44, 0000.
9. Corey, E.J.; Desai, M.C.; Engler, T.A. *J.Am.Chem.Soc.*, **1985** 107, 4339.
10. Paquette, L.A.; Wright, J.; Drtina, G.J.; Roberts, R.A. *J.Org.Chem.*, **1987** 52, 2960.
11. Hudlicky, T.; Radesca, L.; Luna, H.;Anderson, III, F.E. *J.Org.Chem.*, **1986** 51, 4746.
12. Attah-Poku, S.K.; Chau,F.; Yadav, V.K.; Fallis, A.G. *J.Org. Chem.*, **1985** 50, 3418; see also ref. 9 above.
13. Paquette, L.A.; Roberts, R.A.; Drtina, G.J. *J.Am.Chem.Soc.*, **1984** 106, 6690.
14. Hudlicky, T.; Short, R.P. *J.Org.Chem.*, **1982** 47, 1522.
15. Hudlicky, T.; Lovelace, T.C. *Tetrahedron Lett.*, in preparation.
16. Hudlicky, T.; Seoane, G.; Seoane, A.; Frazier, J.O.; Kwart, L.D.; Tiedje, M.H.; Beal, C. *J.Am.Chem.Soc.*, **1986** 108,3755.
17. Hudlicky, T.; Seoane, G.; Lovelace, T.C. *J.Org.Chem.*, **1988** 53, 2094.
18. Seebach, D.; Herradon, B. Tetrahedron Lett. 1987 3791; Ridley, D.; Stratlow, M. *J.Chem.Soc.Chem.Commun.*, **1975** 400.
19. Adams, R.; Van Duuren, B.L. *J.Am.Chem.Soc.*, **1954** 76, 6379; General review: Robins, D.J. *Prog.Chem.Org.Nat.Prod.*, **1981** 41, 115.
20. Fleming, A.; Sinai Zingde, G.; Natchus, M.G.; Hudlicky, T. *Tetrahedron Lett*, **1987** 167.
21. Chiang, H.C.; Woods, M.C.; Nakadaira, U.Y.; Nakanishi, K. *J.Chem.Soc.Chem.Commun.*, **1967** 1201.
22. Brooks, D.W.; Mazdiyasni, H.; Grothaus, P.G. *J.Org.Chem.*, **1987** 52, 192, 3223.

# SYNTHETIC AND MECHANISTIC ASPECTS OF THE CYCLOPROPENE TO VINYLCARBENE REARRANGEMENT

**Mark S.Baird,* Juma'a R.Al-Dulayymi and Helmi H.Hussain**
Department of Chemistry, The University of Newcastle,
Newcastle upon Tyne, NE1 7RU, England.

***Abstract*** *The cyclopropenes (6, X = Cl, OMe, Ph) ring open at 0 – 20 °C to give vinylcarbenes (11) which are trapped in inter- or intramolecular reactions in preference to the isomers (10). The rate of cyclopropanation of 2,3-dimethylbut-2-ene in the presence of (15) correlates well with the* $\sigma_I$ *constant of the 4-substituent, with a* $\rho$*-value of −0.79, suggesting the development of positive charge at C-3 in the transition state.*

1,2-Dehalogenation of 1,1,2-trihalocyclopropanes by reaction with one mol. equiv. of methyl lithium leads to 1-halocyclopropenes. In many cases these react with a second equivalent of the reagent under more vigorous conditions by a lithium – halogen exchange, and the resulting lithiocyclopropenes may be used in synthesis;[1] for example, trapping of the corresponding 1-lithio-3,3-dimethylcyclopropene by R-(+)- or S-(−)-methyloxirane leads to the alcohols (1) and (2), which are converted to the optically active methylenefuran (3) or dihydropyran (4) respectively by reaction with bromine:[2]

*A. de Meijere and S. Blechert (eds.), Strain and Its Implications in Organic Chemistry, 117–120.*

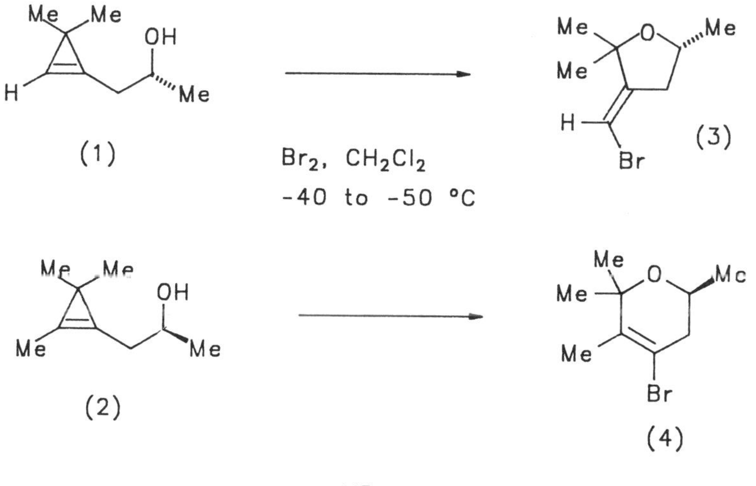

1,2-Dehalogenation of 1,1,2,2-tetrahalocyclopropanes (5, Y = Cl, X = H, OMe, Ar) by reaction with methyl lithium at 0 - 20 °C leads to 1,2-dihalocyclopropenes, eg., (6, X = H, OMe, Ar); the pentachloride (5, X = Y = Cl) also reacts by 1,2-dechlorination to give (6, X = Cl), whereas the bromide (5, X = Cl, Y = Br) apparently undergoes a 1,3-dehalogenation on reaction with two mol.equiv. of methyl lithium at -70 °C, leading eventually to the cyclobutene (7).

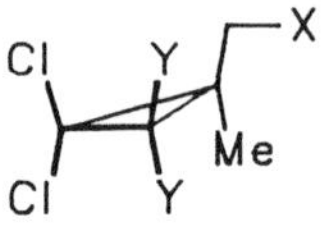

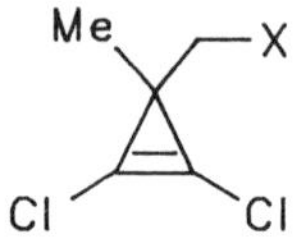

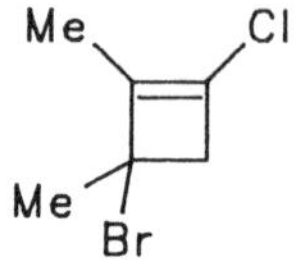

The dimethyl-cyclopropene (6, X = H) reacts rapidly with an added alkyl-substituted alkene at 0 - 20 °C to give a cyclopropane, eg. (8) with 2,3-dimethylbut-2-ene, apparently derived by addition of the vinylcarbene (9); since the cyclopropene is stable for some hours in the absence of the alkene, a reversible cyclopropene – carbene rearrangement is suggested. The conditions required for this rearrangement may be compared with the temperatures of 150 - 180 °C required to ring open many other cyclopropenes, and in particular required for 3,3-dimethyl-cyclopropene and tetrachlorocyclopropene.[3] The carbene (9) is also trapped by electron poor alkenes, adding to dimethyl maleate in a non-stereospecific manner, albeit in low yield; it also inserts into the C – H bond $\alpha$- to oxygen in ethers.[4]

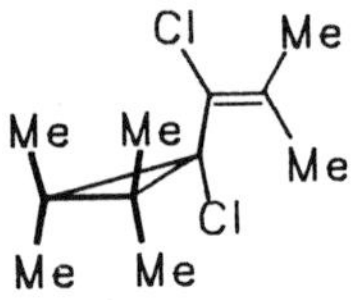

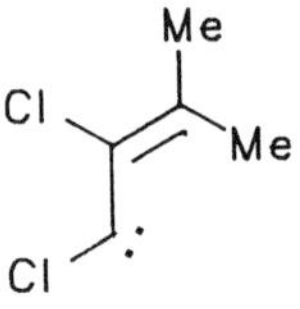

When X ≠ H, the ring-opening of cyclopropenes (6) may in principle produce two stereoisomeric carbenes:

The cyclopropene (6, X = Ph) rearranges in 18 h at 20 °C to give the diene (12); this may be explained in terms of an intramolecular 1,4-hydrogen shift or carbene insertion into a cis-1,4-related C - H bond followed by ring-opening; either process would require the carbene geometry to be (11, X = Ph). Although this may mean that only this carbene is formed, the selectivity may also reflect the greater reactivity of the cis-related C - H bond in (11), combined with a rapid equilibration. However, the same carbene is trapped selectively in intramolecular addition to alkenes. Thus, reaction of the cyclopropene (6, X = Ph) with methyl methacrylate leads predominantly to cyclopropane (13), the structure of which was established by an X-ray crystallographic study. This isomer is apparently derived by trapping of the carbene (11, X = Ph) rather than (10, X = Ph); moreover the ester and alkene substituents are obtained with a cis-stereochemistry. A minor product (ca. 1 : 5) derived by addition of (10, X = Ph) is also thought to have cis-stereochemistry of ester and vinyl groups. The products derived from the corresponding cyclopropenes (6, X = Cl, OMe) reflect an even more selective trapping of the carbenes (11, X = Cl, OMe).[6]

(12)

(13)

The rate of cyclopropanation of 2,3-dimethylbut-2-ene by reaction with (6, X = H, OMe) was very similar in either chloroform or benzene solution, though the reactions did proceed about twice as fast in acetone; in the last case the cyclopropenes reacted with the solvent when no alkene was present, giving products apparently derived by a 1,6-hydrogen shift in an intermediate ylid derived by trapping of the carbene by the carbonyl oxygen. Studies of the relative rates of cyclopropanation of a standard series of alkyl-substituted alkenes by the carbenes derived from (6, X = H) and (6, X = OMe) showed that the reactivities of these carbenes do not give a good linear correlation with dichlorocarbene; however, they do give a linear correlation when compared to each other or to tetrachloroprop-2-enylidene,[6] and are of approximately equal selectivity.

The relative rates of formation of cyclopropanes (14) when (15, Y = H, $CF_3$, Me, OMe) were allowed to ring open in the presence of an excess of 2,3-dimethylbut-2-ene gave a good linear correlation with the $\sigma_I$ constants for the substituents, and corresponded to a $\rho$-value of −0.79. This suggests that the transition state leading to carbenes (11) may involve a considerable charge deficiency at C-3, in agreement with a major contribution from the polar resonance form (16).

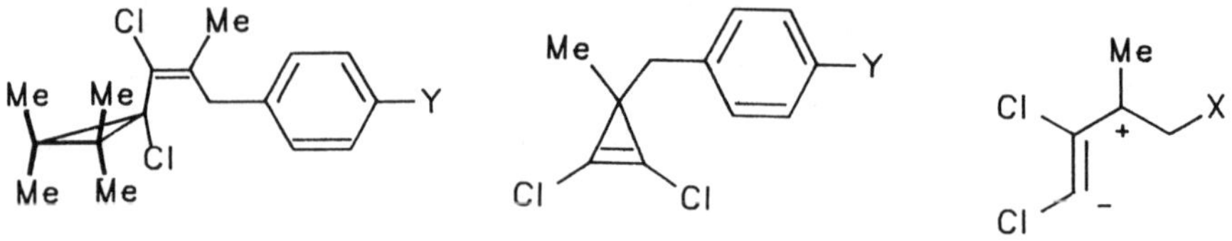

(14)     (15)     (16)

Dehydrochlorination of the derived cyclopropanes provides easy access to two series of allylidenecyclopropanes:

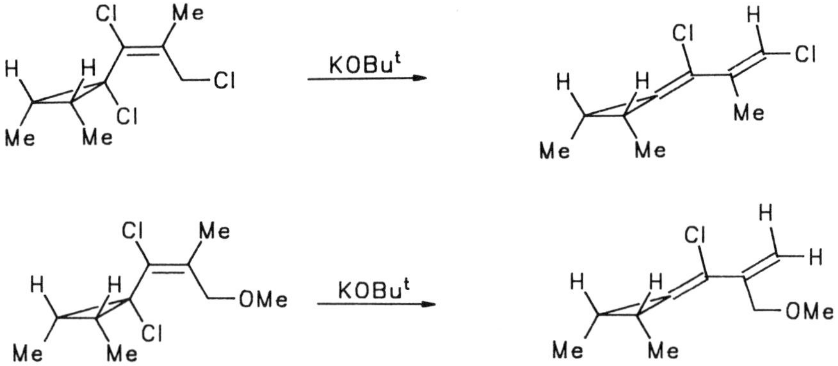

We wish to thank the Government of Iraq and the Arabian Gulf University for the award of grants to J.R.A-D and H.H.H. respectively.

1. M.S.Baird and H.H.Hussain, *J.Chem.Soc.Perkin Trans.I*, 1986, 1845.
2. M.S.Baird and J.R.Al-Dulayymi, unpublished results.
3. M.S.Baird, Synthetic Applications of Cyclopropenes, in Topics in Current Chemistry, Ed. A.deMeijere, 1988, **144**, 138
4. M.S.Baird and H.H.Hussain, unpublished results.
5. J.R.Al-Dulayymi, M.S.Baird and W.Clegg, papers in press.
6. R.Kostikov and A.deMeijere, *J.Chem.Soc.Chem.Comm.*, 1984, 1528.

# GENERATION AND INTERCEPTION OF 1-OXA-2,3-CYCLOHEXADIENE AND 1,2,4-CYCLOHEXATRIENE

Manfred Christl and Martin Braun
Institut für Organische Chemie
der Universität Würzburg
Am Hubland
D-8700 Würzburg
Federal Republic of Germany

ABSTRACT. The cycloadducts 6 and 7 of tricyclo[4.1.0.0$^{2,7}$]-hepta-3,4-diene (5) with styrene and 1,3-butadiene rearrange to unusual products on thermolysis, namely the cycloheptatriene derivatives 9 and 10. 1-Oxa-3,4-cyclohexadiene (20) is generated smoothly from 6,6-dichloro-3-oxabicyclo[3.1.0]-hexane (22) and n-butyllithium. 1-Oxa-2,3-cyclohexadiene (21) is formed from 6-exo-bromo-6-endo-fluoro-2-oxabicyclo-[3.1.0]hexane (30) and methyllithium. In the presence of activated olefins, this reaction provides an efficient route to 28 and 33 - 38, the trapping products of 21. Interestingly, [2+2]-cycloadditions do not take place at the same double bond of 21 as [4+2]-cycloadditions. The reactions of 1,3-cyclopentadiene and indene with bromofluorocarbene afford 6-exo-bromo-6-endo-fluorobicyclo[3.1.0]hex-2-ene (50) and its benzo derivative 45, respectively. On treatment of these compounds with methyllithium in the presence of styrene, the interception products 53 and 47 of 1,2,4-cyclohexatriene (44) and its benzo derivative 43, respectively, are formed in good yields.

## INTRODUCTION

The short-lived intermediate 1,2-cyclohexadiene (2) was generated for the first time by Wittig and Fritze.[1] They utilized a β-elimination pathway and trapped 2 by [4+2]-cycloaddition with 1,3-diphenylisobenzofuran. Moore and Moser[2] discovered a more convenient method for the generation of 2, i. e. the reaction of 6,6-dibromobicyclo[3.1.0]-hexane with methyllithium. Under these conditions, 2 can be



*A. de Meijere and S. Blechert (eds.), Strain and Its Implications in Organic Chemistry, 121–131.*

intercepted by activated olefins. Styrene gives a diastereomeric pair of [2+2]-cycloadducts (1).[3] [2+2]-Cycloadducts, e. g. 3, are also the major products formed from 1,3-butadiene[4] and its methyl derivatives[4,5] but small quantities of [4+2]-cycloadducts, e. g. 15, are observed too, which apparently arise directly from 2 and the diene and not via the thermal rearrangement of the [2+2]-cycloadducts (see below).[4] 1,3-Cyclopentadiene, furan, and 2-methylfuran give [4+2]-cycloadducts exclusively, whereas 1,3-cyclohexadiene produces a mixture of [4+2]- and [2+2]-cycloadducts.[5]

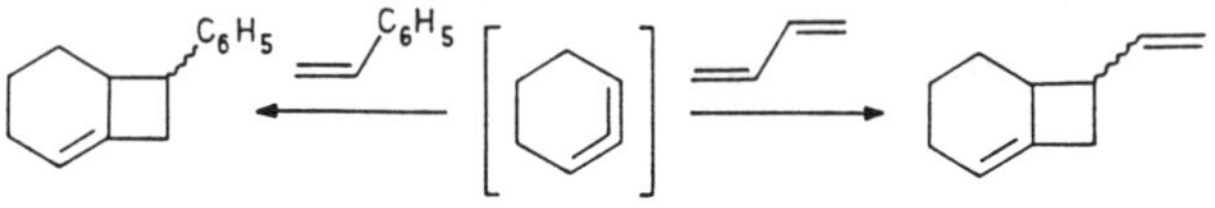

1        2        3

Our lead-in to the chemistry of 1,2-cyclohexadiene and its derivatives was the reaction of dibromocarbene with benzvalene giving rise to the dibromocyclopropane 4.[6,7] Treatment of 4 with methyllithium in the presence of activated olefins of the kind mentioned above affords a series of products, e. g. 6 - 9, formed by interception of tricyclo[4.1.0.0²,⁷]hepta-3,4-diene (5).[8]

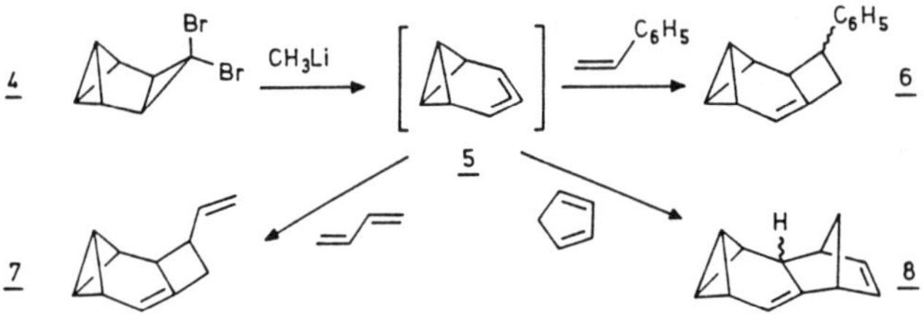

Compounds 6 - 8 behave very differently on thermolysis. Whereas both the diastereomers 8 rearrange to the bicyclo-[3.2.0]hepta-2,6-diene derivatives 11 as expected for homobenzvalenes,[9] 6 and 7 are converted into the cycloheptatrienes 9 and 10, respectively, at unusually low temperatures without the intermediacy of compounds of type 11.[10]

To elucidate the mechanisms of these rearrangements we have studied the thermolyses of the model compounds 1 and 3. The equilibrium between the diastereomers 1a and 1b is established above 140 °C.[11] In this process, diradical 12 is the most likely intermediate, which should also play the key role in the formation of 1 from 2 and styrene. In accordance with the latter hypothesis is the finding that on utilization of (Z)-deuteriostyrene the label in 1a as well as 1b was determined to be evenly distributed over both the 8-positions.[11,12] Thus, a one-step pathway for the reaction of 2 and styrene is ruled out.

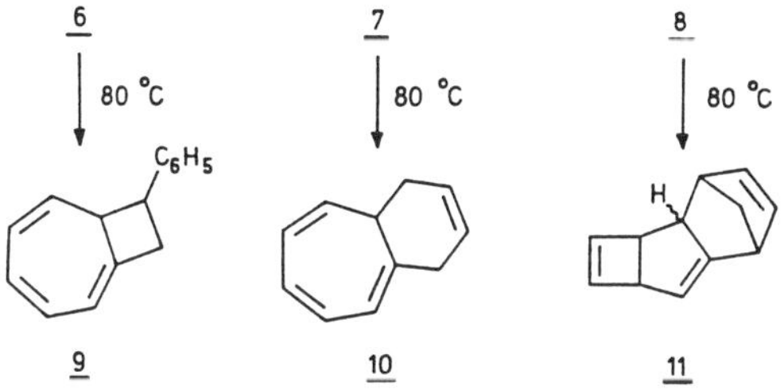

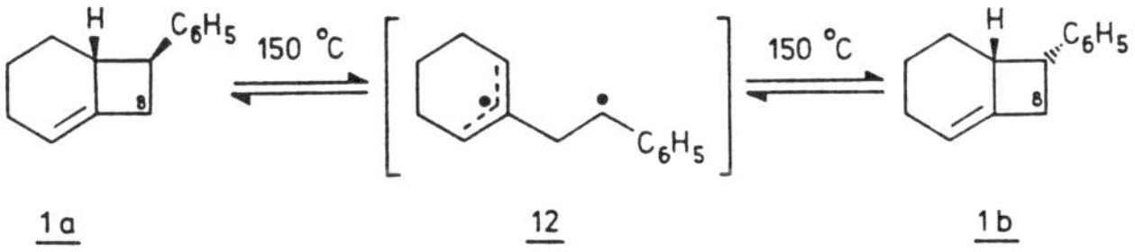

The thermolysis of 3 proceeds in a more complex manner. The interconversion of 3a and 3b does take place, most probably via the diradical 13 having an (E)-substituted exocyclic allyl moiety, but the equilibrium composition is arrived at only very incompletely since there is a leak leading to the formation of the more stable isomer 15. We presume that this vinylcyclobutane-cyclohexene-rearrangement requires diradical 14 as intermediate with a (Z)-substituted exocyclic allyl subunit.[4].

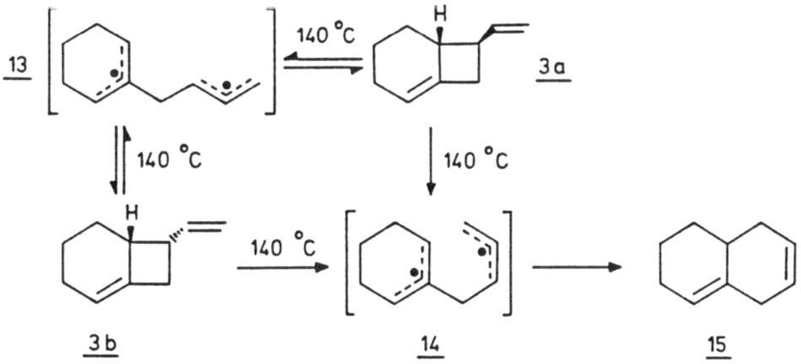

On the basis of these results, mechanisms can be proposed for the rearrangements of 6 to 9 and of 7 to 10. Thermal activation could transform 6 to diradical 16, the homobenzvalenyl radical moiety of which could rearrange to the tropylium radical subunit. Thus generated, the diradical 17

could collapse to 9. The formation of 10 from 7 should proceed analogously via the diradicals 18 and 19, with the allyl moiety in the (Z)-configuration required for the expansion of the four- to the six-membered ring. In agreement with the assumed ease of the rearrangements of 16 to 17 and of 18 to 19, it has been shown recently that the unsubstituted homobenzvalenyl radical transforms to the tropylium radical with an activation energy of 13.4 kcal/mol.[13]

The intermediacy of several further 1,2-cyclohexadiene derivatives, namely bicyclo[3.2.1]octa-2,3-diene,[14] bicyclo[3.2.1]-octa-2,3,6-triene,[15] and a benzanellated derivative of the latter,[16] has been proved by trapping reactions. The properties of these strained allenes and theoretical investigations on them have been summarized.[17] We have generated 1-methyl-1,2-cyclohexadiene from 6,6-dibromo-1-methylbicyclo[3.1.0]hexane and trapped it with styrene[11] and 1,3-butadiene.[4] Until recently, very little was known, however, about hetero derivatives of 2,[17] which is why we undertook to investigate the existence of both the possible oxa derivatives 20 and 21.

1-Oxa-3,4-cyclohexadiene is smoothly generated from 6,6-dibromo- and 6,6-dichloro-3-oxabicyclo[3.1.0]hexane (22) with methyllithium and n-butyllithium, respectively.[18] The interception of 20 succeeded with styrene, 1,3-butadiene, methyl derivatives of the latter, and furan. Styrene provides the diastereomeric [2+2]-cycloadducts 23, whereas 1,3-butadiene and its methyl derivatives give rise to [2+2]-cycloadducts and small amounts of [4+2]-cycloadducts such as 24 and 25, respectively. In contrast, furan affords the endo [4+2]-cycloadduct exclusively. The vinylmethylene-cyclo-

butane derivatives of type 24 rearrange to isochromenes of
type 25 on heating at 165 °C.

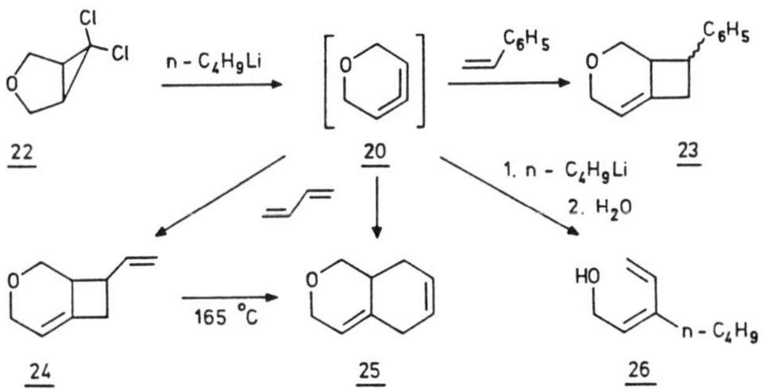

A new type of process was discovered when 20 was generated in the absence of a sufficiently active olefin. The central carbon atom of the allene moiety is then attacked by the organolithium reagent causing the ring opening in an $S_N2'$-type substitution with the oxygen atom serving as leaving group. We have obtained 3-substituted 2,4-pentadien-1-ols analogous to 26 with a methyl, ethyl, n-propyl, neopentyl, phenyl, 2-methoxyphenyl, 2-thienyl, or a 3-thienyl group.

RESULTS AND DISCUSSION

The conversion of 22 into 20 and the trapping of the latter encouraged us to treat the known 6,6-dichloro-2-oxa-bicyclo[3.1.0]hexane (27)[19] with n-butyllithium in the presence of styrene. Indeed, two diastereomers of 28, seemingly [2+2]-cycloadducts of 1-oxa-2,3-cyclohexadiene (21), were formed, but the yield was only 7%. The product mixture consisted mainly of polystyrene and 1-chloro-1-phenylhexane (29). It is well known that n-butyllithium polymerizes styrene.[20] The initial step produces 1-phenylhex-1-yllithium, which sequentially attacks further styrene molecules but in the presence of 27 undergoes a competing reaction, namely the lithium-chlorine-exchange leading to 29 and the carbenoid precursor of 21. Obviously, the desired reaction of 27 with n-butyllithium, generating 21, proceeds considerably more slowly than that of 22 producing 20.

Hence, we searched for another source of 21. The dibromo compound analogous to 27, from which 21 could be liberated with the less nucleophilic methyllithium, cannot be isolated due to rapid rearrangement.[19] However, the fluoride ion is

27  28  29

a less favorable leaving group. Thus, we treated 2,3-dihydrofuran with dibromofluoromethane and sodium hydroxide in the presence of triethylbenzylammonium chloride. After hydrolytic work-up we isolated the desired exo-6-bromo-endo-6-fluoro-2-oxabicyclo[3.1.0]hexane (30) in 25% yield and the acetal 32, which also results from the reaction of chlorofluorocarbene with 2,3-dihydrofuran.[19] Apparently, the diastereomer 31 is formed in addition to 30 and converted into 32 by sequential rearrangement and hydrolysis.

30  31  32

Fluorocompound 30 proved to be an efficient precursor of 1-oxa-2,3-cyclohexadiene (21), which could be generated smoothly by reaction of 30 at -25 °C with methyllithium in ether. In the presence of the trapping reagents styrene, 1,3-butadiene, 2,3-dimethyl-1,3-butadiene, furan and 2,5-dimethylfuran, products with the expected structures 28 and 33 - 38 were formed. The ratio of the diastereomeric styrene adducts 28 (54% yield) was determined to be close to 1 : 1, whereas the ratio of the corresponding adducts of 2 was about 3.5 : 1 in favor of the exo-isomer 1a in the reaction at -25 °C.[11] 1,3-Butadiene afforded a 2 : 1-mixture (80% yield) of the [2+2]-cycloadduct 33 consisting largely of one stereoisomer, probably the exo-compound, and the [4+2]-cycloadduct 36. 2,3-Dimethyl-1,3-butadiene yielded a 7 : 1 : 1-mixture (57%) of the two diastereomeric [2+2]-cycloadducts 34 and the [4+2]-cycloadduct 37. Furan and 2,5-dimethylfuran took up 21 to form the [4+2]-cycloadducts 35 (31%) and 38 (37%), respectively.

It is interesting to note that [2+2]- and [4+2]-cycloadditions of 21 proceed with a different chemoselectivity. While the former take place at the enol ether double bond, in the case of the latter the double bond more remote from the oxygen reacts exclusively. We take these results as evidence for different mechanisms. If the [4+2]-cycloadditions were one-step processes, i. e . true Diels-Alder reactions, on the basis of the frontier orbital concept, the more electron-rich enol ether double bond would be the less active one toward the electron-rich diene in agreement with experiment. According to the above discussion, we consider the [2+2]-cycloadditions as two-step processes with the diradicals 40 as intermediates. Whether the collapse of 40 occurs involving positions 2 or 4 to give 39 and 41, respectively, is a matter of relative reaction rates, since the equilibrium between 39 and 41 should not be mobile under the reaction and work-up conditions. Due to the radical-stabilizing property of an alkoxy substituent,[21] the spin density within the allyl radical subunit of 40 should be larger in position 2 than in position 4 favoring the ring closure in position 2 and giving rise to 39.

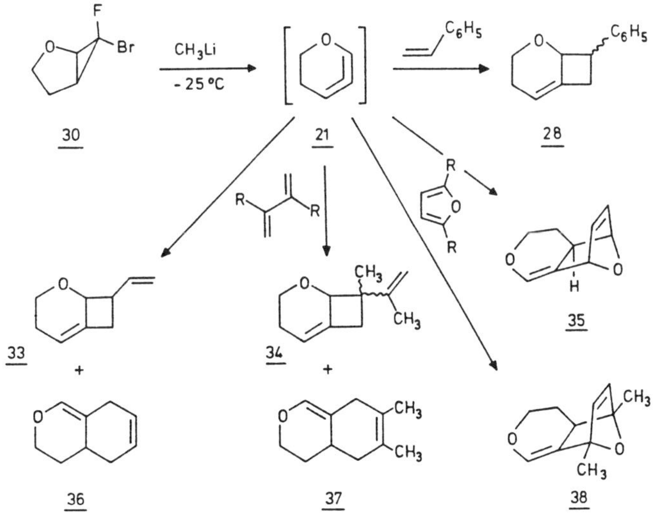

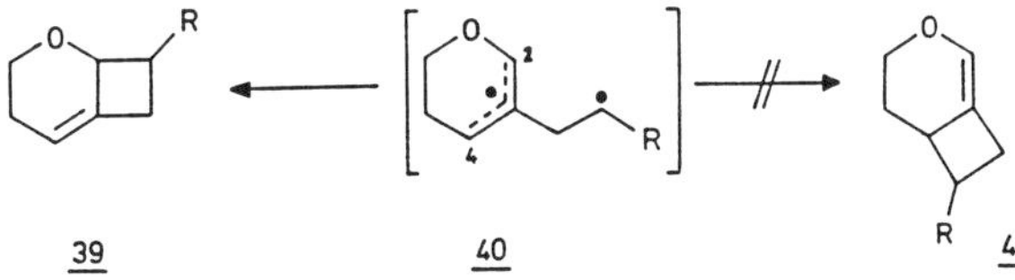

The kinetic stability of 30 encouraged us to investigate whether additional derivatives of 2 could exist. In 1987, Miller and Shi[22] reported the generation of the allenic isomer **42** of the dimethylnaphthalenes by a β-elimination route and the interception of **42** with 1,3-diphenylisobenzofuran. We were interested whether the geminal methyl groups in **42** would be a necessary condition for a sufficiently long lifetime to allow trapping reactions. After all, in the case of the unsubstituted analogue **43** of **42**, a rapid base-induced isomerization to naphthalene is conceivable since the methylene group is doubly activated for deprotonation due to its allylic and benzylic position. In the benzene isomer **44**, the methylene group has doubly allylic character and thus should be similarly acidic as 1,4-cyclohexadiene, which is readily deprotonated with potassium amide in liquid ammonia.[23]

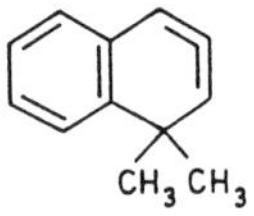

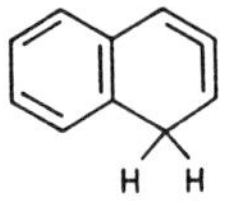

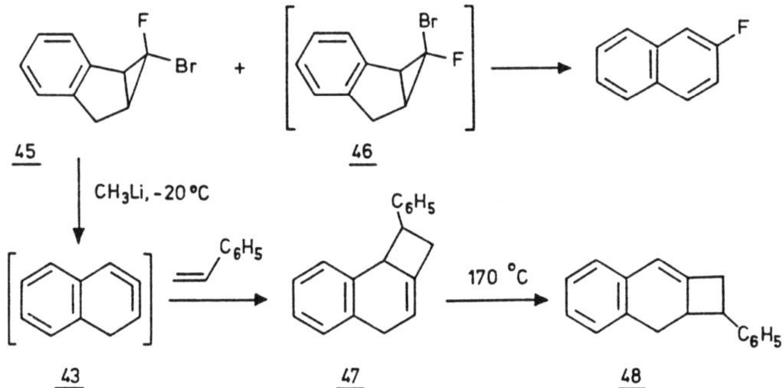

It follows from the above discussions that dihalocarbene adducts of indene and 1,3-cyclopentadiene should be appropriate precursors of **43** and **44**. The dichlorocarbene adducts of indene[24] and 1,3-cyclopentadiene[25] are known, but they convert into 2-chloronaphthalene and chlorobenzene readily. Since n-butyllithium is required to generate allenes from dichlorocyclopropanes but attacks the favorable trapping reagent styrene,[20] we again applied the fluorine trick. From the reaction of indene with bromofluorocarbene we obtained the exo-bromo-endo-fluoro compound **45**, albeit in 8% yield. To a considerable extent 2-fluoronaphthalene was formed originating from the endo-bromo-exo-fluoro isomer **46**. Treatment of **45** with methyllithium in the presence of styrene afforded a 63% yield of the [2+2]-cycloadduct 47 of the allenic naphthalene isomer 43. On heating at 170 °C for one hour, 47 rearranged to the alternative [2+2]-cycloadduct 48.

Most probably, 43 takes up styrene to give diradical 49, which collapses kinetically affording 47 due to the higher spin density in position 1 of 49 relative to position 3. At higher temperatures, the bond making in 49 to give 47 becomes reversible, and the kinetically unfavorable but more exergonic ring closure involving position 3 of 49 leads to 48.

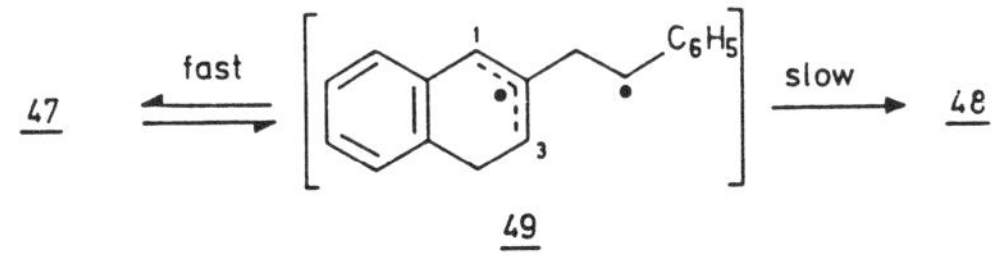

Analogous to the synthesis of 45, we prepared exo-bromo-endo-fluorobicyclo[3.1.0]hex-2-ene (50) (13% yield) by treatment of 1,3-cyclopentadiene with dibromofluoromethane and sodium hydroxide in the presence of triethylbenzylammonium chloride. The reason for the low yield is in part the formation of the endo-bromo-exo-fluoro compound 51, which is converted rapidly into fluorobenzene. Reaction of 50 dissolved in styrene with methyllithium afforded the dihydrophenylcyclobutabenzene 53 in 66% yield. We propose that at first 1,2,4-cyclohexatriene is generated, which attacks styrene to produce diradical 52 containing a pentadienyl radical subunit. Interestingly, 52 is not subject to an intramolecular hydrogen transfer, which would yield 1,2-diphenylethane, but collapses via the position 3, where the spin density of the pentadienyl radical moiety is higher than in position 1.

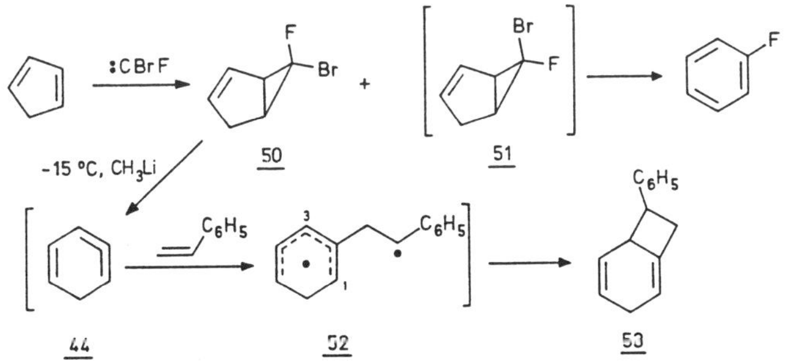

In summary, we have shown that the properties of geminal bromofluorocyclopropane derivatives are unique in providing synthetic access to certain 1,2-cyclohexadienes. First, by addition of bromofluorocarbene to the corresponding cyclopentenes exo-bromo-endo-fluorobicyclo[3.1.0]hexane derivatives are formed. These do not readily rearrange at room temperature because of the strength of the carbon-fluorine bond, which prevents the dissociation into a fluoride ion and a carbocation. Second, the bromine atom is replaced smoothly by a lithium atom under mild conditions, i. e. on treatment with methyllithium, which is followed by the α-elimination of lithium fluoride to give the cyclopropylidene and then, by ring enlargement, the cyclic allene. This methodology allows the generation of 1-oxa-2,3-cyclohexadiene (21) and its trapping by several activated olefins. Also, the benzene isomer 1,2,4-cyclohexatriene (44) and its benzo derivative 43, which is an isomer of naphthalene, do not aromatize under these conditiones and can be intercepted by [2+2]-cycloadditions with styrene.

REFERENCES AND NOTES

1) G. Wittig, P. Fritze, Liebigs Ann. Chem. 711 (1968) 82.
2) W. R. Moore, W. R. Moser, J. Am. Chem. Soc. 92 (1970) 5469.
3) W. R. Moore, W. R. Moser, J. Org. Chem. 35 (1970) 908.
4) M. Christl, M. Schreck, Angew. Chem. 99 (1987) 474; Angew. Chem. Int. Ed. Engl. 26 (1987) 449.
5) A. T. Bottini, L. L. Hilton, J. Plott, Tetrahedron 31 (1975) 1997.
6) M. Christl, G. Freitag, G. Brüntrup, Chem. Ber. 111 (1978) 2307.
7) Experiments for the isolation of 4 may lead to an explosion: C. Herzog, R. Lang, D. Brückner, P. Kemmer, M. Christl, Chem. Ber. 119 (1986) 3027. Appropriate safety precautions are indispensable.
8) M. Christl, R. Lang, M. Lechner, Liebigs Ann. Chem. 1980, 980.
9) M. Christl, U. Heinemann, W. Kristof, J. Am. Chem. Soc. 97 (1975) 2299.
10) M. Christl, Angew. Chem. 93 (1981) 515; Angew. Chem. Int. Ed. Engl. 20 (1981) 529.
11) M. Christl, M. Schreck, Chem. Ber. 120 (1987) 915.
12) S. Harnos, S. Tivakornpannarai, E. E. Waali, Tetrahedron Lett. 27 (1986) 3701.

13) H.-G. Korth, W. Müller, R. Sustmann, M. Christl, Chem. Ber. 120 (1987) 1257.
14) P. Mohanakrishnan, S. R. Tayal, R. Vaidyanathaswamy, D. Devaprabhakara, Tetrahedron Lett. 1972, 2871. A. T. Bottini, L. L. Hilton, Tetrahedron 31 (1975) 2003.
15) M. Balci, W. M. Jones, J. Am. Chem. Soc. 103 (1981) 2874, and references cited therein.
16) M. Balci, M. Harmandar, Tetrahedron Lett. 25 (1984) 237.
17) R. P. Johnson in Molecular Structure and Energetics (J. F. Liebman, A. Greenberg, Eds.), Vol. 3, VCH Publishers, Deerfield Beach, Florida (1986), p. 85.
18) M. Schreck, M. Christl, Angew. Chem. 99 (1987) 720; Angew. Chem. Int. Ed. Engl. 26 (1987) 690.
19) E. V. Dehmlow, K. Franke, Liebigs Ann. Chem. 1979, 1456.
20) A. Echte in Methoden der Organischen Chemie (Houben-Weyl), 4th Edn., Vol. E20,2 (H. Bartl, J. Falbe, Eds.), Thieme, Stuttgart (1987), p. 962.
21) R. Merényi, Z. Janousek, H. G. Viehe in Substituent Effects in Radical Chemistry (H. G. Viehe, Z. Janousek, R. Merényi, Eds.), D. Reidel Publishing Company, Dordrecht (1986), p. 301.
22) B. Miller, X. Shi, J. Am. Chem. Soc. 109 (1987) 578.
23) G. A. Olah, G. Asensio, H. Mayr, P. v. R. Schleyer, J. Am. Chem. Soc. 100 (1978) 4347.
24) W. E. Parham, H. E. Reiff, P. Swartzentruber, J. Am. Chem. Soc. 78 (1956) 1437.
25) M. S. Baird, D. G. Lindsay, C. B. Reese, J. Chem. Soc. C 1969, 1173.

# ON THE RELATIONSHIP BETWEEN STRAIN AND CHEMICAL REACTIVITY OF TORSIONALLY DISTORTED CARBON-CARBON DOUBLE BONDS

Kenneth J. Shea
Department of Chemistry
University of California
Irvine, California 92717

ABSTRACT. In an effort to develop relationships between the strain of a carbon-carbon double bond and its chemical reactivity, a series of bridgehead alkenes containing torsionally distorted bridgehead double bonds have been synthesized. A combination of computational, spectroscopic and structural studies have provided information regarding the manner and degree of strain in these molecules. In addition, the rates of epoxidation of these double bonds have been obtained. The relationship between strain energy, reaction exothermicity and other spectroscopic parameters to the rate of epoxidation are examined in an effort to identify the chemical manifestations of strain.

Introduction

Although it is generally assumed that strained molecules are more reactive than unstrained molecules, quantitative relationships between strain and reactivity are rare.[1] We have a long-standing interest in molecules that contain torsionally distorted carbon-carbon double bonds. This structural feature is embodied in molecules such as *trans*-cycloalkanes 1, bridgehead alkenes 2, and meso-bridgehead dienes 3. In an effort to develop relationships between the strain associated with a twisted carbon-carbon

1

2    3

double bond and its chemical reactivity, a study was undertaken involving the rates of reaction of a series of bridgehead alkenes 2 and bridgehead dienes 3 with the electrophilic reagent m-chloroperbenzoic acid (MCPBA).

## Synthesis of Bridgehead Alkenes and Dienes

The strain energy associated with torsionally distorted carbon-carbon double bonds is relatively small. For example, *trans*-cyclooctene and bicyclo[3.3.1]nonene, two molecules that exist at the limits of isolatability, have steric energies that are relatively modest, i.e., 22[2] and 24 kcal/mol,[3] respectively. This is a consequence of the relatively "soft" bending potentials associated with carbon-carbon double bonds.[4] The relatively low strain energy of these compounds has important implications with regard to the synthesis of compounds of this type. This can be appreciated by consideration of the accompanying table of reaction enthalpies. A number of common pericyclic reactions have sufficient thermodynamic driving force to be used for the synthesis of molecules with torsionally distorted double bonds, even those that exist at the limits of isolatability. Although pericyclic reactions are not often thought of as synthetic entries into strained molecules, the thermodynamic evidence suggests otherwise.

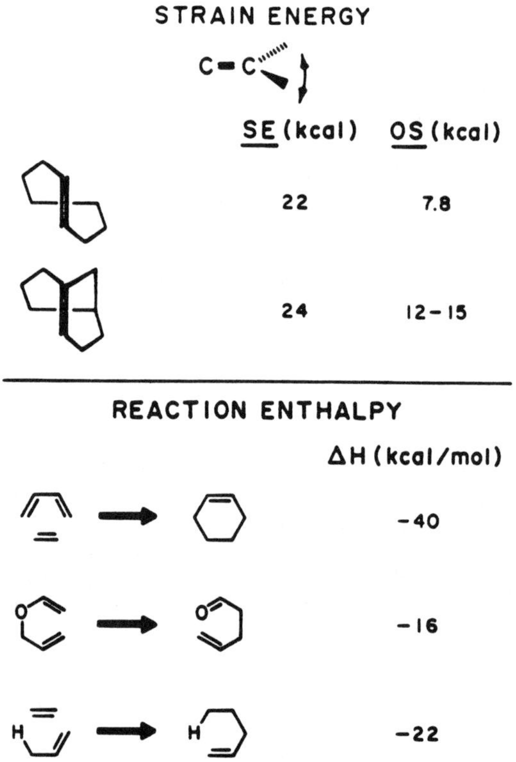

### STRAIN ENERGY

| Compound | SE (kcal) | OS (kcal) |
|---|---|---|
| *trans*-cyclooctene | 22 | 7.8 |
| bicyclo[3.3.1]nonene | 24 | 12–15 |

### REACTION ENTHALPY

| Reaction | ΔH (kcal/mol) |
|---|---|
| butadiene + ethylene → cyclohexene | −40 |
| vinyl enol ether → dihydropyran | −16 |
| 1,3,5-hexatriene → 1,3-cyclohexadiene | −22 |

We have been successful in developing a number of pericyclic reactions that serve as useful synthetic entries into bridgehead alkenes 2 and bridgehead dienes.[5] In the first illustration of this chemistry, the Type 2 intramolecular Diels-Alder reaction[6,7,8,9] has been employed for the synthesis of a homologous series of bridgehead alkenes (Figure 1). The conditions for the cycloadditions typically involve passage of the triene precursor through an atmospheric pressure flow pyrolysis apparatus. Temperatures for the cycloadditions range from 325-450°C with a contact time of 5-15 seconds.

TYPE 2 INTRAMOLECULAR DIELS-ALDER

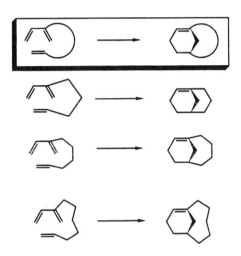

A second strategy utilizes a [3.3] sigmatropic rearrangment of *cis*-1,2-divinylbicycloalkane precursors. Cope rearrangement results in formation of a series of meso-bridgehead dienes that span a considerable range of steric energy.[10,11]

## [3.3] SIGMATROPIC REARRANGEMENTS

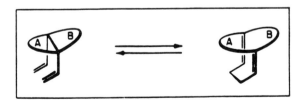

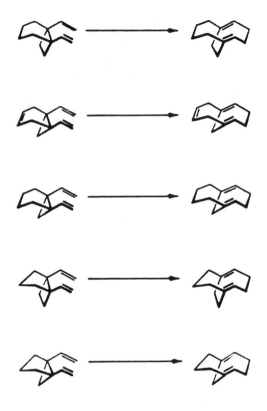

The steric energies for the two types of bridgehead alkenes are given in the table below. The steric energies were calculated by the MM2 force field.[12,13] In addition, values for the olefin strain (OS) are also included. These values are arrived at by taking differences in steric energy of the bridgehead alkene and the corresponding saturated bicyclic alkane as suggested by Schleyer and coworkers.[14,15] The OS is a measure of the release (or gain) in steric energy upon saturation of the bridgehead double bond. This value ranges from 13.8 kcal/mol in bicyclo[3.3.1]nonene to a negative 11.2 kcal/mol for the bicyclo[4.4.2]dodecane derivative. Such negative olefin strain is associated with hyperstable bridgehead alkenes,[14] molecules that show a net <u>reduction</u> in steric energy upon introduction of the bridgehead double bond.

With a quantitative measure of steric energy for each molecule in hand, a study of the epoxidation of the bridgehead alkenes with MCPBA was undertaken.

The epoxidation, illustrated in the figure, exhibits second order kinetics in $CH_2Cl_2$ resulting in formation of a single bridgehead epoxide product. The bridgehead dienes followed a consecutive second order kinetic pattern, giving rise in all but one case, to clean formation of monoepoxide followed by a much slower reaction to form bisepoxide ($k_1 >> k_2$). Rate constants at 0°C were obtained either titrimetrically or spectrophotometrically.

## STRAIN ENERGY

| Structure | Steric Energy (MM2) (kcal/mol) | $OS_1$ (kcal/mol) | $OS_2$ (kcal/mol) |
|---|---|---|---|
|  | 31.7 | 13.8 | – |
|  | 25.0 | 0.5 | – |
|  | 23.9 | -1.7 | – |
|  | 29.4 | -2.1 | -11.22 |
|  | 24.2 | 2.9 | 5.3 |
|  | 36.1 | 2.7 | -6.1 |
|  | 31.2 | 6.7 | 1.5 |

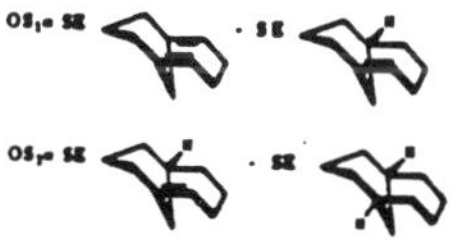

$OS_1 = SE$ – $SE$

$OS_2 = SE$ – $SE$

## EPOXIDATION OF BRIDGEHEAD ALKENES

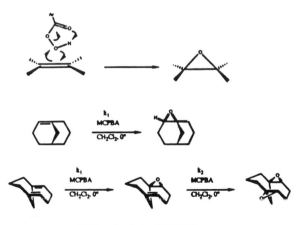

$1/[\text{Alkene}]_t = k_1 t + 1/[\text{Alkene}]_0$

$[\text{Alkene}]_0 = [\text{MCPBA}]_0$

The rate data, for epoxidation of the bridgehead alkenes and bridgehead dienes, together with reference alkenes are given in the table below. The reactivity differences are substantial and range over a factor greater than 20,000. The bridgehead olefins and bridgehead dienes comprise an isosteric series of trisubstituted torsionally distorted double bonds.

## SUMMARY OF KINETIC DATA

| structure | $k_{rel}(0°)$ | $k(0°)$ |
|---|---|---|
|  cyclohexene | 1 | 0.0223 |
|  cyclooctene | 2 | 0.0449 |
|  norbornene | 1.54 | 0.0343 |
|  methylcyclopentene | 24.1 | 0.537 |
|  methylcyclohexene | 12.2 | 0.273 |
|  bridgehead alkene | 4570 | 102 |
|  bridgehead alkene | 135 | 3.02 |
|  bridgehead diene | 570 | 12.7 |
|  bridgehead diene | 9650 | 215 |
|  bridgehead diene | 53.8 | 1.2 |
|  bridgehead diene | 23200 | 518 |

In an effort to uncover a relationship between the kinetic and structural findings, several attempts at correlating this data were undertaken. Although the structural changes resulting from epoxidation are modest compared with those that arise from hydrogenation, the changes that do take place (pyramidalization and C-C bond lengthening) are on the structural continum from bridgehead alkene to alkane. We considered a plot of log $k_{rel}(0°C)$ vs OS (kcal/mol). This is shown in the figure below. As can be seen, there is no correlation between these two parameters. Indeed, we have been

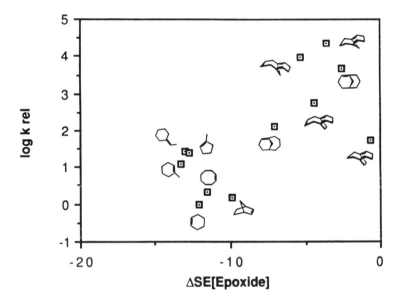

unable to establish a simple strain energy related property that correlates with this data. One interesting finding, however, is an apparent relationship between chemical reactivity and ionization potential of the bridgehead alkenes. Several selected values for the IP of the series of bridgehead dienes are given in the figure below. The splitting of the two $\pi$ energy levels exhibits a smooth progression that is due to a combined through space and through bond interactions of the two $\pi$ bonds. Plots of IP vs log $k_{rel}(0°C)$ produce a reasonably straight line (log $k/eV=1/10^{3.82}$).

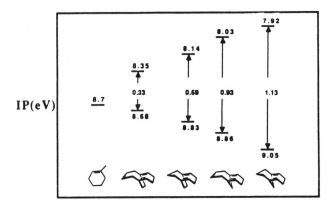

Efforts are continuing to evaluate the generality of this relationship and, in particular, the deviations from the correlation that result with the more highly strained bridgehead alkenes.

**Acknowledgment**

I would like to thank my coworkers in this effort, Dr. Art Greeley, Mr. Jang-S. Kim and Professor Don Aue of UC-Santa Barbara and to acknowledge the National Science Foundation for their financial support.

**References**

1. (a) Sterling, C. J. M. *Tetrahedron*, **1985**, *41*, 1613.
   (b) Sterling, C. J. M. *Pure and Appl. Chem.*, **1984**, *56*, 1781.

2. Rogers, D. W.; Voitkengerg, H.; Allinger, N. L. *J. Org. Chem.*, **1978**, *43*, 360.

3. Lesko, P. M.; Turner, R. B. *J. Am. Chem. Soc.*, **1968**, *90*, 6888.

4. (a) Ermer, O.; Lifson, S. *J. Am. Chem. Soc.*, **1973**, *95*, 4121.
   (b) Allinger, N. L.; Sprague, J. T. *J. Am. Chem. Soc.*, **1972**, *94*, 5734.
   (c) Ermer, O. *A spekte von Kraftfeldrechnungen*, W. B. Verlag: Munich, **1981**.

5. Shea, K. J. *Tetrahedron*, **1980**, *36*, 1683.

6. Shea, K. J.; Wise, S. *J. Am. Chem. Soc.*, **1978**, *100*, 6519.
7. Shea, K. J.; Wise, S. *Tetrahedron Lett.*, **1979**, 1011.

8. Shea, K. J.; Beauchamp, P. D.; Lind, R. *J. Am. Chem. Soc.*, **1980**, *102*, 4544.

9. Shea, K. J.; Wise, S.; Burke, L. D.; Davis, P. D.; Gilman, J. W.; Greeley, A. C. *J. Am. Chem. Soc.*, **1982**, *104*, 5708.

10. Shea, K. J.; Greeley, A. C.; Nguyen, S.; Beauchamp, P. D.; Wise, S. *Tetrahedron Lett.*, **1983**, *24*, 4173.

11. Shea, K. J.; Greeley, A. C.; Nguyen, S.; Beauchamp, P. D.; Aue, D. H.; Witzeman, J. S. *J. Am. Chem. Soc.*, **1986**, *108*, 5901.

12. Burkert, V.; Allinger, N. L. *Molecular Mechanics*, ACS Monograph 177, American Chemical Society, Washington, DC, **1982**.

13. Maier, W. F.; Schleyer, P. v. R. *J. Am. Chem. Soc.*, **1981**, *103*, 1891.

14. McEwen, A. B.; Schleyer, P. v. R. *J. Am. Chem. Soc.*, **1986**, *108*, 3951.

# PHOTOINDUCED SINGLE ELECTRON TRANSFER FROM STRAINED RINGS

Professor Paul G. Gassman
Department of Chemistry
University of Minnesota
Minneapolis, Minnesota 55455 USA

ABSTRACT. Evidence is presented for the existence of one-electron carbon-carbon bonds. Molecules containing this very reactive moiety, which is generated photochemically by single electron transfer, have been studied in detail.

## 1. Introduction

Subsequent to the discovery of the transition metal complex promoted rearrangement of quadricyclane to norbornadiene in our laboratories and independently in the laboratories of Hogeveen and Volger at Dutch Shell, many individuals throughout the world proceeded to evaluate the quadricyclane - norbornadiene interconversion as a possible basis for the development of a cell for the storage of solar energy. One of the labs which investigated this phenomenon was

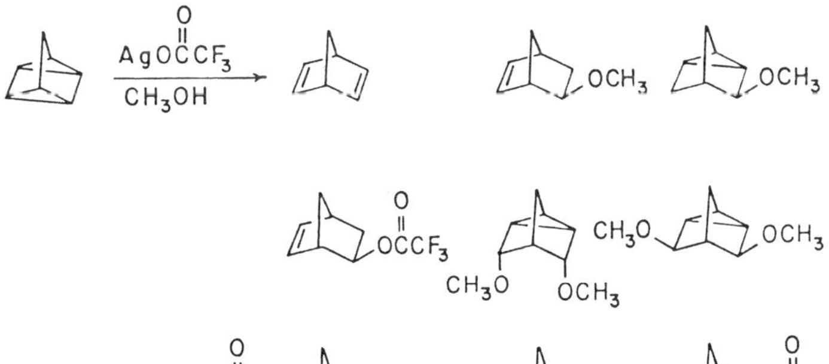


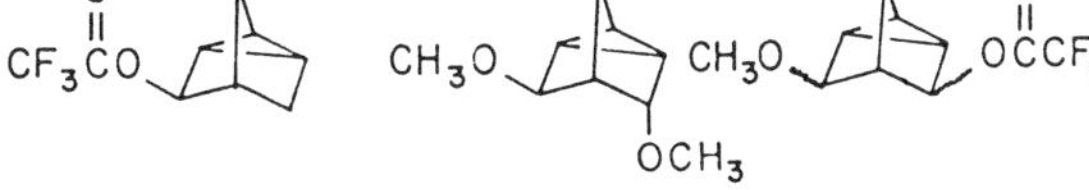


G. F. Koser and J. N. Faircloth, J. Org. Chem., 41, 583 (1976)

that of Professor G. Koser at the University of Akron. Koser and Faircloth found that silver trifluoroacetate in methanol readily isomerized quadricyclane, with norbornadiene being the major product. However, in addition to norbornadiene these workers found numerous other products. Some of these products merely involved the addition of methanol or trifluoroacetic acid to the $C_7H_8$ skeleton. What intrigued us was the formation of a series of products in which two equivalents of



*A. de Meijere and S. Blechert (eds.), Strain and Its Implications in Organic Chemistry, 143–167.*

nucleophile had been added to the basic skeleton. This obviously involved a two-electron oxidation. Silver metal was found in a quantity equivalent to the amount of oxidation products formed in the reaction. This raised the question in our minds as to whether there was a relationship between transition metal promoted rearrangement of strained molecules and the ease of oxidation of these compounds.

Examination of the literature indicated that an energy of 2.5-3.2 V versus a saturated calomel electrode was needed to oxidize unstrained alkanes. Unstrained alkenes required 2.0-2.6 V of applied potential relative to a saturated calomel electrode. In the case of unstrained alkanes, one observed oxidation of the carbon-hydrogen bond while with the unstrained alkenes, the electron was removed from the π-orbital associated with the carbon-carbon double bond. With this as background, we examined the oxidation of quadricyclane and found that it could be oxidized at 0.91 V versus a saturated calomel electrode. Thus, this strained

 $E_{1/2}$ = 0.91 V vs SCE

hydrocarbon gave up an electron at approximately 2 V less than most saturated hydrocarbons. This indicated to us that the HOMO of quadricyclane resided at an energy about 45 kcal per mol higher than the HOMO of most unstrained saturated hydrocarbons. With this in mind, we examined a large series of strained hydrocarbons and found that, depending on strain incorporated within the molecule, irreversible oxidation potentials ranged from approximately 0.5 V to 2.0 V. In every

| Compound | $E_{1/2}$ (V) | Compound | $E_{1/2}$ (V) |
| --- | --- | --- | --- |
|  | 0.91 |  | 1.62 |
|  | 1.23 |  | 1.67 |
|  | 1.50 |  | 1.91 |
|  | 1.73 |  | 2.12 |
|  | 1.74 |  |  |

case that we examined, we found that the oxidation was a two-electron oxidation which involved the cleavage of a carbon-carbon bond, not a carbon-hydrogen bond oxidation as is observed with unstrained hydrocarbons. Examination of a plot of $E_{1/2}$ versus gas phase ionization potentials shows a very linear relationship. To review this point, unstrained alkanes oxidize very readily with cleavage of a C-H bond; unstrained alkenes experience oxidation of the carbon-carbon double bond

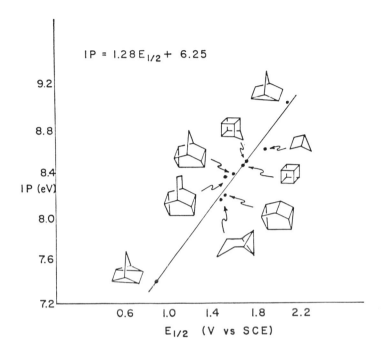

while strained alkanes routinely undergo cleavage of a carbon-carbon bond under oxidative conditions. This two-electron cleavage of the carbon-carbon bond presumably occurs in a step-wise fashion with the initial oxidation involving

Generalized Oxidation Half-wave Potentials

|  | $E_{1/2}$ vs SCE | Bond Oxidized |
|---|---|---|
| Unstrained alkanes | 2.5 – 3.2 | C-H |
| Unstrained alkenes | 2.0 – 2.6 | $\pi$ |
| Strained alkanes | 0.4 – 2.1 | C-C |

removal of an electron from the highest occupied molecular orbital (HOMO) followed by nucleophilic attack of solvent and removal of a second electron to form a second cationic center.

## 2. Photoinduced Single Electron Transfer

We were very interested in examining this mechanistic process with the hope of interrupting it at the first stage. Relatively little was known about cation radicals in which no delocalization of the active centers was possible. A survey of the literature indicated to us that it might be possible to stop the process after

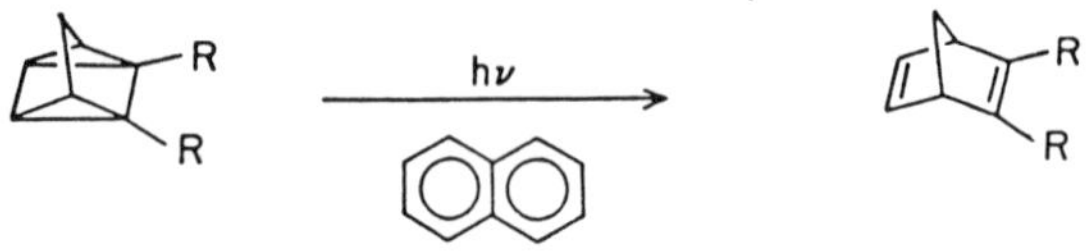

S.L. Murov, R.S. Cole, and G.S. Hammond, J.A.C.S., 90, 2957 (1968).

G. Jones, II, S.-H. Chaing, W.G. Becker, and D.P. Greenberg, JCS, Chem. Comm., 681 (1980).

removal of a single electron. Obviously, this cannot be done electrochemically, nor can it be done through the addition of any macroscopic amounts of an oxidant. Fortunately, the work of Murov, Cole, and Hammond gave a preliminary indication that a single electron transfer process might be viable. Their experimental

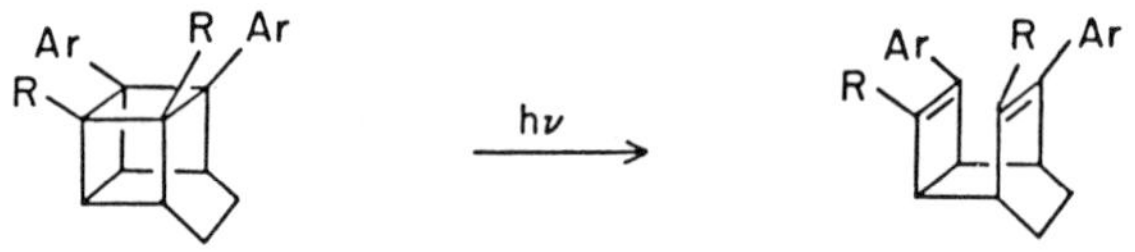

K. Okada, K. Hitsamitsu, and T. Mukai, JCS, Chem. Comm., 941 (1980).

T. Mukai, K. Sato, and Y. Yamashita, J.A.C.S., 103, 670 (1981).

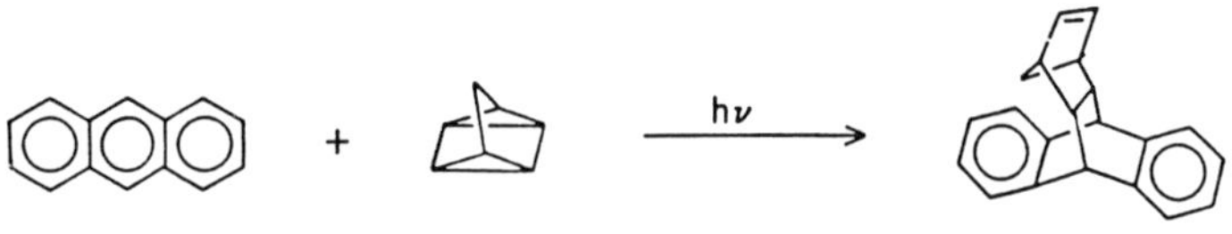

T. Sasaki, K. Kanematsu, I. Ando, and O. Yamashita, J.A.C.S., 98, 2686 (1976).

work, which provided this indication, involved the isomerization of quadricyclane to norbornadiene on irradation in the presence of naphthalene. Subsequent to the initiation of our work, Jones and coworkers established the single electron transfer mechanism for the photoinduced isomerization of quadricyclane to norbornadiene with a series of photosensitizers. At the start of our investigation, the only other pertinent result in the literature was that of Sasaki, who had observed that quadricyclane added to anthracene on irradiation. During the course of our own work, two additional reports appeared from Mukai's laboratory which indicated that aryl groups when attached to strained rings could act as photosensitizers for single electron transfer processes.

3. Fluorescence Quenching

Our own studies started with an investigation of the ability of a series of strained hydrocarbons to quench the fluorescence of naphthalene. As indicated,

| Compound | $\log k_q$ | $E_{1/2}$ (V) |
|---|---|---|
|  | 9.51 | 0.91 |
|  | 8.01 | 1.23 |
|  | 7.32 | 1.50 |
|  | 7.33 | 1.73 |
|  | 7.15 | 1.74 |
|  | 6.99 | 1.62 |
|  | 6.86 | 1.67 |
|  | 6.32 | 1.91 |
|  | — | 2.12 |

strained hydrocarbons are quite effective at quenching naphthalene fluorescence. A range of $10^3$ is observed in the rate constants for fluorescence quenching with the rate decreasing in parallel to the increase in the $E_{1/2}^{ox}$ of the strained hydrocarbon. A reasonable correlation can be obtained for this data. More

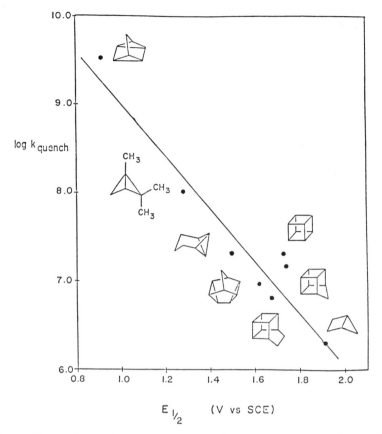

definitive evidence for a single electron transfer process was obtained when our

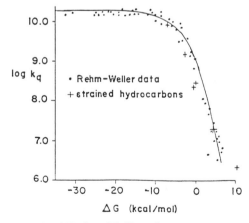

data was plotted on a standard Rehm-Weller plot. As shown, the data points for strained hydrocarbons fall very nicely on the Rehm-Weller plot. This provides excellent evidence that the strained hydrocarbons are quenching the fluorescence of naphthalene via a single electron transfer process.

## 4. The One-Electron Carbon-Carbon Bond

Our initial investigation of productive photochemistry of strained hydrocarbons involved tricyclo[4.1.0.0$^{2,7}$]heptane (Moore's hydrocarbon). When this hydrocarbon was irradiated in the presence of naphthalene, we observed very effective quenching of the naphthalene fluorescence. Unfortunately, no products were observed. Instead, we found only unreacted starting material. This suggested that the initially generated cation radical - anion radical pair underwent back electron transfer to regenerate starting materials rather than collapsing with

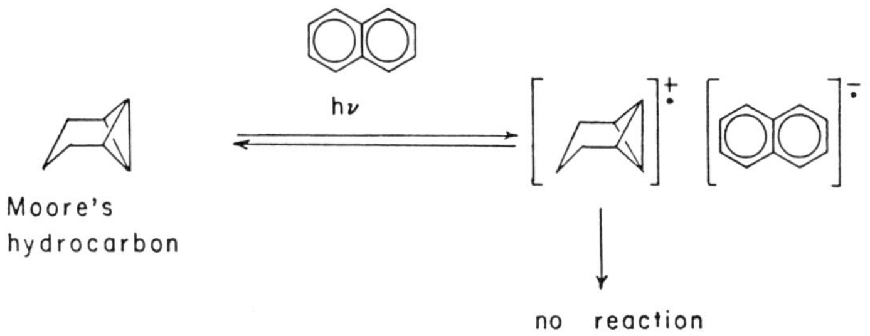

Excited State Reduction Potentials

| | $E_{1/2}$ red | E* Absorpt. | E* Emission | $E_{1/2}$ red* |
|---|---|---|---|---|
| 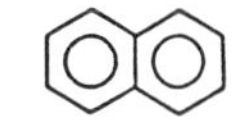 naphthalene | -2.56 V | 3.99 eV | 3.74 eV | 1.43 V |
| 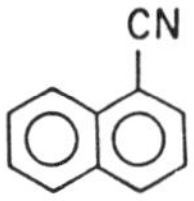 1-CN-naphthalene | -2.00 V | 3.80 eV | 3.69 eV | 1.80 V |
| 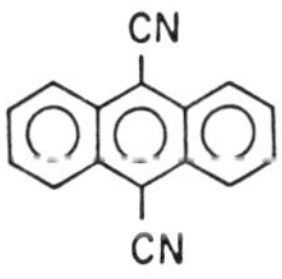 1,4-diCN-naphthalene | -0.82 V | 2.90 eV | 2.70 eV | 2.08 V |

solvent to give product. Since the electron transfer process to form the cation radical - anion radical pair should be slightly endothermic, it is not surprising that back electron transfer might be quite efficient. In its excited state, naphthalene has an $E_{1/2}$ as an oxidant of 1.43 V, while Moore's hydrocarbon requires 1.50 V for oxidation. As a result, we decided to examine more powerful excited state oxidants, namely, 1-cyanonaphthalene and 9,10-dicyanoanthracene, which have $E_{1/2}$ as excited state oxidants of 1.80 V and 2.08 V, respectively. Thus, either of these cyanoaromatics should be sufficiently powerful to remove an electron from Moore's hydrocarbon in an exothermic process. When Moore's hydrocarbon was irradiated in benzene in the presence of 1-cyanonaphthalene, hereafter shown

as 1-CN, we observed the dimer illustrated. Clearly, a central and a side bond of the Moore's hydrocarbon had been cleaved. Electrochemical oxidation in methanol gave the acetal shown while metal catalyzed rearrangement with a wide variety of transition metal complexes generated 3-methylenecyclohexene. We had established previously that the transition metal catalyzed process to yield 3-methylenecyclohexene had occurred via a metal complexed carbene. A mechanistic possibility was that we were generating a homoallylic carbene photochemically and that the observed product which we obtained in 95% yield was merely the result of carbene dimerization. This seemed unlikely to us since homoallylic carbenes bearing a hydrogen on the allylic position are known to hydrogen migrate to form conjugated dienes which great ease. Thus, it would not be reasonable to expect the designated homoallylic carbene to have a sufficient lifetime for dimerization in 95% yield. An alternate hypothesis, which merited consideration, was the possibility that the homoallylic carbene underwent hydrogen migration to give 3-methylenecyclohexene. Through a complex process, 3-methylenecyclohexene could then react with the same homoallylic carbene to form the observed product. This hypothetical path was shown to be inoperative by the addition of small amounts of deuterated 3-methylenecyclohexene to the reaction mixture followed by product examination after short reaction time. No deuterium label was incorporated into the observed triene and, as a result, any intermediacy of 3-methylenecyclohexene could be ruled out.

The first true insight into the mechanistic process involved in this single electron transfer photochemistry came when the process was examined in nucleophilic solvents. Irradiation in methanol gave addition of methanol across the central bond of the bicyclobutane moiety of Moore's hydrocarbon. This occurred in a highly stereospecific manner. Irradiation in tetrahydrofuran - water gave the

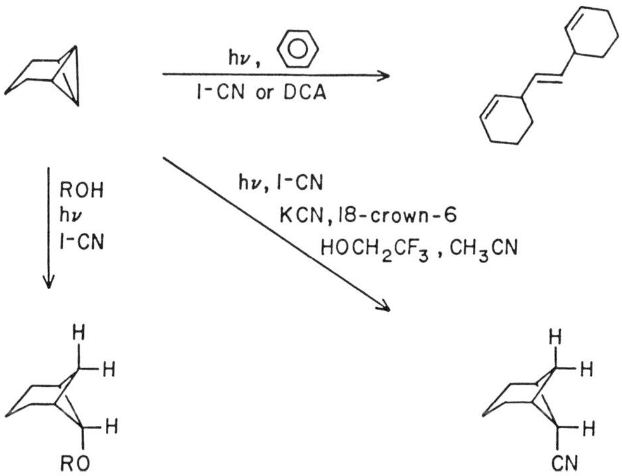

corresponding alcohol, again, in extreme stereochemical purity. Lastly, irradiation in acetonitrile containing 1-cyanonaphthalene, potassium cyanide, 18-crown-6, and 2,2,2-trifluoroethanol gave 6-cyanobicyclo[3.1.1]heptane in excellent stereochemical purity.

Irradiation of 1-methyltricyclo[4.1.0.0$^{2,7}$]heptane in tetrahydrofuran - water containing 1-cyanonaphthalene gave anti-Markovnikov addition of water across the central bond of the bicyclo[1.1.0]butane moiety. Similar irradiation in methanol

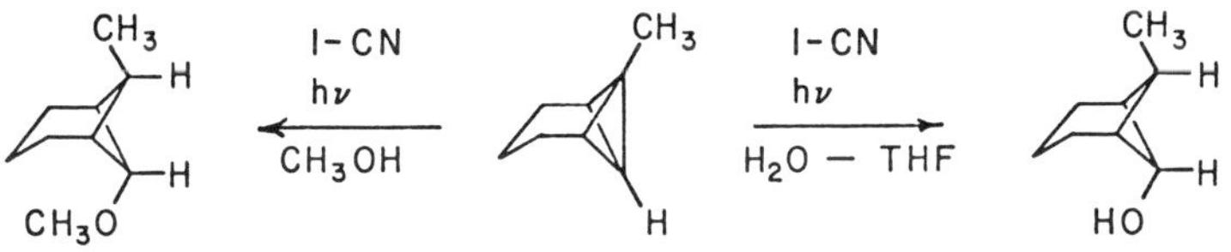

gave the methyl ether which would result from addition of methanol across the central bond of the bicyclo[1.1.0]butane moiety. Both of these reactions occurred in a clean, anti-Markovnikov manner. Stereochemistry of the hydroxyl and methoxyl groups was established with ease. The exo-hydrogen couples with the bridgehead hydrogens with a coupling constant of approximately 6 Hz while the corresponding endo-hydrogen has a coupling constant of approximately 0 Hz. Thus, observation of the carbinol hydrogen shows it to be a triplet and establishes the stereochemistry of the hydroxyl groups. Establishing the stereochemistry of the methyl group was somewhat more difficult. The 300 MHz H$^1$ NMR plot shows the region from 400 to 600 Hz of the proton spectrum of the methyl ether. At the top is the normal spectrum while in the center is the spectrum with the methyl group decoupled. Thanks to the wonders of computers, we were able to subtract these and demonstrate that the hydrogen on the carbon adjacent to the methyl group

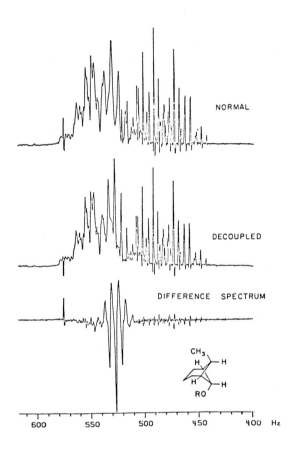

couples with the bridgehead hydrogens by 5.9 Hz. Thus, only the stereochemistry shown is possible.

When the monomethylated derivative of Moore's hydrocarbon was irradiated in benzene with either 1-CN or dicyanoanthracene as the photosensitizer, three products were obtained. The first was a triene analogous to the only product obtained from the parent hydrocarbon. In addition, two formal dimeric reduction products were observed. The structure of the more symmetrical reduction product was established by single crystal X-ray analysis, while the structure of the less symmetrical reduction product was established through comparison of its NMR with that of the more symmetrical reduction product. The ORTEP drawing (minus hydrogens for clarity) of the more symmetrical reduction product is shown. It is

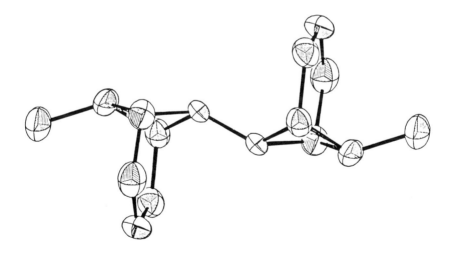

interesting to note that the trimethylene bridge is essentially flat and that the cyclobutane moiety is slightly puckered.

Mechanistically, the dramatic effect of methyl substitution on the dimerization reaction is intriguing. All irradiations are carried out in Pyrex utilizing approximately 2 mol % of the photosensitizer. The strained hydrocarbon is totally transparent in the region of the spectrum above 300 nm. Thus, the absorbed photoenergy is utilized to excite the 1-CN and to convert it into an oxidant which

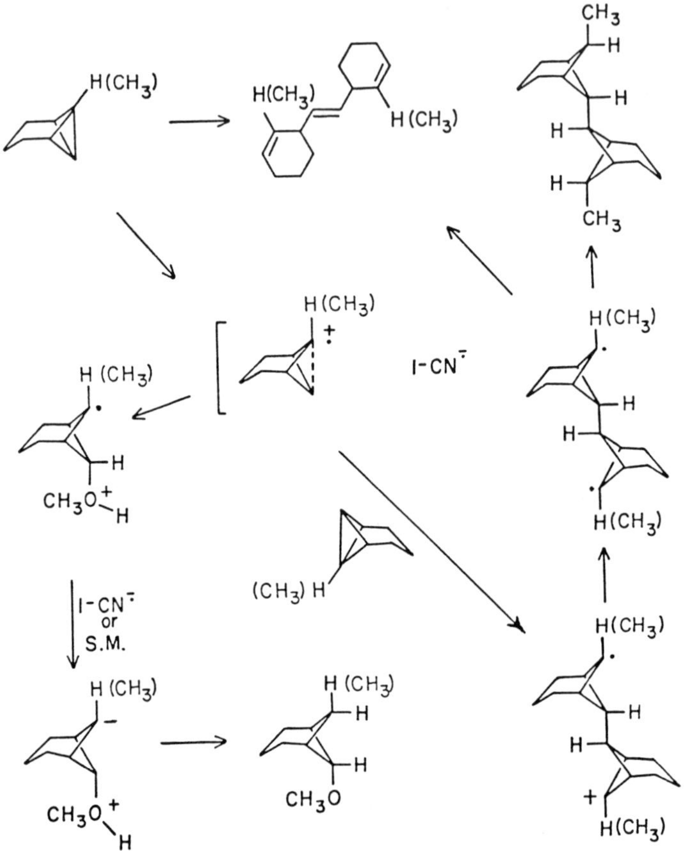

removes an electron from the strained hydrocarbon to generate a cation radical - anion radical pair. In the presence of nucleophiles, attack occurs on the cation radical to yield a distonic cation radical which, by either hydrogen transfer followed by back electron transfer, or back electron transfer followed by proton transfer yields the observed products of nucleophilic addition. In the absence of a nucleophile, the most electron rich species present in solution is Moore's hydrocarbon itself. Dimerization and back electron transfer can generate the diradical illustrated. When the parent hydrocarbon is irradiated, the diradical is secondary at both radical centers. Simple electron shift will then provide the observed triene which was formed in 95% yield. In contrast, when the methyl substituent is present on the starting material, both centers are tertiary. As a result, they are longer lived and have the ability to abstract a hydrogen. Once a hydrogen is abstracted, a second hydrogen will be transferred and the reduction products will be observed. The main question to be answered at this point is "what is the source of the hydrogen." It was unlikely that these hydrogens came from the solvent (benzene). Irradiation in deuterated benzene confirmed this judgment, since no deuterium was incorporated into the reduction products. Thus, the hydrogens involved in the reduction were coming either from the starting material or, more likely, from the triene. The triene has two sites which are doubly allylic. Both of these positions bear a hydrogen which should be readily abstractable. In order to see whether the triene was the source of the hydrogens involved in the reduction, the product ratio was monitored with time. As shown, the ratio of saturated dimer to triene varied considerably as a function of the progress of the reaction. Initially, the triene was present in large quantity with the reduced dimers being there in only trace amounts. As the reaction proceeded, the triene itself became a reactant and the

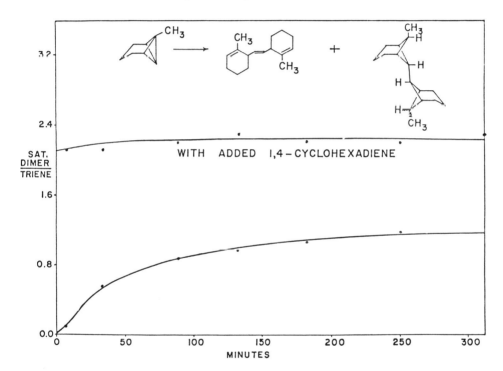

amount of reduced dimer increased. Removal of a hydrogen atom from the triene resulted in a highly delocalized radical which we believe lead to considerable polymerization of the triene. Thus, the overall yield for this reaction was relatively low. In an attempt to avoid this complication, we added 1,4-cyclohexadiene to the reaction mixture. As shown on the plot, the addition of this hydrogen source had a dramatic effect on the product ratio. In addition, it had a major effect on the yield.

Whereas the reaction in the absence of 1,4-cyclohexadiene gave only 37% yield of products and a large of amount of untractable material, the addition of 1,4-

| CONDITIONS | DIMER | REDUCED SYMMETRICAL DIMER | REDUCED UNSYMMETRICAL DIMER | TOTAL YIELD % |
|---|---|---|---|---|
| DCA, $C_6H_6$, hv | 17 | 15 | 5 | 37 |
| DCA, $C_6H_6$, hv, 1,4-CYCLOHEXADIENE | 29 | 50 | 17 | 96 |

cyclohexadiene resulted in the disappearance of the untractable material and formation of a 96% overall yield of the three products.

Whereas tricyclo[4.1.0.0$^{2,7}$]heptane and its monomethyl derivative behaved in a fairly straightforward fashion, the dimethyl derivative provided a variety of surprises. Initial attempts to photodimerize the dimethyl derivative using 1-CN in benzene gave no dimer. Instead, only starting materials were obtained even after days of irradiation. Two possible explanations might be put forth. One is that

electron transfer was not occurring (which seemed very unlikely) and the second is that due to the additional methyl group, one simply could not bring the two molecules together to form a dimer due to the steric interaction of the methyl groups. In order to assure ourselves of the effectiveness of electron transfer, we looked at all three derivatives of Moore's hydrocarbon with 1-CN as a sensitizer. The observed rates for fluorescence quenching were 1.2 x 10$^{10}$, 8.7 x 10$^9$, and 2.0 x 10$^{10}$. Effectively, all of the examples indicated a diffusion controlled quenching process. Obviously, dimethyl Moore's hydrocarbon, with an $E_{1/2}$ versus saturated calomel of 1.06 V, should be the best candidate for electron transfer. Indeed, the quenching constants showed that all three derivatives are extremely effective and, hence, that all three derivatives electron transfer to the excited state of 1-CN. In view of these observations, we decided to carry out the irradiation in methanol which would present itself as a smaller attacking agent. When the dimethyl derivative was irradiated in methanol containing 1-CN, we obtained some of the

|  | $k_q$ (L/mol/sec) | $E_{1/2}$ vs SCE (V) |
|---|---|---|
|  [norbornene] | $1.16 \times 10^{10}$ | 1.50 |
| 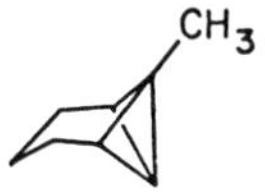 [methyl-norbornene] | $8.66 \times 10^{9}$ | 1.22 |
| 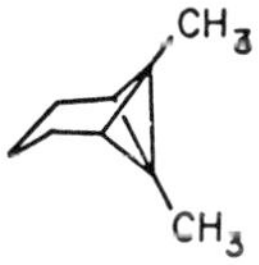 [dimethyl-norbornene] | $2.03 \times 10^{10}$ | 1.06 |

$\tau$ for 1-CN = 18.7 nsec

product which we expected on the basis of earlier experience. In addition, we obtained two new products. One of these differed from the expected product only in the stereochemistry of the methyl group. The other possessed a terminal

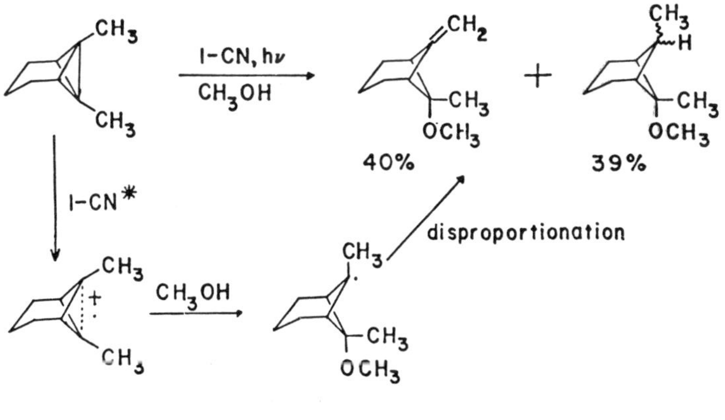

methylene. The ratio of these compounds was 40:39. This looked suspiciously like the result which one might expect from a classic disproportionation reaction between radicals. Thus, it seemed possible that the 1-CN was accepting an electron from the dimethyl derivative to give a cation radical - anion radical pair. Attack of methanol and loss of a proton would give a radical, which, upon disproportionation, could provide the observed results. It should be noted that there was a second difference between the dimethyl derivative and the parent in this transformation. This was associated with the photosensitizer which now became a stoichiometric reagent. A third significant difference was observed when the irradiation was carried out in deuterated methanol. While all of the products incorporated the $CD_3$ group, no deuterium was incorporated on the carbon to which the methyl was attached. This indicated that this hydrogen was coming from the methyl group which served as a precursor of the terminal methylene. This

hypothesis was tested through the preparation of the dimethyl derivative in which the two methyl groups were completely deuterated. When this compound was irradiated in nondeuterated solvent, the products were shown to be $\underline{d}_5$ (of necessity) and $\underline{d}_7$. Thus, the atom which was added to the carbon to which the methyl group was attached came from the methyl group of a second molecule.

In order to explain this behavior, we need to revert back to a consideration of the total mechanistic scheme. In the examples where the photochemistry of the parent hydrocarbon and the monomethyl derivative were examined, we believe that the initially formed cation radical - anion radical pair existed as a "tight" ion pair. Nucleophilic attack at this stage could result in the formation of a distonic cation radical - anion radical pair. Back electron transfer could then generate an anion which would be susceptible to electrophilic attack and lead directly to the observed products. In transversing from the parent hydrocarbon to the dimethyl derivative,

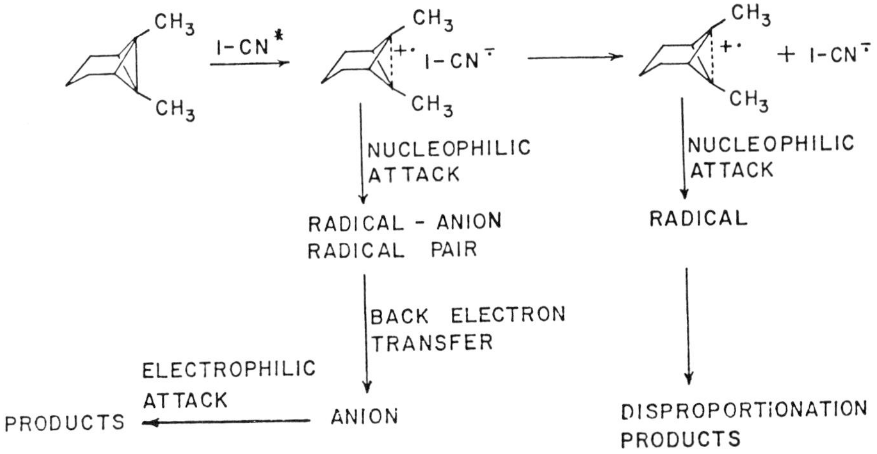

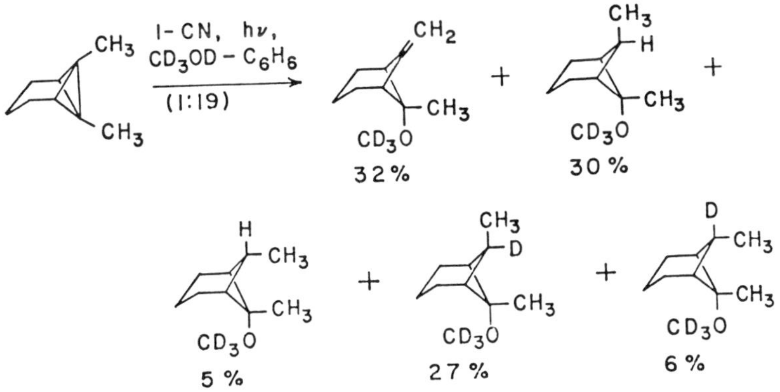

a change in oxidation potential of 0.44 V occurs. This means that the cation radical generated from the dimethyl derivative will be substantially more stable than a cation radical generated from the parent hydrocarbon. In this case, we believe that it might be possible for the cation radical - anion radical to undergo dissociation in the polar solvent methanol. Attack of methanol on the cation radical could then eventually lead to a radical which would undergo disproportionation. This type of mechanistic scheme is extremely difficult to document. In an attempt to provide support for this hypothesis, we chose to change the polarity of the solvent. This was done by using benzene-methanol in a 95:5 ratio. The methanol which we used in this experiment was completely deuterated. In this manner, after irradiation in the presence of 1-CN, we observed five products. Obviously, the material bearing only the terminal methylene would contain the $CD_3O$ group and no other deuterium. In contrast, we were able to establish that the two isomers in which methanol had been added across the central bond of the bicyclo[1.1.0]butane moiety were approximately 1:1 in the ratio of hydrogen to deuterium on the carbon to which the methyl group was attached. This indicated that, through a change in solvent polarity, we were able to revert at least part of the reaction path to the same mechanistic process as was involved with the parent hydrocarbon and the monomethyl derivative. We believe that this is directly attributable to a decrease in the ability of the cation radical and anion radical to undergo separation in solution.

The last question which we asked relative to the dimethyl derivative was whether a classical disproportionation reaction between two radicals was involved or whether a radical was abstracting a hydrogen from the starting material in a chain process. This question could be answered through any process that would vary radical concentration and, at the same time, provide competitive paths for the radical. In order to achieve this set of circumstances, we carried out the irradiation of the dimethyl derivative in the presence of 1-CN and 1,4-cyclohexadiene in methanol. If the radical were reacting with the starting material, the product ratios of **1:2** would not be dependent on the concentration of radical in solution. Obviously, the concentration of the radical in solution can be varied by varying the

| $I_{REL}$ | TIME | 1/2 |
|---|---|---|
| 16 | 1 H | .26 |
| 8 | 2 H | .21 |
| 4 | 4 H | .18 |
| 2 | 8 H | .15 |
| 1 | 16 H | .13 |

light intensity. As a result, the relative intensity of the irradiation was varied from 16 to 8 to 4 to 2 to 1 with simultaneous increase in time of irradiation from 1 to 2 to 4 to 8 to 16 h. Thus, all solutions received the same amount of irradiation. However, when the ratio of **1:2** was examined, it was found that the ratio varied by a factor of 2 over this range of irradiation intensity. Thus, the product ratio was unequivocally shown to be a function of the radical concentration. This supported the concept of a classical radical - radical disproportionation reaction.

I hope that by this point, I have convinced you that one can take a strained hydrocarbon, such as Moore's hydrocarbon, and through selective electron transfer to an excited state photosensitizer generate molecules containing one-electron carbon-carbon bonds. Having arrived at this point, it now became of interest to explore additional chemistry of molecules containing one-electron carbon-carbon bonds.

### 5. Molecular Rearrangement

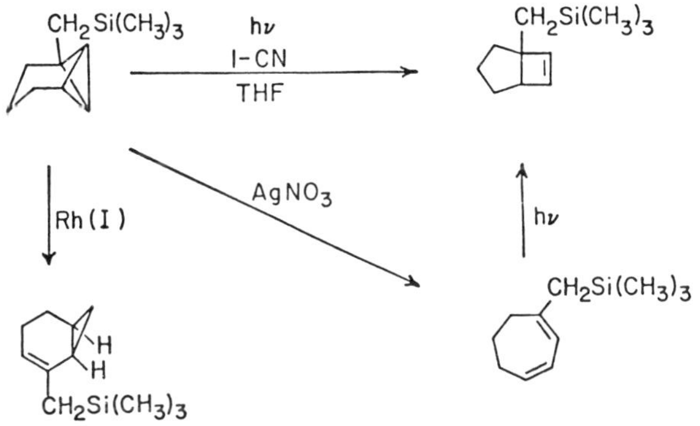

Of particular interest was the question of whether one could generate situations in which cation radicals, such as those found in molecules containing one-electron carbon-carbon bonds would undergo facile rearrangement. We observed that when 2-trimethylsilylmethylbicyclo[4.1.0.0$^{2,7}$]heptane was irradiated in tetrahydrofuran in the presence of 1-CN we obtained an 88% yield of 1-trimethylsilylmethylbicyclo[3.2.0]hept-6-ene. This compound could be prepared independently through the direct irradiation of the appropriate cycloheptadiene which could be made by silver nitrate induced rearrangement of the starting substituted derivative of Moore's hydrocarbon. Interestingly, rhodium promoted rearrangement of this compound gave only the norcarene derivative shown. There are obvious questions which need to be answered. The first is whether the steric bulk of the group in the 2-position promoted the rearrangement. Interestingly, when the tertiary butyl derivative shown was irradiated in methanol, we obtained an excellent yield of the addition of methanol across the central bond of the bicyclo[1.1.0]butane moiety. It could also be asked whether the presence of silicon in the molecule provided some unique property which would promote such rearrangement. As illustrated, the trimethylsilyl group in the 2-position had little effect on the photochemistry. Addition of both methanol and water occurred across the central bond of the bicyclo[1.1.0]butane moiety in high yields. Thus, it would appear that the trimethylsilylmethyl group is unique. This was anticipated because the trimethylsilylmethyl

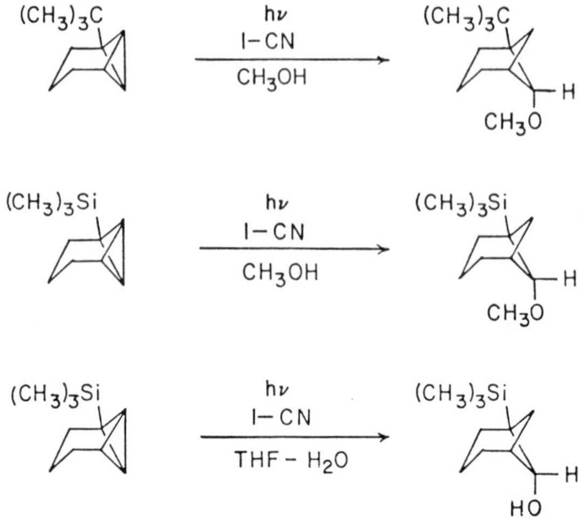

group is an extremely fine stabilizer for cations. The real question which needs to be answered is how does one get from the initially formed cation radical - anion radical pair to the rearranged product. We believe that it might be possible for "leakage" to occur from one cation radical to another. We could also ask whether it might be possible for the starting material to go directly to a cation radical in

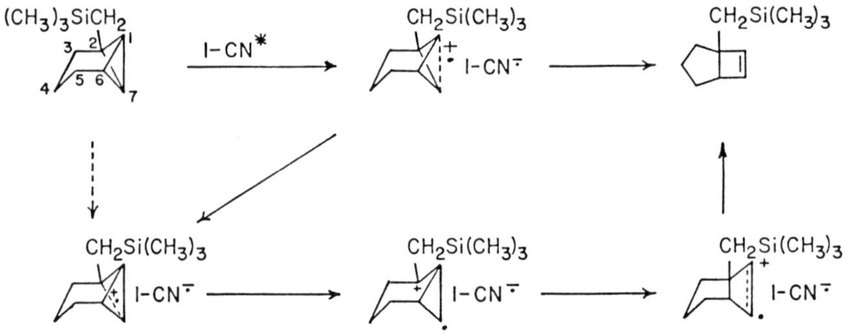

which the removal of an electron did not occur from the central bond of the bicyclo[1.1.0]butane moiety but rather from a side bond. We think this is improbable, but we cannot rule it out at this time. If electron "leakage" could occur, one could envision an opening of the cation radical to generate a distonic cation radical in which the unpaired electron was in an orbital which approached being orthogonal to the orbital associated with the cationic center. Should this be the case, a simple cyclopropylcarbinyl cation to cyclobutyl cation rearrangement would generate the cation radical which is expected to be associated with the observed product. Williams and coworkers have proposed that such a transformation can

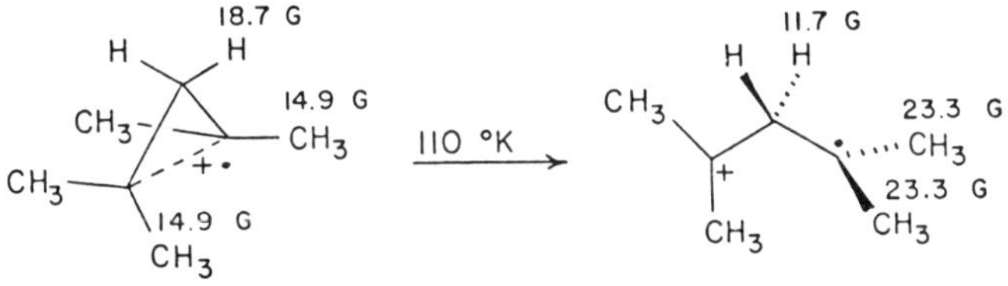

X.-Z. QIN, L.D. SNOW, F. WILLIAMS,
J. AM. CHEM. SOC. 1984, 106, 7640.

occur and Roth and Schilling have discussed this possibility from a hypothetical point of view on several occasions. In order to explore this possibility further, we decided to examine other substituents in the 2-position of tricyclo-[4.1.0.0$^{2,7}$]heptane. It was obvious that we needed a substituent that was more

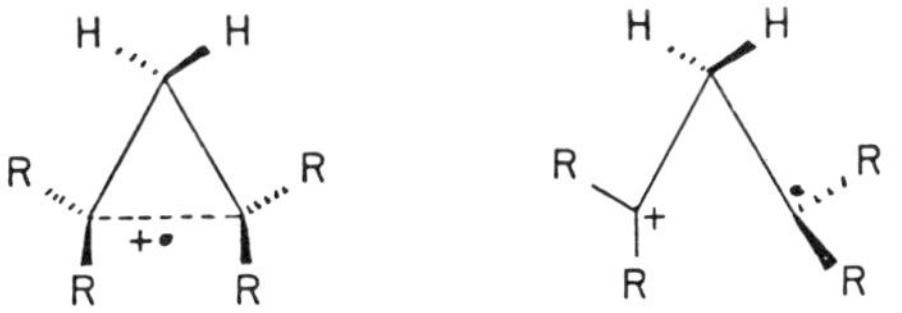

H. D. ROTH and M. L. M. SCHILLING
J. AM. CHEM. SOC. 1980, 102, 7956.
J. AM. CHEM. SOC. 1983, 105, 6805.
CAN. J. CHEM. 1983, 61, 1027.

carbocation stabilizing than *tertiary*-butyl but less carbocation stabilizing than trimethylsilylmethyl. If this could be achieved, we felt that we might be able to have an intermediate cation radical which could chose between the two possible reaction paths depending on conditions. We felt that the methylcyclopropyl substituent would be approximately right in terms of its ability to stabilize a neighboring carbocation. This turned out to be a very good choice. When this derivative of Moore's hydrocarbon was irradiated in methanol, we obtained 5% of the rearranged olefin and 22% of the product involving addition of methanol across the central bond of the bicyclo[1.1.0]butane moiety. Both of these products were photolabile under the reaction conditions. It was obvious that the cyclobutene was the most photolabile since the quantum yields for the two products were 0.01 and 0.02. When the same irradiation was carried out for a short reaction time in the presence of 1 mol % of 1,4-cyclohexadiene, we observed a 26% yield of the cyclobutene and a 53% yield of the methanol adduct at 13% photoconversion.

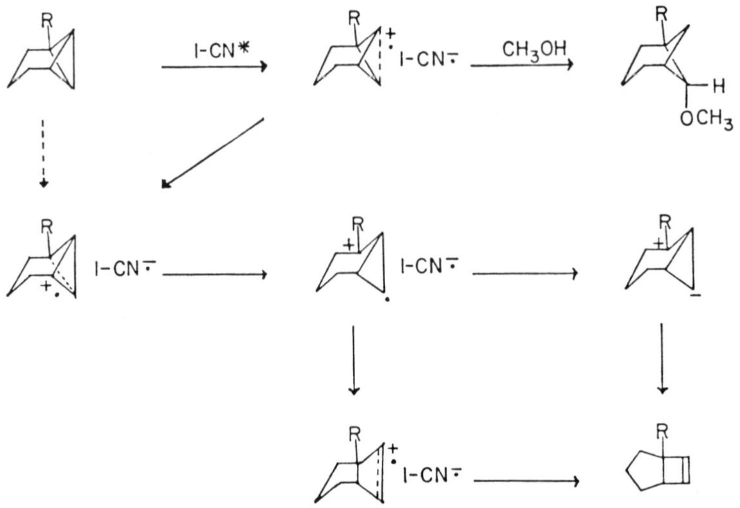

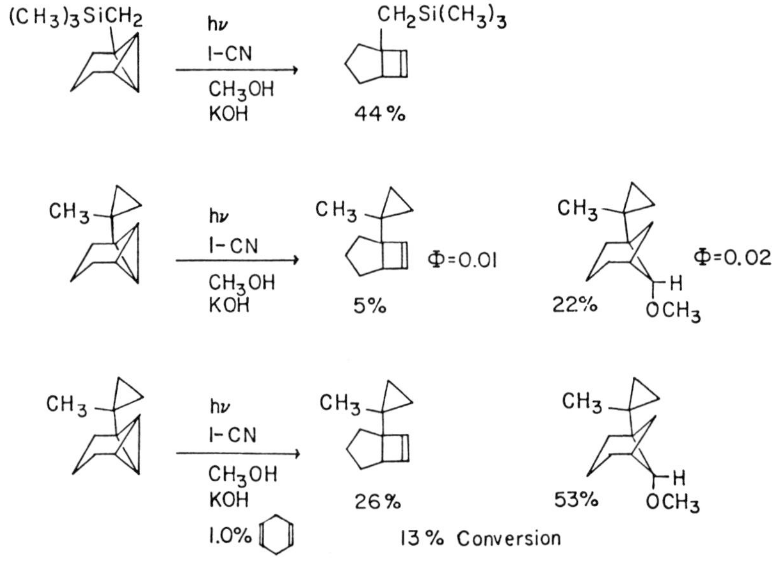

Obviously, these yields are in good agreement with the quantum yields observed above. When the irradiation was carried out in tetrahydrofuran, only the rearranged product was observed. Thus, an intermediate (or intermediates) is generated which permit a competition between rearrangement and nucleophile attack.

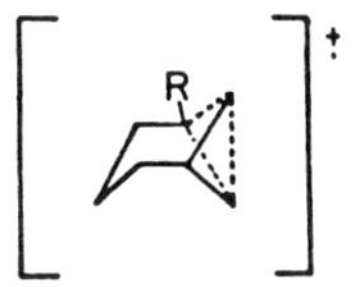

$$\left[ \begin{array}{c} R \\ \diagup\!\!\!\diagdown \end{array} \right]^{\ddagger}$$

Whether this involves two discreet intermediates or a single intermediate with a highly delocalized structure cannot be determined on the basis of present experimental data.

### 6. Photodesilylation

The last example of a single electron transfer promoted reaction which I want to discuss with you involves 1-trimethylsilyltricyclo[4.1.0.0$^{2,7}$]heptane and 1,7-ditrimethylsilylbicyclo[4.1.0.0$^{2,7}$]heptane. Initial experiments with the monotrimethylsilyl derivative gave two products. When 1-CN was used as a sensitizer, the

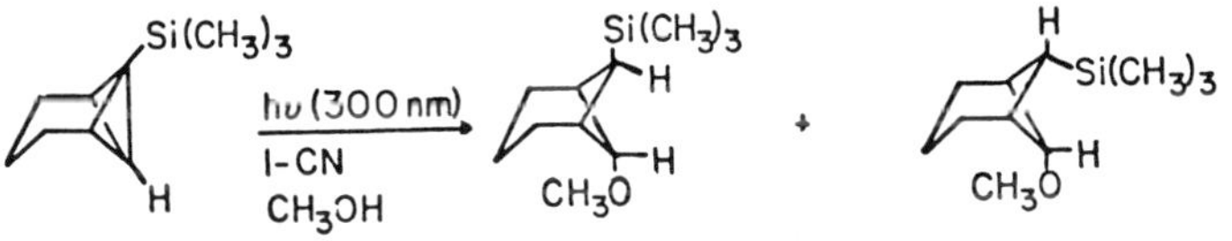

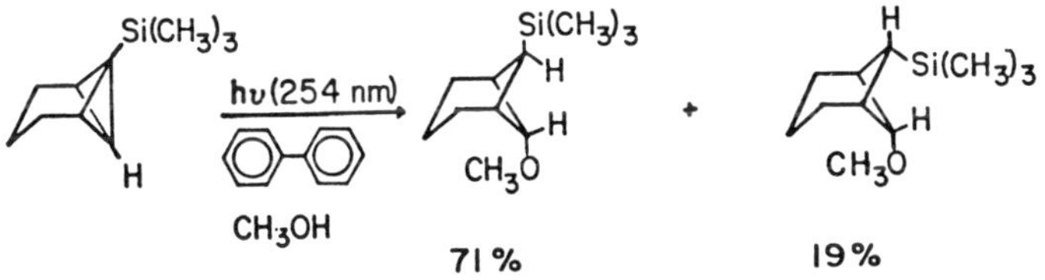

1-CN was consumed at a very rapid rate and very low yields of the products were observed. In contrast, when biphenyl was used as the electron acceptor photosensitizer, we observed 71% yield of the major product and 19% yield of the minor methanol adduct. These two compounds differed only in the stereochemistry of the trimethylsilyl group. When the ditrimethylsilyl derivative was irradiated using biphenyl (BP) as a photosensitizer, facile desilylation occurred to give the monotrimethylsilyl derivative with a quantum yield of 0.10. This, in turn, was carried on to give the two products previously observed from this intermediate with quantum yields of 0.18 and 0.04. Overall, we believe that a very complex process is involved. When the monotrimethylsilyl derivative was irradiated in the presence of biphenyl-$\underline{d}_{10}$ in completely deuterated methanol, we observed greater than 99%

incorporation of deuterium on the carbon to which the trimethylsilyl group was

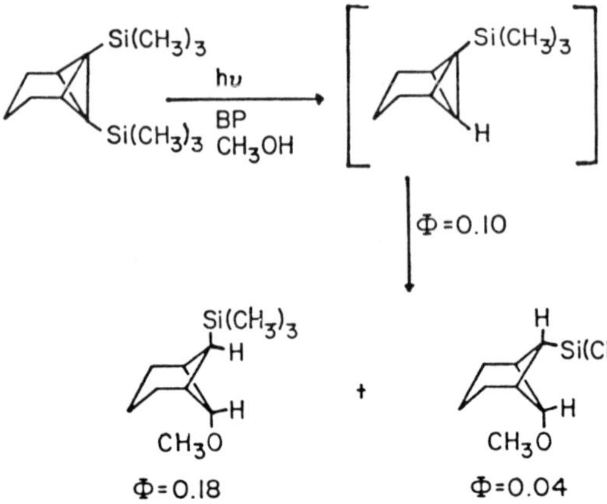

attached. The same was true with the analogous photochemistry of the ditrimethylsilyl derivative. The major surprise was associated with the incorporation of only

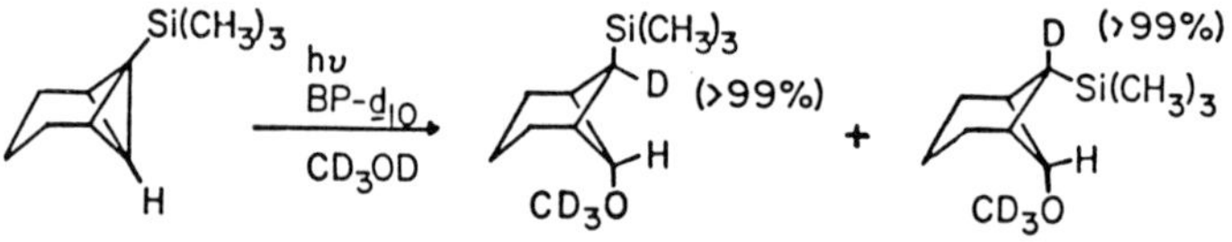

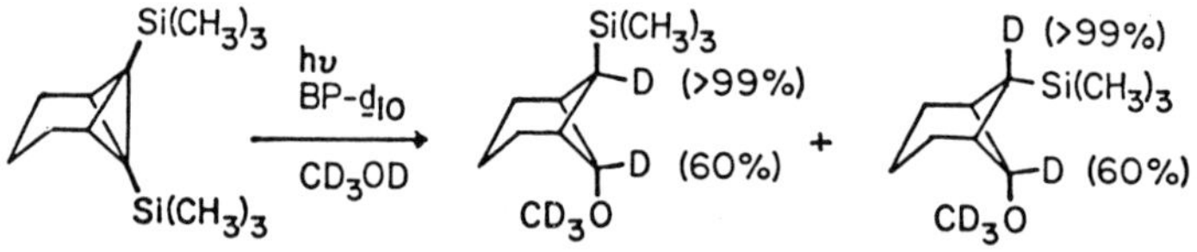

60% deuterium on the carbon to which the methoxyl group was attached. Experiments currently in progress indicate that this is a very complex photochemical process which involves a number of different reaction paths.

In summary, I need to thank a group of very excellent coworkers who were responsible for the experimental results which I have described to you today. Our initial studies in this area were carried out by Dr. Ryohei Yamaguchi, who is currently an assistant professor at Kyoto University. He was followed in these studies by Dr. Kurt Olson, who is currently a research chemist with Union Carbide. Kurt, in turn, was followed by Dr. Bruce Hay, who is currently a research chemist at Charles Pfizer & Co. Most recently, this area was studied by Dr. Steven Husebye, who has also joined the research effort at Union Carbide. This work was supported by a grant from the National Science Foundation and I wish to thank them for their support and you for your attention.

## 7. References

'The Ease of Oxidation of Highly Strained Polycyclic Molecules,' P. G. Gassman, R. Yamaguchi, and G. F. Koser *J. Org. Chem.* **1978**, *43*, 4392.

'The Electrochemical Oxidation of Strained Hydrocarbons,' P. G. Gassman and R. Yamaguchi *J. Am. Chem. Soc.* **1979**, *101*, 1308.

'Photoexcitation of Nonconjugated, Strained, Saturated Hydrocarbons. Relationship between Ease of Oxidation and Quenching of Naphthalene Fluorescence by Saturated Hydrocarbons,' P. G. Gassman, K. D. Olson, L. Walter, and R. Yamaguchi *J. Am. Chem. Soc.* **1981**, *103*, 4977.

'Electron Transfer from Highly Strained Polycyclic Molecules,' P. G. Gassman and R. Yamaguchi *Tetrahedron* **1982**, *38*, 1113.

'Photochemistry of Saturated Hydrocarbons. Mechanistic Changes as a Function of Methyl Substitution in the Photosensitized Reactions of the Tricyclo-[4.1.0.0$^{2,7}$]heptyl System,' P. G. Gassman and K. D. Olson *J. Am. Chem. Soc.* **1982**, *104*, 3740.

'Photoinitiated Electron Transfer Reactions. The Radical Cations of Bicyclo-[1.1.0]butane Derivatives,' H. D. Roth, M. L. M. Schilling, P. G. Gassman, and J. L. Smith *J. Am. Chem. Soc.* **1984**, *106*, 2711.

'Alkyl Group Migration in Photoinduced Cation Radical Reactions,' P. G. Gassman and B. A. Hay *J. Am. Chem. Soc.* **1985**, *107*, 4075.

'The Reaction of 1,2,2-Trimethylbicyclo[1.1.0]butane with Excited State 1-Cyanonaphthalene,' P. G. Gassman and G. T. Carroll *Tetrahedron* **1986**, *42*, 6201.

'The Synthesis and Photoinduced Electron-Transfer Promoted Isomerization of 7,7-Dimethyl-*trans*-bicyclo[4.1.0]hept-3-ene,' P. G. Gassman and K. Mlinarić-Majerski *J. Org. Chem.* **1986**, *51*, 2397.

'Mechanistic Insight into the Photoinduced Rearrangement of the Tricyclo-[4.1.0.0$^{2,7}$]heptyl Skeleton,' P. G. Gassman and B. A. Hay *J. Am. Chem. Soc.* **1986**, *108*, 4227.

# CHEMICAL BEHAVIOUR OF CATION RADICALS DERIVED FROM STRAINED MOLECULES IN SOLUTION. SPIRO-ACTIVATION IN THE CLEAVAGE OF SOME ARYLCYCLOPROPANES UPON TREATMENT WITH TCNE OR DDQ

Shinya Nishida
Department of Chemistry
Faculty of Science
Hokkaido University
Sapporo, Hokkaido 060
Japan

ABSTRACT. We have observed that certain cyclopropanes, activated by a fluorene group linked in a spiro fashion to the cyclopropane, and by efficient cation-stabilizing groups in a geminal manner, are highly reactive in the thermal reactions with TCNE or DDQ. The reaction with TCNE produces a five-membered [$\sigma 2+\pi 2$] adduct, whereas the product isolated in the reaction of methyl-substituted substrate with DDQ is a Diels-Alder adduct after ring-cleaving dehydrogenation. Since no such reactions take place in the reactions of the corresponding diphenyl derivatives, we believe that the spiro-activation is concerned with an initial SET. Either a planar biphenyl moiety or a constrained bisected structure in these arylcyclopropanes would lower the oxidation potential of the substrate to make the thermal SET energetically feasible. Recent results with regard to the spiro-activation are presented and discussed.

## 1. INTRODUCTION

Methodology to construct a five-membered ring is an interesting and important synthetic problem in preparative organic chemistry. One of the attractive strategies will be a cycloaddition between a C-3 unit with an ethylenic linkage. In 1970, Martini and Kampmeier (1) have realized this pathway by using a cyclopropane derivative as the C-3 component. They heated a mixture of 1,1-diphenylcyclopropane with TCNE at 125°C for several days to obtain a cyclopentane derivative as one of the reaction products. Relatively recently, we (2) uncovered the fact that some spiro-linked arylcyclopropanes undergo the same type of reaction very readily. Namely, 1,1-dicyclopropyl, spiro-activated cyclopropane reacts with TCNE at room temperature. Since the reaction is limited to occur in the spiro-linked substrates, we proposed that the reaction may involve a rate-controlling single electron transfer (SET) process at the very early stage of the reaction. The facts that photostimulation is effective in the reaction of not-spiro-activated substrates (diphenyl derivatives) and alkyl substituents on the three-membered ring exert substantial effects on the reaction rate support the conclusion. Successful trapping of the intermediate by methanol indicates the polar nature of the present reactions.

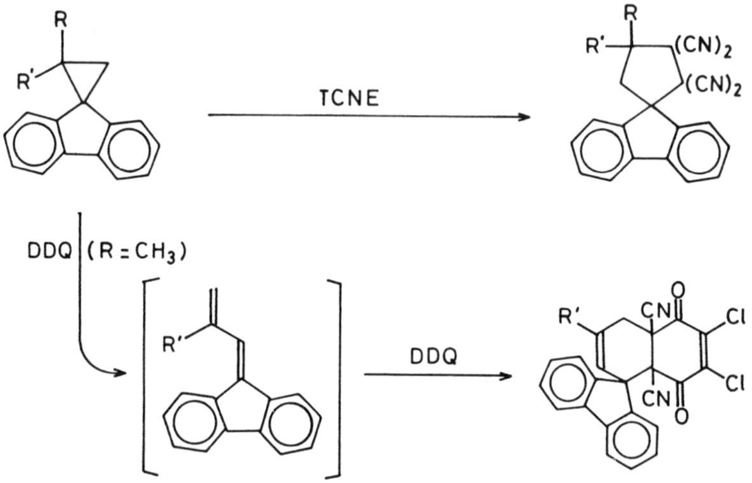

The spiro-activation exerts also a dramatic influence on the reaction of vinylcyclopropanes with TCNE. Previously, we (3) have demonstrated that most of the vinylcyclopropanes react with TCNE in a [$_\pi 2+_\pi 2$] manner to give a cyclobutane derivative, whereas certain tetrasubstituted compounds react with TCNE at their cyclopropane sigma bond in a [$_\sigma 2+_\pi 2$] fashion to produce a vinylcyclopentane derivative (type I reaction and type II reaction). Introduction of a spiro-linked fluorene group to one of the cyclopropane rings in di- or tricyclopropylethylenes, which are known to undergo the type I reaction, not only accelerates the reaction but also results in a total change of the reaction course from the type I to the type II. The results are consistent with the previous supposition that the type II reaction takes place via an SET (3).

In the present paper, we wish to present some additional results with regard to the spiro-activation and discuss the results in terms of the thermal SET initiation.

## 2. RESULTS AND DISCUSSION

### 2.1 Reaction of 1,2-Dicyclopropyl Derivatives with TCNE

In the type I reaction, we have noted that the effect of the substituents depends very much upon the way of the substitution pattern. Namely, one-sided substituents greatly enhance the reactivity of the substrate, whereas the evenly substituted compounds exhibit quite low reactivity. For example, 1,1-dicyclopropylethylene reacts with TCNE almost instantly at room temperature, but the reaction of 1,2-dicyclopropylethylene with TCNE requires several hours at 100°C in a polar solvent such as nitromethane. The same was true in the reactions of spiro-activated dicyclopropylcyclopropane. When the pendant cyclopropyls are substituted at a carbon of the central cyclopropane in a geminal manner, the $[_\sigma 2+_\pi 2]$ type cycloaddition completes almost instantly. However, the reaction of vicinal dicyclopropyl derivatives with TCNE requires several days at 100°C to realize the same type of the cycloaddition.

With regard to the cation-stabilizing ability of the pendant cyclopropyl group, we have good reasons to believe that the cyclopropyl group will be as effective as or more effective than the combined effects of two methyls (4). Accordingly, we expected that vic-dicyclopropyl derivatives may exhibit the reactivity similar to or greater than gem-dimethyl compounds. However, the reverse was true. Thus, the results appear to suggest that not only the electronic effect but also the steric effect are operating in the present reaction. The steric acceleration will be in accordance with the reaction scheme in which the SET process is considered as rate-controlling.

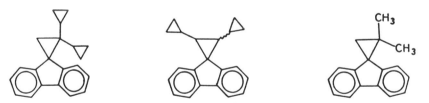

### 2.2 Reaction of 1-Cyclopropyl-1-methyldibenzo[d,f]spiro-[2.4]heptane: Dehydrogenation and Fragmentation

The reaction of the title compound with DDQ proceeded smoothly at room temperature. From the reaction mixture, we obtained not only the dehydrogenation-Diels-Alder product but also 9-methylenefluorene in 56% and 19% yield, respectively. The formation of the latter product (fragmentation product) is of considerable interest since it is a fragmentation product of the parent cyclopropane. If the fragmentation occurs also via the SET, the other half of the fragments will be a cation radical of cyclopropylethylidene (a carbene cation radical). Although we have made enormous efforts to uncover the fate of the $C_5$ fragment, we have been unsuccessful to isolate any products derived from it. In the methanol quenching experiments, the dehydrogenation was totally quenched but the fragmentation was not. Accordingly, we expected that methanol might trap the $C_5$ fragment, but the attempts were also fruitless.

If our supposition that the fragmentation is also taking place via the SET is correct, we should observe in the mass spectra that the peak corresponding to 9-methylenefluorene is dominant. In fact, a m/z = 178 peak ($C_{14}H_{10}^+$) is the base peak in the mass spectra of 1,1-dicyclopropyl and 1-cyclopropyl-1-methyl derivatives (Table 1). It appears that our supposition is substantiated, but the studies with the deuterated compounds showed that the situation is more complex. Namely, the 178 peak was still the base peak in the corresponding dideuterio compounds, although the m/z = 180 peak gained a considerable intensity. Accordingly, there must be at least two important fragmentation pathways to give $C_{14}H_{10}^+$. The first one will be the fragmentation to give methylenefluorene, but the second, more important pathway, in which the labeled cyclopropane methylene group is not incorporated, should exist. We postulate that the second pathway will involve an expansion of the pendant cyclopropyl group to form a cyclohexene derivative, which undergoes a retro-Diels-Alder reaction to afford the unlabeled $C_{14}H_{10}^+$ fragment. In fact, we frequently observed that the 178 peak was the base peak in many related compounds carrying a cyclopropyl group as a substituent at the C-1 position of the spiro-linked fluorene derivatives.

The mass spectral results are thus in accordance with the supposition that 9-methylenefluorene is produced in the fragmentation of the cation radical of the spiro-activated 1-cyclopropyl-1-methyl derivatives. In solution, the fragmentation <u>via</u> the cyclohexene derivative is of minor importance since the deuterated methylenefluorene obtained in the reaction of the ring-dideuterio substrate with DDQ retained practically the same deuterium content as that the

Table 1. Mass Spectral Data of Spiro-Activated Cyclopropanes and Their Deuterated Compounds: Relative Intensity of Some Representative Peaks at 70 eV.

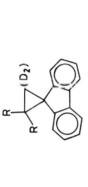

| R,R' | △,△ [a] | | △,CH₃ [b] | | △,CD₃ [c] | | CH₃,CH₃ [d] | |
|---|---|---|---|---|---|---|---|---|
| | $H_2$ | $D_2$ | $H_2$ | $D_2$ | $H_2$ | $D_2$ | $H_2$ | $D_2$ |
| m/z | | | | | | | | |
| 165 | 45 | 52 | 45 | 60 | 43 | | 65 | 43 |
| 178 | 100 | 100 | 100 | 100 | 100 | | 24[e] | 11[f] |
| 180 | ... | 40 | ... | 60 | 3 | | 1 | 21 |

[a] More than 99% deuterated.  [b] Ca. 88% deuterated.  [c] Ca. 82% deuterated.  [d] More than 99% deuterated.  [e] The base peak at m/z = 205.  [f] The base peak at m/z = 207.

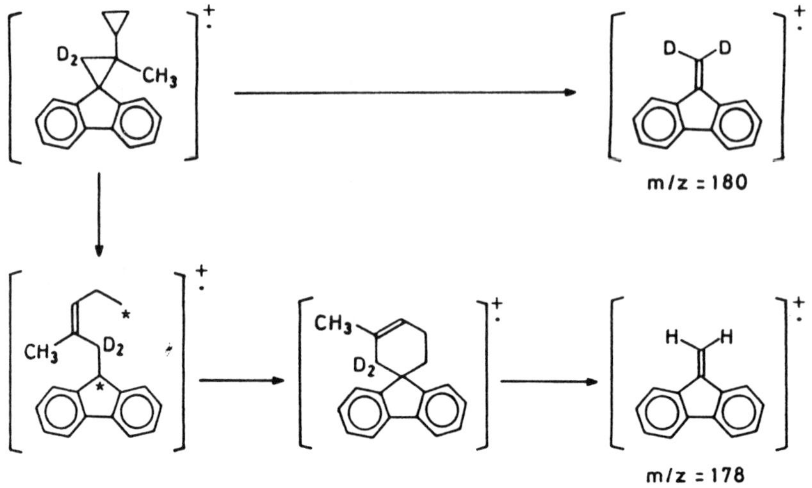

starting substance held.

9-Methylenenfluorene was obtained only in the reaction of the 1-cyclopropyl-1-methyl derivative. Reasons why 1,1-dicyclopropyl and 1,1-dimethyl derivatives did not produce 9-methylenefluorene in reasonable yields are unclear at present.

2.3 Reactions of Other Spiro-Linked Substrates with TCNE

In order to uncover the factors upon which the spiro-activation is based, we examined the reactions of spirocyclopropanes with less extended conjugating systems. Both spiro-indene and spiro-indane derivatives reacted with TCNE at room temperature, but they took totally different reaction pathways. Namely, the indene derivative reacts in a modified type I manner to give an extensively rearranged product whereas the indane derivative takes the $[_\sigma 2+_\pi 2]$ reaction pathway. Apparently, the monobenzo derivative exhibits the reactivity similar to that of the dibenzo derivative.

Further removal of the benzo group to a spiro[2.4]heptadiene derivative resulted in the occurence of the Diels-Alder reaction, as expected. The corresponding totally saturated compound exhibits no reactivity toward TCNE. Thus, the presence of a $\pi$ bond system is essentially required in the present reaction. This is in accordance with the SET initiation mechanism for these reactions. The electrochemical data and modes of the reaction examined here are summarized in Table 2.

Table 2. Oxidation Potential and Reaction Mode of Some Representative Spiro-Activated Cyclopropanes

| | | | | | |
|---|---|---|---|---|---|
| $E_p^{ox}$ (V)[a] (vs SCE) | 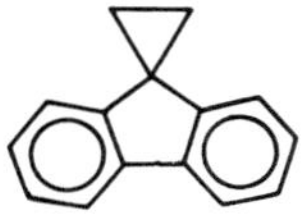 1.65 | 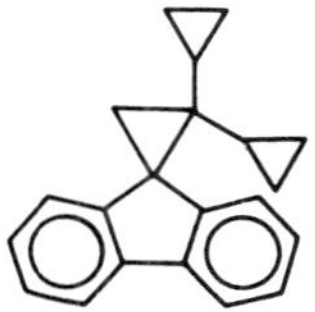 1.33 | 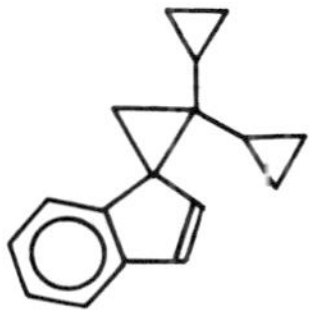 1.46 | 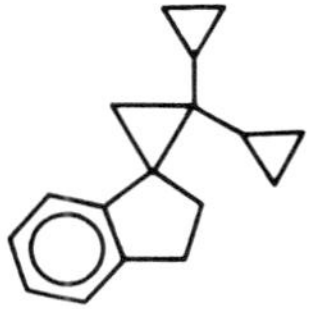 1.20 |  1.41 |  1.67 |
| $\Delta G_{ET}$ (kcal/mol)[b] | 33 | 26 | 29 | 23 | 27 | 33 |
| Reaction with TCNE | ..... | Type II | Type I with rearr. | Type II | Diels-Alder | no reaction |

[a] The oxidation was irreversible and the $E_p^{ox}$ value varied with the sweep rate. Apparently, some follow-up reactions disturbed the oxidation wave. However, the $E_p^{ox}$ value listed above was reproducible within ± 0.03 V under a standard set of conditions with the sweep rate of 400 mV/sec.

[b] An approximate free energy change calculated by: $\Delta G_{ET} = 23[E^{ox} - E^{red}]$.

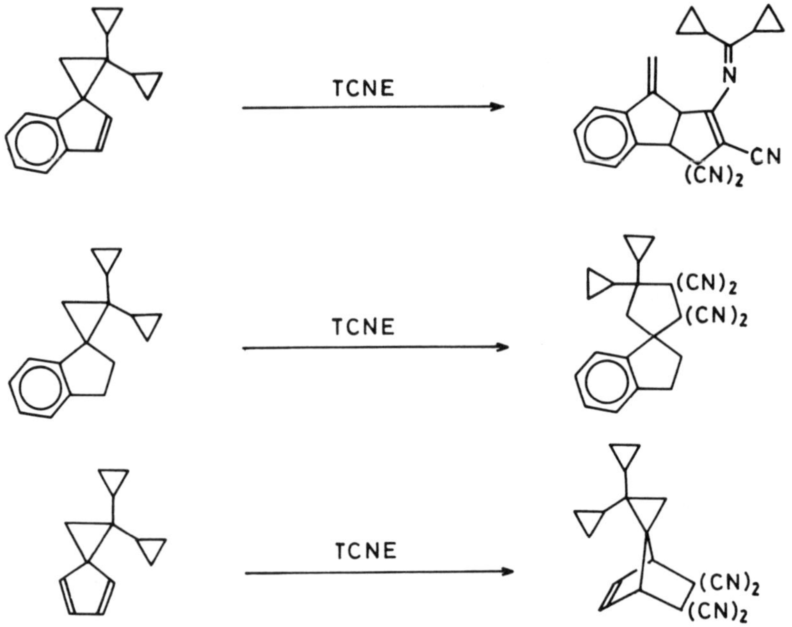

## 3. CONCLUSION

All results presented here support the previous conclusion that these reactions are initiated by rate-controlling SET. The spiro-activation is concerned with the ready oxidizability of the substrate in contact with strongly electron-demanding TCNE or DDQ. It might be caused by either the presence of a planar biphenyl moiety or a fixed, bisected geometry of the spiro-linked phenylcyclopropanes.

## 4. REFERENCES

(1) Martini, T.; Kampmeier, L. A. Angew. Chem., Int. Ed. Engl. 1970, 9, 236.
(2) Nishida, S.; Murakami, M.; Mizuno, T.; Tsuji, T.; Oda, H.; Shimizu, N. J. Org. Chem. 1984, 49, 3428. Murakami, M.; Tsuji, T.; Oda, H.; Nishida, S. Chem. Lett. 1987, 863. Nishida, S.; et al. submitted to J. Org. Chem.
(3) Nishida, S.; Moritani, I.; Teraji, T. J. Chem. Soc., Chem. Commun. 1970, 501; Ibid. 1971, 36; J. Org. Chem. 1973, 38, 1878. Kataoka, F.; Nishida, S. Chem. Lett. 1980, 1115. Tsuji, T.; Nishida, S. Acc. Chem. Res. 1984, 17, 56.
(4) See for example: Saunders, M.; Rosenfeld, J. J. Am. Chem. Soc. 1970, 92, 2548. Brown, H. C.; Gnedin, B. G.; Takeuchi, K.; Peters, E. N. Ibid. 1975, 97, 610. Tidwell, T. T. In "The Chemistry of the Cyclopropyl Group", Rappoport, Z., Ed.; Wiley: New York, 1988; Part 1, Chapter 10.

# THE CYCLOPROPYL GROUP IN STUDIES OF ENZYME-CATALYSED REACTIONS

C.J. Suckling, University of Strathclyde, Department of Pure
and Applied Chemistry, 295, Cathedral St., Glasgow G1 1XL
Scotland, U.K.

ABSTRACT. The cyclopropylcarbinyl radical ring opening reaction has been applied to investigate the mechanism of many enzyme-catalysed reactions especially oxidation-reduction reactions. The advantages and limitations of this approach reflect a balance between accessibility of cyclopropane-containing substrates to enzymes and the intrinsic reactivity of substituted cyclopropylcarbinyl radicals. When both factors are taken into account, penetrating insights into the mechanisms of enzyme-catalysed reactions can be obtained by studies of the products of enzyme-catalysed reactions on cyclopropane-containing substrates.

## 1. INTRODUCTION: PROBES OF ENZYME MECHANISMS

In common with many reactions involving catalysts, the investigation of the mechanism of enzyme-catalysed reactions poses distinctive problems. Normally sharp probes of reaction mechanism are blunted by the fact that the reaction takes place enclosed within the active site of the enzyme, a region that is shielded from the open-environment of the solvent by association/dissociation equilibria. This paper discusses the use of substrate analogues containing the cyclopropyl group as probes of the mechanism of action of enzymes. It is one approach by which the mechanism disguised by the binding steps can be unmasked, especially in oxidation-reduction reactions. The cyclopropyl group must be thoughtfully applied, and before discussing examples of its use, it will be valuable to consider alternative approaches to the determination of mechanisms of enzyme-catalysed reactions using probe molecules and the necessary background of physical organic chemistry.

In many enzyme-catalysed reactions association/dissociation equilibria are rate determining, or at least partially rate determining. Hence normal kinetic approaches are compromised because the step in which the substrate undergoes a transformation of structure into an intermediate or product may be inaccessible to kinetics. In all cases, the mechanism of the reaction can be looked at from two points of view, that of the enzyme, and that of the substrate. From the point of view of the enzyme, it is important to be able to describe the structure of the catalytic site in atomic terms, to identify those groups that are involved in transforming the substrate, and finally to account for the function of the whole protein in catalysing the reaction. The most widely accepted current view is that the catalytic action of enzymes is due to the stabilisation of transition states along the catalysed reaction path [1]. These transition states connect the behaviour of the enzyme with the intrinsic chemical properties of the substrate. As an adjunct to studies with the enzymes themselves, model studies of related non-enzymic reactions have often been undertaken to define the intrinsic chemical reactivity of the substrate. Such studies are valuable to provide a mechanistic framework within which the enzyme catalysed reaction can be fitted [2]. It is never a trivial matter to establish the correspondence between a model reaction and its enzymic prototype especially since it is often necessary to use modified substrates in order to obtain a sufficiently rapid reaction for measurement. Unfortunately in modifying the substrate it is easy to introduce changes into the substrate to modify the mechanism of the reaction and hence invalidate the comparison. Examples of this will be discussed later.

In a reaction away from an active site, an intermediate may be detectable by kinetics or by spectroscopy. However at the active site, the lifetime of many intermediates is likely to be so short that even fast spectroscopic methods will have limited use at normal temperatures and kinetic methods are compromised as we have seen. There is thus a need for methods that can report the nature of an intermediate on the reaction path without the intervention of spectroscopic of kinetic methods. Typically such methods involve reporter groups introduced into normal substrates for the enzyme.

Reporter groups were introduced many years ago particularly in the context of chromophoric or fluorophoric groups. For example, spectrofluorimetric properties could reveal the environment of an active site, and in favourable circumstances, could allow distances between groups in proteins to be measured [3]. Similarly, spin probes gave information on the dynamics of compounds at the active sites of enzymes [4]. Although much useful information has come from such studies, there has always been the drawback that the probes are large groups that will themselves determine at least part of the enzyme-substrate interaction perhaps in so doing disguising an important mechanistic feature. Also such studies tend to be static since the probes usually only report the initial substrate binding step. Recently, advances in nmr technology have made it possible to use isotopic substitution as a mechanistic probe with both substrates and substrate analogues; in the case of substrate analogues the formation of intermediate analogues could be observed [5] and at lower temperatures, the progress of reactions can be monitored [6]. Although low temperature studies (cryoenzymology) extend the use of nmr probes, it is necessary to be sure that the kinetics of the enzyme-catalysed reaction have not been severely altered by the temperature or the use of the solvent mixture required for sub zero degree studies.

In many cases, the putative intermediates of interest have lifetimes too short to permit detection by a spectroscopic method. It is necessary to devise a probe molecule with a particularly rapid response. This is the field in which the cyclopropyl group comes into its own [7]. Shortly after Ingold demonstrated that the rate of ring opening of the cyclopropylmethyl radical exceeded $10^8$ $s^{-1}$ at 25°C [8] it was realised that this reaction offered an excellent opportunity to probe for radical intermediates in enzyme-catalysed reactions, intermediates that would be expected to be very reactive and to have short lifetimes. Naturally, radical intermediates had been proposed in many biological oxidation-reduction reactions and the chemistry of the cyclopropyl group made it possible to investigate them from a new direction. A further attraction of the cyclopropyl group, of course, is its small size. This makes it reasonable to expect that a cyclopropyl-modified substrate will be still accepted by the enzyme: unlike the large spectroscopic probes mentioned earlier, the substitution of a cyclopropane will not dominate the interaction with an enzyme.

For a modified substrate, even bearing such a small change as a cyclopropane, it is still important to establish how the modified substrate binds to the enzyme. The normal kinetic measurements to establish substrate properties by measuring $K_m$ and $v_{max}$ should be carried out and the competition between the normal substrate and a modified substrate measured, if possible. The latter experiment can only be done if an established substrate responds to a different analysis technique from that to which the cyclopropane-containing compound responds as may be the case if the 'normal' substrate is a specially designed chromophoric or fluorophoric molecule and the cyclopropane containing compound lacks these features. Whatever the measurements possible, it is important that some efforts are made to characterise the enzyme-probe interaction. Without this work, the experiments are little better than an uncontrolled model study. However in unusual situations, it is possible that the probe molecule will reveal alternative, and hitherto unknown chemistry that the enzyme can catalyse. When this occurs, the bonus results offer a new dimension to the understanding of the behaviour of the protein. What is not acceptable, however, is to extrapolate the mechanism observed with a probe molecule uncritically to that of the normal substrate reaction.

Another requirement is also mandatory. It is necessary to know that the cyclopropane-containing compound is a kinetically competent probe. In other words, one must have some idea of the rate of the step to be probed and know that the rate of ring opening of the expected intermediate cyclopropylalkyl radical is faster than the rate of the critical step. This is particularly important because it is known that the rate of ring opening of cyclopropylalkyl radicals is very sensitive to substituents on the radical bearing carbon atom (see below). Further, since the ring opening reaction is reversible, it is important to choose a probe in which the equilibrium position lies in a direction that will permit meaningful conclusions. Finally, it is desirable to have a view of the way in which the probe molecule interacts at the enzyme's active site since it is possible that the topology of the enzyme may prevent the ring opening of the probe. This is equivalent to saying that the position of equilibrium at the active site favours the closed form even though a radical intermediate may be involved. One subtle way in which such enzymic control could be exerted is through a stereoelectronic effect. It is well known that enzymes can control the direction of a reaction by imposing a particular conformation upon the reacting molecule [9]. Cyclopropylalkyl radical opening too is known to be under stereoelectronic control as has been shown in the steroid series [10] and this too must be considered where possible in interpreting an enzyme catalysed reaction's mechanism. Happily, in many cases, most of these criteria can be satisfied and reasonable conclusions can be drawn as will be discussed below.

The cyclopropyl probe has been applied to studies of many enzyme-catalysed reactions and their models over the last 10 years. In this paper, they will be discussed in increasing order of molecular complexity of the probe molecule beginning with molecules bearing only alkyl substituents, progressing through cyclopropylmethanols and cyclopropyl aldehydes and ketones, cyclopropylamines and cyclopropanols, and concluding with $\alpha,\alpha$-difunctional derivatives (aminoacids). This survey will encompass monooxygenases (cytochrome P-450, dopamine $\beta$-hydroxylase and lipoxygenases), NAD-dependent dehydrogenases, thiamine pyrophosphate-dependent enzymes, monoamine oxidase, PQQ-dependent enzymes, ethylene biosynthesis, coenzyme $B_{12}$ mediated rearrangements, and penicillin biosynthesis following the same sequence.

## 2. STRUCTURAL EFFECTS ON THE RING OPENING OF CYCLOPROPYLALKYL RADICALS

### 2.1 Substituent Effects

Only rarely are the substrates of enzyme-catalysed reactions alkanes that correspond to Ingold's seminal observation [8] and hence it is important to measure the rates of ring opening of substituted cyclopropylalkyl radicals to calibrate the probe system. The reaction conditions required for measurement are typically in organic solvents using photochemical or thermal initiation methods; the probe molecule used for calibration will often bear protecting groups to direct the radical-forming step in the appropriate direction, substituents that will not be present in the probe molecule used for the enzymic experiments themselves. Accepting these limitations, it is possible to proceed.

Consider firstly the effect of the substitution of electron donating and electron withdrawing groups on the carbon at which the

radical is formed (Table 1) [11]. Although alkoxy substituents have
a relatively small effect, about a 10 fold reduction in rate, the
reaction is still very fast 3, 4. On the other hand, the introduction
of a conjugating electron withdrawing substituent causes a substantial reduction in the rate of ring opening. When both are present,
the ring opening reaction has essentially been halted at room temperature 8, 9. Although esr measurements failed to detect a ring
opened radical in the case of cyclopropaneglyoxylic acid derivatives
8, 9, product studies showed that the ketoester was reduced in the
presence of an initiator with ring opening [12]. The use of such
compounds to probe the mechanisms of enzyme-catalysed reactions at
room temperature or thereby is clearly not a strong test although in
model reactions, they may still be valuable.

Measurements have also been made on some ring substituted
cyclopropylalcohols (5, 6) and on some bicyclic cyclopropanes (10).
In both cases, the rate of ring opening is measurably increased;
such compounds are perhaps the most sensitive cyclopropane-
containing probes available. Two contributions to the increased
rates of ring opening can be considered. Firstly, the substituted
rings have greater intrinsic strain because of the non-bonded
interactions between the substituents. Secondly, the derived
homoallyl radicals are themselves more highly substituted and
correspondingly more stable than the reactant radical. For example,
in 6 ring opening affords a tertiary radical from a secondary radical
and in 12, a secondary benzylic radical is formed. Heteroanalogues
of the cyclopropylalkyl radical, the cyclopropylammonium and
-oxonium radicals also have a significant enzyme chemistry. As with
ring substitution, rates of ring opening are substantially increased
[11b].

2.2 Equilibria in radical reactions

One of the reasons for the comparative neglect of radical chemistry
was no doubt the view that complex organic molecules could not react
cleanly by radical pathways. Insufficient attention to the rates of
reactions was certainly a contributing factor but also a failure to
realise that radical reactions can exist as equilibria played a part.
It is important that even the principal probe reaction of this paper,
the cyclopropylalkyl radical ring opening, is an equilibrium reaction
in which the ring opened position is heavily favoured [13]. As was
hinted earlier, this equilibrium position is sensitive to the structure
of the molecule. Thus a bicyclic system such as nortricyclane 11
forces the reacting partners to remain in close proximity; consequently the equilibrium favours the closed ring.

2.3 Alternative Probes to Cyclopropane-containing Compounds

Cyclopropanes are not the only molecules to be sensitive to the
induction of radicals in their vicinity. As was mentioned above,
radical cyclisations can also take place and these have been
considered as a mechanistic probe for enzyme catalysed reactions

Table 1    Stability of substituted cyclopropylmethyl radicals and heteroanalogues measured by esr spectroscopy.

| # | Radical structure | k s$^{-1}$ | Comment | Reference |
|---|---|---|---|---|
| 1 | 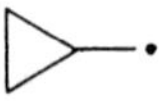 cyclopropylmethyl• | $2.2 \times 10^8$ | | 8 |
| 2 | 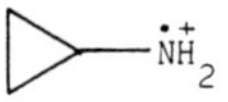 cyclopropyl-$\overset{+}{\text{N}}\text{H}_2$• | – | only ring opened radical observed down to $-193°C$ | 11b |
| 3 | 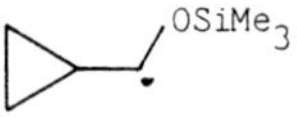 cyclopropyl-CH(•)-OSiMe$_3$ | $1.7 \times 10^7$ | | 11a |
| 4 | 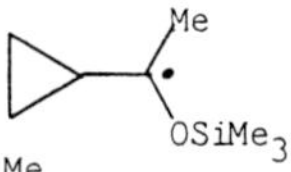 cyclopropyl-C(•)(Me)-OSiMe$_3$ | $2.4 \times 10^7$ | | 11a |
| 5 | 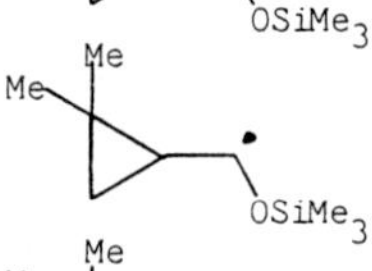 Me,Me-cyclopropyl-CH(•)-OSiMe$_3$ | – | only ring opened radical observed down to $-153°C$ | 11c |
| 6 | 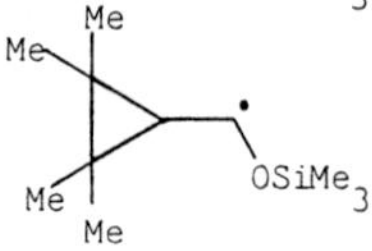 Me,Me,Me,Me-cyclopropyl-CH(•)-OSiMe$_3$ | – | only ring opened radical observed down to $-153°C$ | 11c |
| 7 | 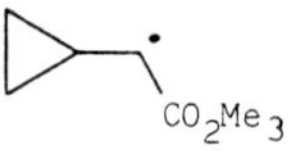 cyclopropyl-CH(•)-CO$_2$Me | – | ring opening not detected up to $7°C$ | |
| 8 | 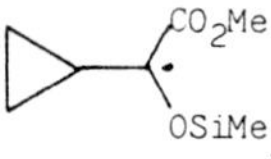 cyclopropyl-C(•)(CO$_2$Me)-OSiMe$_3$ | – | ring opening not detected up to $110°C$ | 11c |
| 9 | 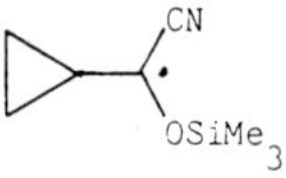 cyclopropyl-C(•)(CN)-OSiMe$_3$ | – | ring opening not detected up to $32°C$ | 11c |
| 10 | 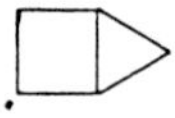 bicyclic radical | $> 10^9$ | | 11d |
| 11 | 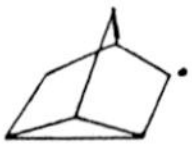 norbornyl-type radical | | equilibrium unfavourable to ring opening | 13 |
| 12 | 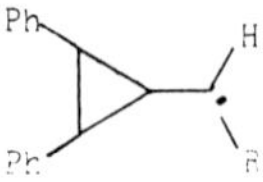 Ph,Ph-cyclopropyl-CH(•)-H, R | $\sim 10^{10}$ | | 11e |

[14]. Their rate constants are also large but a cyclisation reaction demands more conformational freedom at the enzyme's active site than a ring opening or fragmentation [15]. Radical isomerisations also offer possibilities [14,16]. For example, the isomerisation of a cis alkene to a thermodynamically favoured trans alkene via a radical intermediate could serve as a probe for radicals. A class of compounds that undergoes fragmentation as a response to radical formation is α-haloketones through the ketyl radical anion [17]. Particular care must be applied when using this probe, however, because the substitution by the halogen atoms may change the redox potential of the ketone, a fact that could influence the mechanism of a reaction. In other words, the easier reduction permitted by the haloketone may promote a single electron transfer path. The fastest rate that such a probe is likely to respond to is at least $10^5$ $s^{-1}$ [17]. To achieve the greatest possible rigour, all of these probes must also satisfy the criteria discussed in section 1 for the probe molecule to be an acceptable substitute for the substrate. In particular, because α-haloketones are potent alkylating agents, it is essential to demonstrate that the enzyme itself has not been deleteriously modified by alkylation. For these reasons, the cyclopropyl probe has been found to have the widest applicability to the study of enzyme-catalysed reactions.

## 3. STUDIES OF OXYGENASE MECHANISMS

### 3.1 Cytochrome P-450

One of the first examples of the use of the cyclopropyl probe in the study of a mechanism related to enzymic catalysis was by Groves as part of his studies of model systems for cytochromes P-450 [16]. At that time, the nature of oxidising species coordinated to iron and to other metals in haem complexes was not understood. Groves found that the manganese(III) complex of tetraphenylporphyrin oxidised bicyclo[4.1.0]heptan in the presence of iodosobenzene [16]. The product mixture contained 32% ring opened products but the cyclopropane ring was intact in the major part. The result can be understood if the oxidation reaction proceeds via hydrogen abstraction followed by radical recombination in what has become known as the 'oxygen rebound' mechanism (Figure 1). The proportion of products will depend upon the rate of the rebound step in comparison with the rate of ring opening of the cyclopropylalkyl radical (Table 1); clearly in this case, the rates are comparable. This study provided a chemical background in which the mechanism of the enzyme-catalysed reaction could be understood.

The first indications of a radical mechanism from a cyclopropane derivative for cytochrome P-450 came with a study on the oxidation of cyclopropylbenzylamines [18]. The inhibition of cytochrome P-450 by these compounds was interpreted in terms of electron transfer from nitrogen to the iron oxoporphyrin complex leading to an aminium radical cation that rapidly underwent ring opening as has now been

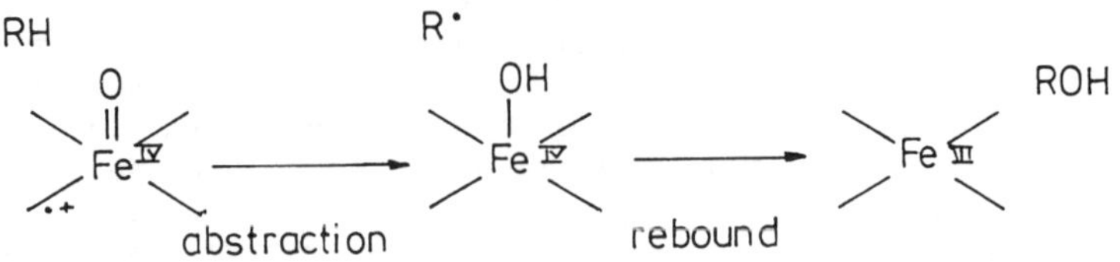

Figure 1  The oxygen-rebound step in reactions catalysed by cytochrome P-450

demonstrated spectroscopically (Table 1) [11b]. Inhibition probably takes place by the reaction of the ring opened radical with the porphyrin ring. This, together with results from other probes [19] strongly indicated that a radical pathway was open to the enzyme. Redox reactions are, however, notoriously sensitive in mechanism to the redox potential of the substrate. Indeed, it has been argued that oxidation by cytochrome P-450 with various substrates is initiated by attack at the most oxidisable atom [20].

Cytochromes P-450 include many enzymes, the non-specific liver enzymes studied above and some highly specific ones that are involved in steroid biosynthesis. In one of the first examples of the use of cyclopropyl probes in the study of such specific enzymes, we looked at cholesterol 7α-hydroxylase using 5,6-methanocholesterol as the probe 13 [21] (Figure 2). To ensure that there was no

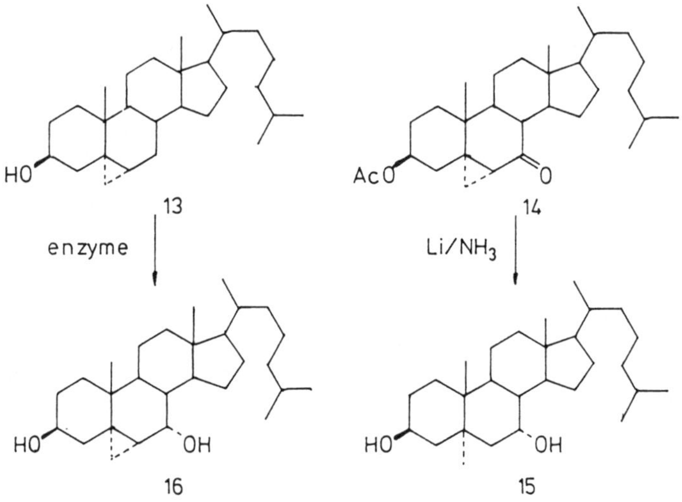

Figure 2  Probes of the mechanism of action of cholesterol 7α-hydroxylase

impediment to the ring opening reaction, we reduced the corresponding 7-ketosteroid 14 with lithium in ammonia following

Dauben [10]. The diol 15 was obtained quantitatively. However, in
the enzyme-catalysed oxidation, the only identifiable product was the
unopened diol 16; the cyclopropyl probe 13 was an adequate substrate and enzyme inhibition was not observed. In the absence of
the results described above, it would be tempting to suggest that
this result implied that an insertion mechanism operated. However
this enzyme, in common with most steroid metabolising enzymes, has
very precise substrate requirements, even minor alterations of the
structure of the side-chain causing major changes in rate of
oxidation. Hence it is more likely that cholesterol 7α-hydroxylase
holds the substrate so tightly that there is no time for the cyclopropane ring to open before the rebound step occurs.

Ortiz de Montellano has obtained results consistent with this
interpretation using a non-specific cytochrome P-450 [22]. He found
that cyclopropylmethane was oxidised without ring opening; this
implies that the rate of the rebound step must be at least $10^8$ $s^{-1}$.
However, when the more reactive substrate, bicyclo[2.1.0]pentane was
used, a degree of ring opening was detected together with a small
proportion of inactivation of the enzyme. Clearly the rates of
rebound and ring opening were comparable in this reaction which
provides an estimate of the rate of rebound as $10^9$ $s^{-1}$. It is also
interesting that enzyme inhibition was only observed when a cyclopropylalkyl ring opening occurred. Enzyme inhibition often
accompanies ring opening of cyclopropyl alkyl probes.

The above studies of the mechanism of cytochrome P-450 concern
aliphatic hydroxylation. Such enzymes are also capable of hydroxylating suitable aromatic substrates and in an attempt to study the
mechanism of this reaction, we investigated the oxidation of arylcyclopropanes [23]. Cyclopropylbenzene was a good substrate for
rat liver cytochrome P-450 but the only detectable product of
oxidation was benzoic acid. Neither cyclopropyl- nor alkylphenols
were detectable. This unexpected reaction was also observed in
model reactions including Fenton's reagent and Groves' system [16].
Such an oxidation, of course, cleaves a C-2 fragment from the probe
and in an attempt to discover the fate of the remaining carbon atoms,
we used 1,2-diphenylcyclopropane as a substrate. With this
molecule, which was a good competitive substrate for ethoxycoumarin,
the normal assay substrate, benzoic acid was again detected, but also
phenylacetaldehyde. A possible explanation of this reaction involves
an addition of the active oxygen complex to the cyclopropane ring in
a manner analogous to that prescribed for alkene epoxidation followed
by further oxidation and cleavage of the ring opened substrate
(Figure 3a). This reaction does not involve a cyclopropylalkyl
radical and no inhibition was observed in accordance with the above
generalisation.

Oxidation of quadricyclane, a bicyclopropyl compound (Figure 3b)
has also been observed to occur with net addition to the cyclopropyl
ring [24]. A mechanism involving initial electron transfer was
proposed. The absence of ring opening in many cases of cytochrome
P-450 and its models has encouraged the search for more sensitive
probes. Although the situation for aliphatic hydroxylation has been

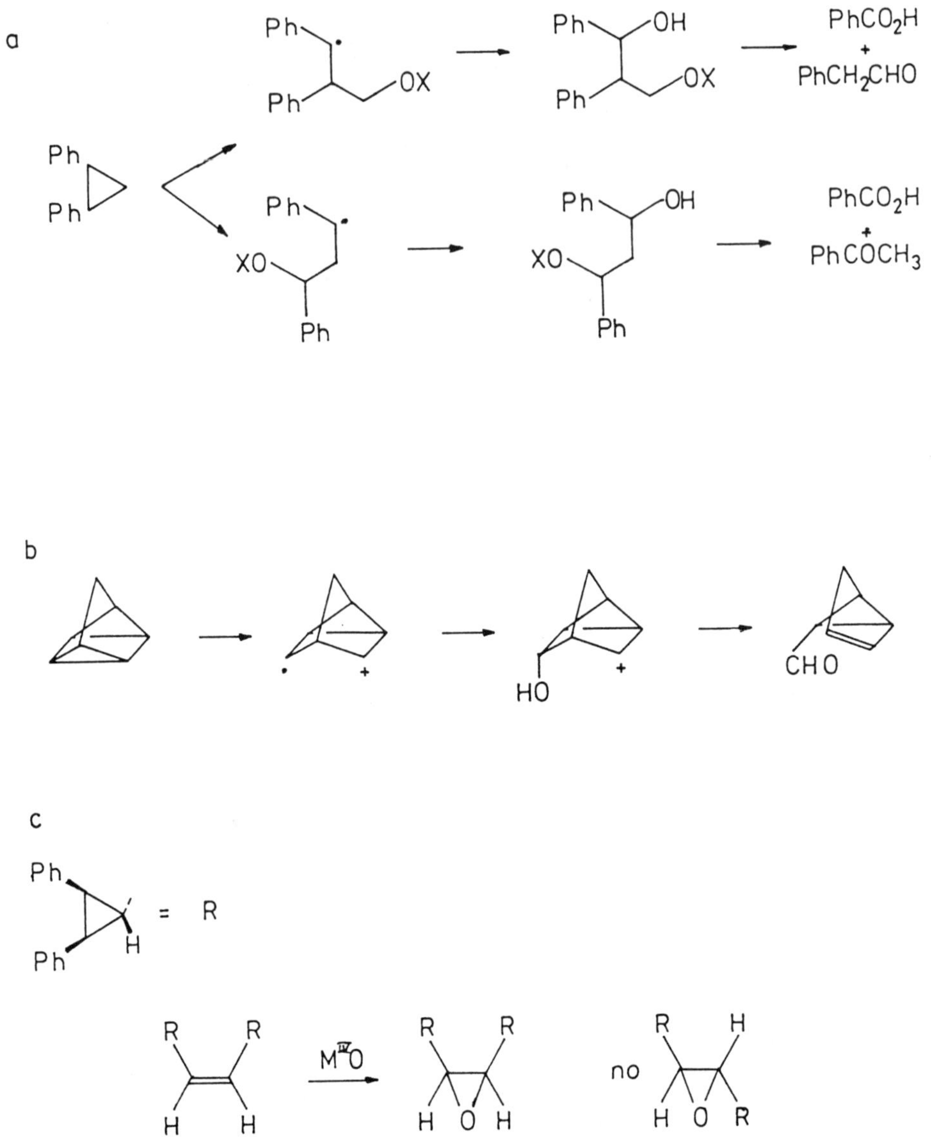

XO = Cytochrome P-450 bearing an active oxygen atom.

MO = Oxygen bearing tetrakis(pentafluorophenyl)porphinatoion(III)-chloride - the oxygen source was pentafluoroiodoso benzene.

Figure 3   Reactions of highly substituted cyclopropanes with cytochrome P-450 and model systems.

clarified by the work of Ortiz de Montellano [24], alkene epoxidation, another characteristic reaction of cytochrome P-450, has not yet been definitively studied. The closest approach has been by Bruice [11e] who investigated the oxidation reaction shown in Figure 3c; This study is based upon the highly reactive probe 12. Ring opening was not detected in this reaction suggesting that direct addition to the alkene does not occur. However, a radical cation intermediate was considered a possibility. It will be interesting to see how this probe behaves in reactions with cytochrome P-450.

3.2 Other Monoxygenases

The cyclopropyl probe has also been used in studies of monoxygenases using different systems to activate oxygen. For example the enzyme from *Methylococcus capsulatus* oxidised cyclopropane to cyclopropanol and methylcyclopropane to cyclopropane methanol [25] and no enzyme inhibition was reported. Unlike cytochrome P-450, this oxidase failed to oxidise 1-methyl-1-phenylcyclopropane but toluene and ethyl benzene were oxidised to afford a mixture of the 1-phenylalcohols and 4-substituted phenols. If a radical mechanism is appropriate for this enzyme, a rapid radical recombination step must also occur. However the occurrence of an NIH shift suggested that addition to C=C as well as insertion into C-H is possible with this enzyme. In contrast to this reaction, the oxidation of 4-cyclopropylmethylphenol by dopamine β-hydroxylase leads to inhibition analogous to that observed with 4-allylphenol [26]. Some hydroxylation was observed for both substrate analogues. Recently experiments have been undertaken that use the cyclopropyl probe to investigate enzymes that have not been isolated and characterised, for example in the bacterial decomposition of phosphonates [27]. Oxidation of cyclopropylmethyl organophosphonates led to partial ring opening.

3.3 Lipoxygenases

Lipoxygenases are of key importance in the biosynthesis of prostaglandins and leucotrienes from polyunsaturated fatty acids. The inhibition by alkynes has been extensively studied and the reactions can be interpreted to follow radical mechanisms in which an alkynyl radical undergoes rearrangement to an allyl radical with which the enzyme reacts [28a] (Figure 4a). The possibility that weakly bonded organoiron intermediates might be involved has been considered and one approach to this problem used a cyclopropylalkyl probe 17 [28b]. The probe 17 inhibited soybean lipoxygenase whilst undergoing conversion into the intact product 18 together with small quantitites of the ring-opened products 19 and 20. The authors argued that this result supports an organoiron intermediate for which other information was available [28b]. However, even if radicals long enough lived to undergo ring opening are not on the main reaction pathway, such results clearly indicate a radical mechanism from both the structures of the products and the observation of enzyme inhibition. It is doubtful that the probe **17** is an especially reactive compound with respect to ring opening because the putative intermediate radical is doubly allylically stabilised whereas the product radical is only singly allylically stabilised. Nevertheless, it is possible that a weak organoiron intermediate cleaves to form a radical pair from which the ring-opened products are derived; such a mechanism has been discussed in the context of β-lactam antibiotic biosynthesis (see below).

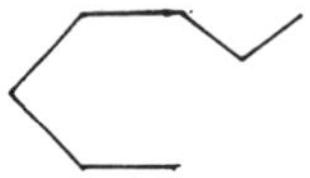

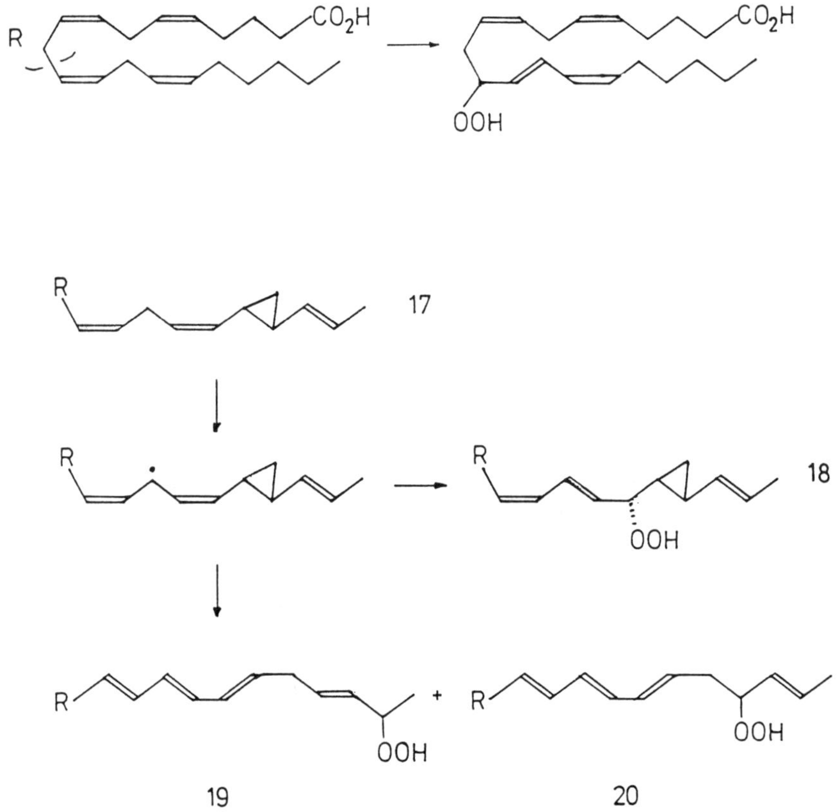

Figure 4   Lipoxygenase inhibition and mechanisms.

## 4. NAD Dependent Dehydrogenases

### 4.1 Model Studies

Just as in the case of cytochromes P-450, model studies have made a significant contribution to our understanding of the mechanisms of action of NAD-dependent dehydrogenases. Although the contribution of the enzymes to catalysis is critical and still not well understood in all its aspects, the chief question that cyclopropyl probes have helped to settle concerns the nature of the hydrogen transfer step. Since the thought provoking review of Hamilton in 1971 [29] the discussion has centred round single electron transfer mechanisms and direct hydrogen transfer reactions (Figure 5). That pyridinium salts and dihydropyridines are capable of reacting by either mechanism is indisputable. It has proved more difficult to model the oxidation by pyridinium salts than the reduction by dihydropyridines. However, alkoxide anions can be oxidised by pyridinium salts [30] and also undergo the common nucleophilic addition reaction. On the other hand, examples in which good one electron acceptors such as haloarylketones and halonitroalkanes undergo fragmentation of the

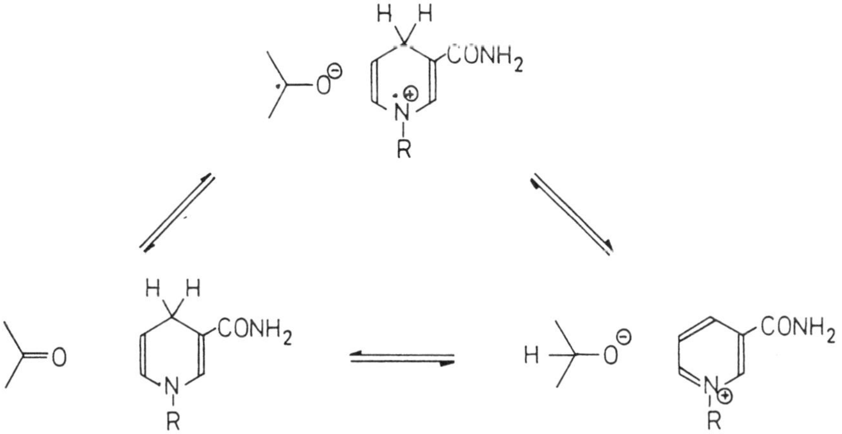

Figure 5    Pathways for oxidation/reduction by nicotinamides and dihydronicotinamides.

radical ions generated on interaction with model dihydropyridines have been known for many years [31]. But such compounds are not close relatives of typical natural substrates which are aldehydes and ketones (including aryl derivatives). The critical problem for model studies therefore, is to define the behaviour of compounds as closely related to the natural substrates as possible with respect to reduction by dihydropyridines.

Since simple aldehydes and ketones are too unreactive for such a reduction in the absence of the enzyme, a suitable cyclopropane-containing substrate must be designed. Following extensive

Japanese studies in which radical mechanisms had been advocated, Pandit's group [32] and our own [12] approached the design by including in the substrate an additional donor function that could coordinate with a catalytic Lewis acid cation analogous to the zinc of the active site of alcohol dehydrogenase (Figure 6). That such substrates were sensitive to ring opening under radical conditions was demonstrated by both groups using tri-n-butyltin hydride in the presence of an initiator at 80-100°C ring opened and dimeric products

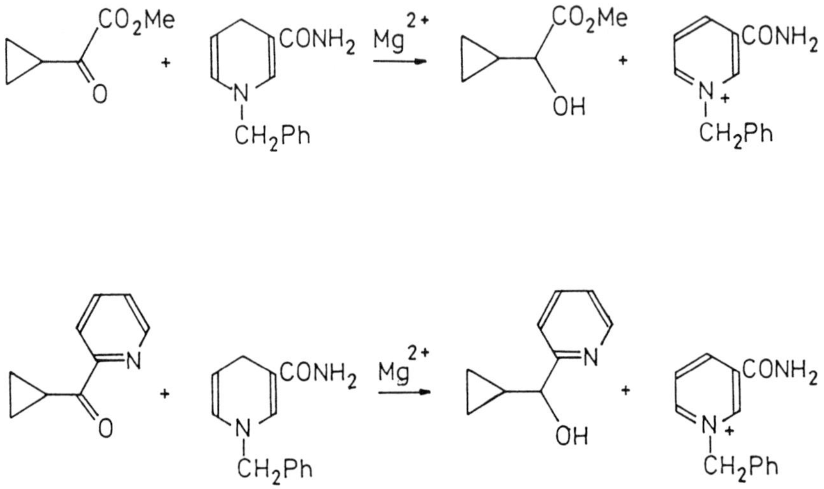

Figure 6    Model reactions for nicotinamide-dependent dehydrogenases.

characteristic of the ring opening of cyclopropylalkyl radicals was demonstrated. In contrast, when dihydropyridines, either Hantsch ester or N-benzyl-1,4-dihydronicotinamide, were used at a similar temperature, no ring opened products attributable to a radical mechanism were observed. Unless a remarkably tenacious dihydropyridine-metal substrate complex were formed in these reactions, these results strongly suggest that the reaction of dihydropyridines with simple ketones avoids radical paths. Indeed, one of these substrates, cyclopropane glyoxylic acid ester, is a close analogue of pyruvic acid which is, of course, a substrate of the NAD-dependent enzyme, lactate dehydrogenase.

4.2 Enzymic Studies

As I argued in the introduction, the key to obtaining useful information on enzyme mechanisms using probe molecules is that the probe molecule should be as closely as possible related to the substrate. For two NAD-dependent dehydrogenases, horse liver alcohol dehydrogenase (HLADH) and lactate dehydrogenase (LDH) this is easily accomplished by the incorporation of a cyclopropyl group into typical substrates, cycloalkanols in the former case, and lactic acid in the latter [12,33]. In order to investigate the mechanism of the reactions with the probes, the kinetic properties of the probes were first established and it was shown that the probes were all competent substrates for the enzymes. Secondly, preparative scale incubations of probe with enzyme were carried out recycling the cofactor with an appropriate reagent. For HLADH, the principal probe was bicyclo[4.1.0]heptan-7-methanol 21 and the corresponding aldehyde 22; for LDH, cyclopropaneglycollic acid 23 and cyclopropane glyoxylic acids were used 24. In neither case were ring opened products characteristic of cyclopropylalkyl radical chemistry observed. At first sight, this result suggests that radical intermediates are not involved, especially in view of the clear cut observations of the model studies [32,33]. However, it is necessary to examine the situation more critically.

For HLADH, it was known that the rate of hydrogen transfer was of the order of 100 s$^{-1}$ and hence the cyclopropyl probe should easily be fast enough to intercept a radical intermediate (cf. Table 1) provided that the enzyme does not impose any unexpected constraints upon the reaction. One observation that was significant was that the primary alcohol and aldehyde probes caused inhibition of the enzyme irreversibly, the rate of inhibition to normal turnover being approximately 0.01. Since no typical radical products were observed, we felt that this inhibition reaction was due to a nucleophilic ring opening catalysed by the Lewis acid zinc ion at the active site. To test this, the more reactive *gem*-dimethyl probes 25 and 26 (cf. Table 1) were submitted to enzyme-catalysed oxidation [11c]. It was anticipated that these compounds would be very much poorer inhibitors because the site for nucleophilic attack would be sterically hindered. This expectation proved correct, the comparable ratios being 0.00024 for 25 and 0.000021 for 26. Once again ring opening was not observed. It remained to test whether HLADH bound the probe molecules in a conformation unfavourable for ring opening. Without an X-ray crystal structure of an enzyme-probe complex, it is impossible to answer this question with certainty. However a molecular graphics analysis showed that the enzyme permitted the stereoelectronic arrangement required firstly for hydrogen transfer and secondly for ring opening, and also accommodated the most bulky ring opened product, that from 26 [33]. For this type of substrate, therefore, there is no evidence to support a radical mechanism; a radical intermediate would have to have a lifetime of $<10^{-9}$s. Other potential probes of this mechanism have not been subjected to such detailed scrutiny with regard to their interactions with the enzyme.

Recently, Tanner has used arylhaloketones as probes of the mechanism of action of HLADH [17]. In contrast to the aliphatic probes, elimination reactions characteristic of radical intermediates were observed in model reactions using this type of probe [17,35]. In the enzyme-catalysed reaction, however, with monohaloacetophenones, stereospecific formation of the secondary alcohols was observed, a result consistent with direct hydrogen transfer. Unfortunately, the kinetic properties of these probes with respect to the enzyme were not reported; with a reactive alkylating agent such as an α-haloketone, some enzyme inhibition would have been expected. Further, the rate of elimination of halide has been estimated at $10^5$ $s^{-1}$; this probe is clearly less sensitive than the cyclopropyl probes. A ketyl radical formed at the active site would almost certainly react to give acylated coenzyme or inhibited enzyme since the rate of hydrogen transfer has been shown to be as slow as 100 $s^{-1}$. What Tanner's work has confirmed, however, is that the mechanism by which dihydropyridines react is substrate dependent [35]. Such a situation is not unprecedented recalling the properties of cytochromes P-450 and their models, and it is reasonable that a spectrum of mechanisms should exist for reductions by NADH depending upon the nature of the electron acceptor [36].

With LDH, however, despite the model reactions and the clean enzyme-catalysed redox reaction, there is a genuine problem. Unlike the simple alkoxy-substituted cyclopropylmethyl compounds, the rate of ring opening of the hydroxy-carboxylate radical formed from 23 could not be measured by flow esr methods [11c] and, as reported in Table 1, the rate of ring opening of the analogous ester radical was also slow. It is therefore not possible to be so certain about the mechanism of the LDH reaction although the model reactions show what the essential chemical character of the reactants is. The most balanced conclusion that can be reached today is that although dihydropyridines will reduce appropriate substrates by single electron transfer mechanisms, for biologically significant substrates in enzyme-catalysed reactions there is no evidence to support a radical mechanism.

## 5. THIAMINE PYROPHOSPHATE-DEPENDENT ENZYMES

It is appropriate to follow a discussion of the mechanism of the nicotinamide-dependent enzyme LDH which has pyruvate as a substrate with consideration of other pyruvate-metabolising enzymes that have been studied using the same probe, cyclopropylglyoxylate. Walsh and his colleagues investigated the reactions of pyruvate:ferredoxin oxidoreductase, pyruvate decarboxylase, pyruvate oxidase, and pyruvate dehydrogenase [37]. The first of these reactions was of particular interest because it is believed that this enzyme functions through a sequence of two one-electron oxidations of the pyruvate-thiamine adduct via a long-lived free radical (Figure 7, 27) derived from hydroxyethylthiamine pyrophosphate, the

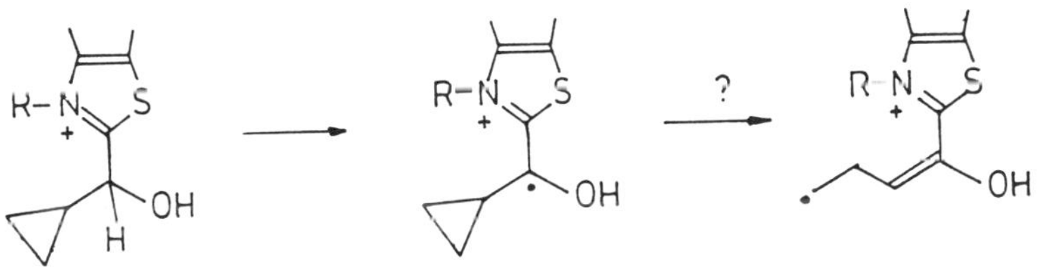

Figure 7  Possible radical generating steps in oxidations of thiamine pyrophosphate adducts.

decarboxylation product of the pyruvate-thiamine adduct. The probe, cyclopropylglyoxylate, was characterised as a competent substrate for all four enzymes and in no case was enzyme inactivation observed. Further, no ring opened products were detected. In the case of the ferredoxin utilising enzyme, an obligate one electron pathway exists and the radical derived from the coenzyme-substrate adduct has been detected by esr [38]. Ring opening might reasonably have been anticipated in this case. However, in view of our results on the capto-dative stabilisation of cyclopropylalkyl radicals (Table 1), it is perhaps not surprising that the three membered ring survived. The C=N bond of the thiazolium salt is a good electron acceptor through the normal mechanism of action of this coenzyme and an additional donor substituent is provided by the hydroxyl group. Thus essentially ideal arrangements are provided for the stabilisation of the radical intermediate in this reaction. Although it is difficult to rule out the possibility that an enzyme inhibits ring opening through constraints at its active site, in this case, the electronic stabilisation of the radical seems the most likely explanation for the failure to observe ring opened products. In these studies, pyruvate decarboxylase acted as a positive control because it processes pyruvate nonoxidatively and ring opening would not be expected. However radical mechanisms are possible for both the other two enzymes, pyruvate dehydrogenase and pyruvate oxidase. The experiments with cyclopropylglyoxylate cannot rule out such mechanisms for the reasons mentioned above.

## 6. AMINE OXIDASES

Monoamine oxidases (MAOs) were amongst the first enzymes to be subjected to the cyclopropyl probe because of interest in the mechanism of action of the antidepressant, tranylcypromine, which is a phenylcyclopropylamine (Figure 8, 28) and an inhibitor of MAO. Although initial experiments were interpreted in favour of oxidation to a cyclopropyliminium cation followed by addition to the C=N forming a relatively stable cyclopropanone hydrate-like adduct, most evidence now clearly points towards a radical mechanism. Following the observations on cyclopropylamines oxidised by cytochromes P-450, Silverman [39] has carried out an extensive series of investigations of the oxidation of a wide range of arylcyclopropylamines by flavin-dependent MAOs. Recalling that a cyclopropyliminium cation radical has a particularly short lifetime (Table 1) [11b] and that electron transfer from nitrogen is the easiest oxidation step in such compounds, it is reasonable to postulate the mechanism shown in Figure 8a for this reaction. The mechanism applies equally to primary and to substituted amines. Once again, inhibition accompanies radical ring opening.

To establish which fragments of cyclopropyl inhibitors were attached to MAO, Silverman studied N-benzyl-1-methylcyclopropylamine and found that only the methyl and cyclopropyl carbon atoms remained attached. This result was interpreted as shown in Figure 8b in which the addition product of flavin and inhibitor is hydrolysed leaving the flavin alkylated by a 3-oxobutyl substituent [40]. An important observation in support of the radical mechanism is that dimethylbenzylamine is a substrate but fails to inactivate MAO [41] thus emphasising the significance of the cyclopropyl group. Recently, a cysteine residue has been implicated as taking part in the formation of a transient intermediate in the inhibition of mitochondrial MAO by 1-phenylcyclopropylamine [42].

## 7. METHANOL OXIDISING ENZYMES

NAD is the cofactor usually employed by enzymes for the oxidation of ethanol and higher alcohols. Recently, methanol metabolising organisms have become of special interest because of their potential contribution to energy production and in the course of studies on such organisms, the importance of a new cofactor known as pyrroloquinolinequinone (PQQ) has become clear (Figure 9, 29). The presence of an ortho quinone substructure immediately suggests that

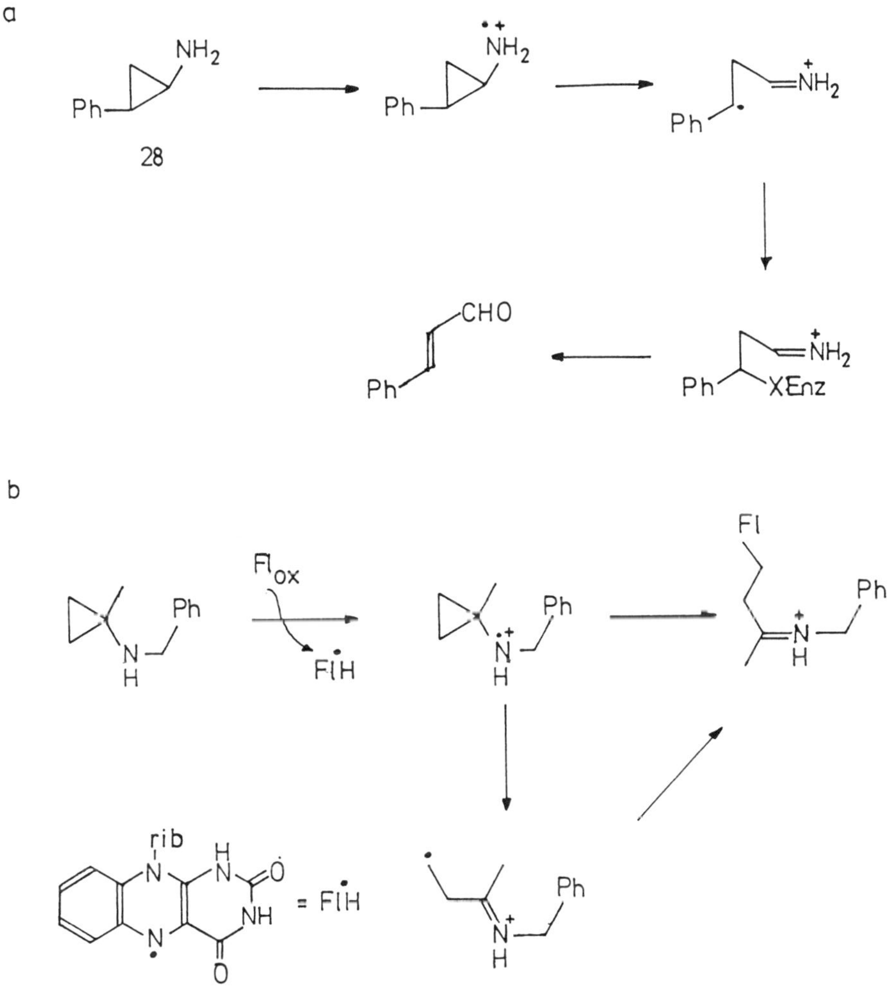

Figure 8   Oxidation of cyclopropylamines by monoamine oxidase.

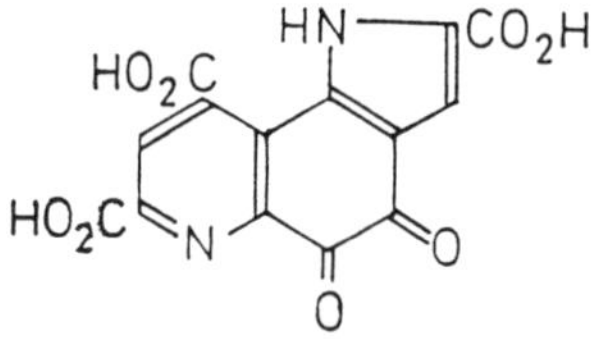

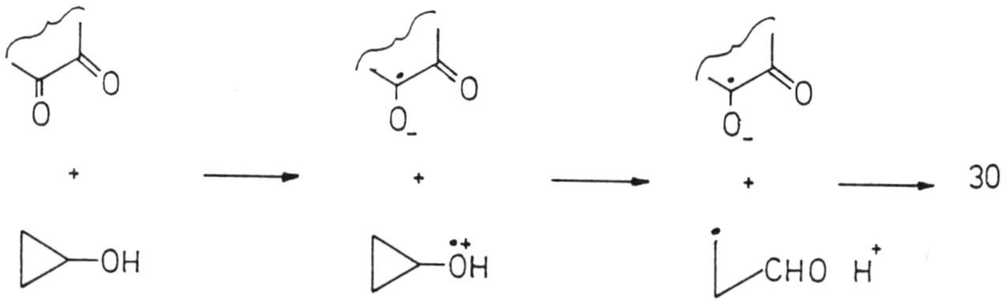

Figure 9  PQQ and the oxidation of cyclopropanol.

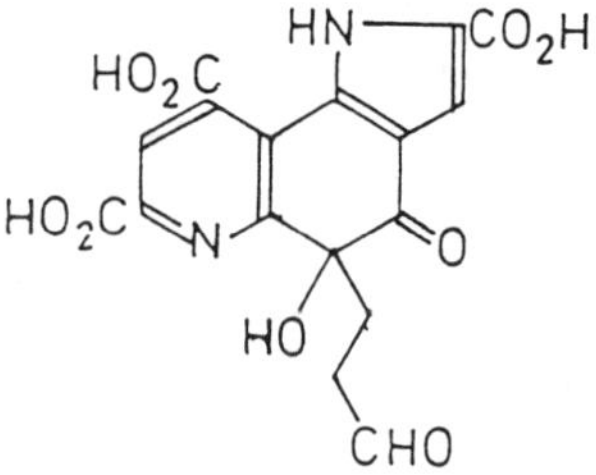

radical mechanisms may be important in the action of this coenzyme. It was therefore natural that cyclopropane-containing molecules should be used as probes of the mechanism of action of PQQ-dependent enzymes. There has been some disagreement about the native oxidation state of PQQ bound to its enzymes and, indeed, several types of PQQ-dependent methanol dehydrogenases have been described. These problems will be resolved by more complete characterisation of the enzymes. Here we concentrate upon the chemical nature of the oxidation step.

Duine has shown that methanol oxidase from *Pseudomonas* BB1 is inhibited by cyclopropanol but oxidises cyclopropylmethanol, cyclobutanol and cyclohexanol as substrates [43]. This enzyme was also capable of oxidising secondary alcohols such as cyclopropanone hydrate and cyclopropanone ethyl hemiketal. In both cases inhibition occurred by modification of PQQ and the structures of the adducts from cyclopropanone hydrate and from cyclopropanol were found to be different. It is therefore unlikely that a direct dehydrogenation of cyclopropanol to cyclopropanone occurs during the normal oxidation cycle. The structure of the adduct has been characterised (30) and its identity makes the formulation of a mechanism possible (Figure 9). One-electron oxidation of cyclopropanol by PQQ leads to a cyclopropyloxonium radical which, by analogy with the corresponding iminium radical, undergoes rapid ring opening. The ring opened radical combines with the radical form of PQQ leading to the observed product. The same structure had been proposed by Abeles [44] as a result of his work on the inhibition of a dehydrogenase from *Methylococcus methanica* by cyclopropanol. He also found that the reaction proceeded without a measurable primary deuterium or tritium isotope effect, a result not inconsistent with the proposed mechanism. Generalising from these and other experiments, Duine has suggested that cyclopropanol is an excellent probe for the existence of PQQ-dependent alcohol dehydrogenases in microorganisms; if a PQQ enzyme is present, it will be inhibited and growth impaired whereas NAD-dependent enzymes will not be inhibited [45].

Abeles has studied a further class of methanol oxidising enzyme in which the prosthetic group is a flavin [46]. Cyclopropanol was again used as a probe and the enzyme was found to be inhibited through the formation of an N-5 adduct. Two possible mechanisms were considered, one in which a cyclopropoxy radical undergoes ring opening before combining with the flavin (cf. PQQ and MAO above), and the other in which cyclopropoxide undergoes ring opening prior to adding to the oxidised flavin. In view of the very slow rate of the latter ring opening [47], Abeles favoured the radical mechanism. Cyclopropylmethanol was a substrate and not an inhibitor of the enzyme. This strongly suggests that the enzyme's initial move is to abstract an electron from oxygen rather than a hydrogen atom from the methylene group. If the latter were the case, ring opening and enzyme inhibition would be expected, neither of which was observed.

It is notable in these reactions that enzyme-bound oxidants, sufficiently powerful to abstract an electron from an alcohol oxygen atom appear to exist.  The formation of oxygen radicals in cytochrome P-450 catalysed reactions has not been well demonstrated (see above) but peroxidase, another haemoprotein, has been shown to be inhibited by cyclopropanone hydrate [48].  The reaction mechanism is presumed to be similar to that of the methanol oxidising enzymes namely, the formation of an oxygen centred radical, ring opening, and recombination of the ring opened radical with the coenzyme leading to an inhibited enzyme.

## 8. ETHYLENE BIOSYNTHESIS

Ethylene is an important hormone in the ripening of fruit and its biosynthesis has naturally attracted considerable attention.  It is derived by oxidative fragmentation of aminocyclopropane carboxylic acid (ACC) and clues as to the likely mechanism have arisen from both radiochemical labelling [49] and model studies [50].  Both identified cyanide as the product bearing the nitrogen atom and the electrochemical experiments suggested a radical mechanism.  The susceptible site of the substrate is once again the amino group oxidation of which to the usual radical cation will again initiate the ring opening reaction.  As will be seen in Figure 10 a further oxidation is required before fragmentation to ethylene occurs.  Thus the flavoprotein amine oxidase mechanism described above is a likely precedent.  This mechanism is also in accord with stereochemical studies provided that the first acyclic radical has a long enough lifetime to undergo rotation at the active site leading to loss of stereochemical integrity.

Ingenious experiments of Pirrung using two cyclopropane rings in tandem have added support to the mechanism proposed above [51]. Bearing in mind that the ethylene producing enzyme also accepts substituted aminocyclopropane carboxylic acids, Pirrung suggested that the amino acid **27** should undergo oxidation and ring opening to afford a cyclopropylmethyl radical **28**.  This in turn would cleave to afford a bishomoallyl radical **29** which would then inhibit the enzyme.  Time dependent inhibition in mung bean hypocotyl segments was observed together with substrate protection.  Radical chemistry is clearly important in the biosyntheis of ethylene.

## 9. COENZYME B12 MEDIATED REARRANGEMENTS

The mechanism by which coenzyme $B_{12}$ directs an intriguing group of molecular rearrangements has been a topic of continuous interest in bioorganic chemistry for more than 20 years.  In the early studies, interest was heightened because of the lack of chemical precedent for the 1,2-shifts of both hydrogen and a heavy atom that occur in these reactions (Figure 11a).  The lability of the cobalt-carbon bond of the coenzyme was clearly an important factor; it was known, for example, that homolytic cleavage to give a cobalt(II) complex and an organic radical was possible and such dissociation had been observed in

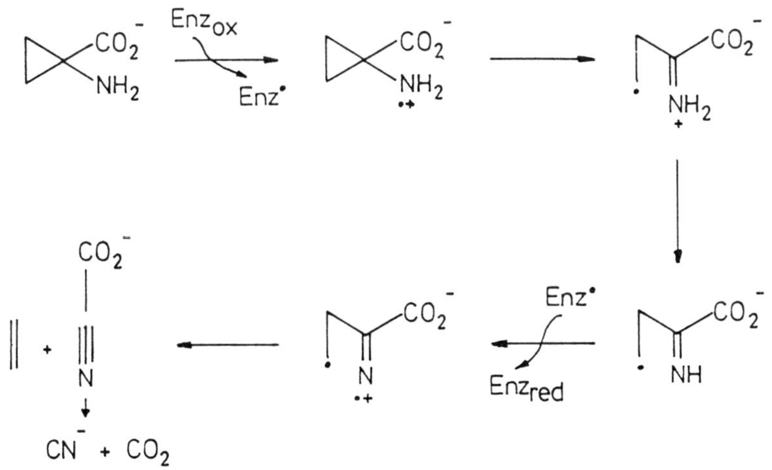

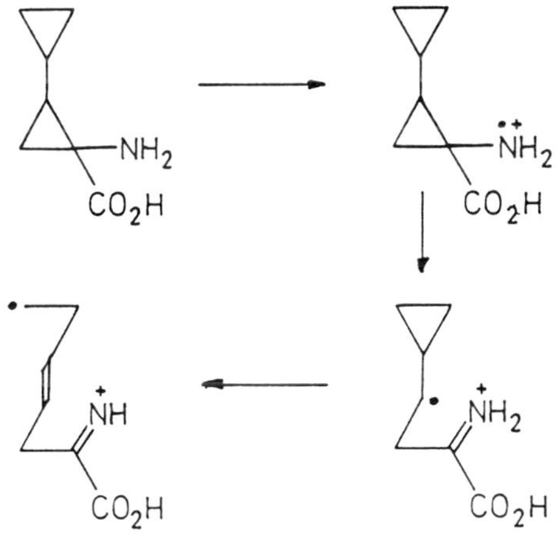

Figure 10   Ethylene biosynthesis and its inhibition by a bis-cyclopropylamino acid.

enzyme-substrate complexes. The connection between radical chemistry and the mechanism of action of $B_{12}$ was strengthened when it was found that the reaction of hydroxyl radicals with ethylene glycol led to the formation of acetaldehyde radicals (Figure 11b) [52]. There were thus several possible mechanisms for consideration for the rearrangement process once a substrate radical had been formed. Two mechanisms involved electron transfer but the chief contenders were a radical rearrangement or a mechanism involving a new organocobalt bond formed by combination of the substrate derived radical with the cobalt. π-Complexes were also considered. The closest approach to a resolution of this problem has come through studies of model systems that relate to rearrangements in which carbon atoms undergo migration by Golding [53].

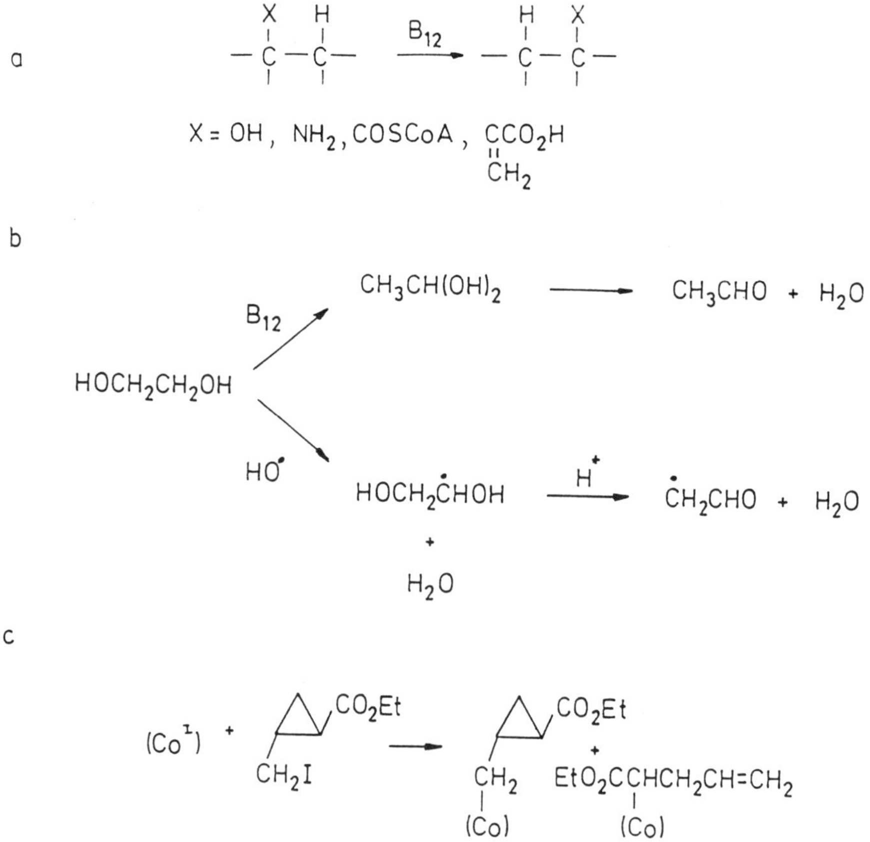

a

$$-\underset{|}{\overset{X}{C}}-\underset{|}{\overset{H}{C}}- \xrightarrow{B_{12}} -\underset{|}{\overset{H}{C}}-\underset{|}{\overset{X}{C}}-$$

$X = OH, NH_2, COSCoA, \underset{\underset{CH_2}{\|}}{C}CO_2H$

b

$HOCH_2CH_2OH \xrightarrow{B_{12}} CH_3CH(OH)_2 \longrightarrow CH_3CHO + H_2O$

$HOCH_2CH_2OH \xrightarrow{HO^\bullet} HOCH_2\overset{\bullet}{C}HOH + H_2O \xrightarrow{H^+} \overset{\bullet}{C}H_2CHO + H_2O$

c

$(Co^I) + \underset{CH_2I}{\triangle\!\!\!-\!\!CO_2Et} \longrightarrow \underset{\underset{(Co)}{|}}{\underset{CH_2}{\triangle\!\!\!-\!\!CO_2Et}} + EtO_2C\underset{\underset{(Co)}{|}}{C}HCH_2CH=CH_2$

34        35        36

<u>Figure 11</u>    Rearrangements catalysed by coenzyme $B_{12}$ and some model reactions.

Cyclopropane chemistry can be used to study the rearrangement catalysed by α-methyleneglutarate mutase and Golding [53] set out to investigate whether the cobalt remained attached to the organic substrate during rearrangement or whether a radical free from cobalt was involved using the model compounds shown in Figure 11c.  The appropriate probe for a model reaction is a cyclopropyl cobaloxime such as 35: this was prepared from the corresponding iodide 34 and cobaloxime(I) with partial ring opening to 36 indicative of a radical substitution pathway.  The more hindered cis-isomer of 34 led exclusively to the ring-opened product, a result that is in keeping with our experience in the ring opening of substituted cyclopropylalkyl radicals [11].  Both isomers of the iodide 34 underwent ring opening upon radiolysis of 77 K thus confirming the sensitivity of the chosen probe to the formation of a radical.  In contrast to this reaction, the cyclopropane substituted cobaloxime 35 did not undergo rearrangement and ring opening.  This important result indicates that the cobalt-carbon bond must break for rearrangement to occur and hence that the cobalt is not a direct partner in the rearrangement process.  It has, in Golding's words [53], a 'spectator' and not a 'conductor' role.  The model result thus suggests that in the enzyme-catalysed reaction, an alkyl cobalamin would be too stable to undergo rearrangement and assigns a role to the deoxyadenosine formed in the reaction, namely to prevent recombination of the rearranging organic substrate radical to afford a stable organocobalt species.  In the case of α-methyleneglutarate mutase, it was inferred that the rearrangement takes place via a cyclopropylalkyl radical [53].

## 10. PENICILLIN BIOSYNTHESIS

Baldwin has considered for many years that the formation of the thiazolidine ring of penicillin is a radical reaction on the basis of model studies, but more recently from the structures of products of allylic rearrangement in cyclisations of alkenyl tripeptides by the enzyme isopenicillin N synthetase [54].  He has also used the cyclopropyl probe to investigate the reaction [55] and clear evidence for ring opening was obtained (Figure 12).  Thus probe 36 underwent cyclisation with partial ring opening to 37 and with probe 38, the only isolated product was ring opened (39).  This result is consistent with the more favourable ring opening for the radical derived from 38 (primary → primary) compared with that from 36 (secondary → primary).  As in the case of lipoxygenase mechanism, the possibility that the radical ring opening is a side reaction peculiar to the probe cannot be ruled out on the basis of these results.  Organoiron intermediates may still be involved but sufficient time must exist at least for a radical pair to form and for ring opening to occur.  It will be interesting to see how finely the distinctions between the possible mechanisms can be drawn from further experiments.

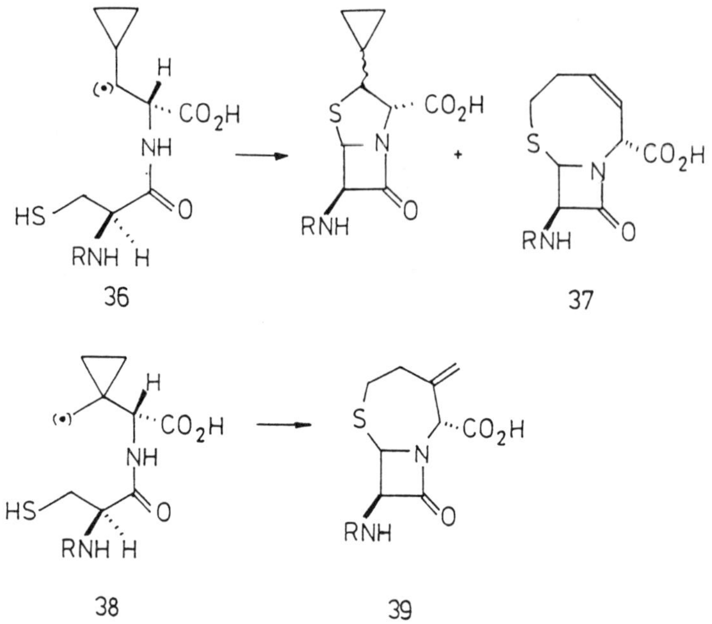

[Structures 36, 37, 38, 39]

Figure 12    Cyclopropyl peptides as probes of the mechanism of penicillin biosynthesis.

## 11. CONCLUSIONS

Although the cyclopropane group has a very extensive chemistry, it is not the only small ring with significant biological activity.  Its heteroanalogues have long been known to inhibit enzymes, sometimes beneficially as in chemotherapeutic nitrogen mustards, [56], sometimes deleteriously as episulphonium mustard gas [57] and in the mutagenicity of 1,2-dihaloethanes through conjugation with glutathione [58].   Epoxides too can be used in cancer chemotherapy and as probes of enzyme active sites [59].   Apart from the vast literature on antibiotic β-lactams, relatively little work has been done on four membered rings.   Isolated cases of the use of such compounds in mechanistic studies exist.   Cyclobutadiene has been generated at the active site of cytochrome P-450, which it inhibits by alkylating the haem [60].   1-Amino-1-phenylcyclobutane has been used to complement the cyclopropane-based studies of MAO [61].   Biologically active oxetanes have recently been discovered [62].   Four-membered rings are well worth further study because they offer alternative substituent patterns and an alternative symmetry to that provided by the cyclopropane group.   There remains much that small rings can reveal in enzyme chemistry and many uses to which their reactivity can be put [63].

REFERENCES

1. M.I. Page in 'Enzyme Mechanisms', eds. M.I. Page and A. Williams, Royal Society of Chemistry, London, 1987, p.1.
2. C.J. Suckling, Chem. Soc. Reviews, 1984, 13, 97.
3. Y. Kanaoka, Angew. Chem. Int. Edn. Engl., 1977, 16, 137.
4. H.M. McConnell and B.G. McFarland, Quart. Rev. Biophys., 1970, 3, 91.
5. T.C. Liang and R.H. Abeles, Biochemistry, 1987, 26, 7603.
6. K.F. Geoghan, A. Galdes, R.A. Martinelli, B. Holmquist, D.S. Auld, and B.L. Vallee, Biochemistry, 1983, 22, 2255; J.P.G. Malthouse, M.P. Gamscik, A.S.F. Boyd, N.E. Mackenzie, and A.I. Scott, J. Am. Chem. Soc., 1982, 104, 6811.
7. C.J. Suckling, Angew. Chem. Int. Edn. Engl., 1988, 27, 537.
8. B. Maillard, D. Forrest, and K.V. Ingold, J. Am. Chem. Soc., 1976, 98, 7024.
9. V.C. Emery and M. Akhtar in 'Enzyme Mechanisms', eds. M.I. Page and A. Williams, Royal Society of Chemistry, London, 1987, p.345; P. Haake, ibid., p.390.
10. W.G. Dauben, L. Schutte, R.E. Wolf, and E.J. Deviny, J. Org. Chem., 1969, 34, 2512; W.G. Dauben and R.E. Wolf, ibid., 1970, 35, 374.
11a. D.C. Nonhebel, C.J. Suckling, and J.C. Walton, Tetrahedron Lett., 1982, 23, 4477; b. X.-Z. Quin and F. Williams, J. Am. Chem. Soc., 1987, 109, 595; c. D. Laurie, E. Lucas, D.C. Nonhebel, C.J. Suckling, and J.C. Walton, Tetrahedron, 1986, 42, 1035; d. C. Jamieson, J.C. Walton, and K.U. Ingold, J. Chem. Soc., Perkin Trans. 2, 1980, 1366; e. A.J. Castellino and T.C. Bruice, J. Am. Chem. Soc., 1988, 110, 1313.
12. D.C. Nonhebel, S.T. Orszulik, and C.J. Suckling, J. Chem. Soc., Chem. Commun., 1982, 1146.
13. A. Effio, D. Griller, K.U. Ingold, A.L.J. Beckwith, and A.K. Serelis, J. Am. Chem. Soc., 1980, 102, 1734.
14. S.K. Chung and S.U. Park, J. Org. Chem., 1982, 47, 3197.
15. A.L.J. Beckwith, C.J. Easton, T. Lawrence, and A.K. Serelis, Aust. J. Chem., 1983, 36, 545; C. Chatgilialoglu, K.U. Ingold, and J.C. Scaiano, J. Am. Chem. Soc., 1981, 103, 7739.
16. J.T. Groves, W.J. Kruper Jr., and R.C. Haushalter, J. Am. Chem., Soc., 1980, 102, 6375.
17. D.A. Tanner and A.R. Stein, J. Org. Chem., 1988, 53, 1642.
18. R.P. Hanzlik and R.H. Tullman, J. Am. Chem. Soc., 1982, 104, 2048; T.L. MacDonald, K. Zirvi, L.T. Burka, P. Peyman, and F.P. Guengerich, ibid., 2050.
19. J.T. Groves and D.V. Subramanian, J. Am. Chem. Soc., 1984, 106, 2177.
20. F.P. Guengerich and T.L. MacDonald, Accts. Chem. Res., 1984, 17, 9.
21. L. Brown, W.J.S. Lyall, C.J. Suckling, and K.E. Suckling, J. Chem. Soc., Perkin Trans. 1., 1987, 595; D.J. Houghton, S.E. Beddows, K.E. Suckling, L. Brown, and C.J. Suckling, Tetrahedron Lett., 1986, 27, 4655.

22. P.R. Ortiz de Monellano and R.A. Stearns, J. Am. Chem. Soc., 1987, 109, 3415.
23. K.E. Suckling, C.G. Smellie, I.E. Ibrahim, D.C. Nonhebel, and C.J. Suckling, FEBS Letters, 1982, 145, 179; C.J. Suckling, D.C. Nonhebel, L. Brown, K.E. Suckling, S. Seilman, and C.R. Wolf, Biochem. J., 1985, 232, 199.
24. R.A. Stearns and P.R. Ortiz de Montellano, J. Am. Chem. Soc., 1985, 107, 4081.
25. H. Dalton, B.T. Golding, B.W. Waters, R. Higgins, and J.A. Taylor, J. Chem. Soc., Chem. Commun., 1981, 482.
26. P.F. Fitzpatrick and J.J. Villafranca, J. Am. Chem. Soc., 107, 5022.
27. J.W. Frost, S. Low, M.L. Cordeiro, and D. Li, J. Am. Chem. Soc., 1987, 109, 2166.
28.a E.J. Corey and R. Nagata, Tetrahedron Lett., 1987, 28, 5391;
   b E.J. Corey and R. Nagata, J. Am. Chem. Soc., 1988 submitted.
29. G.A. Hamilton, Prog. Bioorg. Chem., 1971, 1, 113.
30. S. Shinkai, H. Era, T. Tsuno, and O. Manabe, Bull. Chem. Soc. Jpn., 1984, 57, 1435; A. Shirra and C.J. Suckling, Tetrahedron Lett., 1975, 3323.
31. S. Shinkai in 'Enzyme Chemistry, Impact and Applications', ed. C.J. Suckling, Chapman & Hall, London, 1984, p.46.
32. L.H.P. Meyer, J.C.G. van Niel, and U.K. Pandit, Tetrahedron, 1984, 40, 5185.
33. I. MacInnes, D.C. Nonhebel, S.T. Orszulik, and C.J. Suckling, J. Chem. Soc., Perkin Trans. 1, 1982, 1146.
34. R.J. Breckenridge and C.J. Suckling, Tetrahedron, 1986, 42, 5665.
35. D.A. Tanner and A. Kharrat, J. Org. Chem., 1988, 53, 1646.
36. S. Fukuzumi, N.Nishizawa, and T. Tanaka, J. Org. Chem., 1984, 49, 3571; M.F. Powell and T.C. Bruice, J. Am. Chem. Soc., 1983, 105, 1014; A. Sinha and T.C. Bruice, ibid., 1984, 106, 7291; J.W. Bunting and N.P. Fitzgerald, Canad. J. Chem., 1985, 65, 655.
37. D.J. Livingston, S.L. Shanes, R. Gerris, and C.T. Walsh, Bioorg. Chem., 1987, 15, 358.
38. L. Kerscher and D. Osterhelt, Eur. J. Biochem., 1981, 116, 595, TIBS, 1982, 7, 374.
39. R.B. Silverman in 'Topics in Medicinal Chemistry', ed. P.J. Leeming, Royal Society of Chemistry, London, 1988, p.73.
40. R.B. Silverman and R.B. Yamasaki, Biochemistry, 1984, 23, 1822.
41. R.B. Silverman and P.A. Zieske, Biochemistry, 1985, 24, 2128.
42. R.B. Silverman and P.A. Zieske, Biochem. Biophys. Res. Commun., 1986, 135, 154.
43. M. Dijkstra, J. Frank Jzn., J.A. Jongenjan, and J.A. Duine, Eur. J. Biochem., 1984, 140, 369.
44. T. Mincey, J.A. Bell, A.S. Mildvan, and R.H. Abeles, Biochemistry, 1981, 20, 7502; C. Parkes and R.H. Abeles, ibid., 1984, 23, 6355.
45. J.A. Duine personal communication.
46. B. Sherry and R.H. Abeles, Biochemistry, 1985, 24, 2595.
47. A. Thibblin and W.P. Jencks, J. Am. Chem. Soc., 1979, 101, 4963.
48. J.S. Wiseman, J.S. Nichols, and M. Kolpak, J. Biol. Chem., 1982, 257, 6328.

49. R.M. Adlington, J.E. Baldwin, and B.J. Rawlings, J. Chem. Soc., Chem. Commun., 1983, 290; G.D. Reiser, F.-J. Wang, N.E. Hoffmann, S.F. Yang, H.W. Liu, and C.J. Walsh, Proc. Natl. Acad. Sci. USA., 1984, 81, 3059.
50. M.C. Pirrung and G.M. McGeechan, J. Org. Chem., 1983, 48, 5143.
51. M.C. Pirrung, J. Am. Chem. Soc., 1983, 105, 7207; M.C. Pirrung and G.M. McGeechan, Angew. Chem. Int. Edn. Engl., 1985, 24, 1044.
52. B.C. Gilbert, J.P. Larkin, and R.O.C. Norman, J. Chem. Soc., Perkin Trans. 1, 1972, 794.
53. R.M. Dixon, B.T. Golding, S. Mwesigyne-Kibende, and D.C. Ramakrishna Rao, Phil. Trans. Roy. Soc. London Ser. B, 1985, 311, 531.
54. J.E. Baldwin, R.M. Adlington, M.J.C. Crabbe, T. Homoto, and C.J. Schofield, Tetrahedron, 1987, 43, 4217.
55. J.E. Baldwin, R.M. Adlington, B.P. Domayne-Hayman, G. Knight, and H.-H. Ting, J. Chem. Soc., Chem. Commun., 1987, 1161.
56. 'Molecular actions and targets for cancer chemotherapeutic agents', eds. A.C. Sartorelli, E. Lazo, J.R. Bertino, Bristol Meyers Cancer Symposium Series, Academic Press, New York, 1981.
57. D.E. Hathaway, 'Molecular Aspects of Toxicology', Royal Society of Chemistry, London, 1984, p.204.
58. D. Reichert, Angew. Chem. Int. Edn. Engl., 1981, 20, 135.
59. for example, R.M. Pollack, R.H. Kayser, and C.L. Bevins, Biochem. Biophys. Res. Commun., 1979, 91, 783; J.M. Penning, Biochem. J., 1985, 226, 469.
60. R.A. Stearns and P.R. Ortiz de Montellano, J. Am. Chem. Soc., 1985, 107, 234.
61. R.B. Silverman and P.A. Zieske, Biochemistry, 1985, 25, 341.
62. H. Nakamura, S. Hasegawa, N. Shimada, A. Fujii, T. Takita, and Y. Iitaka, J. Antibiot., 1986, 39, 1626; S. Omura, M. Murata, N. Imamura, Y. Iwai, and H. Tanaka, J. Antibiot., 1984, 37, 1324.
63. C.J. Suckling, Proceedings of 10th International Symposium on Medicinal Chemistry, Budapest, 1988, in press.

# INTRAMOLECULAR [2+2] CYCLOADDITONS OF KETENES AND OLEFINS

Beat Ernst*, Alain de Mesmaeker, Hans Greuter and
Siem J. Veenstra
Central Research Laboratories
CIBA-GEIGY Ltd., CH-4002 Basel, Switzerland

ABSTRACT. The synthetic potential of the intramolecular ketene cycloaddition is presented. It proves to be a versatile method for the
synthesis of a variety of bi- and tricyclic cyclobutanones. The major
side reaction, namely the dimerization of the ketene is prevented by
the introduction of conformational restrictions. This has been
achieved by an additional double bond or by an annulated ring system.
Following this strategy C-analogues of a penem and a carbapenem as
well as linear and angular annulated triquinanes are obtained in high
yield.

## 1. INTRODUCTION

More than 75 years have elapsed since Staudinger discovered in 1911
the ability of ketenes to react with olefins to yield cyclobutanones
[1].
    The prototype of this reaction is mentioned in a Du Pont patent
[2], where ketene and ethylene react at 200° under a pressure of
1000 atm. to form cyclobutanone 1. Continuous addition of monomeric
ketene to a supply of ethylene under reaction conditions at a rate
nearly equivalent to the rate of cycloaddition was envisioned as the
most efficient procedure for formation of cyclobutanone 1 instead of
ketene dimer.

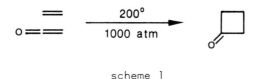

scheme 1

To avoid these extreme reaction conditions and to develop ketene
cycloadditions into a useful synthetic tool many improvements were
made. In intermolecular cycloadditions this is done by increasing the

electrophilicity of the ketene or the nucleophilicity of the ketenophile, as demonstrated by the following literature examples:

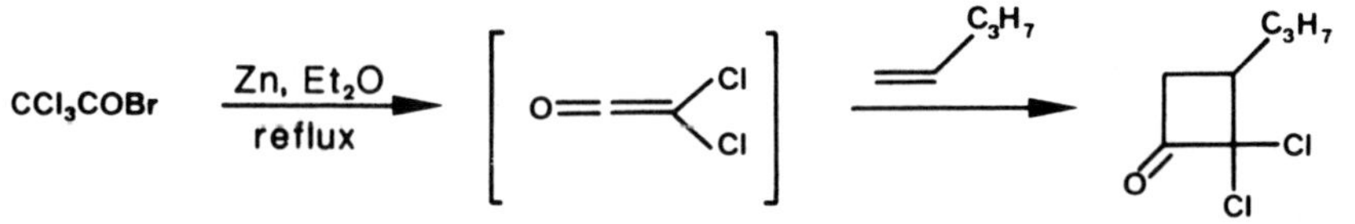

scheme 2

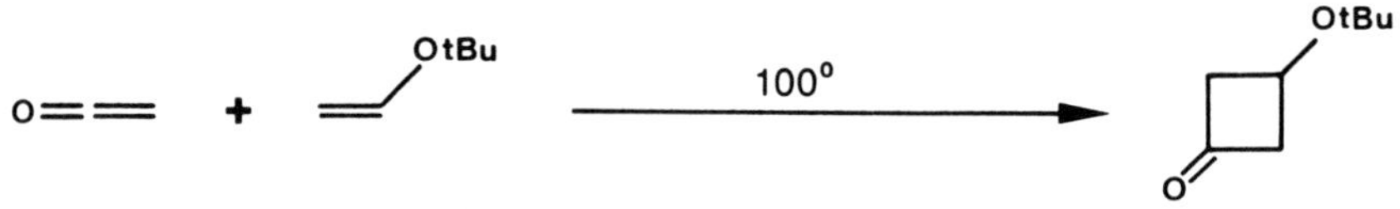

With the more electrophilic dichloroketene instead of ketene itself the cycloaddition takes place at 35°C to yield the cyclobutanone 2 [3]. On the other hand the cycloaddition can be carried out under mild conditions by using more nucleophilic olefins like vinylethers [4].

A possibility to decrease the free energy of activation is provided by intramolecular processes. Thus connecting the ketene and the ketenophile over a short chain the degree of freedom for the two reacting functionalities is reduced. From this a less negative activation entropy and therefore a smaller $\Delta G$ results.

$$\Delta G^{\ddagger} = \Delta H^{\ddagger} - T\Delta S^{\ddagger}$$

scheme 3

In principle this situation is similar to an enzymatic reaction. Functional groups which under the conditions of the organic synthesis are extremely unreactive, react in the enzyme catalysed reaction under mild conditions. This is at least partially due to the perfect positioning of the reacting functionalities in the enzyme-substrate-complex.

This kind of proximity of the unactivated ketene and olefin in the norbornene derivative 4 is responsible for the cycloaddition to take place under mild conditions [5].

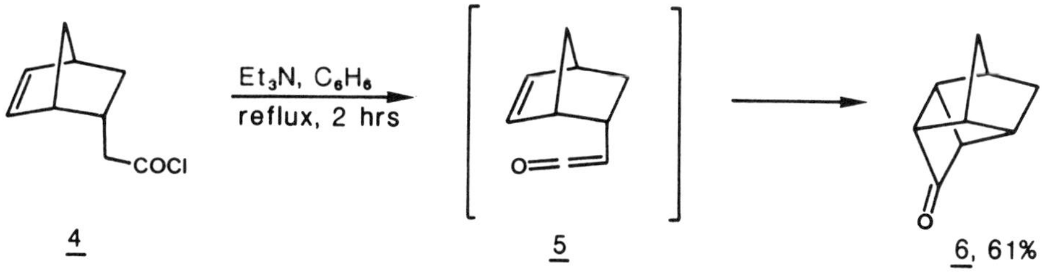

scheme 4

In view of this early result surprisingly few examples of such intramolecular thermal [6] and photochemical [7a] cycloadditions have been reported and no systematic study was undertaken until we initiated our research in this field.

We started our investigation by reexamination of a published example where the ketene and the ketenophile are connected to each other over a short chain without further geometrical restrictions [6c].

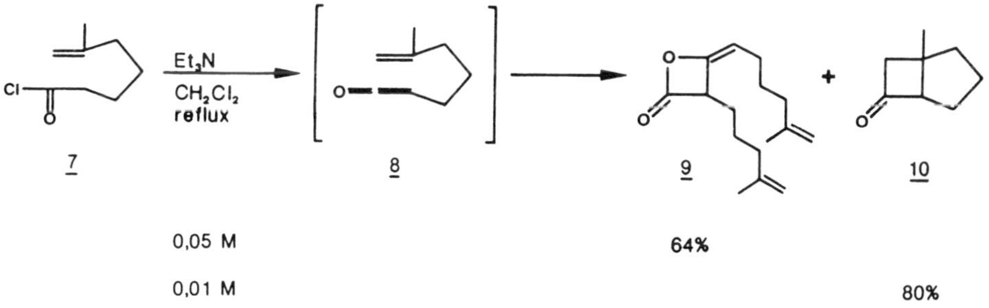

scheme 5

When NEt$_3$ was added to a 0.05M solution of the acid chloride 7 over a period of 24 hrs the ketene dimer 9 was formed almost exclusively. By running the cycloaddition reaction five times more diluted the desired intramolecular cycloaddition product 10 was formed in high yield (scheme 5).

But even under these high dilution conditions dimers 13 or 16, respectively, were formed as major products in the case of the desmethyl-derivative 11 or the chain extended derivative 14 (scheme 6).

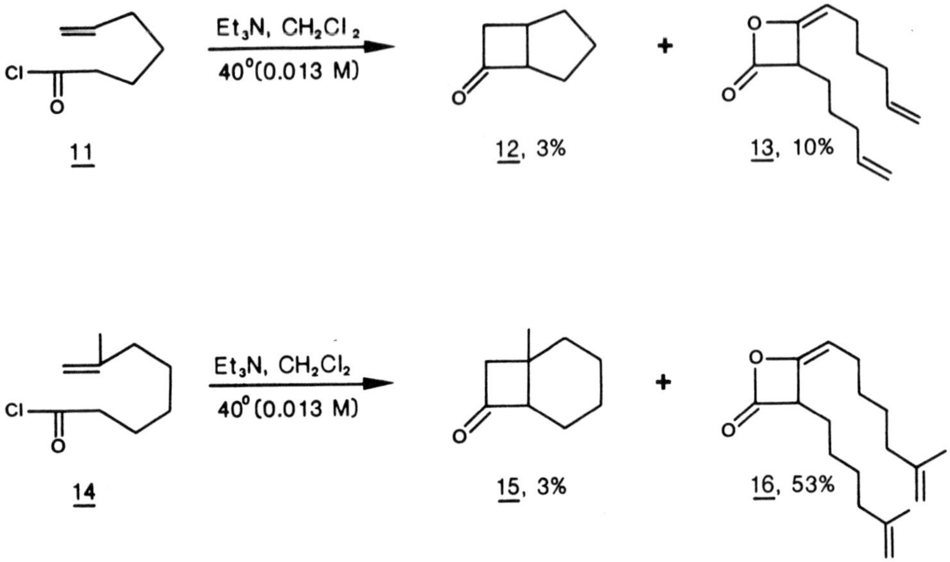

scheme 6

When, as in the first example, the nucleophilicity of the ketenophile is reduced only neglectable amounts of the intramolecular cycloadditon product 12 are formed.

In the second example a dramatic difference in the formation of the [3.2.0] and the [4.2.0]-system respectively is observed. Starting with the acid chloride 14 the ketene dimer 16 is formed as the major product. The desired cyclobutanone 15 was isolated only in traces.

The tendency of ketenes to dimerise as well as their ability to undergo [2+2] cycloadditions with unsaturated substrates is due to the combination of electrophilic character at C(1) and nucleophilic character at C(2) [8].

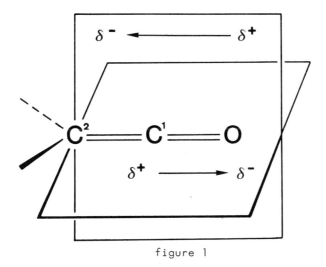

figure 1

For us this implied the question of what kind of modifications would suppress the formation of ketene dimers and hence allow the intramolecular cycloaddition at preparatively reasonable concentrations.

## 2. INTRAMOLECULAR [2+2] CYCLOADDITION OF VINYLKETENES AND OLEFINS

From frontier orbital considerations [9] it becomes obvious that we can not expect an important electronic contribution going from ketene to vinylketene. The introduction of an additional double bond however is creating some geometrical restrictions and increases the entropic factor. Therefore the $\Delta G$ for the intramolecular cycloaddition relative to the $\Delta G$ for the dimerisation should be reduced.

Vinylketenes can be generated easily from α,β-unsaturated acid chlorides by 1,4-dehydrohalogenation [10]. In a first, reversible step an acyl ammonium complex 18 is formed, which acidifies the γ-position. In a second step deprotonation leads to the vinylketene 19.

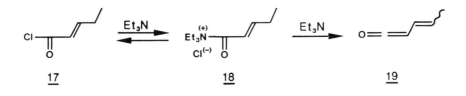

scheme 7

A problem that has to be avoided is the formation of E/Z-isomers in the elimination step. A solution is offered by fixing two equi-

valent groups at the γ-position (cf. section 2.1.) or by regioselective deprotonation (cf. section 2.2.).

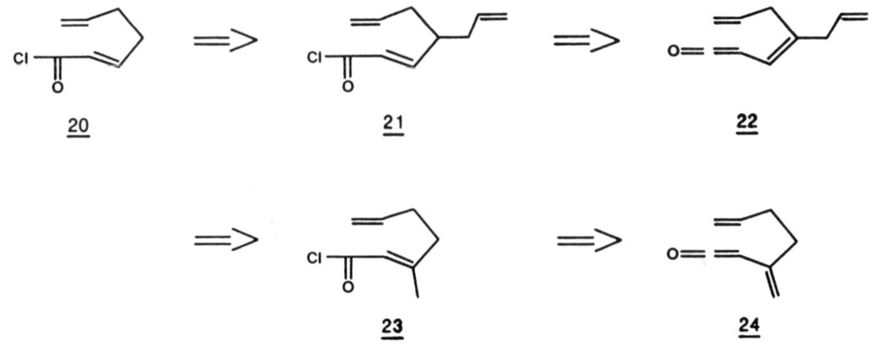

scheme 8

## 2.1. Vinylketenes with two equivalent groups in the γ-position

On treatment of 0.3 molar solution of acid chloride 21 in chloroform with triethylamine the vinylketene 22 was formed. It readily
cyclised to give the bicyclic cyclobutanone 27 in 83 % yield [11].
Dimers or oligomers were not detected [12]. An even more concentrated
solution (0.5 M) was used in the case of the methyl derivative 28
which gave the corresponding cyclobutanone 30 in 84 % yield [11].
Vinylketenes can therefore be used in intramolecular cycloadditions at
preparatively reasonable concentration. The formation of ketene dimers
can be neglected.

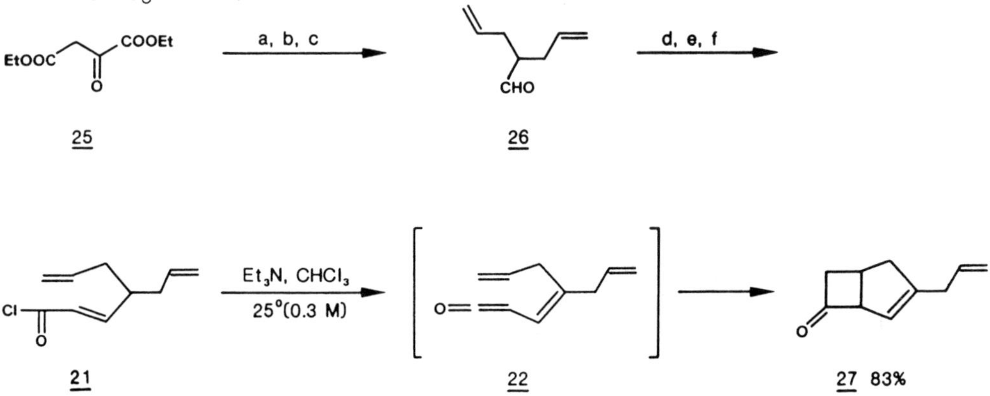

a = $CH_2$=$CHCH_2$Br, $K_2CO_3$, DMF (85 %); b = LiAlH$_4$, Et$_2$O (94 %); c = PCC,
CH$_2$Cl$_2$ (69 %); d = Ph$_3$PCHCOOEt, C$_6$H$_6$ (93 %); e = KOH, EtOH/H$_2$O (95 %);
f = SOCl$_2$ (95 %).

scheme 9

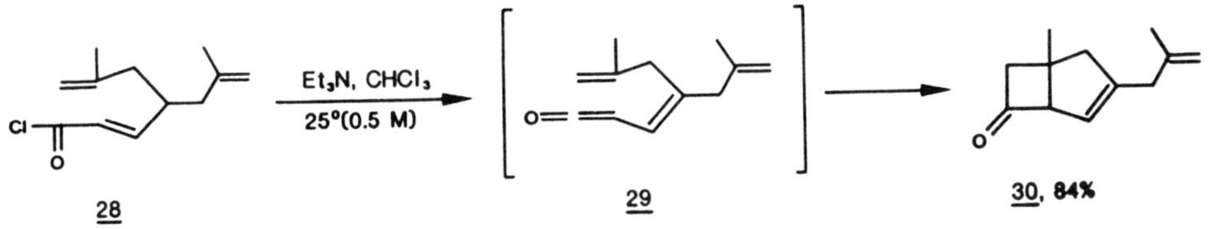

scheme 10

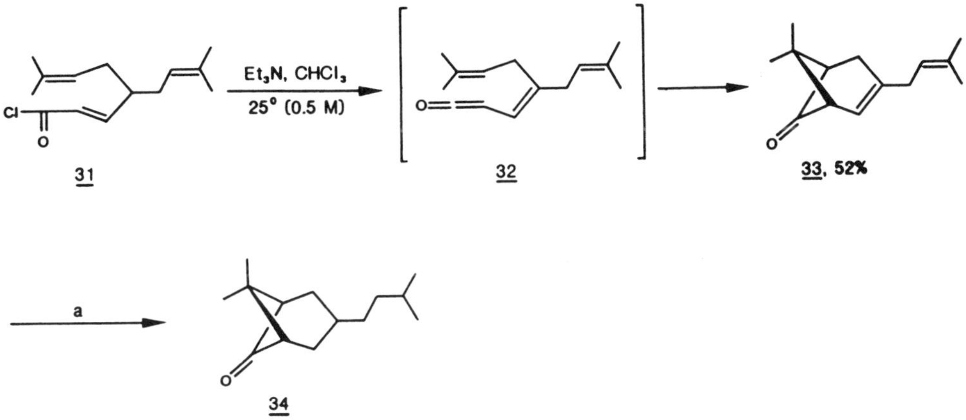

scheme 11

a = H$_2$, Pd/C, EtOH (82 %).

Interestingly the bis-prenyl acid chloride 31 gave the bicyclic structure 33 with inverted regiochemistry. This was easily proven by hydrogenation of 33 giving the symmetrical compound 34. The regiochemistry of [2+2] cycloadditions can be rationalized by HOMO/LUMO considerations. The most important interaction of a ketene undergoing cycloaddition to a double bond is the interaction of the HOMO of the ketenophile with the LUMO of the ketene. Regiochemistry is therefore determined by the fact that the larger lobe of the HOMO of the olefin overlaps with the larger lobe of the LUMO of the ketene. Calculations [12] clearly show that the largest LUMO coefficient is on the central atom of the ketene and therefore it is this atom which will become bonded to the carbon atom having the larger coefficient in the HOMO of the olefin.

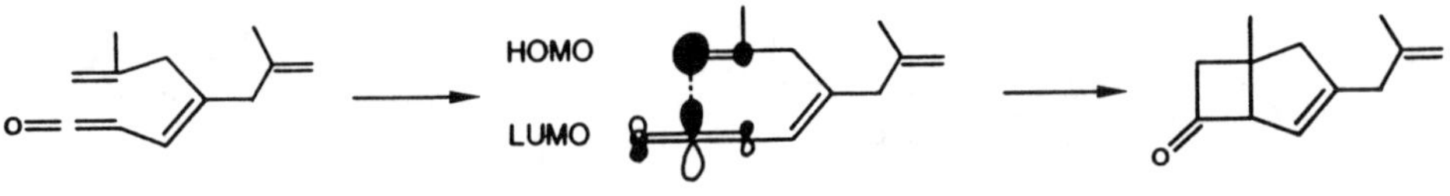

scheme 12

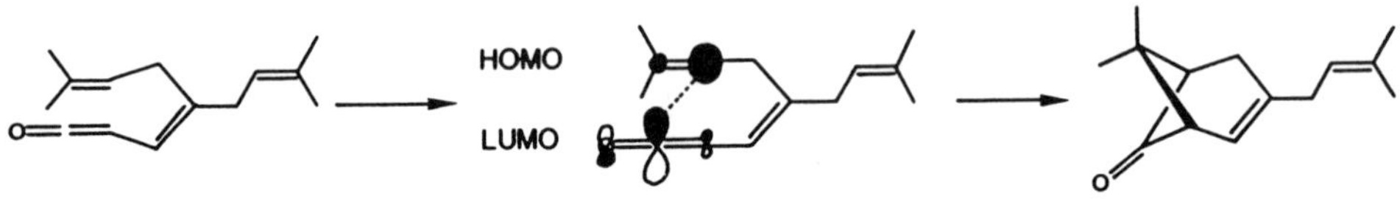

In the first example (cf. 29 ⟶ 30) the larger coefficient in the HOMO of the olefin is on the terminal carbon and therefore the [3.2.0] bicyclic system 30 is formed. In the second example (cf. 32 ⟶ 33) the larger coefficient is on the opposite terminus of the double bond, which gives rise to the formation of the [3.1.1] bicyclic system 33.

2.2. Vinylketenes by regioselective deprotonation

Starting from the acid chloride 35 which is easily available from geraniol the problem of E/Z-isomerism of the double bond was solved by regioselective deprotonation. By treatment with $NEt_3$ the kinetically controlled vinylketene was formed. Cycloaddition led to the [3.1.1] bicyclic product 36.

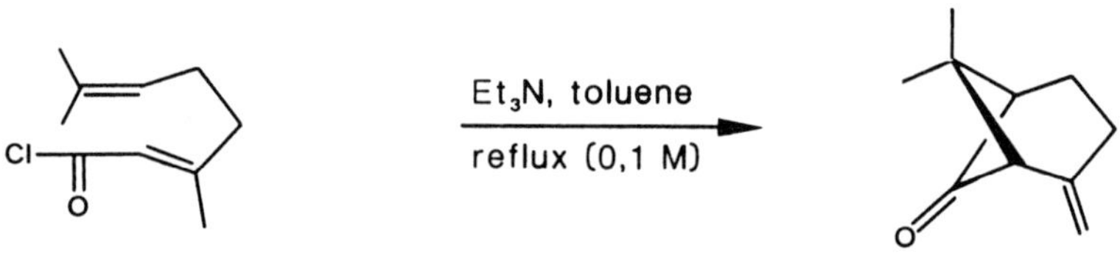

scheme 13

This example was independently investigated by B. Snider [13] who applied this principle to the syntheses of natural products, e.g. chrysanthenone, β-pinene, β-cis-bergamotene, β-trans-bergamotene

or copaene. In an alternative approach Corey [14] started from the
β,γ-unsaturated acid chloride to realise the synthesis of
(+)-β-trans- bergamotene.

An application of the intramolecular [2+2] cycloaddition of
vinylketenes to olefins is the synthesis of C-analogues of β-lactam
antibiotics. The discovery of non-classical β-lactams [15] in recent
years necessitated a reevaluation of the structural features which are
required for biological activity. A new class with possible antibiotic
activity are the C-analogues of β-lactams. This concept was already
proposed in the literature [16], but only in one case a weak
antibacterial activity has been found [16b]. Our first approach was to
mimic a highly active β-lactam by its exact C-analogue.

One of these highly active β-lactams is thienamycin (37). The
challenge of its synthesis has attracted the attention of many industrial and academic researchers [17]. Many analogues of this carbapenem have been prepared, some of which show similar antibiotic activity. One such example is 38 - a carbapenem synthesized by Ciba-Geigy
researchers [18] - which differs from thienamycin only in having the
much simpler side chain at C-3. Consequently, we decided that the
corresponding cyclobutanone 39 should be our first target.

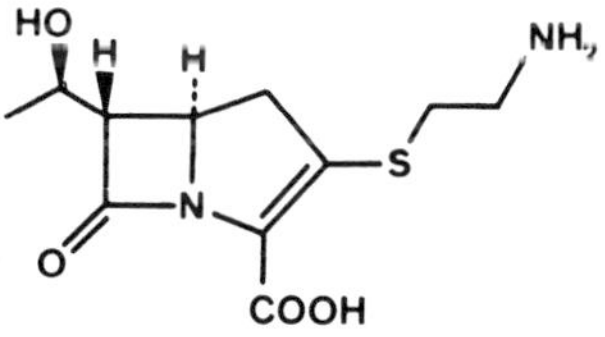

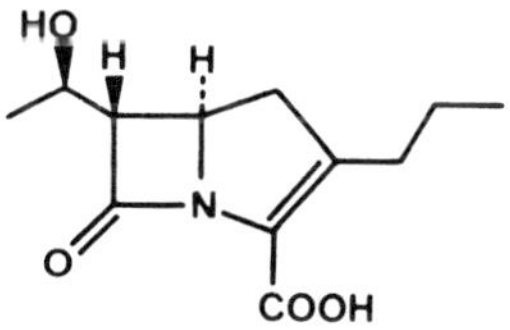

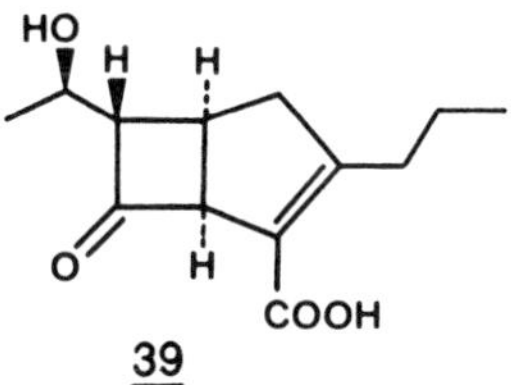

scheme 14

The bis-allylated citraconic acid derivative 41 could be prepared
by standard procedures from aldehyde 26.

The next step was the preparation of the corresponding acid
chloride 42. This had to be done under neutral conditions, otherwise
cyclic anhydride formation could not be prevented.

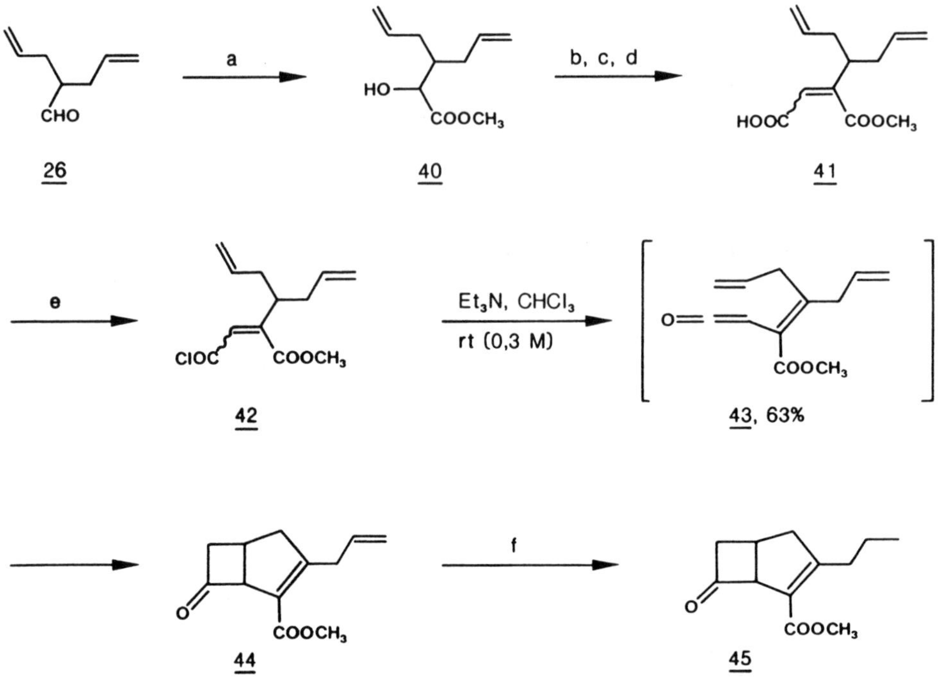

scheme 15

a = NaCN, Na$_2$S$_2$O$_5$; HCl, MeOH (61 %); b = CrO$_3$, H$_2$SO$_4$ (93 %);
c = Ph$_3$PCHCOOtBu, C$_6$H$_6$ (95 %); d = CF$_3$COOH, 25°C (87 %); e = TBDMSCl,
imidazol; (COCl)$_2$, cat. DMF; f = H$_2$, Pd/C, MeOH (93 %).

For this purpose we used Wissner's method [19]. Conversion to the t-butyldimethylsilylester followed by treatment with oxalylchloride and catalytic amounts of DMF produced the very labile acid chloride 42. It was not isolated but immediately treated with NEt$_3$ in chloroform to form vinylketene 43, which cyclised at room temperature in a 63 % overall yield starting from 41.

Finally the double bond in the side chain was selectively hydrogenated under standard hydrogenation conditions to yield 45.

The hydroxy ethyl side chain was introduced successfully by applying the Mukaijama reaction (sequence a) as well as with a boron aldol reaction (condition b).

Both methods led to a mixture of aldol products in good yields, although with different selectivities. The silyl enol ether of 45 was prepared without any difficulty. Tin tetrachloride catalyzed the aldol reaction with acetaldehyde. After removal of the silyl groups with hydrogenfluoride-urea complex two aldol products syn-46 and anti-46 were obtained and separated in 19 % and 66 % yield, respectively. The stereochemistry was proven by X-ray analysis of the ketal of anti-46.

45 syn - 46 anti - 46

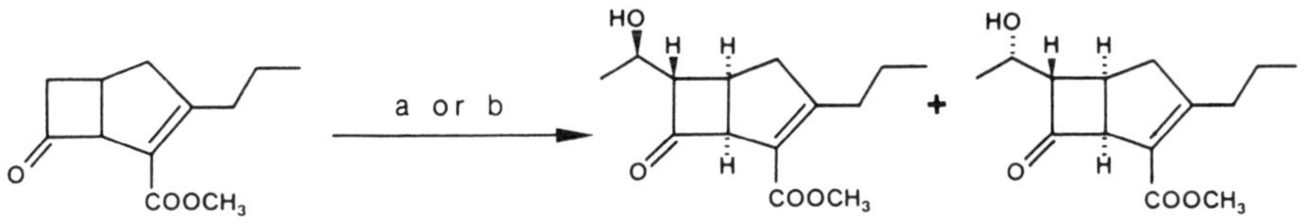

| | | syn - 46 | anti - 46 | |
|---|---|---|---|---|
| a = | 1. TBDMSOTf, $Et_3N$, 25° | | | |
| | 2. $SnCl_4$, $CH_3CHO$, $CH_2Cl_2$, -78° | 19% | 66% | |
| | 3. $HF \cdot H_2NCONH_2$, $CH_3CN$ | | | |
| b = | 1. $iPr_2NEt$, $Bu_2BOTf$, $CH_2Cl_2$, -78° | | | |
| | 2. $CH_3CHO$, -78° | 3% | 62% | (18% cis) |

scheme 16

The boron aldol reaction gave an even less satisfactory result in terms of stereoselection although it proceeded very well chemically. Only traces of the desired isomer were formed surprisingly together with a cis aldol product.

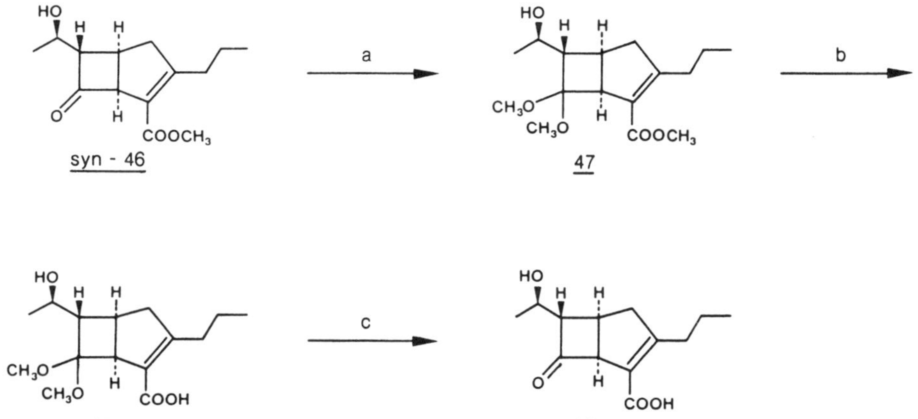

syn - 46 47 48 49

a = $HC(OMe)_3$, MeOH, $(COOH)_2$, rt (58 %); b = LiOH, MeOH, 60°C (86 %); c = 0.1 N HCl, THF (quant.).

scheme 17

To avoid side reactions in the hydrolysis of this relatively hindered ester the cyclobutanone was protected. This was done by ketalization with trimethyl orthoformate in methanol. Then the ester 47 was hydrolysed with LiOH and finally the cyclobutanone 39 was set free with aqueous HCl in THF. Unfortunately this compound did not show any antibacterial or β-lactamase inhibitory activity.

Another possible approach to analogues of β-lactam antibiotics is the synthesis of activated cyclobutanones. If they are able to acylate the target enzymes they might fulfill a function similar to the β-lactams.

For this strategy stands the C-analogue of a carbapenem which was published by Du Pont researchers [16b]. It shows a certain acylating ability in the reaction with benzylamine and it indeed shows weak antibacterial activity. As our second target we defined therefore the penem - analogue 50 of the Du Pont cyclobutanone 49 [16b].

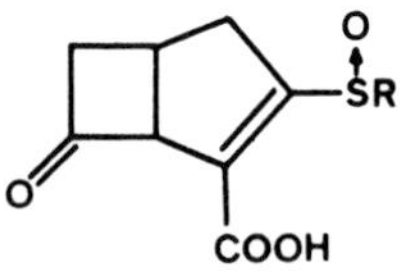

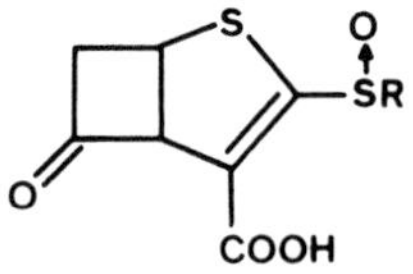

scheme 18

The reaction of vinyl magnesium bromide with sulfur and dibromomethane gave the bis thiovinyl ether 51 in 44 % yield [20]. By condensation with oxalic ester the α-ketoester 52 was obtained.

Olefination and hydrolysis yielded the half ester 53, which represents the starting material for the intramolecular ketene cycloaddition. To transform 53 into the corresponding acid chloride we made use of 1-chloro-N,N,2-trimethylpropenylamine developed by Ghosez [21]. The acid chloride was not isolated but treated in situ with $NEt_3$. The vinylketene 54 thus formed reacted immediately with the electron rich double bond of the thiovinylether. Thereby the skeleton of our target C-analogue was synthesized.

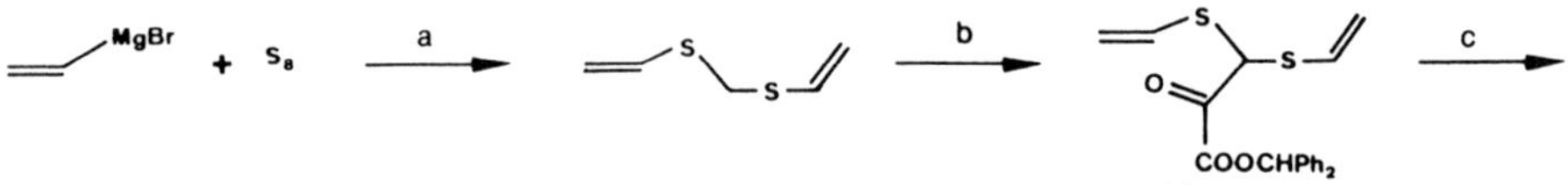

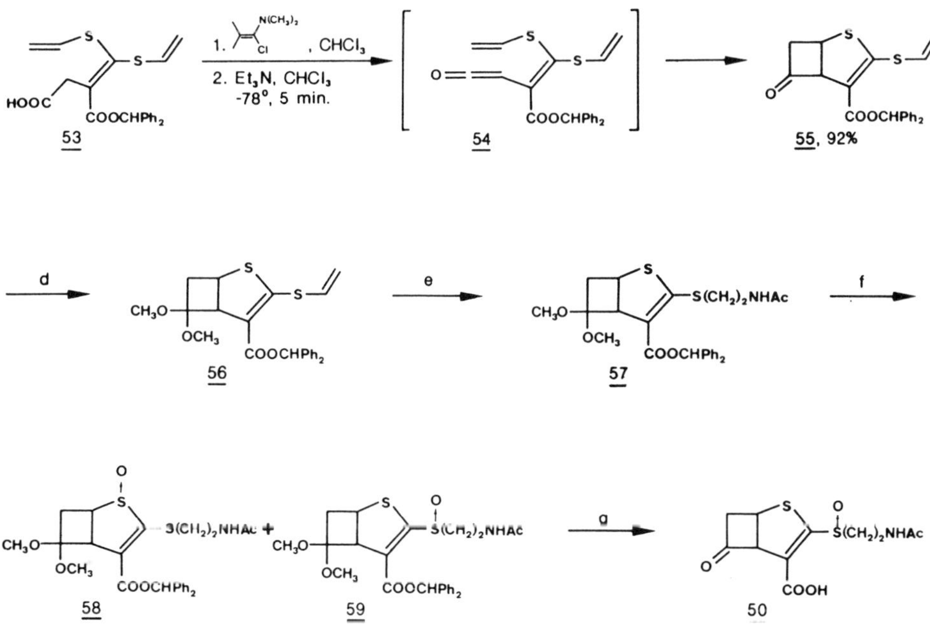

scheme 19

a = THF, 30°C; CH$_2$Br$_2$, 60°C (44 %); b = (COOCHPh$_2$)$_2$, nBuLi, THF (81 %); c = Ph$_3$PCHCOOMe; LiOH, H$_2$O (31 %); d = HC(OMe)$_3$, MeOH, pTsOH, rt (90 %); e = HSCH$_2$CH$_2$NH$_2$, NaH, DMF, rt; Ac$_2$O (66 %); f = mCPBA, CH$_2$Cl$_2$ (58: 51 %; 59: 23 %); g = C$_6$H$_5$OMe, CF$_3$COOH (60 %).

After protection of the cyclobutanone as its ketal ( ⟶ 56) the thiovinyl side chain was functionalized ( ⟶ 57).

The oxidation with mCPBA proceeded without regioselectivity and gave the two sulfoxides 58 and 59, separated by flashchromatography. The remaining steps, namely hydrolysis of the esters and deprotection of the cyclobutanone were accomplished by standard procedures.

Since we have found that, in contrast to the all-C-analogue 39, the cyclobutanone 50 was ring opened very easily with nucleophiles such as benzylamine we tested this intermediate with great expectations. It shows, similar to the C-analogue synthesized by Du Pont [14b], a weak antibacterial activity.

## 3. INTRAMOLECULAR [2+2] CYCLOADDITIONS OF α-SUBSTITUTED VINYLKETENES AND OLEFINS

Another investigation was directed towards the influence of substituents in the α-position of the ketene.

Quite astonishingly we found that starting from the α-chloro-derivative 60 the cycloadduct 61 is formed only in low yield. In case of the methyl-derivative 62 the bridgehead substituted bicyclic compound 64 could not be obtained, even after 60 hrs of reflux in toluene.

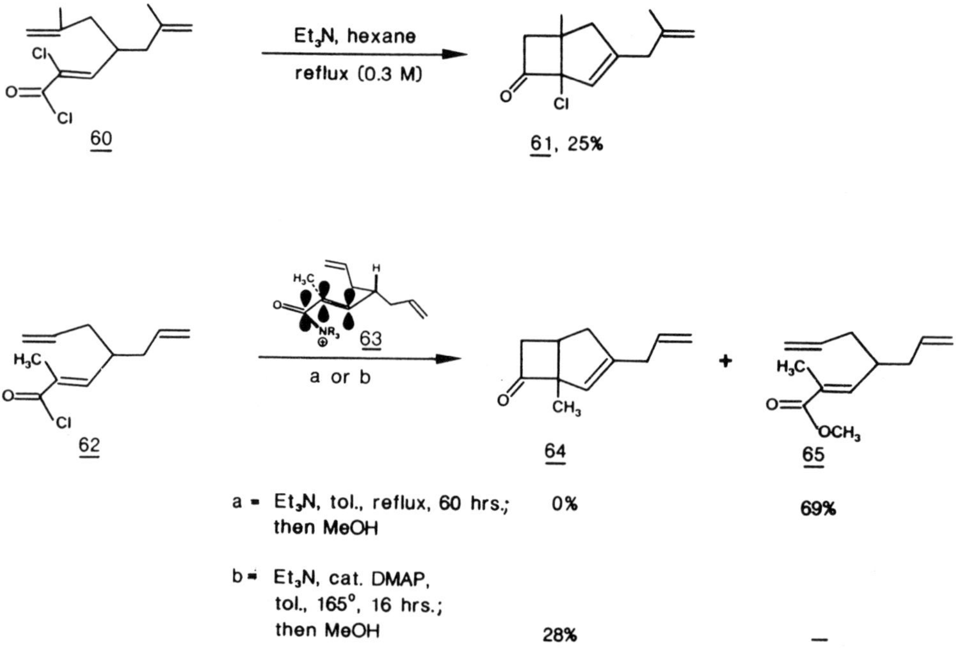

a = Et₃N, tol., reflux, 60 hrs.; then MeOH          0%          69%

b = Et₃N, cat. DMAP, tol., 165°, 16 hrs.; then MeOH             28%          —

scheme 20

When the reaction mixture was quenched with methanol the α,β-unsaturated ester 65 was isolated in 69 % yield. The deconjugation of the double bond which usually takes place in the methanol quenching of vinylketenes [22] was not observed.

The reason for the failure of this reaction is therefore not a problem in the cycloaddition step but in the formation of the ketene itself. This is either due to the difficult formation of the acylammonium complex 63 or to the difficult deprotonation in the γ-position. For the deprotonation the C-H bond has to be orientated parallel to the p-orbitals, but in this conformation severe steric interactions are taking place.

Therefore the 1,4-dehydrohalogenation was tried under harsher conditions and indeed the cycloadduct 64 could be isolated, albeit in low yield, employing DMAP as catalyst.

In order to prove that the 1,4-dehydrohalogenation of the sterically congested system and not the cycloaddition is the rate determining step the vinylketene 69 should be generated by an alternative procedure. Such a possibility is the electrocyclic ringopening of cyclobutenones [23].

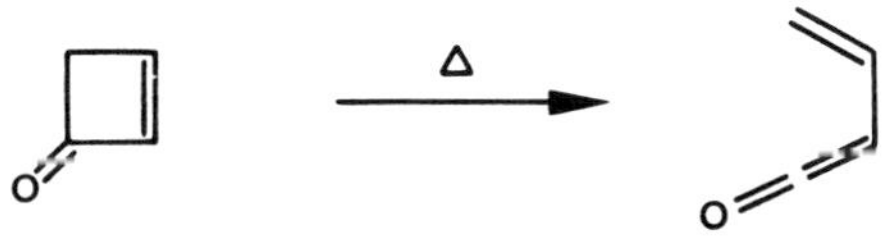

scheme 21

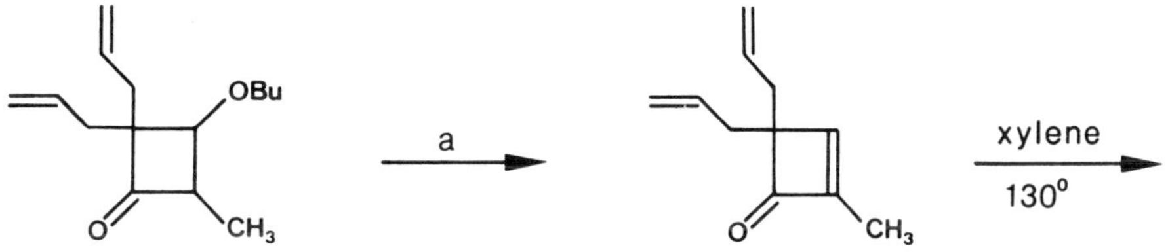

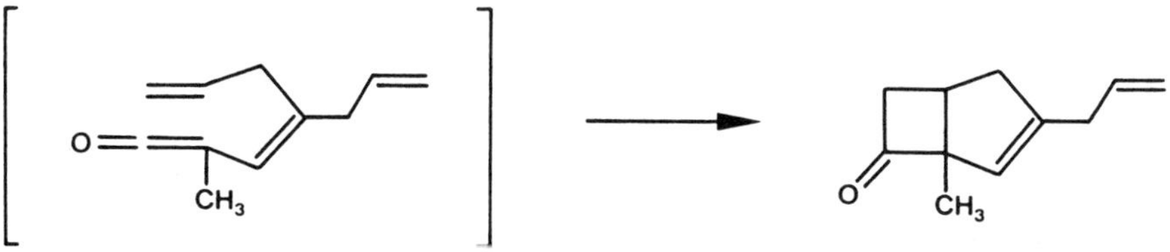

scheme 22

a = Me$_3$SiOTf, CH$_2$Cl$_2$, Et$_3$N (64 %).

The cyclobutenone we used to generate the vinylketene 69 was synthesized starting from cyclobutanone 67 which was easily accessible by an intermolecular ketene cycloaddition. The elimination of butanol was only successful when trimethylsilyltriflate/NEt$_3$ was used. With Al$_2$O$_3$ [24], TiCl$_4$ or BBr$_3$ the desired product was formed only in traces.

Thermolysis of 68 at 130°C gave, as anticipated, the intramolecular cycloadduct 64 of vinylketene 69 in good yield.

## 4. INTRAMOLECULAR [2+2] CYCLOADDITIONS OF KETENES WITH OLEFINS

The increase of geometrical restrictions going from ketenes to vinylketenes made the stereocontrolled construction of bicyclic cyclobutanones a synthetically feasable process.

The necessary conformational rigidity can also be imposed by an annulated ring system. This would lead to tricyclic cyclobutanones. By ring expansion the carbon skeleton of triquinanes would be available.

Triquinanes were intensively studied recently due to their particular structures but also due to the interesting physiological activities of some members of this class of natural products.

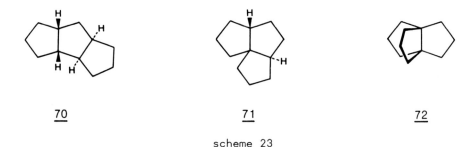

scheme 23

The C-skeleton of the linear annulated triquinane 70 is found e.g. in coriolin [25] or hirsutene [26], the one of the angular annulated triquinane 71 e.g. in isocomene [24] or pentalenene [28] and finally the one of the bisangular annulated triquinane 72 in modhephene [29].

### 4.1. Linear annulated triquinanes [30]

We planned to obtain the tricyclic cyclobutanone 73, a precursor which is easily transformed into the linear annulated cis-anti-cis triquinane 70 by an intramolecular ketene cycloaddition. The cis-substituted cyclopentane 74 would therefore be the appropriate starting material for the cycloaddition.

To avoid configurational problems in the synthesis of 74 an additional double bond was introduced ( ⟶ 75). The starting material for the cycloaddition reaction is therefore the acid chloride 76.

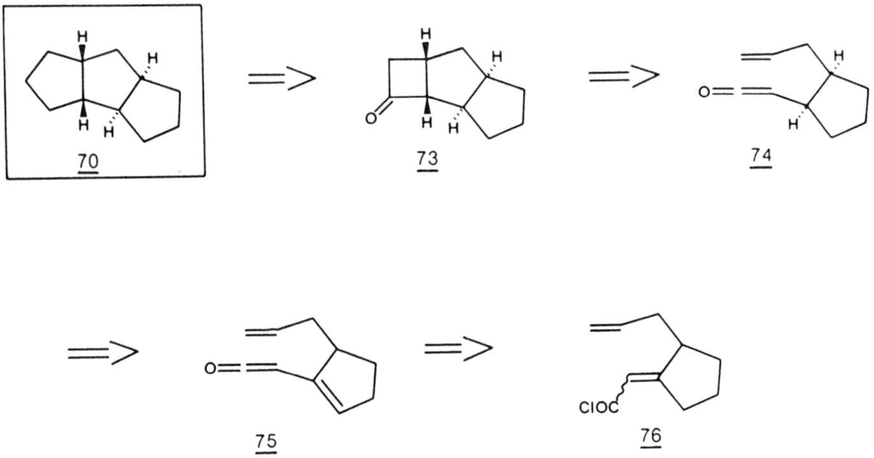

scheme 24

An initial reaction sequence (scheme 25), starting from the allylcyclopentanone 77 [31], yielded mainly the undesired cis-syn-cis fused compound 83. Double bond migration into the five membered ring took place during the Horner-Emmons olefination [32] to the extent of 30 % and even more during alkaline hydrolysis. Consequently the β,γ-unsaturated acid 80B was obtained as the major product. Conversion of the mixture of 78B, 79B and 80B into the corresponding acid chlorides by treatment with 1-chloro-N,N,2-trimethylpropenylamine [20], followed by addition of triethylamine yielded the cyclobutanones 81 and 82. Hydrogenation of this mixture finally led mainly to the undesired cis-syn-cis fused compound 83.

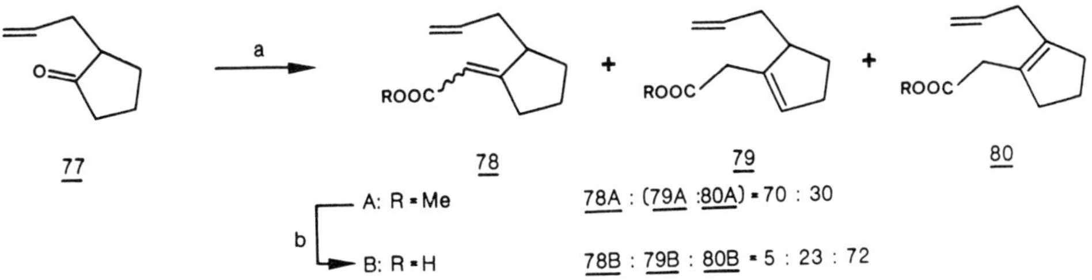

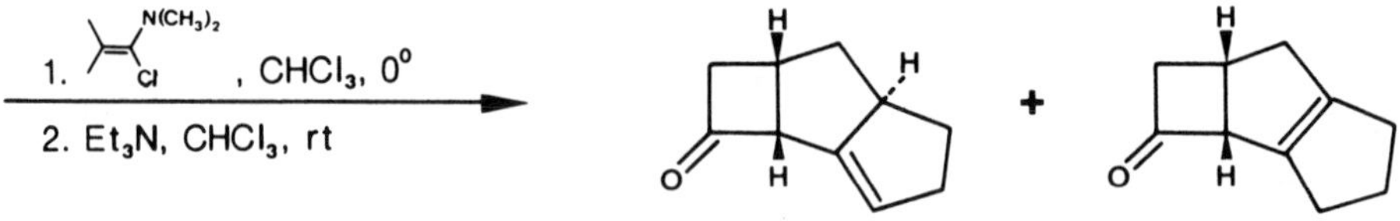

81  75% (81:82 = 25:75)  82

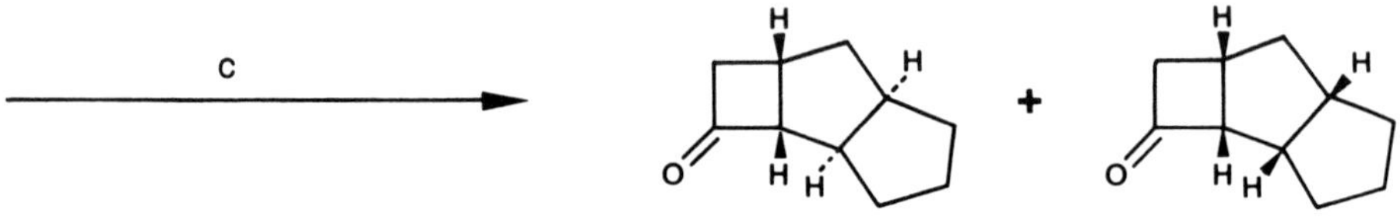

73  65% (73:83 = 25:75)  83

a = (MeO)$_2$POCH$_2$COOMe, NaH, DMF, 60°C (67 %); b = KOH, EtOH/H$_2$O, rt (75 %); c = H$_2$, Pd/C, iPrOH (65 %; 73:83 = 25:75).

scheme 25

A selective synthesis of the compound 73 could be realized by preventing the isomerizations mentioned above. Peterson olefination of 77 furnished exclusively the α,β-unsaturated esters 84, if the reaction mixture was quenched at -25°C with aqueous NH$_4$Cl. The hydrolysis of the tert.butylesters 84 was also proceeding without double bond migration (scheme 26). Interestingly, triethylamine abstracted a proton of C-5 of the acyl chloride 76, with high regioselectivity ( ⩾ 97 %), giving the intermediate vinylketene 75 which cyclized readily at room temperature to the single epimer 81. The stereochemistry was verified on the lactone 85. As expected, the catalytic hydrogenation of the cyclobutanone 81 yielded the cis-anti-cis ring fusion ( ⟶ 73). The ketone 86 with the desired triquinane skeleton was obtained by ring expansion [33].

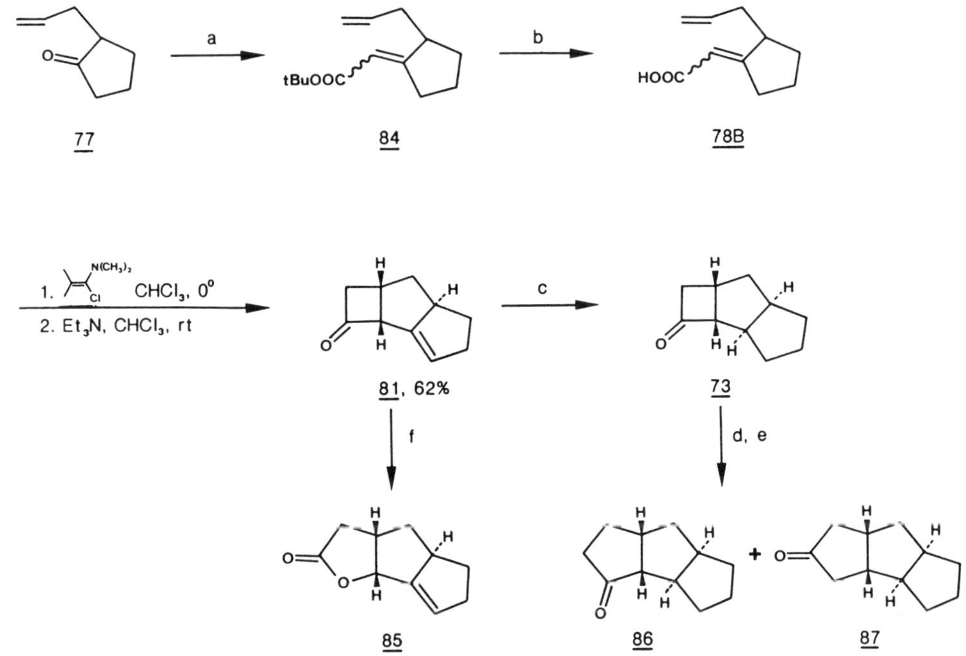

225

a = LICA, Me₃SiCH₂COOtBu, THF, -78°C, 1 hr ⟶ -25°C, 2 hrs; NH₄Cl
aq., -25°C (70 %); b = CF₃COOH, CH₂Cl₂, rt (88 %); Z/E = 67:33);
c = H₂, Pd/C, THF/AcOH (77 %), d = N₂CHCOOEt, BF₃·Et₂O (89 %); e = 4N
HCl, AcOH, reflux (79 %); f = AcOH, H₂O₂ aq. (65 %).

scheme 26

## 4.2. Angular annulated triquinanes [34]

For the synthesis of the angular annelated triquinane 71 we started
either from vinylketene 89 or ketene 90. The former has the advantage
that a synthetically useful olefinic functionality is incorporated in
the molecule.

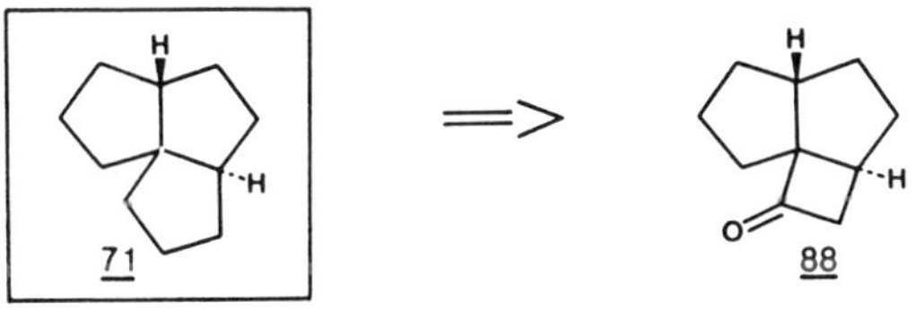

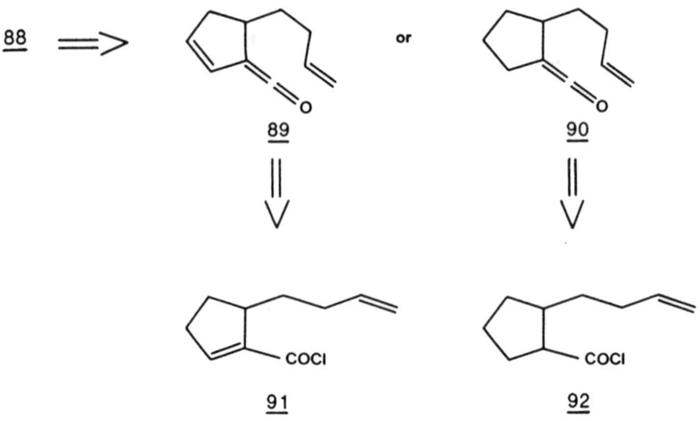

scheme 27

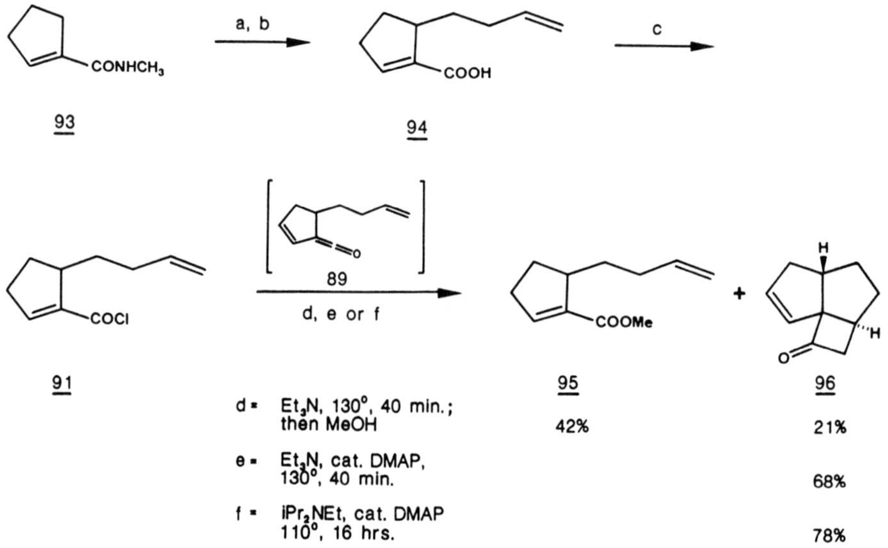

**93** → (a, b) → **94** → (c) →

**91**

+ **89** (d, e or f) →

**95** 42%

+ **96** 21% / 68% / 78%

d = Et₃N, 130°, 40 min.; then MeOH

e = Et₃N, cat. DMAP, 130°, 40 min.

f = iPr₂NEt, cat. DMAP, 110°, 16 hrs.

a = sBuLi, TMED, THF, −100°C; CH₂=CHCH₂CH₂Br [35] (20 %); b = NaOH, HOCH₂CH₂OH, 180°C (75 %); c = Me₂C=CClNMe₂, CHCl₃, 0° (quant.).

scheme 28

Using the α,β-unsaturated acid chloride 91 as ketene precursor suitable experimental conditions proved to be crucial for the success of the cycloaddition.

When 91 was treated under conditions d, (i.e. 130°C for 40 minutes followed by quenching with MeOH) we isolated the α,β-unsaturated ester 95 besides a small amount of cycloadduct 96. It is likely that 95 originates from the acid chloride 91 rather than the vinylketene 89. The latter would produce, upon MeOH quench, at least a partially deconjugated product [22] which actually was not detected.

Presumably, during the generation of vinylketene 89 the formation of an acyl-ammonium complex is the most difficult step, due to steric hindrance.

By use of a catalytic amound of 4-dimethylaminopyridine (DMAP) under otherwise similar conditions (conditions e) the amount of the acylammonium complex should be increased. Indeed a beneficial effect upon rate and yield of the reaction was observed. The yield of 96 could even be improved at somewhat lower temperature using the less volatile diisopropylethylamine (conditions f). To our best knowledge this is the first DMAP catalysis demonstrated in ketene cycloadditions.

The saturated angular annulated cyclobutanone 88 could be obtained similarly. Treatment of acid chloride 92 (3:1 mixture of epimers) gave the cycloadduct 88 in 67 % isolated yield under these optimized reaction conditions.

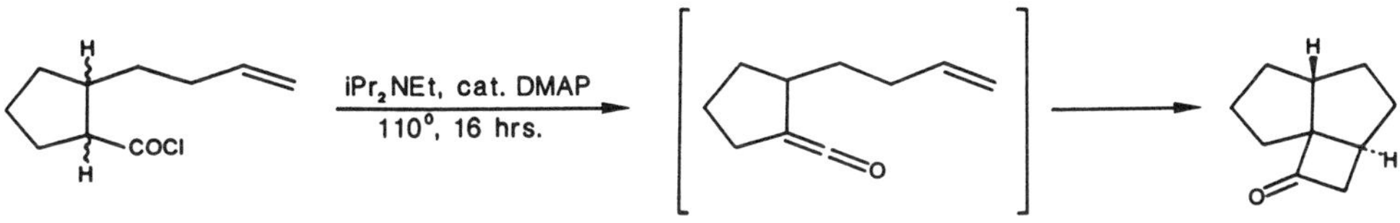

scheme 29

The angular annulated triquinane ring system was finally obtained through the following sequence: Treatment of 96 with ethyl diazoacetate in the presence of $BF_3 \cdot OEt_2$ [33] gave a 8:1 mixture of the isomeric ketoesters 97 and 98, which could be separated on silica gel. Decarboxylation of 97 produced the angular annulated triquinane 99 in 90 % yield.

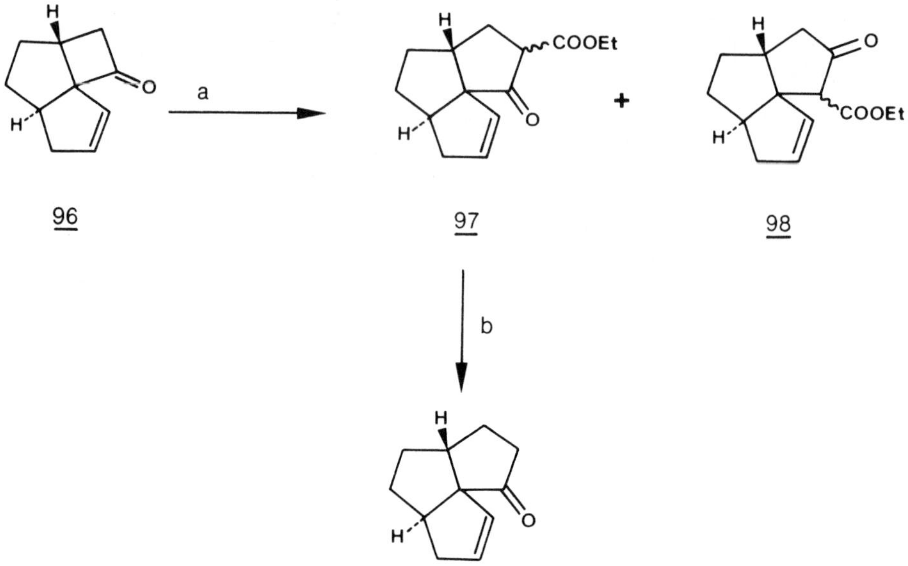

96 → 97 + 98

↓ b

99

a = N$_2$CHCOOEt, BF$_3$·Et$_2$O (97:81 %; 98:10 %); b = 4N HCl, AcOH, reflux (90 %).

scheme 30

The above mentioned methodology opens an elegant and straightforward approach to natural occuring angular annulated triquinanes like isocumene [27], and pentalene [28].

A similar approach to the angular annulated triquinane skeleton was recently published by Yadav et al. [36].

4.3. Bisangular annulated triquinanes

Finally we investigated the intramolecular approach to the bisangular annulated triquinane 72. Again the cyclobutanone 100 is a reasonable precursor. It was planned to synthesize it from ketene 101, which can be generated from unsaturated acid 102.

scheme 31

For a concerted cycloaddition to take place ketene and ketenophile have to approach each other more or less orthogonally. The optimum conformations of ketene 101 were calculated using the semiempirical quantum chemical method AM1 [37]. It was found that all conformations of low energy have parallel arrangement of ketene and ketenophile. The three most stable conformations are shown in scheme 32. An energetically reasonable conformation where the two reacting functionalities approach each other orthogonally is not possible. If therefore the cycloaddition takes place a non concerted mechanism is required.

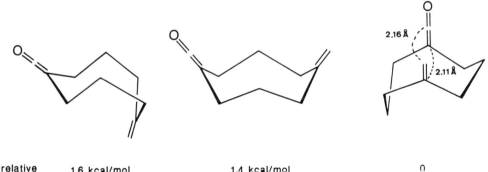

scheme 32

The acid which was obtained from ethyl-2-cyclohexanone carboxylate (103) [38] was transformed again with 1-chloro-N,N,2-trimethylpropenylamin [21] to the acid chloride which was treated in situ with base.

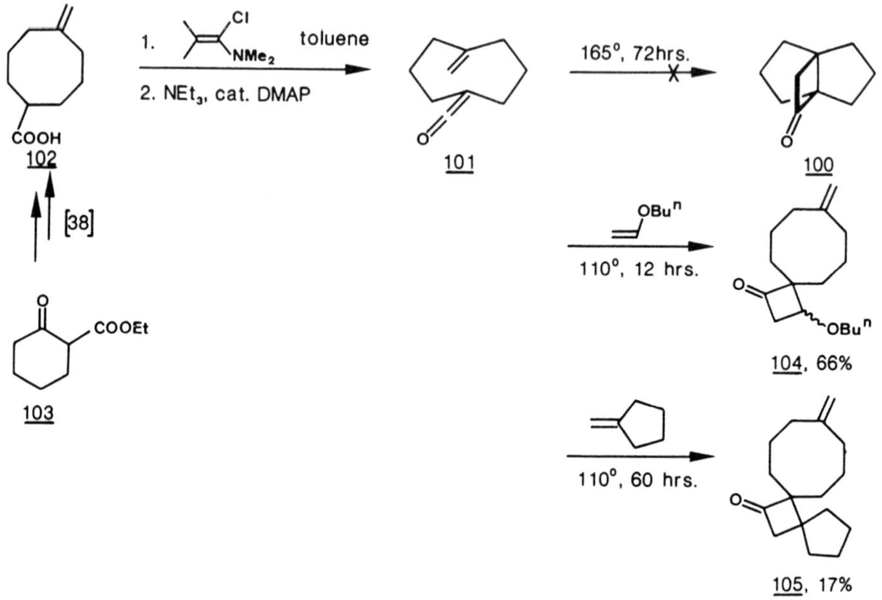

scheme 33

Even under harsh conditions the cycloadduct could not be detected. This raised the following two questions:
First, is the product under these reaction conditions stable and second, is the ketene really generated? The first question was answered by exposing the tricyclic cyclobutanone 100 which was synthesized by an alternative route [39] to the reaction conditions. It proved to be stable. The second question was answered with a trapping experiment. With butylvinylether as well methylene cyclopentane the intermolecular cycloadducts 104 and 105, respectively, were isolated.
These results lead to the conclusion, that an intramolecular [2+2] cycloaddition takes place only when a more or less orthogonal approach of the two reactants is possible.
An example that illustrates this finding is described in scheme 4[5]. There the stereocontrol of the cycloaddition reaction results from an orthogonal approach of ketene and ketenophile. The regioisomer resulting from a parallel approach was not isolated.

## 5. CONCLUSION

Intramolecular [2+2] cycloadditions of vinylketenes (cf. section 2 and 3) and ketenes (cf. section 4) to olefins have proved to be a versatile method for the synthesis of a variety of bi- and tricyclic cyclobutanones. Recently a number of basic investigations in this field but also of applications in natural product synthesis have been published.

Ghosez et al. [11] investigated intramolecular [2+2] cycloadditions of keteniminium salts which are more electrophilic than ketenes and do not dimerize [8]. An application of this strategy is his short synthetic route to prostaglandins [40].

Snider et al. [41] showed that alkenyloxy ketenes, prepared from acid chloride by treatment with triethylamine undergo facile intramolecular [2+2] cycloadditions to give polycyclic cyclobutanones.

A similar approach to polycyclic cyclobutanones was presented by Brady [42] starting from phenoxy ketenes.

In an elegant synthesis of (±)-retigeranic acid Corey [43] incorporated an intramolecular [2+2] cycloaddition of a vinylketene to an olefin as the key step.

By the same type of cycloaddition Oppolzer & Nakao [44] synthesized (±)-protoilludene. In 1986 intramolecular ketene cycloadditions have been reviewed by Reisig [45].

## 6. REFERENCES

[1] H. Staudinger, Chem. Ber. 44 (1911) 521.
[2] H.K. Hall, D.E. Plorde, US-Pat. 3646150 (1968), Du Pont.
[3] W.T. Brady, O.H. Waters, J. Org. Chem. 32 (1967) 3703.
[4] J.B. Sieja, J. Am. Chem. Soc. 93 (1971) 130.
[5] R.R. Sauers, K.W. Kelly, J. Org. Chem. 35 (1970) 3286.
[6] a) J.J. Beereboom, J. Am. Chem. Soc. 85 (1963) 3525; J. Org. Chem. 30 (1963) 4230; b) W.F. Erman, J. Am. Chem. Soc. 91 (1969) 779; c) S.W. Baldwin, E.H. Page, J. Chem. Soc. Chem. Commun. 1972, 1337; d) F. Leyendecker, R. Bloch, J.M. Conia, Tetrahedron Lett. 1972, 3703; e) R.H. Bisceglia, C.J. Cheer, J. Chem. Soc. Chem. Commun. 1973, 165; f) S. Moon, T.F. Kolesar, J. Org. Chem. 39 (1974) 995; g) L. Libit, US-Pat. 4005109 (1974); h) A. Smit, J.G.J. Kok, H.W. Geluk, J. Chem. Soc. Chem. Commun. 1975, 513; i) F. Leyendecker, Tetrahedron 32 (1976) 349; j) T. Sasaki, S. Eguchi, Y. Hirako, J. Org. Chem. 42 (1977) 2981; k) A. Maujean, G. Marcy, J. Cuche, J. Chem. Soc. Chem. Commun. 1980, 92; l) M. Kuzuya, F. Miyake, T. Okuda, Tetrahedron Lett. 1980, 1043 and 2185; m) A. Alder, D. Bellus, J. Am. Chem. Soc. 105 (1983) 6712; n) H. Dhimane, J.C. Pommelet, J. Cuche, Tetrahedron Lett. 26 (1985) 833; o) F. Arya, J. Bouquant, J. Cuche, Tetrahedron Lett. 27 (1986) 1913.
[7] a) W.F. Erman, J. Am. Chem. Soc. 89 (1967) 3828; b) O.L. Chapman, J.D. Lassila, J. Am. Chem. Soc. 90, (1968) 2449; c) O.L. Chapman, M. Kane, J.D. Lassila, R.L. Loeschen, H.E. Wright, J. Am. Chem. Soc. 91 (1969) 6856; d) A.S. Kende, Z. Goldschmit, P.T. Izzo, J. Am. Chem. Soc. 91 (1969) 6858; e) H. Hart, G.M. Lowe, J. Am. Chem. Soc. 93 (1971) 6266; f) D. Becker, M. Nagler, D. Birnbaum, J. Am. Chem. Soc. 94 (1972) 4771; g) Z. Goldschmit, U. Gutman, Y. Bakal, A. Worchel, Tetrahedron Lett. 1973, 3759; h) D. Becker, Z. Harel, D. Birnbaum, J. Chem. Soc. Chem. Commun. 1975, 377; i) S. Ayral-Kaloustian, S. Wolff, W.C. Agosta, J. Org. Chem. 43 (1978), 3314; j) D. Becker, D. Birnbaum, J. Org. Chem. 45 (1980) 570; k) R.E. Ireland, J.D. Godfrey, S. Thaisrivongs, J. Am. Chem. Soc. 103 (1981) 2446; e) E. Lee-Ruff, A.C. Hopkinson, H. Kazarians-Moghaddam, Tetrahedron Lett. 24 (1983), 2067; m) A.G. Schultz, J.P. Dittami, K.K. Eng, Tetrahedron Lett. 25 (1984) 255.
[8] L. Ghosez, M.J. O'Donnell in Pericyclic Reactions, A.P. Marchand, R.E. Lehr, Eds., Academic Press, New York, 1977, Vol. 2, p. 79-140.
[9] a) M. Kuzuya, F. Miyake, T. Okuda, J. Chem. Soc. Perkin Trans. II, 1984, 1471; b) E. Sonveaux, J.M. André, J. Delhalle, J.G. Fripiat, Bull. Soc. Chim. Belg. 94, (1985) 831.
[10] G.B. Payne, J. Org. Chem. 31 (1966) 718.
[11] I. Markó, B. Ronsmans, A.-M. Hesbain-Frisque, S. Dumas, L. Ghosez, B. Ernst, H. Greuter, J. Am. Chem. Soc. 107 (1985) 2192.

[12] I. Fleming in Frontier Orbitals and Organic Chemical Reactions, J. Wiley & Sons, 1976, 143 ff.
[13] a) Y.S. Kulkarni, B.B. Snider, J. Org. Chem. 50 (1985) 2809; b) Y.S. Kulkarni, B.W. Burbaum, B.B. Snider, Tetrahedron Lett. 26 (1985) 5619; c) B.B. Snider, E. Ron, B.W. Burbaum, J. Org. Chem. 52 (1987) 5413.
[14] E.J. Corey, M.C. Desai, Tetrahedron Lett. 26 (1985) 3535.
[15] M.I. Page, Acc. Chem. Res. 17 (1984) 144.
[16] a) E.M. Gordon, J. Pluščec, M.A. Ondetti, Tetrahedron Lett. 22 (1981) 1871; b) G.A. Boswell, Jr., A.J. Cocuzza, US-Pat. 4 505 905 (1982), Du Pont; c) O. Meth-Cohn, A.J. Reason, S.M. Roberts, J. Chem. Soc. Chem. Commun. 1982, 90; d) G. Lowe, S. Swain, ibid. 1983, 1279; e) A.J. Cocuzza, G.A. Boswell, Tetrahedron Lett. 26 (1985) 5363; f) G. Lowe, S. Swain, J. Chem. Soc., Perkin Trans, I 1985, 391; g) G. Lange, M.E. Savard, T. Viswanatha, G.I. Dmitrienko, Tetrahedron Lett. 26 (1985) 1791; h) D. Agathocleous, G. Cox, M.I. Page, ibid. 27 (1986) 1631.
[17] H. Muruyama, T. Hiraoka, J. Org. Chem. 51 (1986) 399 and references cited herein.
[18] Prepared by Dr. P. Schneider, Pharmaceuticals Division, CIBA-GEIGY Ltd., CH-4002 Basel.
[19] A. Wissner, C.V. Grudzinskas, J. Org. Chem. 43 (1978) 3972.
[20] The analogous reaction with vinyl lithium was described: C.L. Semmelhack, I.-C. Chiu, K.G. Grohman, J. Am. Chem. Soc. 98 (1976) 2005.
[21] B. Haveaux, A. Dekoker, M. Rens, A.R. Sidani, J. Toye, L. Ghosez, Org. Synth. 59, 26.
[22] L. Lombardo, Tetrahedron Lett. 26 (1985) 381.
[23] a) J.E. Baldwin, M.C. McDaniel, J. Am. Chem. Soc. 89 (1967) 1537; b) J.E. Baldwin, M.C. McDaniel, ibid. 90 (1968) 6118.
[24] H. Mayr, R. Huisgen, Ang. Chem. 87 (1975) 491.
[25] C. Exon, P. Magnus, J. Am. Chem. Soc. 105 (1983) 2477 and references cited therein.
[26] R.L. Funk, G.L. Bolton, J. Org. Chem. 49 (1984) 5021 and references cited therein.
[27] a) L.A. Paquette, Y.-K. Han, J. Am. Chem. Soc. 103 (1981) 1835; b) P.A. Wender, G.B. Dreyer, Tetrahedron 37 (1981) 4445; c) M.C. Pirrung, J. Am. Chem. Soc. 103 (1981) 82 and references cited therein.
[28] G.D. Annis, L.A. Paquette, J. Am. Chem. Soc. 104 (1982) 4504.
[29] H. Schostarez, L.A. Paquette, J. Am. Chem. Soc. 103 (1981) 722.
[30] A. De Mesmaeker, S.J. Veenstra, B. Ernst, Tetrahedron Lett. 29 (1988) 459.
[31] The allylcyclopentanone 77 was prepared in three steps from cyclopentanone via allylation of the corresponding dimethylhydrazone (overall yield 72 %) in analogy with E.J. Corey, D. Enders, Tetrahedron Lett. 1976, 3.
[32] For deconjugation of $\alpha,\beta$-unsaturated esters during Horner-Emmons and Peterson olefination see a) S.L. Hartzell, D.F. Sullivan, M.W. Rathke, Tetrahedron Lett. 1974, 1403; b) H. Taguchi,

Soc. Jpn. 47 (1974) 2529.
- [33] J.R. Stille, R.H. Grubbs, J. Am. Chem. Soc. 108 (1986) 855.
- [34] S.J. Veenstra, A. De Mesmaeker, B. Ernst, Tetrahedron Lett. 29 (1988) 2303.
- [35] P. Beak, D.J. Kempf, K.D. Wilson, J. Am. Chem. Soc. 107 (1985) 4745.
- [36] J.S. Yadav, B.V. Joshi, V.R. Gadgil, Indian J. Chem., Sect. B 26 (1987) 399.
- [37] The autors thank Dr. D. Poppinger, Zentrale Funktion Forschung, Computer Chemistry, Ciba-Geigy Ltd., CH-4002 Basel for the calculations according to: M.J.S. Dewar, E.G. Zoebisch, E.F. Healy, J.J.P. Stewart, J. Am. Chem. Soc. 107 (1985) 3902; Quantum Chemistry Program Exchange Nr. 506.
- [38] H. Marschall, F. Vogel, Chem. Ber. 107 (1974) 2176.
- [39] Y. Fukuda, T. Negoro, Y. Tobe, K. Kimura, Y. Odaira, J. Org. Chem. 44 (1979) 4557.
- [40] L. Ghosez, I. Marké, A.-M. Hesbain-Frisque, Tetrahedron Lett. 27 (1986) 5211.
- [41] a) B.B. Snider, R.A.H.F. Hui, Y.S. Kulkarni, J. Am. Chem. Soc. 107 (1985) 2194; b) B.B. Snider, R.A.H.F. Hui, J. Org. Chem. 50 (1985) 5167.
- [42] a) W.T. Brady, Y.F. Giang, J. Org. Chem. 50 (1985) 5177; b) W.T. Brady, Y.F. Giang, A.P. Marchand, A.H. Wu, J. Org. Chem. 52 (1987) 3457; c) W.T. Brady, A.P. Marchand, Y.F. Giang, A.H. Wu, Synthesis 1987 395.
- [43] E.J. Corey, M.C. Desai, T.A. Engler, J. Am. Chem. Soc. 107 (1985) 4339.
- [44] W. Oppolzer, A. Nakao, Tetrahedron Lett. 27 (1986) 5471.
- [45] H.-U. Reissig, Nachr. Chem. Tech. Lab. 34 (1986) 880.

# STRAINED IMINIUM SALTS IN SYNTHESIS

Léon Ghosez*, Chen Lian Yong, Benoît Gobeaux, Cathy Houge,
Istvan Marko, Matthew Perry and Hiroyuki Saimoto
Laboratoire de Chimie Organique de synthèse
Université Catholique de Louvain
Place Louis Pasteur, 1 - 1348 Louvain-La-Neuve, BELGIUM

ABSTRACT. Cycloadditions of olefins to keteniminium salts derived from chiral amides provide an excellent method of synthesis of homochiral cyclobutanones. Applications to the asymmetric alkylation of olefins and the synthesis of optically active γ lactones or prostanoids will be discussed.

Strained rings are no longer considered as chemical curios of interest to the structural chemist. The unusual bonding of three- and four-membered rings and the relief of potential energy associated with cleavage of these rings has been successfully used for the construction of complex molecular frameworks (1). Our own interest in this area has been focused on the synthesis and applications of four-membered rings.

Heterosubstituted ketenes such as dichloro-or thioalkylketenes first discovered in our laboratory have been extensively used (2). They are key reagents in a two-step method for the stereocontrolled vicinal

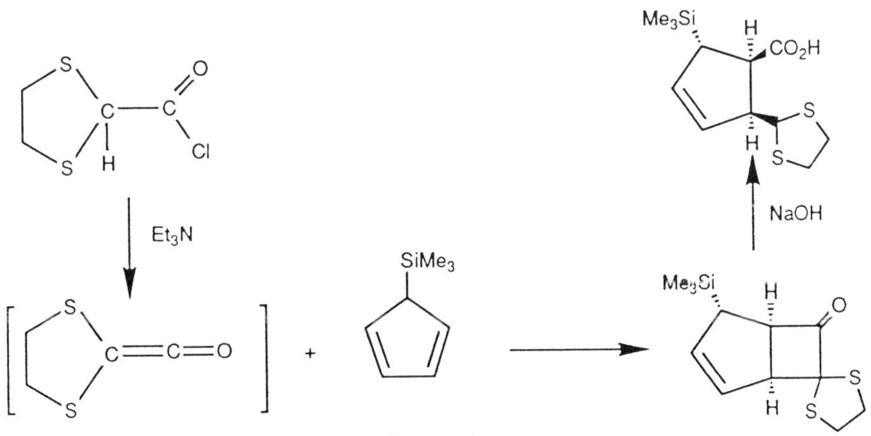


Scheme 1



*A. de Meijere and S. Blechert (eds.), Strain and Its Implications in Organic Chemistry, 235–254.*

alkylation of olefins (Scheme 1)(3). Both cycloaddition and ring opening reactions occur with retention of configuration. When applied to cyclopentadiene, the method provides a short high yield synthesis of functionalized cyclopentene derivatives. The vicinal alkylation of cyclopentadiene with (carbomethoxy)chloroketene has opened a practical route to advanced intermediates in the synthesis of prostaglandin hormones and their analogues (Scheme 2) (4).

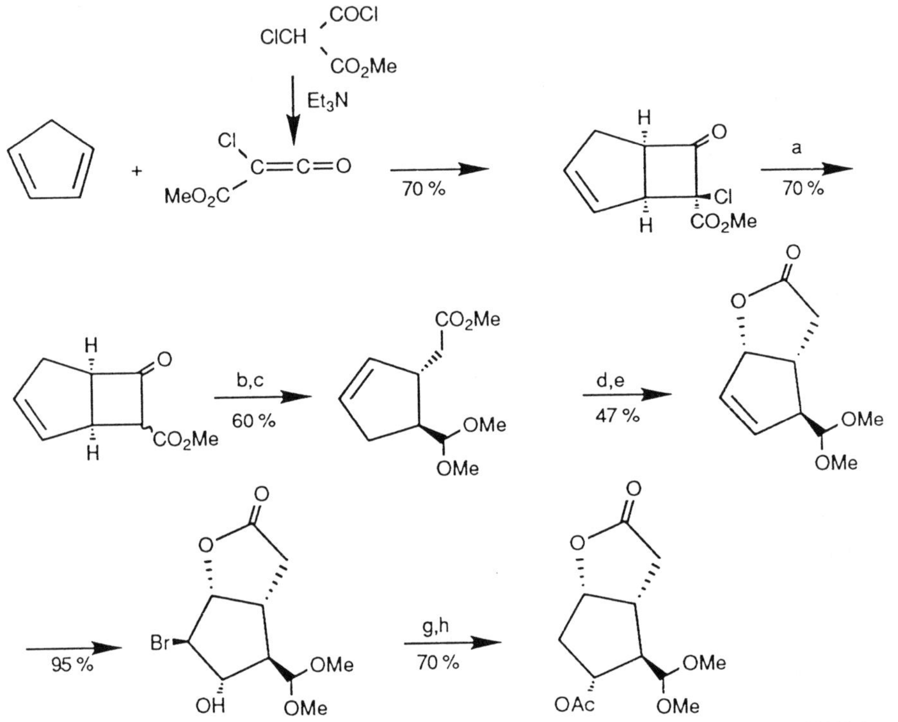

Reagents : a,Bu$_3$SnH ; b,NaBH$_4$ methanol,-70°C then MeONa → 20°C ; c,HC(OMe)$_3$ ; d,KOH then KI$_3$ ; e,DBU ; g,CH$_3$COCl-pyridine ; h,Bu$_3$SnH.

Scheme 2

Obviously, the availability of asymmetric [2+2] cycloadditions would considerably enhance the usefulness of this methodology. Disappointingly the reaction of cyclopentadiene with ketenes 1 or 2 bearing a chiral auxiliary gives poor diastereomeric excesses (<20%) (Scheme 3).

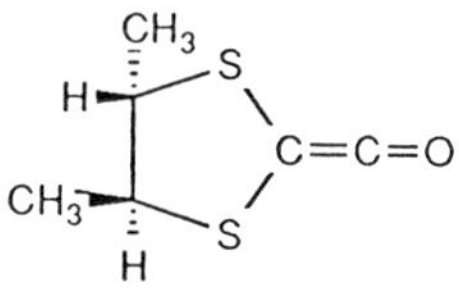

Scheme 3

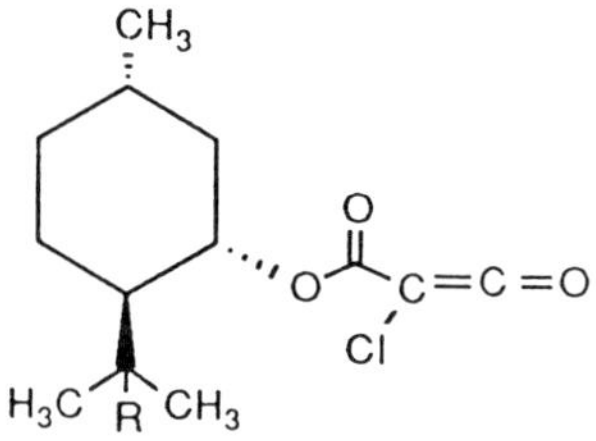

We have then examined a ketene equivalent bearing the chiral auxiliary on the $C_1$ carbon atom where bonding is expected to be the most advanced in the transition state of the cycloaddition. In principle keteniminium salts derived from chiral amines meet this structural modification.

Keteniminium salts bearing no hydrogen at C-2 are readily available from the corresponding α-chloroenamines (5). These are conveniently prepared by phosgenation of tertiary amides followed by treatment of the amide chloride 3 with triethylamine (Scheme 4)(6). When $R^1$ and $R^2$ are

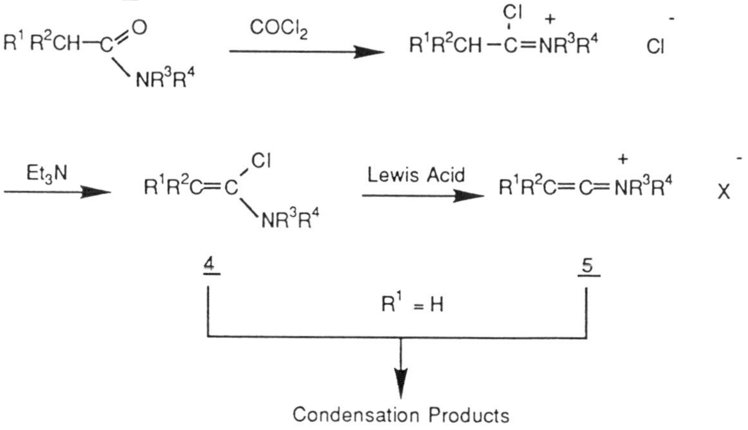

Scheme 4

different from hydrogen, the α-chloroenamine 4 is thermally stable and can be readily purified by distillation.

α-Chloroenamines are ambiphilic molecules: interaction of the nitrogen lone-pair with the carbon-carbon double bond confers nucleophilic character at C-2. However, a 90° rotation around the C-N bond allows an interaction of the nitrogen lone pair with the σ* C-Cl orbital. This promotes the heterocyclic cleavage of the C-Cl bond, yielding the strongly electrophilic keteniminium chloride. When $R^1$ and $R^2$ are different from hydrogen, the nucleophilic character of 4 is weakened: treatment of 4 with Lewis acid ($ZnCl_2$, $TiCl_4$, $SnCl_4$, $Ag^+$) generates keteniminium salt 5 without competitive condensation reactions. However, when $R^1$=H, the same conditions yield products resulting from the reaction of the nucleophilic enamine 4 with the electrophilic iminium salt 5 (5). This undesired reaction can be suppressed by replacing the chlorine substituent by a better leaving group.

The reaction of tertiary amides with triflic anhydride generates α-triflyliminium triflates which have been identified by NMR spectroscopy (7). In the presence of a weak base such as collidine, elimination of triflic acid occurs to yield an α-triflylenamine which spontaneously ionises to give the corresponding keteniminium salt (Scheme 5).

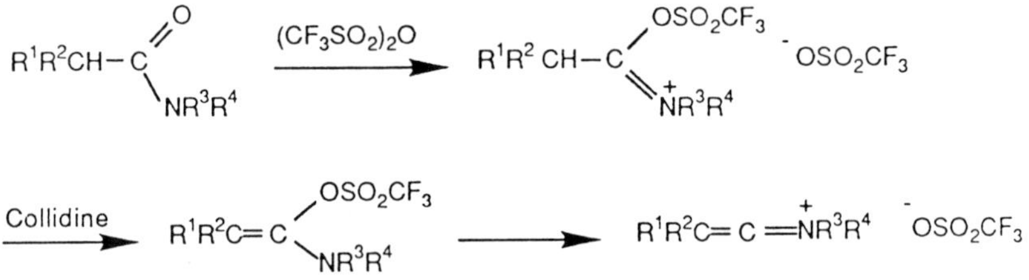

$R^1R^2CH-C(=O)NR^3R^4$
$(CF_3SO_2)_2O$
$R^1R^2CH-C(OSO_2CF_3)=\overset{+}{N}R^3R^4$ $^-OSO_2CF_3$
Collidine
$R^1R^2C=C(OSO_2CF_3)NR^3R^4$
$R^1R^2C=C=\overset{+}{N}R^3R^4$ $^-OSO_2CF_3$


Scheme 5

Keteniminium salts have been trapped in situ by a large variety of olefins (Scheme 6)(5,8). The high reactivity of keteniminium salts is demonstrated by their facile reaction with propylene or cyclohexene at room temperature. The corresponding aldoketenes do not react under those conditions. Good yields of cyclobutaniminium salts 6 or their hydrolysis products 7 have been obtained with several mono- and 1,2 disubstituted olefins.

$R^1R^2CH-C(=O)-NAlk_2$
Olefin
$(CF_3SO_2)_2O$ + Collidine
method B
method A
1. $COCl_2$
2. $Et_3N$
Alk
N +
Alk
$H_2O$
$R^3$
$R^4$
O
$R^1R^2C=C(Cl)NAlk_2$
Olefin + Lewis Acid
method A
6
7


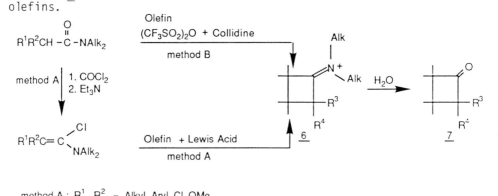

method A : $R^1$, $R^2$ = Alkyl, Aryl, Cl, OMe

method B : $R^1$, $R^2$ = H, Alkyl, Aryl, SR, OMe, F, Cl

Scheme 6

Cycloadditions of keteniminium salts derived from chiral amides have been examined. In model studies, we selected amides derived from derivatives of L-proline. 1-Acetyl-2-(methoxymethyl)pyrrolidine 8 has been converted into the corresponding keteniminium triflate 9 in the presence of an excess of cyclopentene. Hydrolysis ($H_2O$, overnight) yields pure cycloadduct 10. The optical purity of 10 is determined by comparing the optical rotation of 10 with that of an authentic sample of (1R, 5R)-bicyclo[3.2.0]hepten-6-one. The enantiomeric excess in the adduct 10 is 55.4% in favour of the (1S,5S) isomer 10b (Scheme 7)(9).

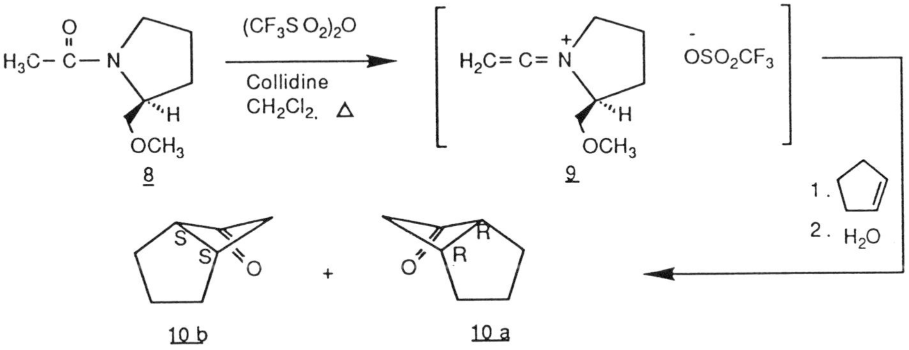

Scheme 7

The diastereoselectivity is even higher for the reaction of cyclopentene with the β-disubstituted keteniminium salt 11 which is formed in situ from the corresponding α-chloroenamine 12 in the presence of zinc chloride (Scheme 8). Hydrolysis of the crude iminium salt 13 yields cyclobutanone 14 of high optical purity (≥ 95%)(9).

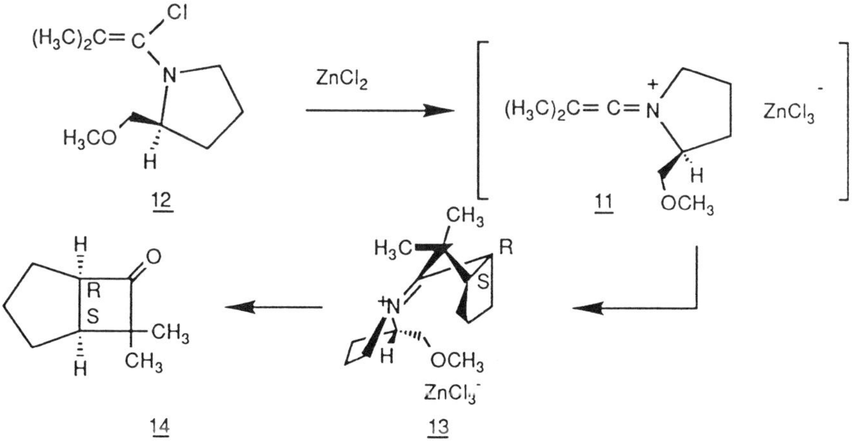

Scheme 8

Configuration of the adduct has been shown to be (1R,5S)(9). Thus, in addition to providing a much higher enantiomeric excess (95% vs 55%) the reaction of cyclopentene with 11 also produces an adduct of opposite configuration to that formed from the unsubstituted keteniminium salt 9.

Excellent diastereoselectivities (≥94%) are also observed with keteniminium salts 15 and 16 derived respectively from a derivative of (-) ephedrine and from (-) 2-methylpyrrolidine (Scheme 9). The latter result indicates that, in contradiction to our earlier expectation, the methoxy group in 11 plays no significant role in the discrimination between the two diastereoisomeric transition states of the cycloaddition.

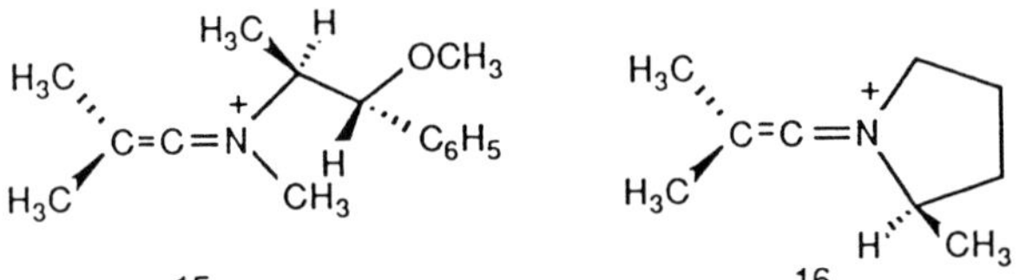

Scheme 9

Diastereoselectivities of the reactions of 11 with cyclohexene (a less rigid cyclic olefin) and with open-chain alkenes are slightly lower (Scheme 10).

The chemical reactivity of cyclobutanones makes these strained ketones exceedingly useful building blocks. For instance they readily undergo Baeyer-Villiger oxidation to yield the corresponding γ-lactones. Using performic acid generated in situ at -20°C, bond migration was highly specific as shown by the conversion of 17 to 19 with a 95:5 regioselectivity. The reaction does not affect the chiral centers. The sequence - asymmetric [2+2] cycloaddition + Baeyer-Villiger oxidation - represents a new and potentially useful synthesis of homochiral γ-lactones and products derived therefrom. This sequence is also a two-step method for the enantiospecific vicinal alkylation of a prochiral olefin.

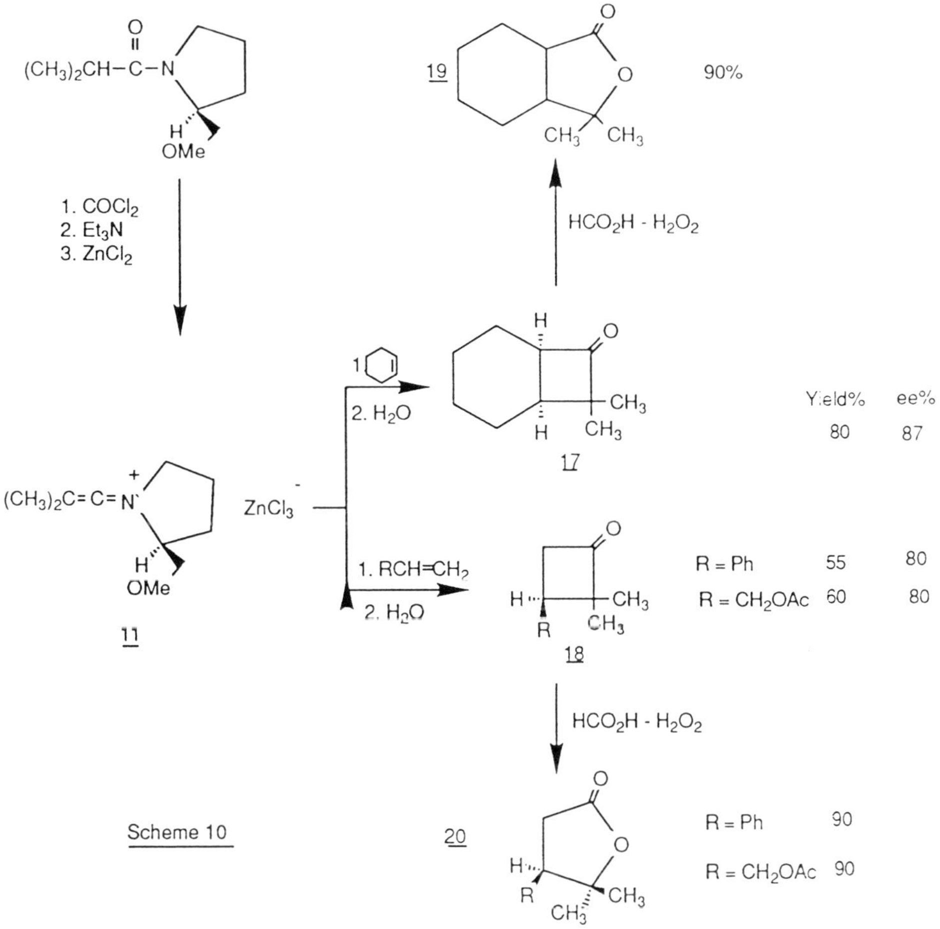

Scheme 10

One might be tempted to simply consider keteniminium salts as activated derivatives of ketenes. The exceptionally low-lying position of the LUMO (Scheme 11) accounts for the higher electrophilic character of keteniminium salts as compared to ketenes (10). Clearly the most important perturbation at the early stages of the reaction results from a transfer of electrons from the olefin HOMO to the keteniminium LUMO which has its highest coefficient at C-1. However, in contrast to ketenes, keteniminium ions have strongly bonding HOMO and, therefore, the secondary interaction between the LUMO of the olefin and the HOMO of the keteniminium ion will be weak, at least in the early part of the reaction profile. Moreover, the coefficient of the HOMO of a keteniminium ion being also higher at C-1, both interactions should lead to a homoallylic cation (Scheme 12). At this stage, bonding at C-2 is not expected to

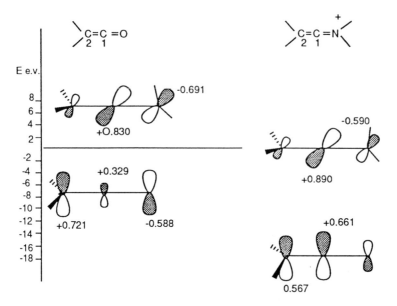

LCAO coefficients of frontier HOMO and LUMO orbitals of ethylene and cumulenes (STO - 3G - RHF)

Scheme 11

occur before rotation around the carbon-nitrogen bond has created a nucleophilic enamine system. It cannot be estimated where, on the energy profile, this rotation will take place.

Scheme 12 : Keteniminium + Olefin

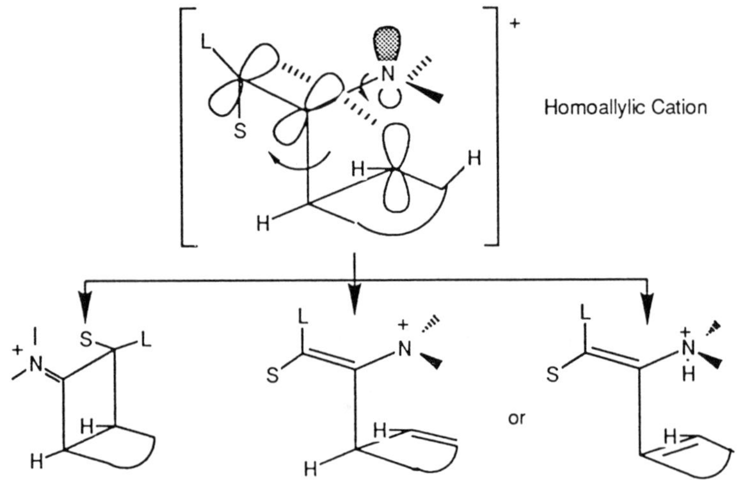

The eight possible approaches which allow an interaction between the π bond of cyclopentene and the π electrons of the carbon-nitrogen bond of the keteniminium ion are shown in Scheme 13. The presence of the methyl groups at C-2 allows to disregard approaches A and C from both the α and β faces. Approches B and D from the β face are also unfavou-

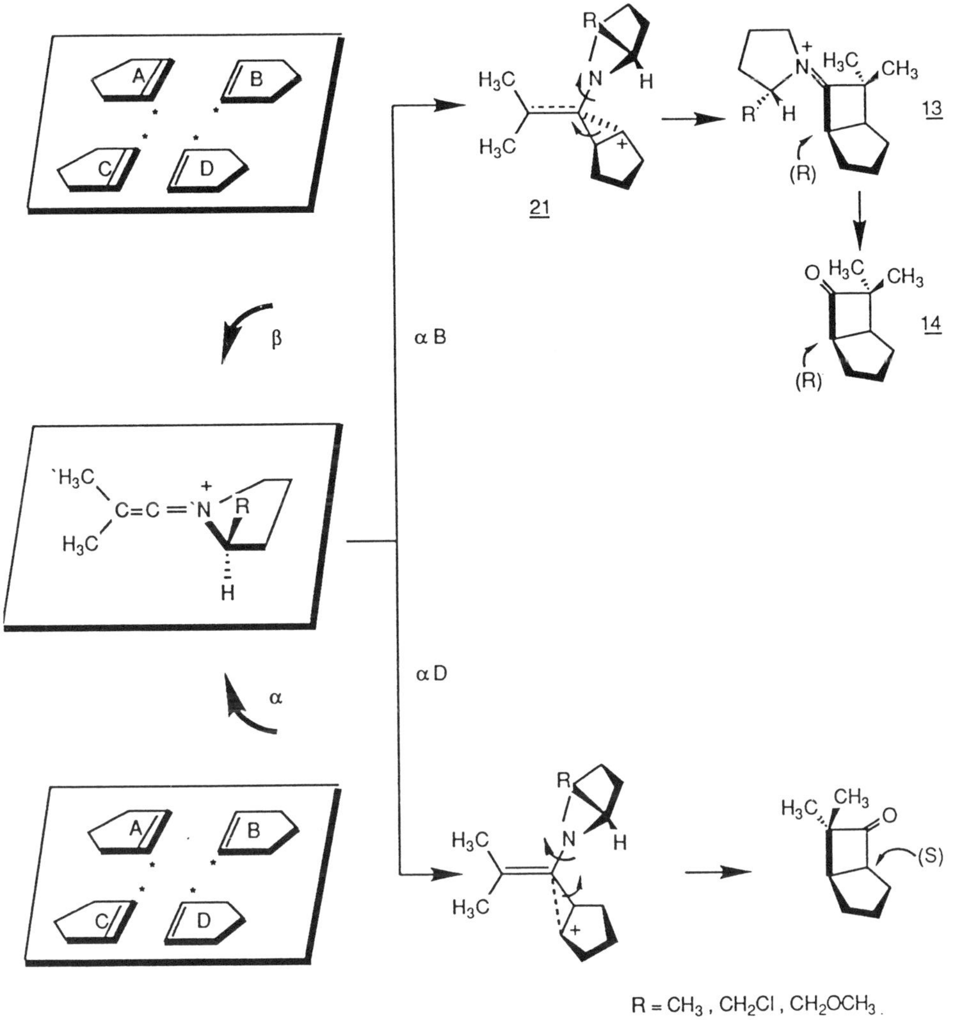

A
B
C
D
β
α
αB
αD
21
13
14
(R)
(S)
R = CH3 , CH2Cl , CH2OCH3


Scheme 13

rable since the olefin would interact with substituent R. However, on the basis of such a rough model, it is not possible to discriminate between α,B and α,D which would lead to enantiomeric adducts. Knowing the absolute configuration of both adducts 13 and ketones 14, it is obvious that the most favourable transition state results from the α,B approach of the two reactants. This leads to the homoallylic cation 21 which, after a rotation around the C-N bond away from the cyclopentane ring, yields 13.

The stereochemistry of the reactions of 11 with cis-and trans but-2-enes has been analysed in great detail (Table 1) (11). The cycloaddition with cis-but-2-ene gives a 1:1 mixture of crude diastereoisomeric adducts. In addition, hydrolysis is accompanied by a substantial cis ⇌ trans isomerization. The reaction of 11 with trans-but-2-ene is totally stereoselective but still some isomerisation is observed upon hydrolysis.

Table 1   Steric Course of Cycloadditions of 11 to cis-and-trans-But-2-ene.

|  | Crude Cyclobutaniminium Salts cis : trans | Yield | Cyclobutanones cis : trans | ee(cis) | ee(trans) |
|---|---|---|---|---|---|
| cis-But-2-ene | 50 : 50 | 64 % | 13 : 87 | 64 % | 13-16 % |
| trans-But-2-ene | 0 : 100 | 52 % | 20 : 80 | 97 % | 94 % |

The facile isomerisation of the cyclobutaniminium salts raises a question about the stereochemistry of the cycloaddition. Does the "wrong isomer" come from the cycloaddition itself or from some uncontrolled isomerization after cycloaddition ? The answer comes from the determination of the enantiomeric excesses for the cis and trans cyclobutanones. If the "wrong" isomer is formed by isomerization after the cycloaddition step, both cis and trans adducts should be obtained with the same enantiomeric excess. This is indeed the case for the cis and trans cyclobutanones originating from trans-but-2-ene and 11. On the other hand, the trans cyclobutanone originating from cis-but-2-ene is formed with a much lower enantiomeric excess then the cis isomer. This clearly indicates that a substantial loss of the original configuration of the olefin is occuring during the cycloaddition process.

This can be readily explained on the basis of the two-step mechanism shown in Scheme 14. A least-hindered approach (see above) of the two reactants leads to intermediate 22cis. Rotations a und b create the enamine system and allow the formation of the second bond. Hydrolysis of cis-adduct 23 leads to the cis-cyclobutanone 24 accompanied by its epimer 25 resulting from the cis ⇌ trans equilibration during the hydrolysis. If this is the sole process producing the trans isomer 25 it should indeed be found with the same enantiomeric excess as the cis-cyclobutanone 24. Since a lower enantiomeric excess is observed for the trans-cyclobutanone 25, a stereochemical leak must have occured.

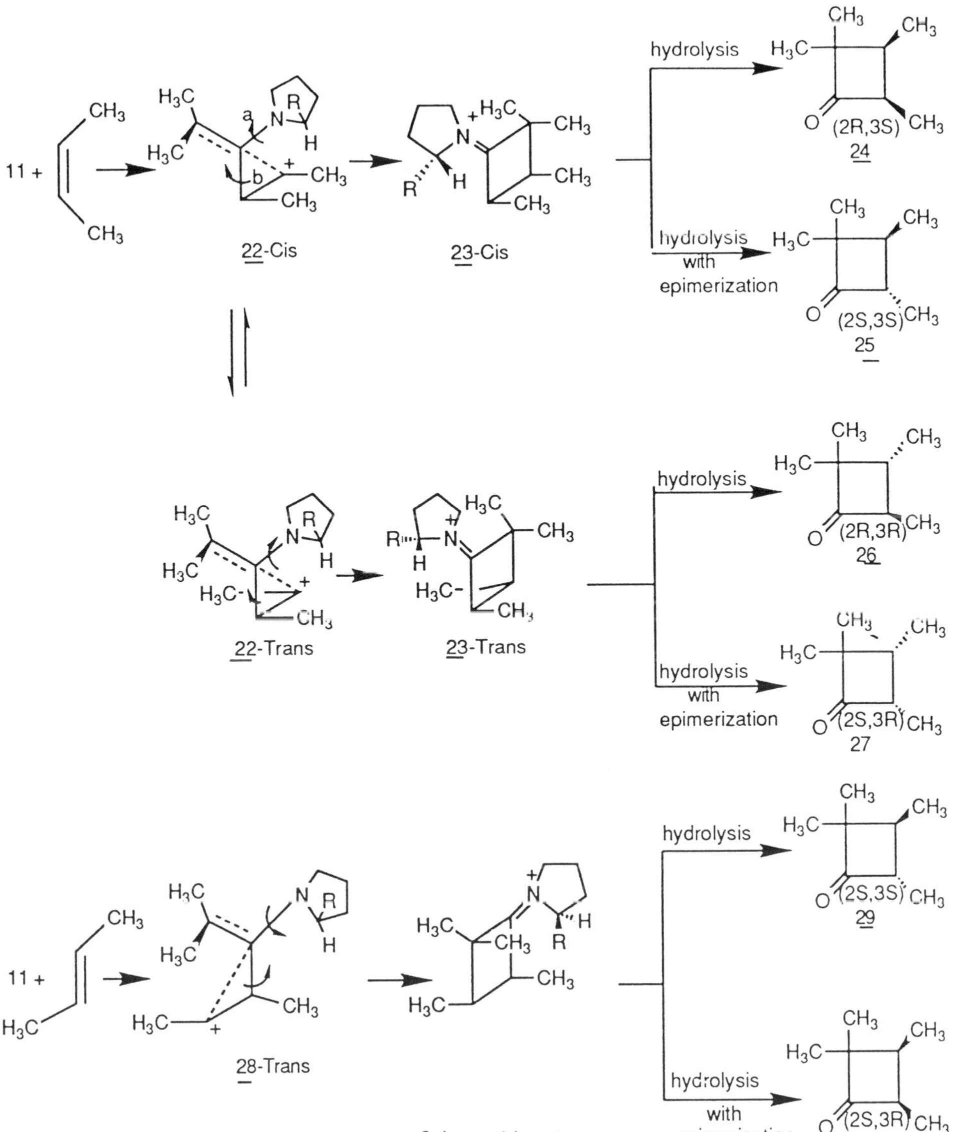

Scheme 14

It was shown that this does not result from the isomerisation of the cis-but-2-ene to trans-but-2-ene under the reaction conditions. However, it can be readily accounted for by some isomerization of intermediate 22 cis to 22 trans. Cyclization of this new intermediate and hydrolysis gives 26, the enantiomer of 25. Since trans→cis isomerization should be less important, the contamination of 24 by its enantiomer 27 should be smaller. The enantiomeric excess for the cis isomer should then be higher than for the trans, as found experimentally.

On the other hand, the reaction of 11 with trans-2-but-2-ene is always exclusively channeled through intermediate 28 trans which should not be expected to readily isomerize to the corresponding cis isomer. Consequently both the trans and cis cyclobutanones 29 and 30 are formed with almost equal enantiomeric excesses.

The intramolecular version of these (2+2) cycloadditions could offer promising routes for the regio- and stereocontrolled synthesis of polycyclic compounds. Studies have been initiated aiming at developing this intramolecular reaction as a synthetic tool (12). The requisite alkenylketeniminium salts 31 are generated from the reaction of unsaturated amides 32 with triflic anhydride in the presence of collidine in refluxing dichloromethane or 1,2 dichloroethane. The resulting cyclobutaniminium salts 33 are directly hydrolysed to the corresponding cyclobutanones (Scheme 15).

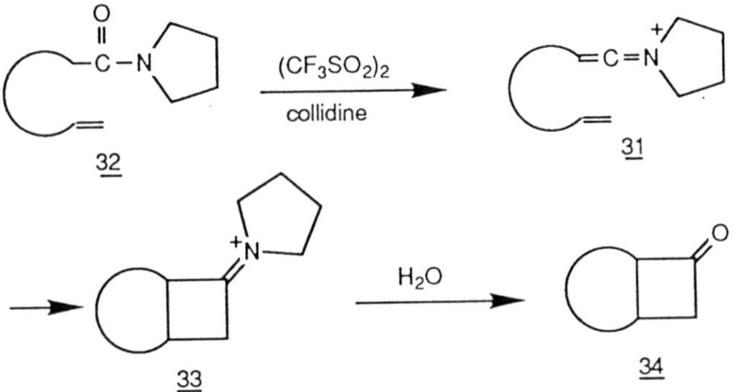

Scheme 15

Good yields of cycloadducts are obtained for a variety of chain lengths including those leading to cyclobutanones fused to a medium ring (Table 2). The reaction gives cis-fused adducts except in case g. Here the trans isomer probably results from the epimerization of the cis adduct under the reaction conditions. The formation of a tricyclic ketone (case g) illustrates the generality of the method and indicates that it could become useful for the construction of spiranic skeletons. It also indicates that the regiochemistry of the cycloaddition is essentially governed by the electronic properties of the double bond : the less substituted terminal carbon atom becomes bonded to the electrophilic $C_1$ atom, a process that gives the most stable carbenium intermediate but also an eight-membered ring.

Similar results have been obtained with unsaturated keteniminium salts derived from α or β-aminoamides (Scheme 16)(13). In all cases the (2+2) cycloadducts are obtained in good yields (Table 3). The position of the nitrogen atom in the chain has little effect upon the reaction. Substrates with no substituent at the double bond (entries a,c,d,g,h) or in which the internal alkene carbon bears a methyl group (entries b,c) yield adducts resulting from attack of the more nucleophilic terminal carbon atom at the C-1 atom of the keteniminium salts. Small amounts of

Table 2    Intramolecular [2+2] Cycloadditions of Keteniminium Salts.

a. 87%
b. 65%
c. 89%
d. 71%
e. 78%
f. 72%
g. 30%
h. 55%

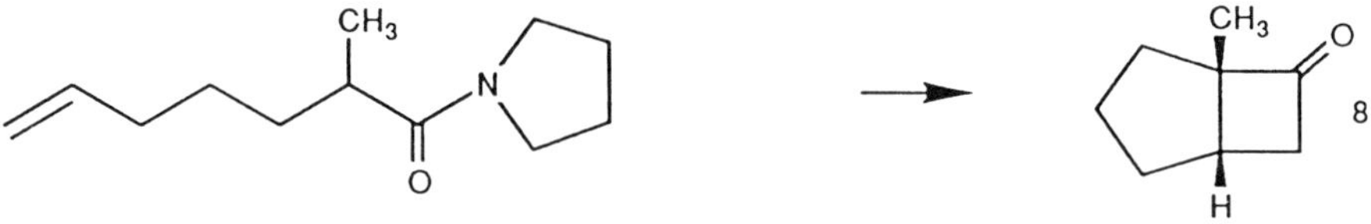

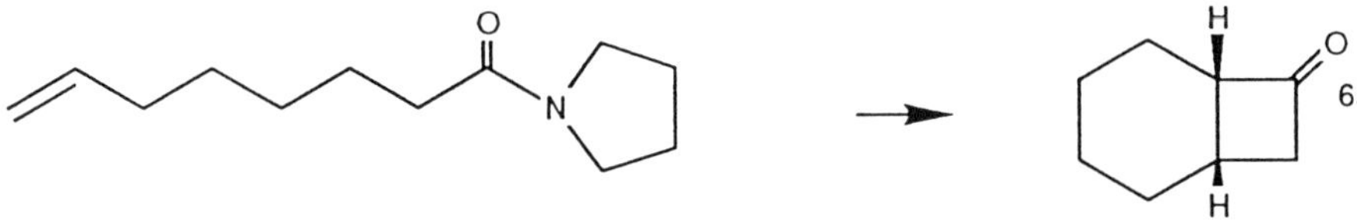

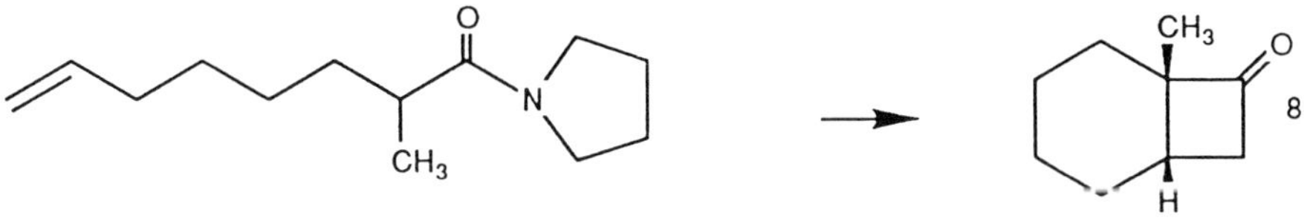

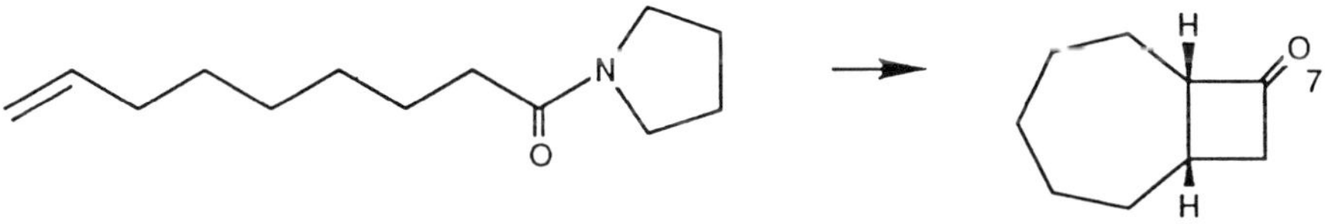

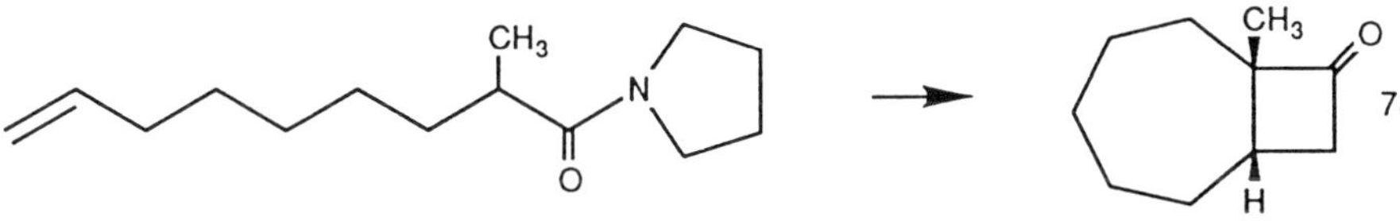

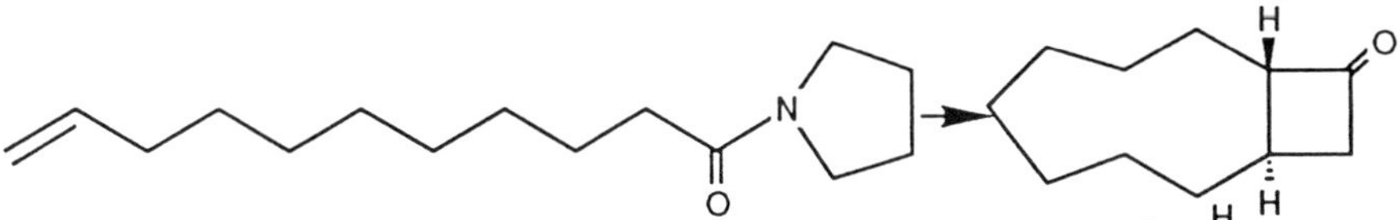

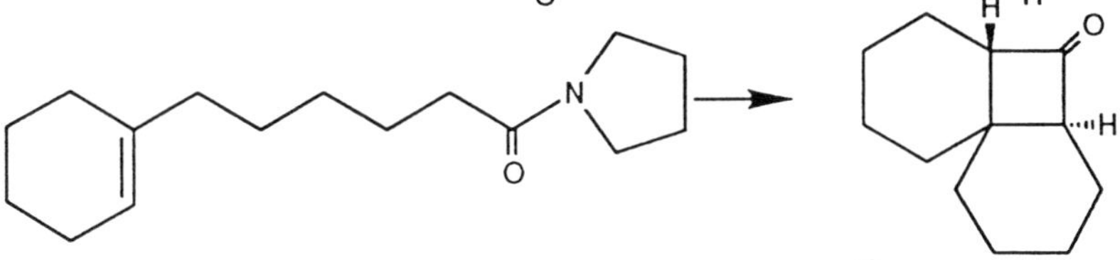

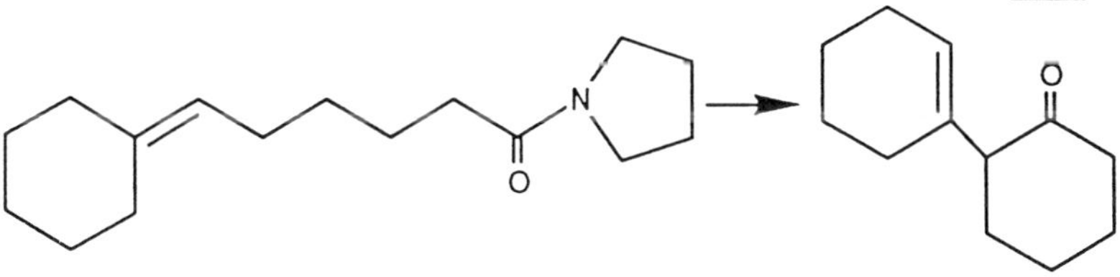

Table 3    Intramolecular Cycloadditions of Keteniminium Salts derived from Aminoacids.

| Entry | Amide | Product |
|---|---|---|
| a | 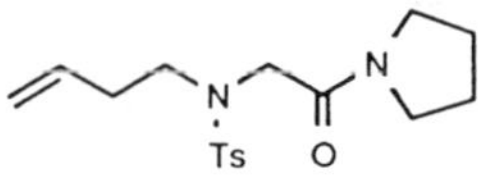 | 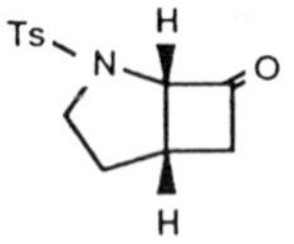 56% |
| b | 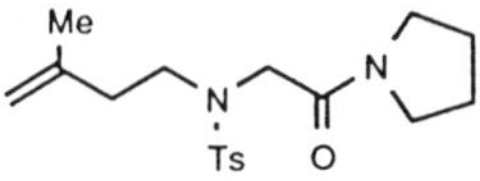 | 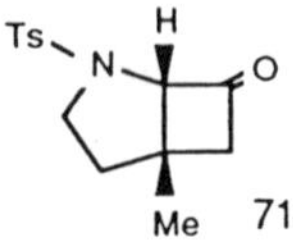 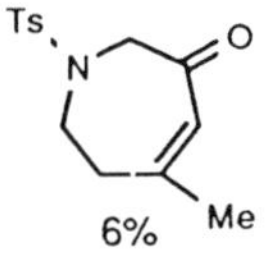 71%, 6% |
| c | 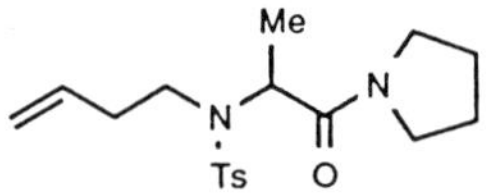 | 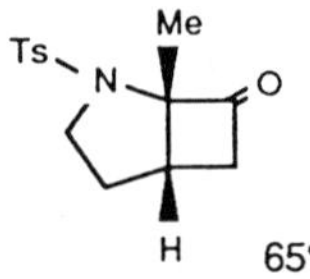 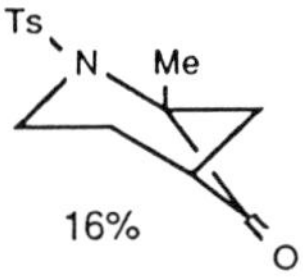 65%, 16% |
| d | 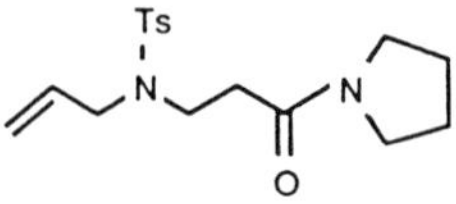 | 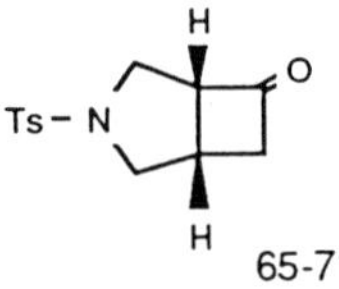 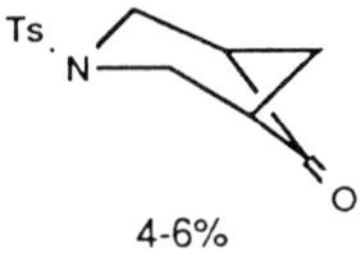 65-71%, 4-6% |
| e | 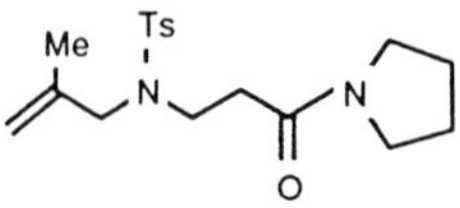 | 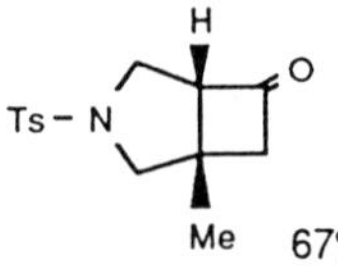 67% |
| f | 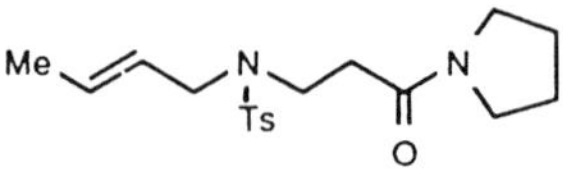 | 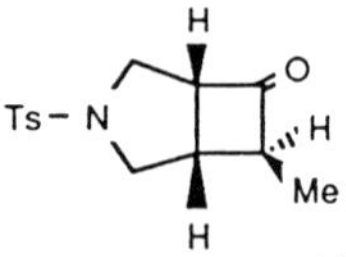 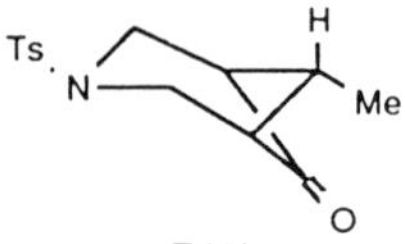 14%, 74% |
| g | 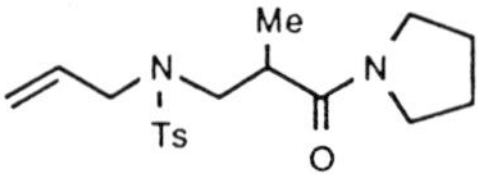 | 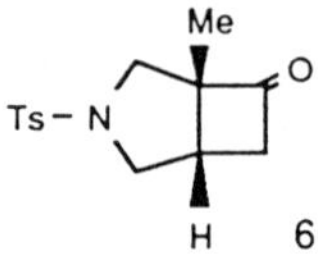 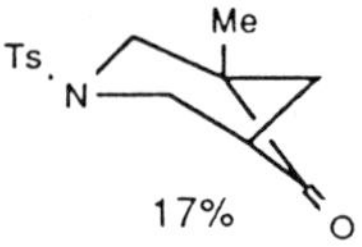 60%, 17% |
| h | 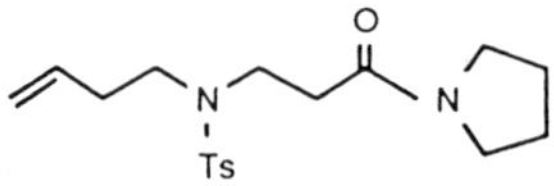 | 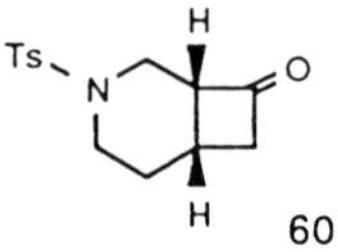 60-65% |

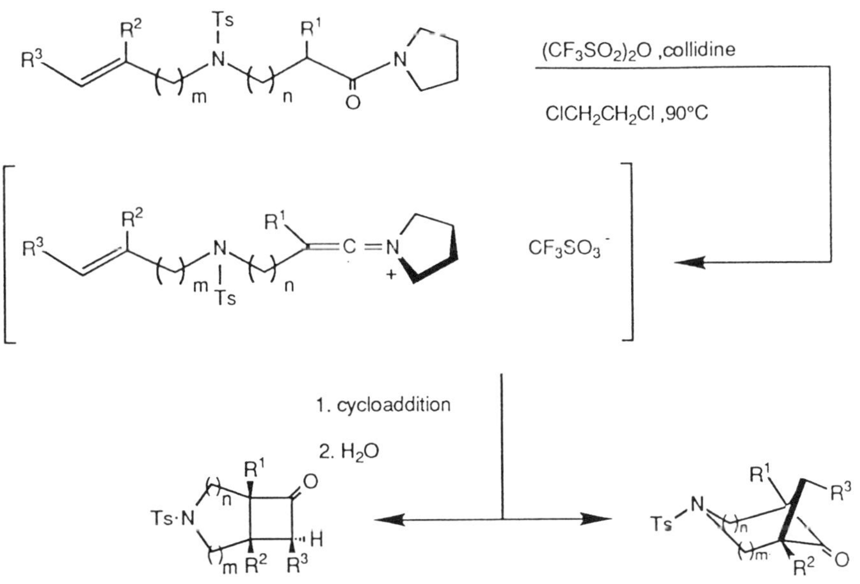

Scheme 16

the regioisomeric adducts are observed in three cases (entries c, d and g). On the other hand, when the terminal olefinic atom bears a methyl group (entry f), the major adduct is 7-methyl-3-azabicyclo [3.1.1.] heptan-6-one. This compound results from attack of the internal alkene carbon on the C-1 atom of the keteniminium salt. Thus, when there is no directing electronic effect, a six-membered transition state is preferred over a seven-membered transition state.

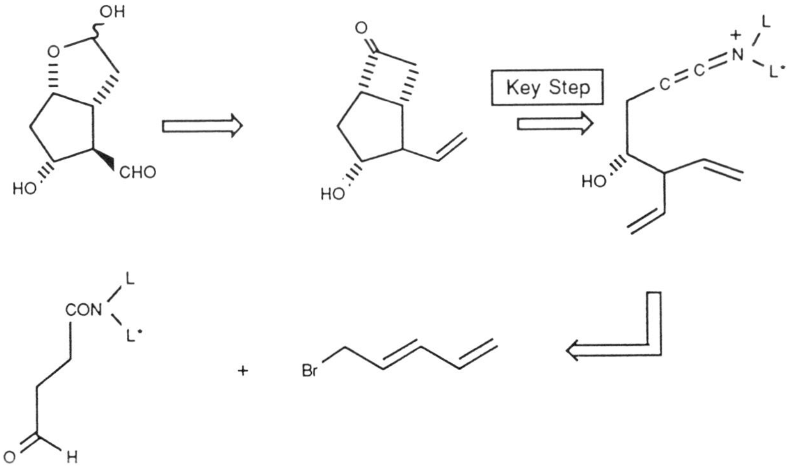

Scheme 17

Scheme 17 describes an application of this strategy to the synthesis of Corey's aldehyde, an advanced intermediate toward protaglandins. The key step is an intramolecular (2+2) cycloaddition of a keteniminium salt.

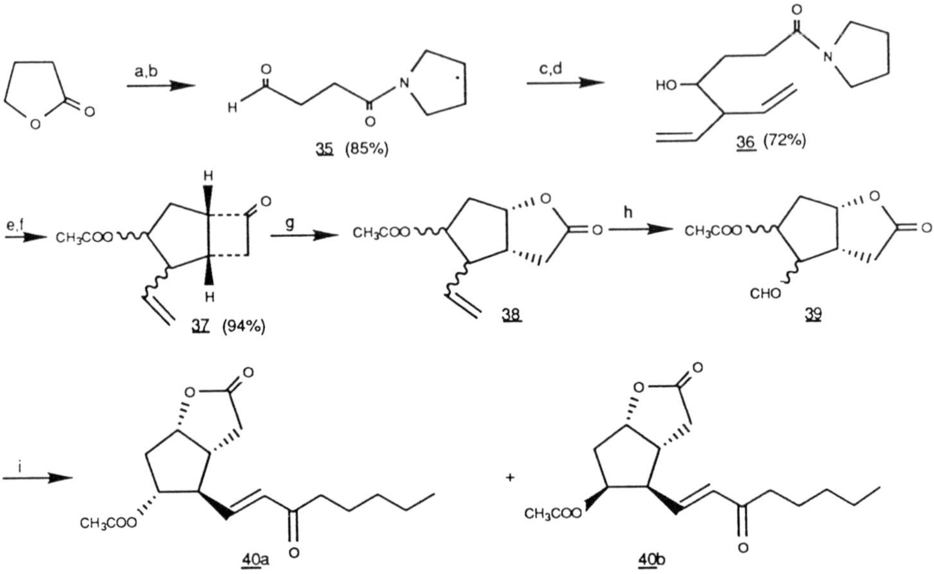

Reagents : a,pyrrolidine, Δ; b,DMSO + oxalyl chloride,Et₃N ; c,pentadienyl bromide + Zn ; d,CH₃COCl,pyridine ; e,Triflic anhydride,collidine ; f,H₂O-CCl₄,90°C ; g,mCPBA,NaHCO₃ ; h,O₃,CH₃OH,-70°C then Me₂S ; i, (EtO)₂P(O)CH₂COC₅H₁₁+NaH,THF,-10°C → 20°C

Scheme 18

In a racemic version of this strategy, the direct precursor 36 of the keteniminium salt is readily obtained in two isolated steps from γ-butyrolactone via 35 (14). The presence of two equivalent olefinic bonds favours the intramolecular reaction and, after hydrolysis and chromatography, a high yield of bicyclo[3.2.0.]heptan-5-one 37 is obtained. As expected, no steric control can be found at C-2 and C-3 and a mixture of four diastereoisomers is obtained. Baeyer-Villiger oxidation of 37 yields lactone 38 resulting from highly (≥ 93%) selective migration of the methine group (15). The crude mixture of diastereoisomeric lactones 38 is directly ozonized to the aldehyde 39 which is used without purification in the Wittig-Horner reaction. Chromatography gives lactones 40a and 40b in 16% and 30%, respectively. Thus the reaction sequence allows the establishment of three stereocenters with high selectivity. Control of the stereochemistry of the side-chain relative to the lactone substituent is the result of equilibration at the aldehyde stage or during the Wittig-Horner reaction. However, the method does not allow control of stereochemistry of the oxygenated functionality which is too remote from the reacting centers in the cycloaddition step.

This convergent strategy offers the possibility of preparing optically active prostaglandins by using keteniminium salts derived from chiral amides. However, intramolecular [2+2] cycloaddition of amide 41 derived from prolinol methyl ether followed by hydrolysis yields bicyclo [3.2.0] heptan-6-one 42 in only 27% enantiomeric excess (Scheme 19).

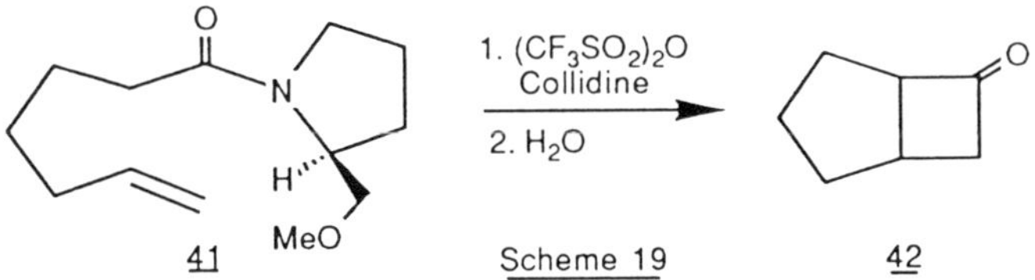

Scheme 19

The use of prolinol derivatives as chiral auxiliaries is thus restricted to keteniminium salts bearing two identical substituents at C-2. When these are different, two diastereoisomeric keteniminium salts 43 and 44 can be formed and the net result is the formation of an adduct of lower optical purity (Scheme 20). In this case a chiral inductor posses-

Scheme 20

sing $C_2$ symmetry is needed. This is spectacularly confirmed by the very high asymmetric induction observed in the intramolecular cycloaddition of keteniminium salts derived from a $C_2$ symmetrical chiral pyrrolidine (16) (Scheme 21). At this stage, this has not yet been applied to the sequence leading to the prostaglandins but it should without any doubt be equally successful.

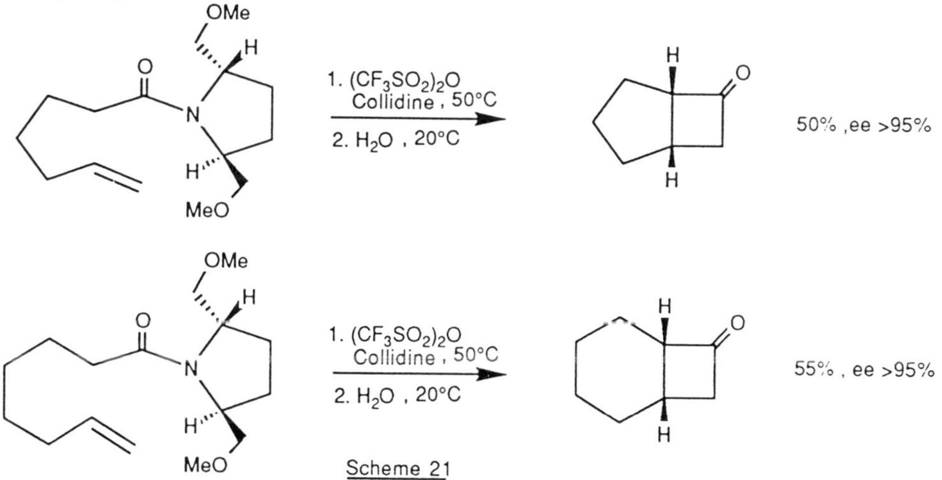

50%, ee >95%

55%, ee >95%

Scheme 21

Acknowledgement : We thank our collaborators who paved the way for this work and whose names have been cited in the references. Dr A.M. Hesbain-Frisque and Prof. E. De Hoffmann are warmely thanked for their help in interpreting NMR and Mass spectra, respectively. This work has been generously supported by the S.P.P.S. (Actions Concertées 86/91-84), F.N.R.S. and I.R.S.I.A.

References

1) (a) Trost, B.M. Top. Curr. Chem. 1986, **133**, 3 (b) Wong, H.N.C.; Lau, K.L.; Tam K.F. ibid. 1986, **133**, 83 (c) Krief, A. ibid. 1987, **135**, 1. (d) Burger, P.; Buch, H.M. ibid. 1987, **135**, 77. (e) Bellus, D.; Ernst, B. Angew. Chem. Int. Ed. Engl. 1988, **27**, 797.

2) Recent Reviews : a) Ghosez, L. in "Stereoselective Synthesis of Natural Products", Bartmann, W.; Winterfeld, E. (Eds) Excerpta Medica, Amsterdam Oxford, 1979, 93. b) Brady, W.T. in "The Chemistry of Ketenes, Allenes and Related Compounds", Patai, S. (Ed), Interscience, New-York 1980, 279. c) Brady, W.T., Tetrahedron 1981, **37**, 2949 (d) Moore, H.W.; Gheorghiu, M.D. Chem. Soc. Rev. 1981, 289.

3) Cossement, E.; Binamé, R.; Ghosez, L. Tetrahedron Lett. 1974, 997; Michel, P.; O'Donnell, M.J.; Hesbain-Frisque, A.M.; Ghosez, L.; Declercq, J.P.; Germain, G.; Van Meerssche, M. Tetrahedron Lett. 1980 **21**, 2577.

4) Goldstein, S.; Vannes, P.; Houge, C.; Hesbain-Frisque, A.M.; Wiaux-Zamar, C.; Ghosez, L.; Germain, G.; Declercq, J.P.; Van Meerssche, M.; Arrieta, J.M. J. Am. Chem. Soc. 1981, **103**, 4616.

5) Review : Ghosez, L.; Marchand-Brynaert, J. Adv. Org. Chem. 1976, **9**, Part 1, 421.

6) Haveaux, B.; Dekoker, A.; Rens. M.; Sidani, A.R., Toye, J.; Ghosez, L. Org. Syn. 1980, **59**, 26.

7) a) Falmagne, J.B.; Escudero, J.; Taleb-Sahraoui, S.; Ghosez, L. Angew. Chem. Int. Ed .Engl. Engl. 1981, **20**, 879. b) Falmagne, J.B. Dissertation U.C.L., 1985.

8) (a) Marchand-Brynaert J.; Ghosez, L. J. Am. Chem. Soc. 1972, **94**, 2870 (b) Schmit, C. Dissertation, Université de Louvain, 1988.

9) Houge, C.; Frisque-Hesbain, A.M.; Mockel, A.; Ghosez, L.; Declercq, J.P.; Germain, G.; Van Meerssche, M.; J. Am. Chem. Soc. 1982, **104**, 2920.

10) (a) Ghosez, L.; O.Donnell, M.J. in "Pericyclic Reactions" Marchand, A.P.; Lehr, R.E. (Eds) Academic Press, New-York, 1977, Vol. II., 79 (b) Sonveaux, E.; André, J.M.; Delhalle, J.; Fripiat, J.G.; Bull. Soc. Chim. Belg. 1985, **94**, 831.

11) Saimoto, H.; Houge, C.; Hesbain-Frisque, A.M.; Mockel, A.; Ghosez, L. Tetrahedron Lett. 1983, **24**, 2251.

12) (a) Marko, I., Ronsmans, B.; Hesbain-Frisque, A.M.; Dumas, S.; Ghosez, L.; Ernst, B.; Greuter, H. J. Am. Chem. Soc. 1985, **107**, 2192. (b) Snider, B.B.; Hui, R.A.H.F.; Kulkarni, Y.S. J. Am. Chem. Soc 1985, **107**, 2194.

13) Gobeaux, B.; Ghosez, L., Heterocycles, Special Issue **28**, accepted for publication.

14) Ghosez, L.; Marko, I.; Hesbain-Frisque, A.M., Tetrahedron Lett. 1986, **27**, 5211.

15) Newton, R.F. and Robert, S.M., Tetrahedron **36**, 2163 (1980).

16) Kawanami, Y.; Ito, Y.; Kitagawa, T.; Taniguchi, Y, Katsuki, T. Yamaguchi, M., Tetrahedron Lett. 1984, **25**, 857.

# THE GENERATION AND REARRANGEMENT OF 2-DIAZOCARBONYL CYCLOBUTANONES: THE FORMATION OF 5-SPIROCYCLOPROPYL-2 (5H) FURANONES

R. D. Miller, W. Theis, G. Heilig, and S. Kirchmeyer
IBM Research Division
Almaden Research Center
650 Harry Road
San Jose, California 95120-6099

$\Delta^{\alpha,\beta}$ Butenolides are valuable synthetic reagents which constitute the active ingredients of many natural products.[1] As a result many synthetic routes to these materials as well as the conjugation extended 5-alkylidene and arylidene derivatives have been reported. In light of this, it is curious that 5-spirocyclopropyl $\Delta^{\alpha,\beta}$ butenolides are, virtually unknown[2] in spite of the fact that they represent a homoconjugated variation of the 5-alkylidene derivatives and as such have considerable synthetic potential. We now report that a variety of 5-spirocyclopropyl $\Delta^{\alpha,\beta}$ butenolides are produced by the pyrolysis of 2-diazocarbonyl cyclobutanones. The reaction conditions are mild, product yields are respectable and the rearrangement appears to be stereospecific.

The diazocarbonyl derivatives are prepared from the corresponding 2-vinyl cyclobutanones in three steps as shown in Scheme I. The starting vinyl cyclobutanones are readily available from a variety of synthetic procedures[3], the most convenient of which is the cycloaddition of in situ generated vinyl ketenes with olefins and dienes.

Scheme I

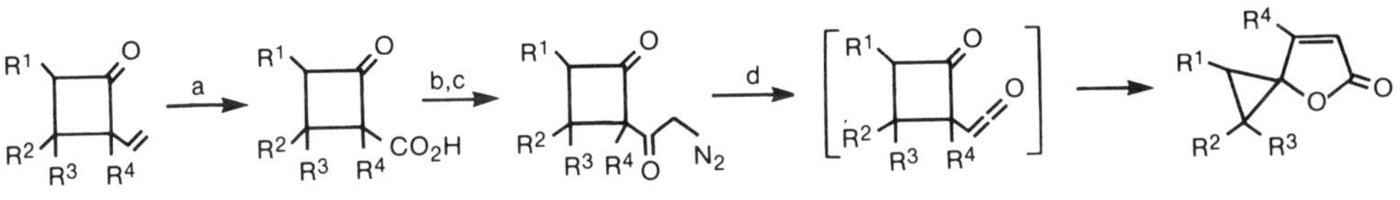


a. $RuCl_3 \cdot H_2O$, $NaIO_4$, $MeCN-H_2O/CCl_4$, 25°, 4h b. $(COCl)_2$, 25°C c. $CH_2N_2$, $Et_2O$, -10°C d. Δ, Xylene

The diazoketones rearranged upon brief heating (20-30m) in xylene to produce the desired 5-spirocyclopropyl $\Delta^{\alpha,\beta}$ butenolides as shown in Table I. The lactone products showed a characteristic high frequency carbonyl band in the IR at $1750-1760\ cm^{-1}$ when measured in nonpolar solvents. The structural assignments were supported by spectral data and confirmed in some cases by single crystal x-ray analysis.

The isolation of epimeric lactone mixtures with the same isomeric composition as the starting diazoketones strongly suggested that the rearrangement was stereospecific. This was confirmed by the transformation of the epimerically pure diazoketones 2e and 2f to the spirolactones 3e and 3f respectively. Rearrangement of other pure epimers gave similar results and stereochemistries. On the basis of these studies, it was deduced that the ring oxygen of the lactone appears in the same



*A. de Meijere and S. Blechert (eds.), Strain and Its Implications in Organic Chemistry, 255–258.*

stereochemical configuration as the original diazocarbonyl moiety in the starting material.

Although it was suspected that α-ketenylcyclobutanones were intermediates in the conversion of the 2-diazocarbonyl cyclobutanones to the spirolactones, the thermal conditions employed prevented the detection of the intermediates. However, irradiation of the diazoketones ($\lambda > 330$ nm) in the presence of alcohols led to the isolation of homologated ketoesters consistent with the formation of the corresponding α-ketenylcyclobutanones. These intermediates were directly spectroscopically detected (IR, 2120 cm$^{-1}$, 1760 cm$^{-1}$) in solutions which had been irradiated at low temperatures ($-40$ to $-50°C$). The α-ketenylcyclobutanones thus generated rearranged cleanly upon warming to the spirocyclopropyl $\Delta^{\alpha,\beta}$ butenolides and the kinetics could be monitored by IR. The rate constants measured were dependent on the ketene structure and varied by more than two orders of magnitude. Similarly there were often considerable differences in the rates of rearrangement between stereoisomeric pairs. The rates were also solvent dependent and increased markedly in more polar solvents.

Table I: Preparation of 5-Spirocyclopropyl $\Delta^{\alpha,\beta}$Butenolides

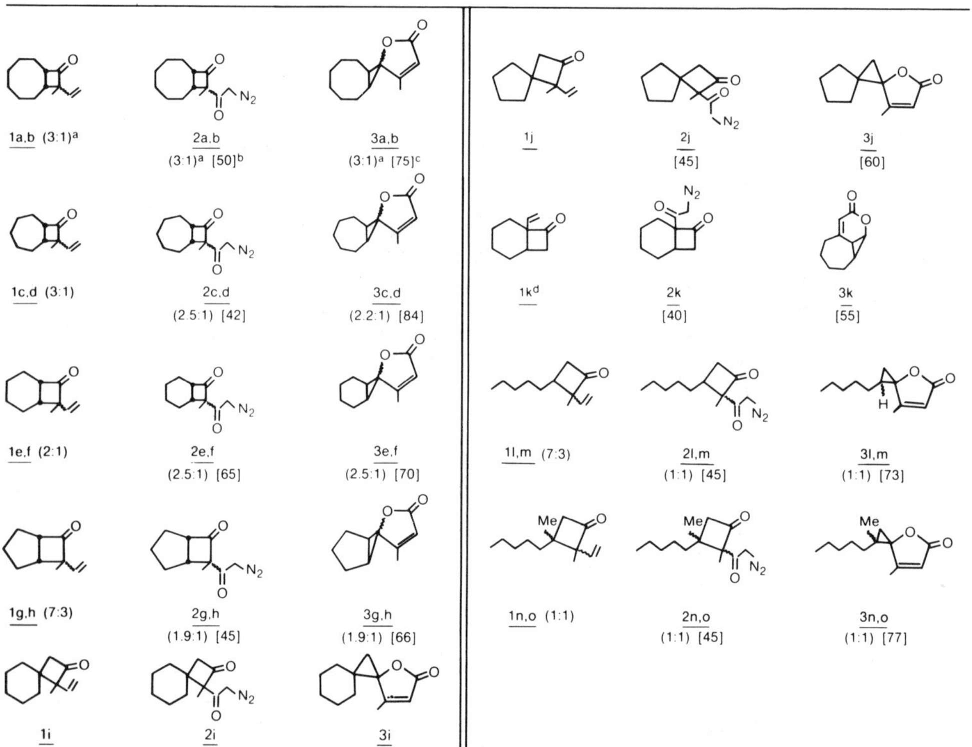

a. Epimer ratios  b. Yields of purified diazoketone based on vinyl cyclobutanone  c. Yields of purified butenolide from diazoketone
d. Prepared as described in footnote 11.

Surprisingly, the variation in the reaction rates with structure seems to be controlled by entropy rather than enthalpy. In this regard, the measured activation enthalpies (ΔH‡) were low and varied only slightly (12-15.5 kcal-mol$^{-1}$) while the corresponding values for the entropies of activation (ΔS‡) were all very negative and varied from − 18 to − 39 eu. In more polar solvents such as chloroform, the enthalpy values again changed very little but ΔS‡ was less negative by as much as 10 eu. The large negative values for the entropies of activation are consistent with a highly ordered and structured transition state.[4] There also appears to be an appreciable electrostatic component to the entropy of activation as evidenced by the significant solvent dependence.[5] This leads us to postulate a cyclic dipolar transition state such 5 shown below for the conversion of 2e to 3e.[6] Consistent with this hypothesis, MNDO calculations also support a dipolar cyclic transition state where there is considerable bond breaking between atoms 2 and 3 and bond formation between atoms 1 and 3.

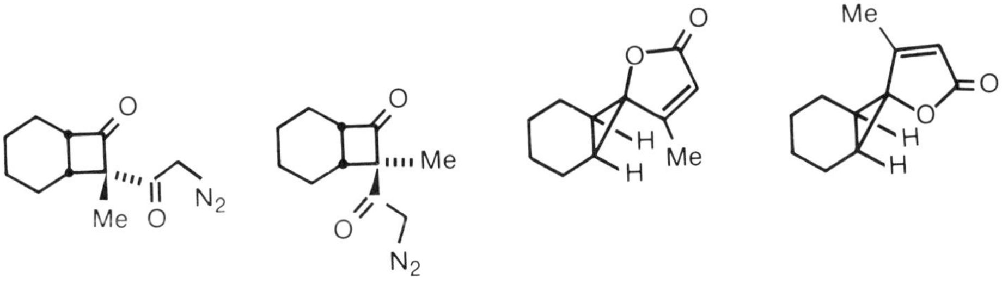

2e    2f    3e    3f

A cyclic structure a such as 5 also rationalizes the large rate difference often observed between epimers in terms of steric crowding between substituents on carbons 2, 3 and 4. An all cis arrangement leads to considerable steric repulsion in the transition state and hence a slower rate of rearrangement relative to those cases where the substituent in position 2 is trans to the substituents in positions 3 and 4.

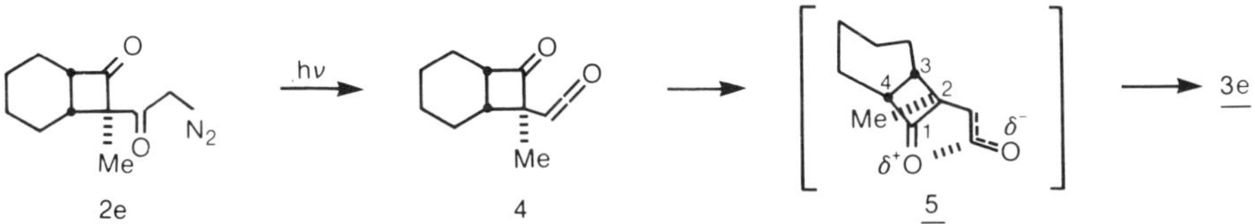

2e    4    5    3e

In summary, we have described the generation of 5-spirocyclopropyl Δ$^{α,β}$-butenolides by the thermal and photochemical rearrangements of 2-diazocarbonyl cyclobutanones which proceeds via an α-ketenylcyclobutanone intermediate. This rearrangement constitutes another synthetically useful example of a electrophilically initiated cyclobutanone rearrangement.

**References**
1. For a comprehensive review of butenolide derivatives see Rao, Y. S., *Chem. Rev.*, 1976 **76**, 625.

2. (a) Breslow, R., Altman, L. J.; Krebs, A.; Eohacsi, E.; Murata, I.; Peterson, R. A.; Posner, J., *J. Am. Chem. Soc.*, 1965, **87**, 1326; (b) Breslow, R.; Altman, L. J. ibid., 1966, **88** 504; (c) Dehmlow, E.V.Z. *Z. Naturforsch.*, 1975, **30b**, 404, (d) Staab, H. A.; Ipaktschi, *J. Chem. Ber.* 1968, **101** 1457 (e) Miller, R. D.; Dolce, D. L. *Tetrahedron Lett.*, 1975, 1831.
3. (a) Matz, J. R.; Cohen, T. *Tetrahedron Lett.*, 1981, **22** 2459. (b) Cohen, T.; Matz, J. R., *Tetrahedron Lett.*, 1981, **22**, 2455; (c) Berge, J. M.; Rey, M.; Dreiding, A. S., *Helv. Chem Acta,* 1982, **65** 2230; (d) Jackson, D. A.; Rey, M.; Dreiding, A. S., *Helv. Chem. Acta,* 1983, **66**,. 2330.
4. Carpenter, B. K. in "Determination of Organic Reaction Mechanism", J. Wiley and Sons, Inc., New York, 1984, Chap. 7.
5. Frost, A. A.; Pearson, R. E. in "Kinetics and Mechanism", J. Wiley and Sons, Inc., New York, 1963, Chap. 7.
6. Reichardt, C. M. "Solvent Effects in Organic Chemistry", *Verlag Chemie,* Weinheim, West Germany, 1979, Vol. 3 Chap. 5 and references cited therein.

# COMPUTATIONAL STUDIES OF PRISMANE, HELVETANE/ISRAELANE AND ASTERANE

E. Ōsawa,* J. M. Rudziński, D. A. Barbiric and E. D. Jemmis[a]
Department of Chemistry, Faculty of Science, Hokkaido University, Sapporo 060, Japan, [a]School of Chemistry, University of Hyderabad, Central Univeristy P.O., Hyderabad 500134, India

ABSTRACT. Lower members of the title series have been calculated with MM2' and AM1 methods in order to gain insights into the strain and structural features of these mostly unknown molecules.

1. INTRODUCTION

The past few years saw a sudden rush of papers on the computation of prismane and related hydrocarbons. Impetuses for these works are the assessment of synthetic possibilities and the evaluation of various computational techniques for the molecules with extremely high strain. We summarize below some of our recent results obtained in this field.

2. PRISMANES

The lower members of [n]prismanes, $(C_2H_2)_n$, n=3 to 14, have been overworked computationally(1-6). Interests are generated by two predictions that have been made based on MM2 calculations(1): [1] systematic variation of C-C bond lengths within and between the n-membered rings, and [2] instability of $\underline{D}_{nh}$ structure in the higher members. After several calculations with higher and higher precision, it appears that ab initio studies by Disch and Schulman(4) finally gave definitive answers to the questions: [1] the bond length variations predicted by the low-level computations are artefacts, and [2] the energy-minimum structures of the [n]prismanes with n ≤ 12 are all $\underline{D}_{nh}$, but the higher members have very low vibrational frequencies near 100 $cm^{-1}$.

Comparison of the results from AM1(2,3) and ab initio(4) calculations seems to support the contention of Dewar(8) that this new semiempirical molecular orbital method performs as good as the ab initio 6-31G* basis set.

3. HELVETANES/ISRAELANES

These trivial names(9) so far referred to the two configurational isomers of [12]prismane wherein cyclobutane rings are all concatenated in all-cis fashion, which can be designated as $(cis)_{12}$. In israelane, the catenation is all-trans, $(trans)_{12}$. In helvetane, one cis-linkage replaces every third trans-lingkage. According to our new notation, helvetane can be designated as $\{(trans)_2(cis)\}_4$.

The previous calculations(3,10) were limited to the two direct descendants of G. Dinsberg's original idea(9). However, the homologous series of these starshaped molecules should provide interesting submolecular environments, especialy regarding the varying congestion in the inner cavity. As a preliminary step, we searched a part of the helvetane series, $\{(trans)_2(cis)\}_n$, n=3-7, using MM2'(11).

Triadhelvetane (n=3, $C_{18}H_{18}$) involves forbiddingly high congestion among the six inner protons and we have so far been unable to optimize the structure. Quadrihelvetane (n=4, $C_{24}H_{24}$) gave the same $D_{2d}$ energy-minimum as obtained with AM1 and MM2(3). The inner congestion is still considerable in quinquhelvetane (**1**, n=5, $C_{30}H_{30}$) and hexadhelvetane (**2**, n=6, $C_{36}H_{36}$): they did not optimize into $D_{nh}$ but ended up with $C_2$ and $D_{3d}$ symmetries, respectively. Heptadhelvetane (**3**, n=7, $C_{42}H_{42}$) is the smallest member of the series that attains the planar $D_{7h}$ structure.

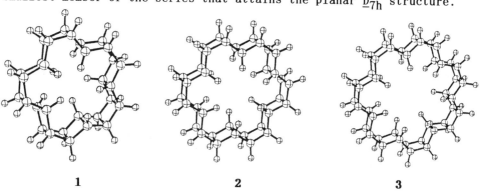

**1**    **2**    **3**

## 4. ASTERANES

The asterane series(12,13), having the general formula **4** [X=$CH_2$, $(C_3H_4)_n$, n≥3], provides novel means of straining a molecule by allowing less and less space between the bow and stern of boat cyclohexane ring in addition to the increasing angle strain similar to that of prismanes. Hence, the series should be, like the helvetanes, a good testing ground for the nonbonded H/H interaction potential function in molecular mechanics(14).

**4**, Asterane, X=$CH_2$
Prismane, X=none

It has been predicted sometime ago that the energy minimum structure of asterane will lose $D_{nh}$ symmetry at higher (n≥8) members based on molecular mechanics calculations(15). Exactly similar to the case of prismanes(1-4), AM1 calculations of asterane series again gave $D_{nh}$ energy minima at least up to octaasterane (n=8). However, the lowest vibrational frequency of octaasterane occurs at 91 $cm^{-1}$.

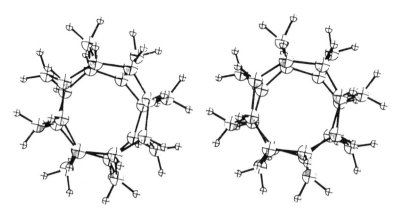

Stereo Pair of [8]Asterane, **4** (X=$CH_2$, n=8), in a Crown Conformation Obtained by Geometry Optimization with MM2'

The nonbonded H/H distances between bow and stern in the AM1-optimized structures of asteranes are significantly smaller than in the MM2'-optimized ones. For example, in hexaasterane (n=6), the distances are 1.707 and 1.832 Å, respectively. This indicates that the proton in MM2' may be still too hard and too large.

1. Reddy, V. P.; Jemmis, E. D. Tetrahedron Lett., **1986**, 27, 3771-3774.
2. Jemmis, E. D.; Rudzinski, J. M.; Ōsawa, E. Chem. Express, **1988**, 3, 109-112.
3. Miller, M. A.; Schulman, J. M. J. Mol. Struct. **1988**, 163, 133-141.
4. Disch, R. L.; Schulman, J. M. J. Am. Chem. Soc. **1988**, 110, 2102-2105.
5. Mehta, G.; Padma, S.; Ōsawa, E.; Barbiric, D. A.; Mochizuki, Y. Tetrahedron Lett., **1987**, 28, 1295-1298.
6. Mehta, G.; Padma, S.; Jemmis, E. D.; Leela, G.; Ōsawa, E.; Barbiric, D. A. Tetrahedron Lett. **1988**, 29, 1613-1616.
7. Ōsawa, E.; Barbiric, D. A.; Lee, O. S.; Kitano, Y.; Padma, S.; Mehta, G. submitted for publication in J. Chem. Soc., Perkin Trans. II.
8. Dewar, M. J. S.; O'Conner, B. M. Chem. Phys. Lett. **1987**, 138, 141-145.
9. Nickon, A.; Silversmith, E. F. 'Organic Chemistry: The Name Game, Modern Coined Terms and Their Origins', Pergamon Press: New York, 1987.
10. Li, W.-K.; Luh, T.-Y.; Chiu, S.-W. Croat. Chem. Acta, **1985**, 58, 1-3.
11. Jaime, C.; Ōsawa, E. Tetrahedron **1983**, 39, 2769-2778.
12. Musso, H. Naturwiss. Rundschau, **1966**, 19, 448; Umschau Wiss. Tech., **1968**, 68, 209.
13. Miller and Schulman(3) called the starshaped helvetane and israelane as "asteranes", apparently without knowledge of the former occupants of the star. See, ref. 9.
14. Murrell, J. N. et al. 'Molecular Potential Energy Functions', John Wiley & Sons: New York, 1984.
15. Ōsawa, E.; Musso, H. Angew. Chem. Int. Ed. Engl. 22, 1-12.

# CYCLOBUTANE-CYCLOPENTANE INTERCONVERSION OF COORDINATED ADDUCTS OF CYCLOHEPTATRIENE AND TETRACYANOETHYLENE VIA THE [2,2]-SIGMAHAPTOTROPIC REARRANGEMENT

Zeev Goldschmidt and Hugo E. Gottlieb
Department of Chemistry
Bar-Ilan University
Ramat-Gan 52100
Israel

ABSTRACT. The reaction of (3,7,7-trimethyl-cycloheptatriene)Fe(CO)$_3$ with tetracyanoethylene gave an equilibrium mixture of the 2+2 and 3+2 isomeric adducts, in a 2:3 ratio. The rate constants for the interconversion process in CDCl$_3$ were evaluated by the spin saturation transfer technique, and in CD$_3$OD by variable temperature line shape analysis. The rate constant for the 3+2 → 2+2 rearrangement in CDCl$_3$ k$_{32}$ 0.19 ± 0.03 s$^{-1}$ with $\Delta G^{\neq}$ 18.37 ± 0.10 kcal/mol at 24°C. In CD$_3$OD k$_{32}$ 1.3 ± 0.6 s$^{-1}$ with $\Delta G^{\neq}$ 17.13 ± 0.27 kcal/mol at 22°C. Other activation parameters were $\Delta H^{\neq}$ 15 kcal/mol, $\Delta S^{\neq}$ −10 e.u. (in CDCl$_3$) and $\Delta H^{\neq}$ 20 kcal/mol, $\Delta S^{\neq}$ +12 e.u. (in CD$_3$OD), respectively. Addition of strong dienophiles, such as (carbomethoxy)maleic anhydride, to the eqilibrium mixture did not interfere with the reaction. This, together with the kinetic data and the small observed solvent effect k(CD$_3$OD)/k(CDCl$_3$) = 6.8, support a mechanism in which isomerization proceeds via a single step pericyclic [2,2]-sigmahaptotropic rearrangement.

## Introduction

Cycloheptatriene (cht) (**1**) enters into the Diels-Alder reaction with tetracyanoethylene (TCNE) by way of its norcaradiene valence isomer **2** (eq. 1).[1] However, the related ($\eta^4$-cycloheptatriene)Fe(CO)$_2$ (**3**), unlike its uncoordinated counterpart **1**, reacts with TCNE to give the 3+2 $\sigma$, $\pi$-allylic adduct **4** as the main primary product. This adduct undergoes in solution a thermally facile rearrangement to the formal 6+2 isomeric complex **5** (eq. 2). The rearrangement has been shown to proceed via the pericyclic [4,4]-sigmahaptotropic ($\sigma\eta$) migration, where both a sigma and a metal bonded group exchange bonding sites across a 4-carbon chain in a single kinetic step.[2]

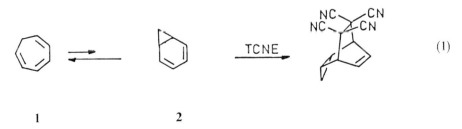

(1)

1    2

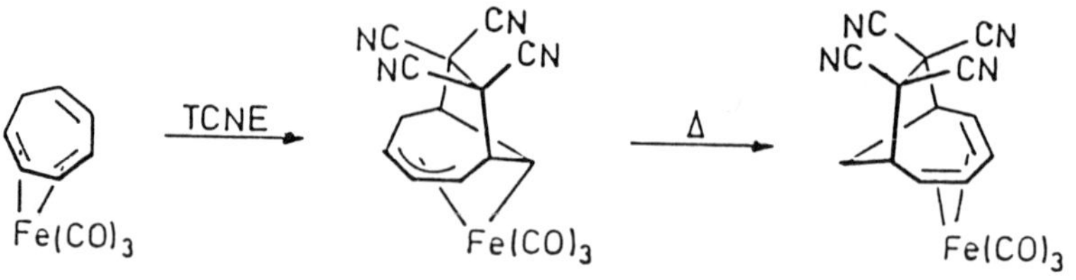

Hereby we report the kinetics of the novel interconversion between the 2+2 (7) and 3+2 (8) adducts of ($\eta^4$-3,7,7-trimethyl-cycloheptatriene)Fe(CO)$_3$ (6) and TCNE, which takes place by way of the [2,2]-$\sigma\eta$ rearrangement (eq. 3).[3] This to our knowledge is the first example in which a tetracyano-substituted cyclobutane-cyclopentane equilibrium takes place in concert.

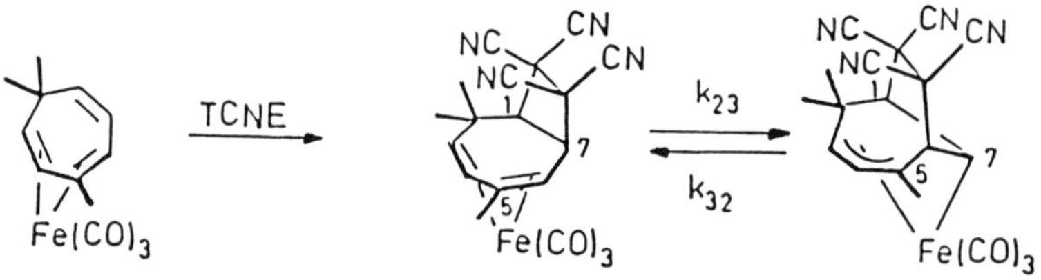

## Results

The reaction of 6 and TCNE was carried out in CHCl$_3$ as described before.[4] The $^1$H NMR spectrum of the recrystallized product in CDCl$_3$ shows the presence of two isomers, 7 and 8, in a ca. 2:3 ratio. The structure of these isomers was determined by a careful analysis of this spectrum. The ratio beteen the isomers remained constant at room temperature or after heating at 50°C for several hours. A similar constant ratio between the isomers was observed in methanolic solutions. Repeated recrystallizations did not change the observed ratio, clearly indicating an equilibrium between the two isomers. An unequivocal confirmation of the presence of this equilibrium resulted from $^1$H{$^1$H} double irradiation experiments which indicated spin saturation transfer (SST) in addition to the expected decoupling. We consequently took advantage of this phenomenon to calculate the kinetic rate constants, using the method developed by Hoffman and Forsen.[5]

In methanol as a solvent, the equilibrium process is faster than in chloroform, causing a significant broadening of the proton lines even at room temperature. Hence, the more convenient line shape analysis technique 6 was utilized, at several different temperatures.

## Kinetics

The exchange rates ($k_{32}$) for the [2,2]-$\sigma\eta$ rearrangement in $CDCl_3$ solutions were measured by the Fourier transform version of the Hoffman-Forsen SST method.[5] Measurements are basd on the exchange process between the two $H_7$ nuclei. In these inversion-recovery experiments the signal of $H_7(8)$ was integrated while $H_7(7)$ was continuously irradiated (Figure 1a). The kinetic data are collected in Table 1.

**Figure 1.** Kinetic measurements by (a) SST and (b) LSA techniques.

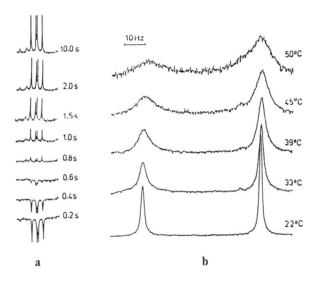

**Table 1.** Kinetic Data for the [2,2]-Sigmahaptotropic Rearrangement by Spin Saturation Transfer ($CDCl_3$).

| T(°K) | $k_{32}$ (s$^{-1}$) | $\Delta G^{\#}$(kcal/mol) | Keq |
|---|---|---|---|
| 297 | 0.19 ± 0.03 | 18.37 ± 0.10 | 3.7 ± 0.7 |
| 302 | 0.30 ± 0.03 | 18.38 ± 0.08 | 3.1 ± 0.7 |
| 307 | 0.48 ± 0.04 | 18.43 ± 0.07 | 3.0 ± 0.7 |
| 316 | 0.97 ± 0.06 | 18.57 ± 0.06 | 3.1 ± 0.7 |

$\Delta H^{\#} = 15.1 \pm 0.6$ kcal/mol $\quad \Delta S^{\#} = -11 \pm 3$ e.u.

The variable temperature line shape analysis (LSA)[6] was employed for the kinetics in CD$_3$OD. The peaks selected for this study were the well isolated 5-methyl signals at δ 2.15 (**8**) and 2.38 (**7**). These appear as two quite sharp singlets at 22ºC, which broaden up as the temperature is raised to 50ºC, without reaching coalescence (Figure 1b). The kinetic data is gathered in Table 2.

**Table 2.** Kinetic Data for the [2,2]-Sigmahaptotropic Rearrangement by Line Shape Analysis (CD$_3$OD).

| T (ºK) | lw$^a$ (Hz) | $k_{32}$ (s$^{-1}$) | $\Delta G^{\#}$ (kcal/mol) |
|---|---|---|---|
| 295 | 0.4 ± 0.2 | 1.3 ± 0.6 | 17.13 ± 0.27 |
| 300 | 0.8 ± 0.2 | 2.5 ± 0.6 | 16.99 ± 0.16 |
| 304 | 1.3 ± 0.2 | 4.1 ± 0.6 | 16.97 ± 0.12 |
| 307 | 1.7 ± 0.3 | 5.3 ± 0.9 | 16.94 ± 0.13 |
| 312 | 3.3 ± 0.4 | 10.4 ± 1.3 | 16.85 ± 0.10 |
| 318 | 5.8 ± 0.5 | 18.2 ± 1.6 | 16.81 ± 0.08 |
| 324 | 10.2 ± 0.7 | 32.0 ± 2.2 | 16.76 ± 0.08 |

$\Delta H^{\#} = 20.7 \pm 0.6$ kcal/mol    $\Delta S^{\#} = 12 \pm 3$ e.u.

$^a$corrected line width, 5-Me (**8**)

**Discussion**

Although there are a few examples reported in the literature in which 2+2 vinylcyclobutane adducts of TCNE with 1,3-dienes rearrange to the isomeric 4+2 (Diels-Alder) adducts in uncoordinated compounds,[7] the present example is to our knowledge the first in which a 2+2 adduct is found to be in equilibrium with its isomeric counterpart (in this case the 3+2 adduct **8**). The 2:3 ratio between the isomers clearly shows that they differ in strain enegy by less than 300 cal/mol.

In order to verify that both the forward and backward reactions occur in a single kinetic step and not by a stepwise cycloaddition-cycloreversion reaction, (carbomethoxy)maleic anhydride (CMA) was added to the equilibrium mixture. None of the expected adducts of (cht)Fe(CO)$_3$ with CMA[8] was detected. Hence, a two step rearrangement in either directions may be excluded.

The present kinetic data are consistent with a concerted mechanism. The rearrangement is evidently first order in both directions. Also we note the low activation energies in both chloroform ($\Delta H^{\#}$ 15.1 kcal/mol) and methanol ($\Delta H^{\#}$ 21 kcal/mol), and the moderate solvent effect, $k_{rel}$ = k(methanol)/k(chloroform) = 6.8, which are typical for pericyclic reactions.

We conclude by outlining the general mechanism of the pericyclic [2,2]-sigmahaptotropic rearrangement, a reaction in which a $\sigma$-bond and a carbon-metal bond interchange about a two carbon chain. In pentacoordinated complexes, this process involves a concerted suprafacial 1,2-sigmatropic shift and a metal (haptotropic)[9] migration, along a symmerical Berry pseudorotation pathway, within the metal coordination sphere. This is shown in Figure 2a for the interconversion of the square pyramidal (sp) butadiene complex and the $\sigma$, $\pi$-allylic trigonal bipyramid (tbp). The molecular orbital correspondance picture, illustrated in Figure 2b, shows the three bonds which participate in the [2,2]-$\sigma\eta$ rearrangement. A more comprehensive analysis of this rearrangement has been recently published elsewhere for the analogous ketene adducts.[3]

**Figure 2.** The [2,2] $-\sigma\eta$ rearrangement along a pseudorotation pathway.
(a) Pseudorotation view in the plane of the paper. (b) Orbital correlation (side view).

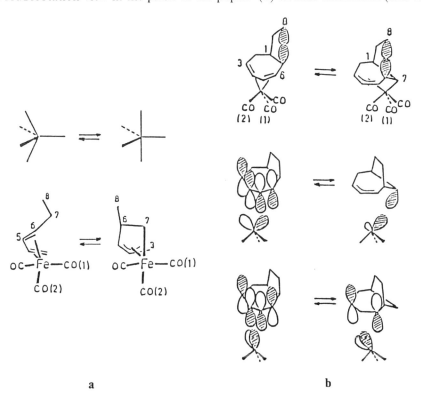

## References

1. N.W. Jordan and I.W. Elliot, *J. Org. Chem.*, **27**, 1445 (1962); G.H. Wahl, *J. Org. Chem.*, **33**, 2158 (1968).
2. Z. Goldschmidt, H.E. Gottlieb, E. Genizi, D. Cohen and I. Goldberg, *J. Organomet. Chem.*, **301**, 337 (1986).
3. Z. Goldschmidt and H.E. Gottlieb, *J. Organomet. Chem.*, **329**, 391 (1987).
4. Z. Goldschmidt and S. Antebi, *J. Organomet. Chem.*, **259**, 119 (1983).
5. R.A. Hoffman and S. Forsen, *Prog. Nucl. Magn. Reson. Spectros.*, **1**, 15 (1966); J.W. Faller in *Determination of Organic Structures by Physical Methods*, J.J. Zuckerman and F.C. Nachod (Eds.), Academic Press, New York, 1973, Vol. **5**; B.E. Mann, *Prog. Nucl. Magn. Reson. Spectros.*, **11**, 95 (1977).
6. I.O. Sutherland, *Ann. Rep. NMR Spectros.*, **4**, 71 (1971); G. Binsch in *Dynamic NMR Spectroscopy*, J.M. Jackman and F.A. Cotton (Eds.), Academic Press, New York, 1968, Chap. **3**, p. 45.
7. F. Kataoka, N. Shimizu and S. Nishida, *J. Am. Chem. Soc.*, **102**, 711 (1980).
8. Z. Goldschmidt, S. Antebi, H.G. Gottlieb, D. Cohen, U. Shmueli and Z. Stein, *J. Organomet. Chem.*, **282**, 369 (1985).
9. N.T. Anh, M. Elian and R. Hoffmann, *J. Am. Chem. Soc.*, **100**, 110 (1978).

# QUEST FOR HIGHER PRISMANES

Goverdhan Mehta
School of Chemistry
University of Hyderabad
Hyderabad 500 134, India.

**Abstract:** Prismanes are a fascinating class of $(CH)_{2n}$ polyhedranes of high symmetry ($D_{nh}$), whose synthetic appeal has been sustained through the prediction and expectancy of novel structural characteristics and unusual chemical reactivity. Among the prismanes, the current favourite for synthesis is [6]-prismane, the heptacyclic, face to face dimer of benzene of $D_{6h}$ symmetry. In our quest for hexaprismane, a novel pentacyclic dimer of benzene and 1,4-bishomohexaprismane "Garudane" (face-to-face dimer of norbornadiene) have been identified as the two advanced precursors and their syntheses have been accomplished through novel strategies. Efforts directed towards the conversion of these precursors to [6]-prismane are described. Some experiments enroute to [7]-prismane and [8]-prismane have been explored.

## INTRODUCTION

The design of strained polycyclic molecules which bear resemblance to familiar geometrical objects and are endowed with high order of symmetry pose a formidable synthetic challenge and hold aesthetic appeal to organic chemists. Among the various polyhedranes, the [n]-prismanes constitute a fascinating class of saturated hydrocarbons made up of even number of methine units of general formula $(CH)_{2n}$ and possessing $D_{nh}$ symmetry, where 'n' is the order of the prismane. Structurally prismanes are composed of two n-membered rings and n-four membered rings, all fused together in cis, syn- manner. The resulting carbocyclic prismatic networks are considerably strained and their creation is a formidable challenge to the experimentalist. At the same time, [n]-prismanes provide a fertile testing ground to the theorist for the application of various levels of molecular structure calculations to predict novel and unusual features. Indeed, there has been a spate of papers recently, dealing with the structure and geometry of prismanes with emphasis on higher prismanes(1a-k). The parameters generated from different calculations ranging from MM2 to ab initio are summarised in Table 1.
    Historically, the prismatic structure 1 (Ladenburg benzene) for an organic molecule was first conceived over a century ago in 1879. How-

ever, the practical realisation of a prismane was achieved only in 1964 by Eaton and Cole through their pioneering effort culminating in the syn-

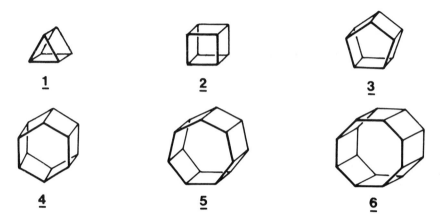

thesis of [4]-prismane 2 (cubane)(2). Nearly, a decade later in 1973 triprismane 1 was synthesised by Katz and Acton(3), to be followed by [5]-prismane 3 (housane) whose synthesis was also accomplished by Eaton et.al.(4). The focus since then has shifted towards higher prismanes, namely [6]-prismane 4, [7]-prismane 5, [8]-prismane 6 and beyond, which present synthetic challenges of increased magnitude. Several groups have recently revealed their intention and endeavours with the objective of [6]-prismane 4 in mind, but only limited progress has been achieved(5). More recently, we at the University of Hyderabad have made considerable headway towards 4 and accomplished the synthesis of secohexaprismane 7(6), the closest, one-bond-away secologue of hexaprismane and of 1,4-bishomo[6]-prismane 8 ("Garudane")(7), a promising precursor of 4 and equally importantly a molecule of interest in its own right. Concurrently, we have also explored some novel synthetic routes to heptaprismane 5 and its seco- and homologues. Herein, we describe the results of experimental efforts made by us in our quest for 4, 5 and 6.

## TOWARDS HEXAPRISMANE 4

The major strategic concern in the construction of the prismatic frameworks is the identification of methodology for efficient construction of multiple cyclobutane rings. In conceptualising a synthetic route to [6]-prismane 4, we recognised at the very outset the pivotal role of the photochemical 2+2-cycloadditions and ring contraction of α-halocyclopentanone moieties as the key cyclobutane generating reactions. Consequently, two sets of precursors from which 4 could be approached were considered and these are shown in Scheme 1. From each of these precursors, hexaprismane can be generated through reactions that result in the formation of two or more cyclobutane rings in a single step. From the Scheme 1, we selected the pentacyclic dimer of benzene 9 and face-to-face dimer 8 (1,4-bishomohexaprismane) of norbornadiene as the most promising precursors of hexaprismane. Since, neither 8 nor 9 or any of their derivatives were known

Table 1 : Optimized Geometries and Heats of Formation of [n]-Prismanes by molecular mechanics,[a] semi-empirical[b] and <u>ab initio</u> methods.[c]

| | | n | 3 | 4 | 5 | 6 | 7 | 8 | 9 | 10 |
|---|---|---|---|---|---|---|---|---|---|---|
| | | θ | 60 | 90 | 108 | 120 | 128.6 | 135 | 140 | 144 |
| (C-C)[e] Å | | MM2 | 1.535 | 1.562 | 1.571 | 1.574 | 1.576 | 1.579 | 1.583 | 1.586 |
| | | MNDO | 1.554 | 1.571 | 1.580 | 1.585 | 1.589 | 1.591 | 1.592 | 1.592 |
| | | AM1 | 1.571 | 1.577 | 1.593 | 1.602 | 1.607 | 1.609 | 1.610 | 1.610 |
| SCF | | 3-21G | | 1.577 | 1.582 | 1.579 | | | | |
| | | 6-31G* | | 1.563 | 1.558 | 1.554 | | | | |
| (C-C)[f] Å | | MM2 | 1.516 | 1.562 | 1.562 | 1.537 | 1.530 | 1.532 | 1.537 | 1.527 |
| | | MNDO | 1.552 | 1.571 | 1.559 | 1.547 | 1.540 | 1.536 | 1.535 | 1.535 |
| | | AM1 | 1.541 | 1.577 | 1.546 | 1.529 | 1.524 | 1.521 | 1.521 | 1.522 |
| SCF | | 3-21G | | 1.577 | 1.561 | 1.558 | | | | |
| | | 6-31G* | | 1.563 | 1.552 | 1.551 | | | | |
| HCC deg. | | MM2 | 132.5 | 125.3 | 116.3 | 112.4 | 110.7 | 108.7 | 108.4 | 108.3 |
| | | MNDO | 129.3 | 125.3 | 123.3 | 121.5 | 120.2 | 119.3 | 118.7 | 118.2 |
| | | AM1 | 128.9 | 125.7 | 122.4 | 120.3 | 118.9 | 117.9 | 117.2 | 116.6 |
| SCF | | 3-21G | | 125.2 | 123.2 | 121.1 | | | | |
| | | 6-31G | | 125.2 | 123.3 | 123.0 | | | | |
| HF[g] | | MM2 | 88.1 | 137.5 | 119.6 | 127.7 | 168.6 | 219.1 | 285.3 | 329.3 |
| | | MNDO | 121.9 | 99.1 | 72.2 | 80.4 | 116.4 | 172.7 | 243.8 | 325.1 |
| | | AM1 | 165.0 | 151.2 | 117.7 | 140.0 | 191.0 | 259.2 | 340.6 | 430.9 |
| SCF | | 3-21G | | 145.1 | 115.4 | 160.3 | | | | |
| | | 6-31G* | | 147.1 | 121.2 | 154.5 | | | | |
| RMP | | 6-31G* | 136.4 | 148.5 | 119.6 | 153.1 | 212.0 | 283.1 | 368.7 | |

[a] References 1b-e.  [b] References 1g,h,k.  [c] References 1i-j.
[d] C-C-C angle along the n-membered ring.  [e] Single C-C bond, within ring.
[f] Single C-C bond, ring to ring.  [g] Kcal/mol.

## Scheme 1

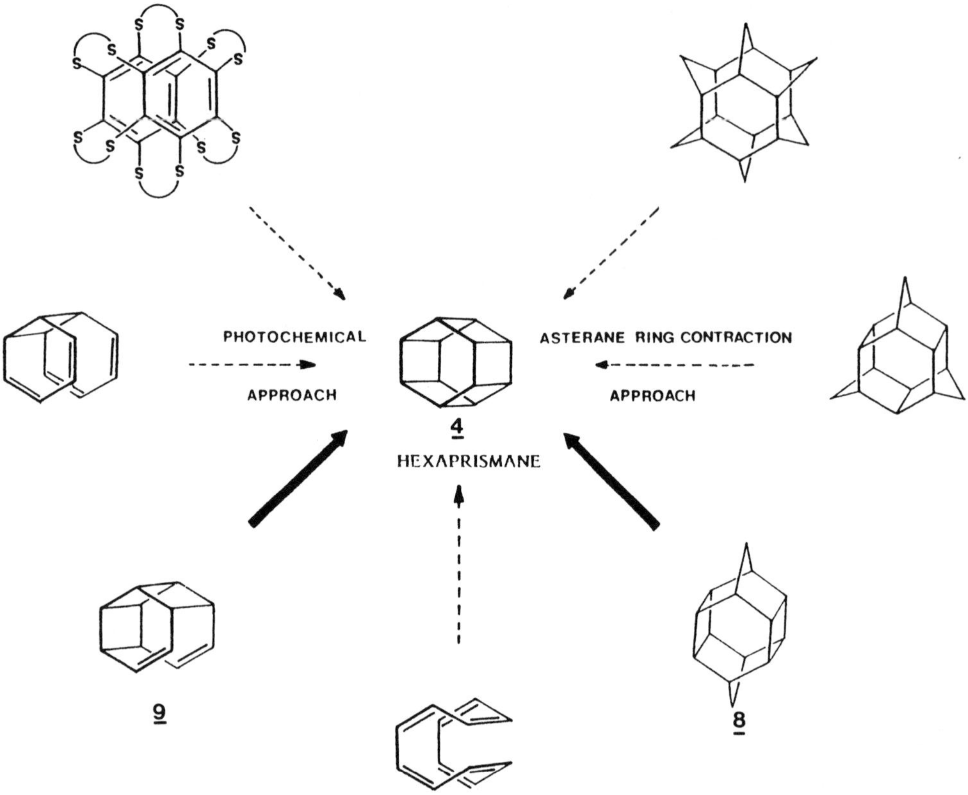

in the literature at the inception of our effort, attention was first centred on their synthesis enroute to 4.

## SECOHEXAPRISMANE 7(6)

Synthesis of 11, a functionalised derivative of 9 was developed from readily available 1,5-cyclooctadiene(COD) and 5,5-dimethoxy-1,2,3,4-tetrachlorocyclopentadiene(DCP) in 13 steps as detailed in Scheme 2. The key concept in this approach was the deployment of COD and DCP as cyclooctatetraene and cyclobutadiene equivalents, respectively. Photochemical intramolecular 2+2-cycloaddition in 11 was expected to deliver the hexaprismane derivative 12. However, 11 → 12 transformation could not be realised under a variety of irradiation regimen. Disappointed by this result, we turned to 10 whose X-ray crystal structure (Fig.1) had revealed the two-double bonds to be within range (<3Å apart) with favourable spatial orientation for an intramolecular 2+2-cycloaddition. It was apparent that 13, obtainable from 10 was also eminently serviceable for conversion to the hexaprismane deri-

Scheme 2

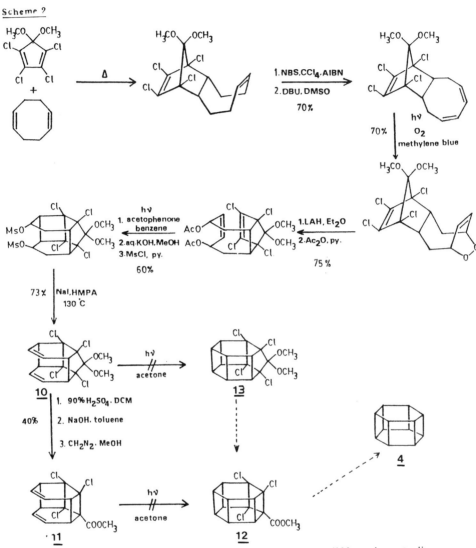

Fig. 1. Perspective view of 10 determined by X-ray diffraction studies.

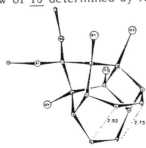

vative 12 through ring contraction protocol. However, once again, the
10 →13 photochemical change could not be induced despite several variations
in the irradiation conditions. We have attributed the failure of 10 → 13
and the 11 → 12 photocycloadditions to the high build-up of strain energy
($\Delta$SE >55 Kcal/mol) during the photocyclisation step(1e). The MM2 calculations, however, predicted that 14 → 15 ( $\Delta$SE 45.93 Kcal/mol) was feasible
on strain energy considerations and the attention was, therefore, turned
to this option.

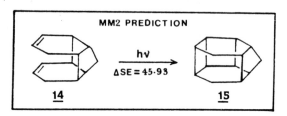

Access to the ring system 14 required regioselective removal of one
of the cyclobutane bonds in the pentacyclic compound 10. This could be
achieved through the preparation of the olefinic mesylate 16 and subjecting
it to diborane mediated fragmentation reaction(8). The resulting tetracyclic
diene 17 on sensitised irradiation took the predicted course and the hexacyclic 2+2-cycloaddition product 18 was realised. Unmasking of the carbonyl
group in 18 set up the Favorskii ring contraction and a 2:5 ratio of secohexaprismane ester 19 and the corresponding Haller-Bauer ester, respectively,
was obtained. Finally, a three step sequence on 19 furnished the hexacyclic
hydrocarbon secohexaprismane 7 as a highly volatile waxy solid, mp > 250°,
and was fully characterised (HRMS, $^1$H & $^{13}$C NMR), Scheme 3.

Scheme 3

While the synthesis of the hydrocarbon closest to hexaprismane 4 was accomplished in 18 steps from cheap abundantly available starting materials, it became quite apparent that the photochemical cycloaddition strategies would not deliver 4 unless 2+2-photoclosure in either 10 or 11 could be successfully executed. This meant overcoming the $\Delta\overline{SE}$ barrier of ~55 Kcal/mol for the photocyclisation and no strategy to circumvent this problem could be readily devised. The attention was, therefore, switched to explore the alternative strategy of asterane ring contraction to hexaprismane, Scheme 1.

## 1,4-BISHOMOHEXAPRISMANE 8 ("GARUDANE")

Among the three asterane-type precursors of hexaprismane depicted in Scheme 1, the 1,4-bishomohexaprismane 8 (now christened "Garudane")(9) was chosen as the initial objective. This heptacyclic, true face-to-face 2+2 dimer of norbornadiene of $D_{2h}$ symmetry is not only promising in the context of 4 but is an architecturally unique molecule of interest. Ideally, 8 should be accessible through the union of two norbornadiene moieties and indeed many efforts in this direction have been made over the past two decades employing a variety of metal catalysts(10). However, unfavourable strain energy and entropy considerations have thwarted such attempts and only dimer 20 has been encountered among other products, Scheme 4. We, therefore, contemplated on a stepwise but rational synthetic design for 8. From the various strategic options available for attaining 8, the one conceptualised in Scheme 5, appeared most attractive. This involved a formal $2C_5$ (1,3-cyclopentadiene) + $C_4$ (cyclobutadiene) union through thermal 4+2 and photochemical 2+2-cycloaddition processes. Imparting practical shape to this theme required identification of a cyclobutadiene equivalent that could twice function as a 2 π component in the 4+2 cycloadditions, control of stereochemistry to facilitate the intramolecular 2+2-photocycloaddition and lastly functional group adjustments to the hydrocarbon level. Synthetic considerations and previous literature precedences identified the "2,5-dibromobenzoquinone between the two cyclopentadienes" as the stratagem for achieving 1,4-bishomohexaprismane, Scheme 5.

Scheme 4

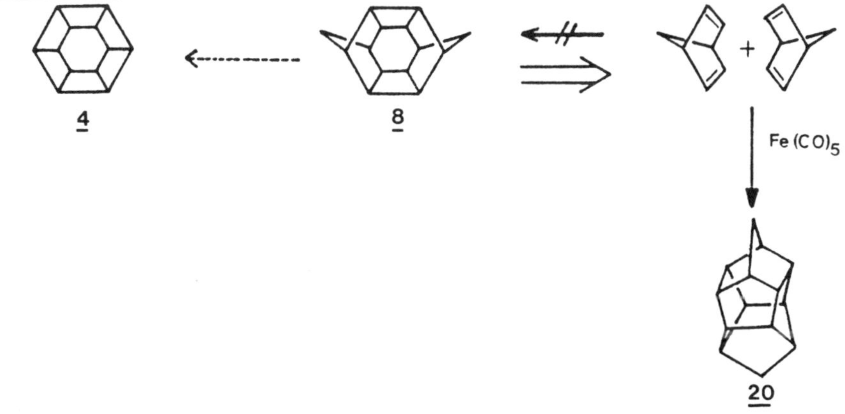

## Scheme 5

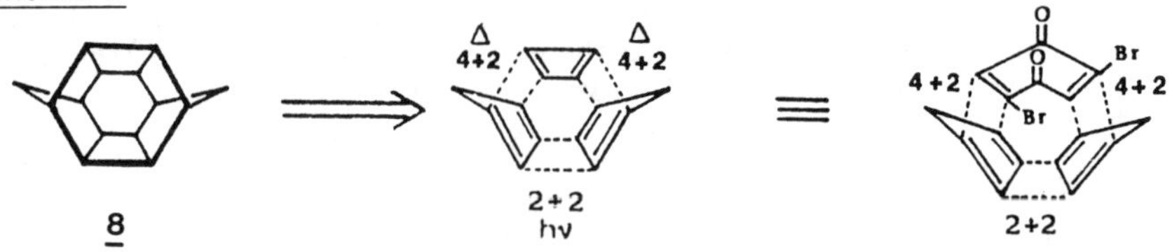

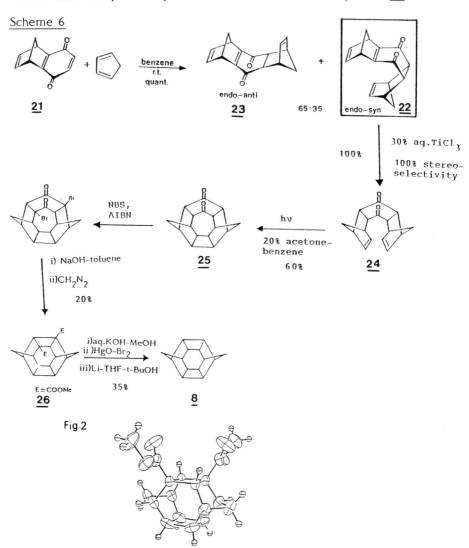

The successful synthesis of 8 is delineated in Scheme 6. The readily available(11) but previously overlooked norbornenobenzoquinone 21 on Diels-

Fig. 2

Alder reaction with cyclopentadiene furnished a 65:35 mixture of endo, syn-22 and endo, anti-23 adducts, respectively, in quantitative yield. The enedione moiety in 22 was regio- and stereoselectively reduced with aq.TiCl$_3$ reagent(12) to furnish the endo, syn, endo-24 product. This readily underwent the projected intramolecular 2+2-photoclosure to give the key compound a 1,4-bishomo-6-seco-[7]-prismane dione 25, which apart from being the key precursor of 8 is also potentially serviceable for further elaboration to [7]- and [8]-prismane analogues. Further evolution of dione 25 to "garudane" required two ring contractions and these were effected through two α-brominations and a single shot double Favorskii rearrangement to give diester 26. Routine functional group manipulations now delivered the target molecule 8 exhibiting a 3 line $^{13}$C NMR spectrum ($\delta$ 43.0, 36.3, 33.5) and a 3 line $^1$H NMR spectrum ($\delta$ 2.38 br, 2.0 br, 1.2 dd, J=1.3 Hz). As an unambiguous proof of the structure of "garudane", a single crystal X-ray structure determination on 1,3-dimethoxycarbonyl derivative 26 was carried out. A perspective drawing is shown in Fig. 2 and the detailed structure has been discussed elsewhere(13). Interestingly, 8 broadly retains the trends in bond lenghts and bond angles exhibited by the parent norbornane.

While the successful synthesis of the parent 1,4-bishomohexaprismane 8 had been achieved, further elaboration of this to hexaprismane 4 required access to suitably functionalised derivatives of 8. In particular, entry was sought to tetraketone 27 either directly or in a stepwise manner to effect a total of four Favorskii ring contractions as indicated in Scheme 7.

Scheme 7

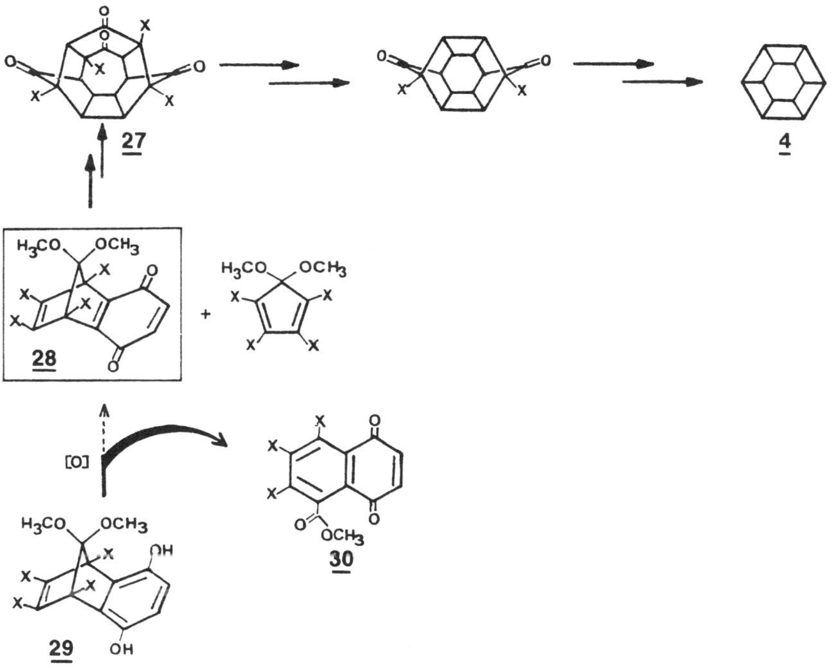

The key tetraketone 27 could, in turn, be assembled from masked 7-keto-norbornenobenzoquinone 28 and 5,5-dimethoxy-1,2,3,4-tetrachlorocyclopentadiene as shown. For this purpose, considerable effort was devoted towards the synthesis of the quinone 28 from the readily available 29. However, oxidation of 29 employing a variety of diverse oxidants only led to the naphthoquinone ester 30 and no trace of the desired quinone 28 could be detected. Attempts towards in situ trapping of 28 too have not frutified so far. Consequently, we decided to pursue the next best alternative and investigated the reaction or norbornenobenoquinone 21 with DCP. The endo, syn-adduct 31 was duly obtained although this time as a minor product (endo, syn : endo, anti = 23 : 77) of cycloaddition. Regio- and stereoselective reduction of the enedione moiety and sensitised irradiation led to the formation of the caged dione 32. Attempts to convert 32 to 33 have not succeeded so far, particularly the step involving installation of α-bromo substituents in preparation for the Favorskii ring contraction, Scheme 8. However, attempts in this direction are being continued presently.

Scheme 8

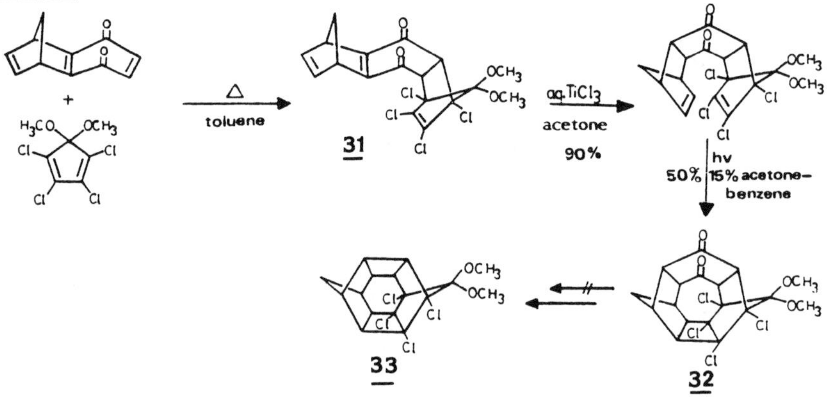

## TOWARDS [7]-PRISMANE AND ITS SECO- AND HOMOLOGUES

No synthetic work aimed at heptaprismane has appeared in literature so far(1f). The heptacyclic dione 25 and its functionalised derivative 32 described above possess a bishomo-seco-[7]-prismane skeleton and, therefore, are potentially useful for elaboration to [7]-prismane 5 or its close analogues.

Several lines of thought were pursued to convert 25 to a 1,4-bishomo[7]-prismane 34 as shown in scheme 9. These included an attempted direct pinacolic coupling to give 35 which did not succeed. Alternatively, the dione 25 was transformed to the diolefin 36 in a 4 step sequence involving C-C bond fragmentation. Attempted 2+2-photoclosure of 36 to give bishomo-heptaprismane 34 failed again, perhaps due to the unsurmountable strain energy barrier ($\Delta SE$ = 63.4 Kcal/mol), Scheme 9.

Another set of reactions were performed to gain entry into the homo-seco-[7]-prismane 37. In one approach, the functionalised dione 32 was transformed to 38 to set up a Favorskii ring contraction. However, alkali

## Scheme 9

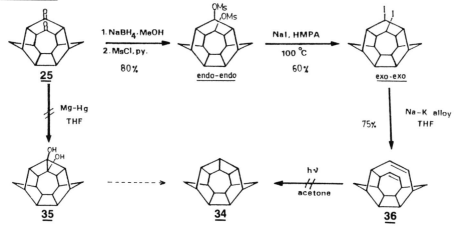

treatment on 39 gave the fragmented diolefin 40 whose 2+2-photoclosure too could not be effected. In another lead, the dione 32 was deacetalised to 41 but its base induced reaction led to intractable products, Scheme 10.

## Scheme 10

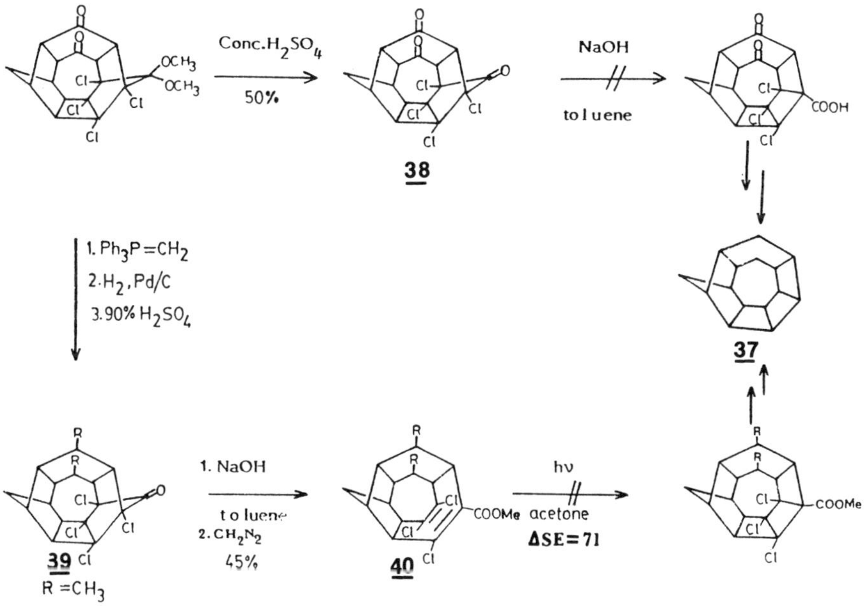

An attempt was also made to create the architecturally beautiful 1,3,5-trishomo[7]-prismane framework 41. In this endeavour a homo-Norrish rearrangement on 42 was unsuccessfully explored. An alternate route to 41 via the olefin 43 got diverted to the lactone 44, Scheme 11. However, further manipulations to gain entry to 41 are currently underway.

Scheme 11

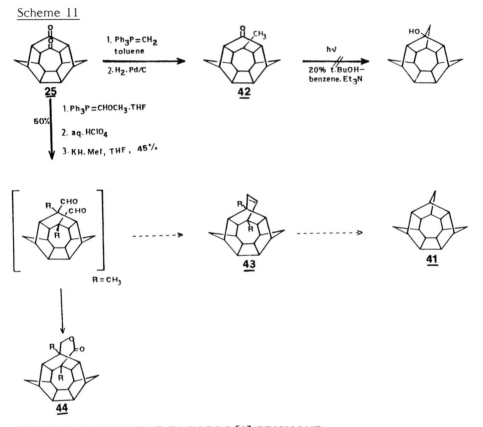

## PROBING EXPERIMENT TOWARDS [8]-PRISMANE

The heptacyclic dione 25 on ring expansion could be converted to 45 (or regioisomer) which is formally a [8]-prismane analogue. However, 25 proved to be extremely resistant towards ring expansion. Even Baeyer-Villiger oxidation on 25 could not be effected beyond 46.

## ACKNOWLEDGMENT

The research described here is the outcome of dedicated and skilfull experimental work of exceptional merit carried out by my graduate student Ms. S. Padma. Support for our research effort has been provided by the University Grants Commission through Special Assistance and COSIST programmes.

## REFERENCES

1. (a) V.I. Minkin, R.M. Minyaev, Zh. Org. Khim., 1981, 17, 221. (b) N.L. Allinger, P.E. Eaton, Tetrahedron Lett., 1983, 3697. (c) R.M. Martin, unpublished results quoted in ref. 4b. (d) V.P. Reddy, E.D. Jemmis, Tetrahedron Lett., 1986, 3771. (e) G. Mehta, S. Padma, E. Osawa, D.A. Barbiric, Y. Mochizuki, ibid, 1987, 28, 1295. (f) G. Mehta, S. Padma, E.D. Jemmis, G. Leela, E. Osawa, D.A. Barbiric, ibid, 1988, 29, 1613. (g) W.P. Dailey, ibid, 1987, 47, 5787. (h) R. Engelke, P.J. Hay, D.A. Kleier, W.R. Wadt, J. Am. Chem. Soc., 1984, 106, 5439. (i) M.A. Miller, J.M. Schulman, J. Mol. Struct., 1988, 163, 133. (j) R.L. Disch, J.M. Schulman, J. Am. Chem. Soc., 1988, 110, 2102. (k) E.D. Jemmis, J.M. Rudzinski, E. Osawa, Chem. Expres, 1988, 3, 109.
2. P.E. Eaton, T.R. Cole, J. Am. Chem. Soc., 1964, 86, 3157.
3. T.J. Katz, N.J. Acton, ibid, 1973, 95, 2738.
4. (a) P.E. Eaton, Y.S. Or, S.J. Branca, ibid, 1981, 103, 2134. (b) P.E. Eaton, Y.S. Or, S.J. Branca, B.K.R. Shankar, Tetrahedron, 1986, 42, 1621.
5. (a) P.E. Eaton, U.R. Chakraborty, J. Am. Chem. Soc., 1978, 100, 3634. (b) G. Mehta, A. Srikrishna, M.S. Nair, T.S. Cameron, W. Tacreiter, Ind. J. Chem., 1983, 22B, 621. (c) H. Higuchi, K. Takatsu, T. Otsubo, Y. Sakata, S. Misumi, Tetrahedron Lett., 1982, 671. (d) N.C. Yang, M.G. Horner, 1986, 543. (e) V.T. Hoffmann, H. Musso, Angew. Chem. Int. Engl., 1987, 26, 1006.
6. G. Mehta, S. Padma, J. Am. Chem. Soc., 1987, 109, 2212.
7. G. Mehta, S. Padma, J. Am. Chem. Soc., 1987, 109, 7230.
8. J.A. Marshall, G.L. Bundy, J. Am. Chem. Soc., 1966, 88, 4291.
9. The protruding bridges ("wings") in 8 are reminiscent of "Garuda" (Sanskrit), the mythological Hindu demi-god, part-bird, part-man, see: Encyclopaedia Britanica, Micropaedia 1981; Vol. IV, p.425. According to the von Baeyer system of nomenclature : heptacyclo-[9.3.0.0$^{2,5}$.0$^{3,13}$.0$^{4,8}$.0$^{6,10}$.0$^{9,12}$]-tetradecane.
10. T.J. Chow, Y.S. Chao, L-K, Liu, J. Am. Chem. Soc., 1987, 109, 797 and previous references cited therein.
11. (a) O. Diels, K. Alder, Chem. Ber., 1929, 62, 2337. (b) R.C. Cookson, R.R. Hill, J. Hudec, J. Chem. Soc., 1964, 3043.
12. L.C. Blaszczak, J.E. McMurry, J. Org. Chem., 1974, 39, 258.
13. E. Osawa, D.A. Barbiric, O.S. Lee, Y. Kitano, S. Padma, G. Mehta, submitted for publication to J. Chem. Soc., Perkin Trans. II.

# PLANARIZING DISTORTIONS IN POLYCYCLIC CARBON COMPOUNDS

R. Keese, W. Luef, J. Mani, S. Schüttel,
M. Schmid, C. Zhang
Institute of Organic Chemistry
University of Berne, Freiestrasse 3
3012 Berne, Switzerland

ABSTRACT. Planarizing distortions of the central $C(C)_4$ fragment in spiro[m.n]alkanes with m,n = 2-4 can be analyzed in a systematic way by symmetry deformation coordinates. As revealed by X-ray structures and MNDO calculations of such compounds, the local distortions around the quaternary carbon atom are dominated by compression and twist. In fenestranes, which are to be considered as spiro compounds with alkylidene-bridges between the two $\alpha,\alpha'$-positions, the deformation of the central $C(C)_4$ fragment is due to twist and angle opening. Various syntheses of all-cis-[5.5.5.5]fenestranes with the intramolecular arene-olefin-photocycloaddition as key step are discussed. Reactions of such compounds as well as attempts to increase the planoid distortions by transforming all-cis-[5.5.5.5]fenestranes into the more strained cis,cis,cis,trans-[5.5.5.5]fenestranes will be reported.

## INTRODUCTION

Synthesis of strained and unusual compounds has fascinated organic chemists since A. v. Baeyer first prepared cyclopropane [1]. This challenge got additional momentum when it became apparent that such compounds have unexpected properties and due to unique reactivity, are synthetically useful intermediates. In addition many of the strained organic compounds serve as models for the study of the nature of the chemical bond [2] and for testing theory by experiment. More recently, the question of planar, tetracoordinate carbon has been discussed [3]. Many structures proposed have been elucidated by quantum chemical methodology [4], few experimental studies have been reported [5]. The energy of planar methane 2 being

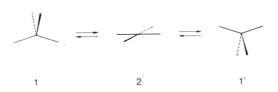

1    2    1'

156.4 kcal/mole higher than that of 1, indicates that tetracoordinate carbon with a planar configuration may only exist under special structural conditions if at all.

Considering the configuration of the $C(C)_4$ fragments in organic compounds containing quaternary C-atoms, it is apparent, that their geometry deviates in most cases from $T_d$ symmetry [6]. Using symmetry deformation coordinates to describe angular distortions in bridged spiro[4.4]nonanes, we had observed that the configuration of the central $C(C)_4$ fragments are best described by twist ($S_{2b}$) and compression ($-S_{2a}$) whereas in [5.5.5.5]fenestranes, to be considered as spiro[4.4]nonanes with ethylene groups bridging each of the $\alpha,\alpha'$-positions, two opposite bond angles at the quaternary carbon·atom are larger than the standard tetrahedral angle [7]. In this paper we discuss the structural features of spiro[m.n]alkanes with m,n = 2-4, in which the two rings are connected by alkylidene groups and the special class of [m.n.o.p]fenestranes with m,n,o,p $\leq$ 5. The synthesis of various [5.5.5.5]fenestranes and some of their chemical transformations are also presented.

## ANGULAR DISTORTIONS IN SPIRO ALKANES

In organic compounds, bond angles and bond lengths of $C(C)_4$ fragments can be analyzed with respect to the type and extent of distortions from local $T_d$ symmetry [8]. Usually all deformations contribute to the local configuration of the $C(C)_4$ fragment [7]. Nevertheless it is possible to extract from the total deformation those angular contributions, which are caused by planarizing distortions. The appropriate bond angle deformations can be described as "spread" ($+S_{2a}$), compression ($-S_{2a}$) and twist ($S_{2b}$) (cf. Fig. 1).

"Spread" leads to angle opening of a pair of opposite angles at the expense of the other four and will, in the limit, lead to a square planar structure. Compression is related to closing a pair of opposite bond angles and gives a linear structure. Twist, which may be pictured by a clockwise or counterclockwise rotation of the plane of one $C(C)_2$ fragment with respect to the plane of the other, will transform a $C(C)_4$ structure into a rectangular arrangement. In terms of symmetry deformation coordinates, the type of deformation is specified as $+S_{2a}(E)$, $-S_{2a}(E)$ and $S_{2b}(E)$ with the appropriate components of the displacement vector given by $s_j$ [7]. Polycyclic compounds with central $C(C)_4$ moieties, which show spread resp. compression of opposite bond angles are given by the [5.5.5.5]fenestrane 3 [9] and the functionalized spiro[4.4]nonane 4 [10]. In 7,7-vespirene 5 [10] the contribution of twist to the over all planoid distortion is much larger [11] (cf. Fig. 1). In extention of our earlier results [7], the planarizing distortions of $C(C)_4$ fragments in spiro[m.n]alkanes (m,n = 2-4) with or without alkylidene groups bridging the $\alpha,\alpha'$-positions have been analyzed. The structural data were obtained from X-ray analysis as well as MNDO calculations [12] (Fig. 2). The distortions of the $C(C)_4$ fragments in the bicyclic spiro compounds, related to the principal axis bisecting the two rings (cf. Fig. 1) are essentially caused by compression ($s_{2a}(E) < 0$).

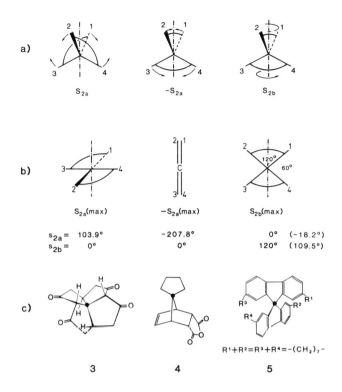

Figure 1. Angular distortions leading from tetrahedral to planar configurations, a) specification of the appropriate symmetry deformation coordinates, b) the limiting structures to be expected from the deformations specified, c) structures with dominating spread 3 $(+S_{2a}(E))$, compression 4 $(-S_{2a}(E))$, and a large proportion of twist 5 $(S_{2b}(E))$.

In the tri- and tetracyclic compounds the angular deformations are related to the principal axis outside the two rings (Fig. 1) and give rise to spread ($s_{2a}(E) > 0$) and twist ($s_{2b}(E) > 0$). According to the angular distortions revealed by X-ray structure analysis spiro[m.n]alkanes with or without alkylidene groups bridging the $\alpha,\alpha'$-positions can be divided into two classes. In relation to the principal axis, compression dominates in spiro compounds without bridges, whereas spread and twist are the main planarizing angular distortions in those tri- and tetracyclic structures, in which the spiro rings are bridged by one or two alkylidene groups.

The bond angle deformation in the spiro[m.n]compounds 6-10, for which m,n = 2,3 (Fig. 2), is due to compression whereas the $C(C)_4$ fragments of the tricyclic and tetracyclic structures 11-13 show considerable spread and twist (Fig. 3). In the "broken windowpane" 11 [13] spread ($s_{2a}(E)$) and twist ($s_{2b}(E)$) amount to $s_{2a}$ = +27.74° resp. 19.65°, whereas the angular distortions in the fenestranes 12 [14] and 13 [5b] are clearly dominated by spread (12: $s_{2a}(E)$ = +28.08°, $s_{2b}(E)$ = 0.83°; 13: $s_{2a}(E)$ = +32.27°, $s_{2b}(E)$ = 9.3°). The analysis of the spirocom-

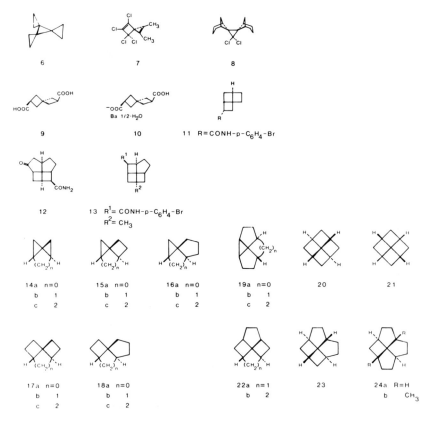

Figure 2. The structures, for which spread ($+S_{2a}(E)$), compression ($-S_{2a}(E)$) and twist ($S_{2b}(E)$) of the central $C(C)_4$ fragments have been analyzed.

pounds 14-18 and the fenestranes 19-24, for which angular distortions were extracted from MNDO calculations, lead to the same conclusions. The compression, indicated by small intraring bond angles in spiro pentane and spiro[3.3]heptane is rather large ($s_{2a}(E) \ll 0$), whereas spread and twist dominate in the tricyclic compounds 14b, 15b, 17b and 14c-18c (cf. Fig. 3b). In 14a-18a the contribution of spread and twist resp. comp-

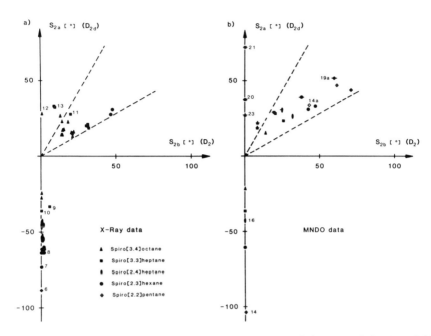

Figure 3a,b. *Planarizing angular distortions ($+S_{2a}(E)$, $-S_{2a}(E)$, $S_{2b}(E)$) extracted from a) data of crystal structures, b) MNDO calculations.*

ression to the over all angular distortions is of secondary importance [12]. Only tricyclo[2.1.0.0$^{1,3}$]pentane 14a, prepared recently by Wiberg [15], may be mentioned. According to our MNDO results, which reproduce the bond angles of the ab initio (6-31G$^*$) calculation for this compound rather well, the secondary, planarizing distortion is due to twist with a minor contribution from spread ($s_{2b}(E) = 43.07°$, $s_{2a}(E) = +33.92°$).

## THE STRUCTURE OF FENESTRANES

A comparison of spread ($S_{2a}(E)$) vs. twist ($S_{2b}(E)$) reveals that the angular distortions of the central $C(C)_4$ fragments in [m.n.o.p]fenestranes with $3 < m,n,o,p \leq 5$ are strongly dominated by spread. Typical examples are given by 20 ($s_{2a}(E) = +37.72°$, $s_{2b}(E) = 0.005°$) and 23 ($s_{2a}(E) = +27.4°$, $s_{2b}(E) = 0.005°$). Significant twist is found in [m.n.o.p]fenestranes with $m,n,o,p \leq 5$, when at least one ring differs in size from the others (Fig. 4). Thus, in [3.5.4.5]fenestrane 19b, spread is very similar to twist ($s_{2a}(E) = +38.85°$, $s_{2b}(E) = 37.85°$), whereas angle opening is larger in 22a ($s_{2a}(E) = +33.09°$, $s_{2b}(E) = 10.23°$) than in 24a ($s_{2a}(E) = +18.87°$, $s_{2b}(E) = 8.13°$). The parent hydrocarbons 22a,

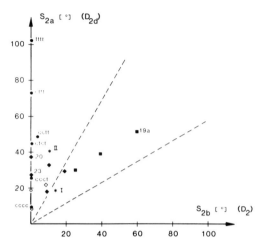

Figure 4. Angular distortions [spread (+$S_{2a}(E)$) vs. twist ($S_{2b}(E)$)] in selected [m.n.o.p]fenestranes.

23 and 24a have yet to be prepared, but syntheses of derivatives viz. 25, 26 and 27 have been reported [5b]. The structural details of the central $C(C)_4$ fragments of the parent compounds 22a and 23 obtained by MNDO calculations, correspond very closely to those obtained by X-ray structure analysis. At present 25, a derivative of [4.4.4.5]fenestrane 22a, is the fenestrane with the largest opposite bond angles (129.2° and 128.2°) [5b].

25      26      27

29      30      31

It is apparent from Fig. 4 that tri- and tetracyclic compounds like 14c-19c and 22-24 have angular distortions, which are related to spread and twist. In general, the angular deformations increase with decreasing size of at least one ring. Increasing planoid deformations in $C(C)_4$ fragments are also found in unsaturated and in stereoisomeric fenestranes (cf. Fig. 5). Whereas spread in the [5.5.5.5]fenestrane 28 is only $s_{2a}(E) = +9.31°$ - corresponding to opposite bond angles 114.93° and 114.9° - $s_{2a}(E) = +27.2°$ in the triene 29. For the tetraene 30, opposite bond angles of 138° and 135° are calculated, leading to $s_{2a}(E) = +44.9°$. In the $C(C)_4$ fragments of all three structures the twist is negligible. The fenestranes discussed so far are formally constructed of four bicyclo[x,y,0]alkane subunits. Stereoisomers of fenestranes are obtained if one allows for cis- as well as trans-fusion in the bicyclic substructures. In this manner, 6 stereoisomers of [5.5.5.5]fenestrane can be constructed [6]. In all-cis-[5.5.5.5]fenestrane 28 the four bicyclo[3.3.0]subunits are cis-fused, whereas in cis,cis,trans,trans-[5.5.5.5]fenestrane 31, two of these substructures are trans-fused. The planarizing distortions in these stereoisomeric [5.5.5.5]fenestranes were related to strain and compared with those of the fenestranes 20, 22a, 23 and 24a, all with cis-fused bicyclo[x,y,0]alkane subunits (Fig. 5). With all-cis-[5.5.5.5]fenestrane as starting point, the over all planarizing distortions increase both with increasing ring contractions - leading to all-cis-[4.4.4.4]fenestrane 20 - and with increasing number of trans-fused bicyclic substructures in [5.5.5.5]fenestrane. Increase of strain is related to increasing spread and hence to further opening of opposite bond angles (i.e. 139° and 139.1° in 31). It remains to be seen, whether compounds with planarizing distortions larger than $s_2(E) = 34,6°$, found for [4.4.4.5]fenestrane 22a and its derivative 25 [5c] can be prepared.

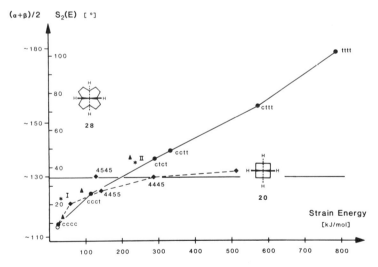

Figure 5  Planarizing distortions ($S_2(E)$) in selected fenestranes vs. strain, calculated from group parameters [16].

In the following section synthetic efforts for preparation of [5.5.5.5]fenestranes will be discussed.

# SYNTHESIS OF [5.5.5.5]FENESTRANES

Syntheses of all-cis-[5.5.5.5]fenestrane and some of its derivatives have been reported [9] [17]. Both approaches are based on functionalized bicyclo[3.3.0]octanes and preserve elements of the target molecule in the synthetic steps. In consideration of preparing stereoisomers other routes to [5.5.5.5]fenestranes with a variety of functionalities in strategic positions were needed. The photoinduced intramolecular metacycloaddition between an arene and an appropriately located double bond offers such possibilities. Two routes to [5.5.5.5]fenestranes with the photoinduced metacycloaddition as the key step have been explored [18] [19].

## a) [5.5.5.5]Fenestranes from a substituted indane

Retrosynthetic considerations suggested that 1-(3-butenyl)-indane would be the key structure for direct formation of [5.5.5.5]fenestrane by photoinduced metacycloaddition. For purposes of selectivity in the photoreaction and the advantageous placement of a methoxy group in the desired fenestrane, 32 was prepared and irradiated. Three photoproducts 33, 34 and 35 were obtained in a ratio of 61.5:30.8:7.7 (Scheme 1).

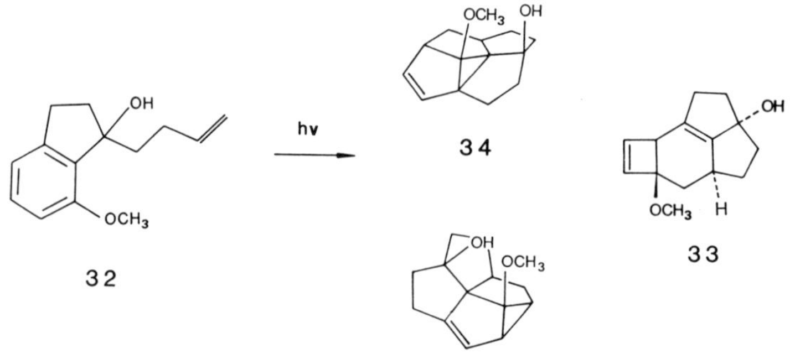

Scheme 1.

Contrary to our expectation, the [5.5.5.5]fenestrane 35 was the minor product, whereas 34 qualifies as [3.5.5.5]fenestrane with an additional ethynilidene bridge. The formation of the major photoproduct 33 merits special consideration. According to detailed structure analysis by advanced NMR methodology, compound 33 has the configuration shown [20]. According to our mechanistic interpretation a photoinitiated [2+2]cycloaddition leads from 32a first to the [4.5.5.6]fenestrane 36, in which the bicyclo[3.2.0]heptane substructure is trans-fused (Scheme 2). Disrotatory opening of 36, which must be highly exothermic, gives the triene 37 from which 33 is formed in a further photoinduced, disrotatory ring closure. Prior to proof of 36 being the primary intermediate in the formation of the major photoproduct, the planarizing distortions have been determined by MNDO calculations. With opposite bond angles of 127°

and 138° and a spread of $s_{2a}(E) = +40°$, 36 presently would qualify as the fenestrane with the largest angle opening (see Fig. 5, II). In contrast, 32b would give the more stable isomer 38, which contains a cis-bicyclo[3.2.0]heptane subunit with bond angles of 119° and 121° (see Fig. 5, I).

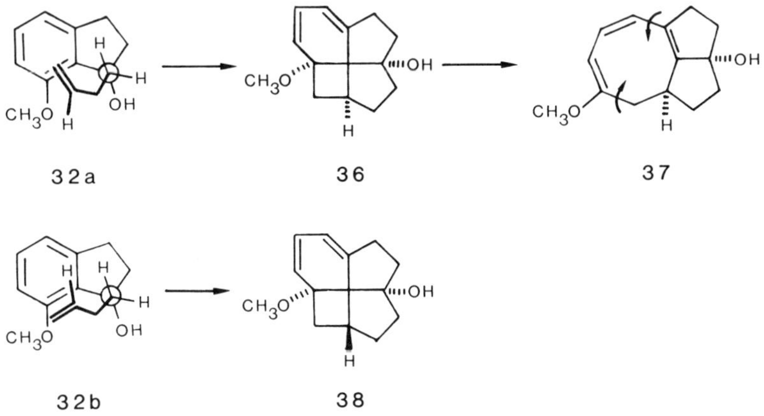

Scheme 2.   32a   36   37   32b   38

The tetracyclic compound 33 is very sensitive to acid and readily gives 39. From this tricyclic structure the [5.5.5.5]fenestrane 42 resp. 28 might be obtained via a transannular reaction of an appropriate intermediate.

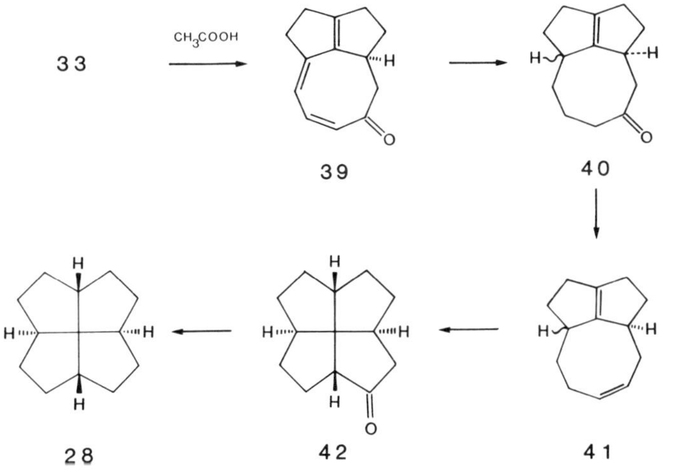

Scheme 3.   33   39   40   28   42   41

Hydrogenation of 39 gave 40 as a mixture of stereoisomers. Reduction of the carbonyl group and elimination gave a mixture of dienes, which was refluxed with formic acid [21]. After hydrolysis and oxidation a pure sample of the ketone 42 could be obtained. Certainly, the selectivity in these reaction sequence can be improved. Meanwhile, reactions of the fenestranes 34 and 35 have been investigated. When 34 was heated to 240°C, the tetracyclic compound 43 is formed via a 1,5-shift of the hydrogen at

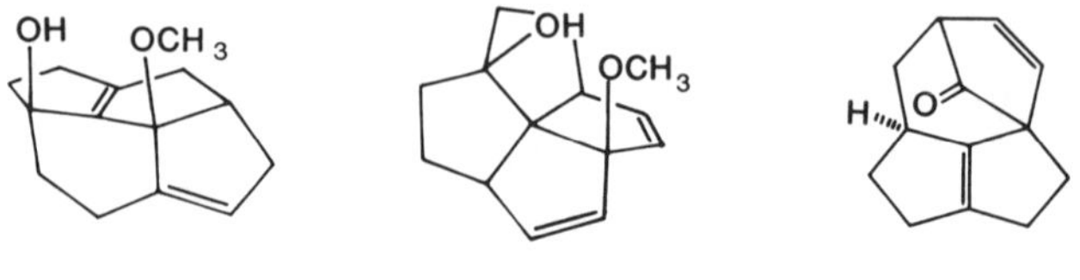

43

44

45

the bridgehead. Much to our surprise, the fenestrane 35 gave not the tetrafunctionalized all-cis-[5.5.5.5]fenestrane 44, but the same tetracyclic compound 43. When 34 was refluxed in dimethylsulfoxide the tetracyclic compound 45 was obtained. Under the same conditions this compound was also obtained from 35.

b) [5.5.5.5]Fenestranes from 5-phenylpentene

We had reported that tetracyclic ketone 50, a derivative of all-cis-[5.5.5.5]fenestrane can be prepared from the substituted 5-phenylpentene 46 (Scheme 4) [18]. Key step in this synthesis was again the photoinduced intramolecular cycloaddition between the phenylring and the double bond and yielded the tetracyclic compound 47. The diazoketone 49, obtained from 48, reacted upon treatment with trifluoroacetic acid to give 50.

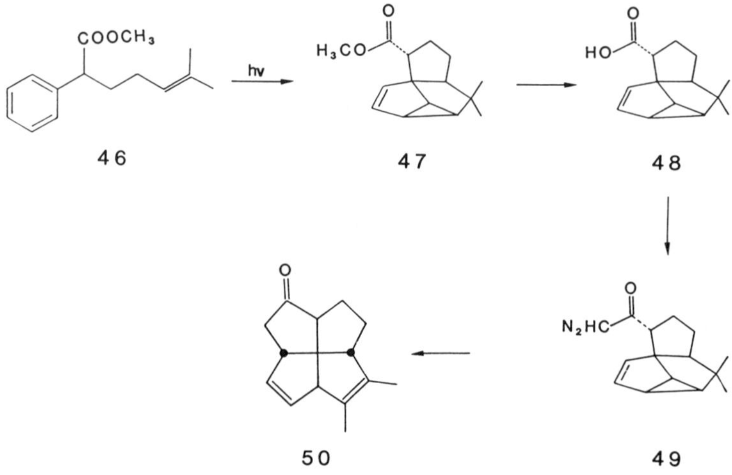

Scheme 4.

46

47

48

50

49

With the aim to enhance the planarizing distortions this fenestrane was used as starting point for further transformations (Scheme 5). The dienones **52** and **53** were obtained in a ratio of 6:1 from the ketone **51** by hydrogenation of **50** and formal dehydrogenation using the well established phenylselenium methodology [22]. Subsequent reduction of the major component **52** gave the allylic alcohols **54** and **55** in a ratio of 4:1. These carbinols could be used for preparation of a stereoisomeric [5.5.5.5]fenestrane or for formation of a fenestrane with a cyclopentadiene substructure.

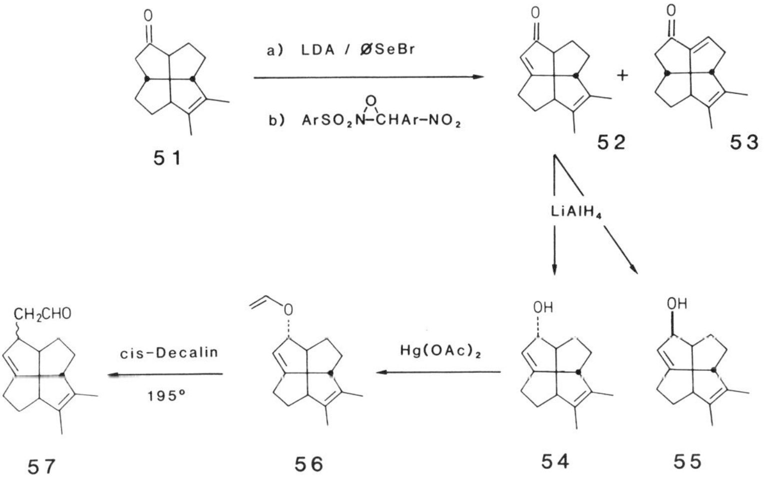

Scheme 5.

When the vinylallylether **56**, prepared by the method indicated (cf. Scheme 5), was heated in decalin to 195°C, only the aldehyde **57**, formed via a [1,3]sigmatropic rearrangement was obtained; the normal product of the Claisen rearrangement to be formed via [3,3]sigmatropic reaction was not observed. This may indicate, that the Claisen rearrangement is not sufficiently exothermic to allow for the formation of a highly strained product. When the mixture of the carbinols **54** and **55** was heated in HMPTA [23] to 245°C only the diene **58** was obtained. Reaction of the carbinol

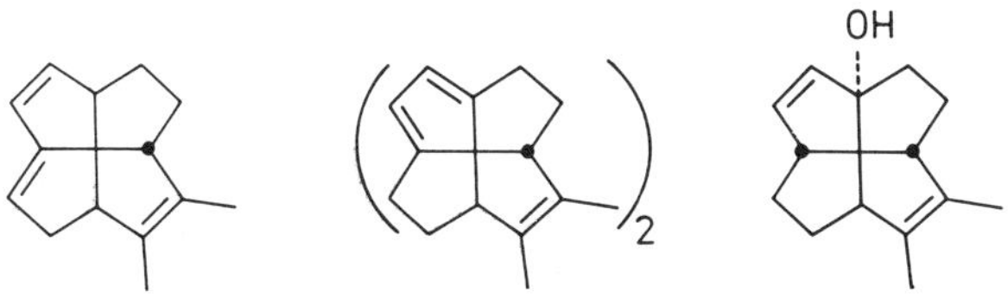

**58**   **59**   **60**

**54** in which the bridgehead hydrogen is in cis-position to the adjacent hydroxy group, with sodium hydride and the anhydride of trifluorosulfonic acid lead to a hydrocarbon **59**, which formally is a dimer of the desired triene. Whether the carbinol **60**, also isolated from this reaction, is formed via 1,2-elimination followed by stepwise 1,4-addition or by $S_N2'$ reaction with inversion remains to be seen. None of these products was obtained, when the stereoisomeric carbinol **55** was submitted to the same reaction conditions [24]. In a further attempt to prepare stereoisomeric [5.5.5.5.]fenestranes the stereochemistry of hydroboration of **54** and **55** was investigated. After oxidative work up the trans-diol **61**

61   62

63   64   65

was obtained from **54**, whereas the cis-diol **62** was formed from **55**. Since both diols are derived from all-cis-[5.5.5.5]fenestrane, attack of $BH_3$ is apparently controlled by the roof-like shape of the bicyclo[3.3.0]oct-1-ene substructure of **54** resp. **55** rather than the neighbouring hydroxy groups in these compounds. However, hydroboration of **64** prepared from **63** by the Mitsunobu procedure [25] yielded upon oxidation a mixture of alcohols, from which the compound **65** could be isolated [26]. Given appropriate substitution of the hydroxy group in **55**, it may be expected that steric hindrance will direct the hydroboration to the appropriate side and eventually give a derivative of cis,cis,cis,trans-[5.5.5.5]fenestrane.

CONCLUSION

Symmetry deformation coordinates can be used for a systematic search of planarizing distortions in $C(C)_4$ fragments. Based on X-ray structure analysis and MNDO calculations the unique structure of fenestranes is due to two opposite angles of the central $C(C)_4$ fragment being significantly larger than those of a regular tetrahedron. Functionalized all-cis-[5.5.5.5]fenestranes can be prepared from readily available starting material with the photoinduced arene-olefin cycloaddition as key step. Further reactions of these compounds lead to unexpected results. Further experimental studies of fenestranes will lead to the limits, beyond which further opening of opposite bond angles is "forbidden".

ACKNOWLEDGEMENT

The support of this work by the Swiss National Science Foundation (project 2.236-0.84 and 2.016-1.86), the Stipendienfonds der Basler Chemischen Industrie, the van't Hoff Fund of the Royal Netherlands Academy of Sciences and the Stiftung zur Förderung der wissenschaftlichen Forschung an der Universität Bern is gratefully acknowledged.

REFERENCES

[1]  a) A. v. Baeyer, *Ber.Dtsch.Chem.Ges.* (1885), **18**, 2277.
     b) K.B. Wiberg, *Angew.Chem.* (1986), **98**, 312; *Int.Ed.* **25**, 312.

[2]  L. Pauling, *The nature of the chemical bond*, Cornell University Press, 3rd. ed., 1960.

[3]  R. Hoffmann, R.W. Alder, C.F. Wilcox, Jr., *J.Am.Chem.Soc.* (1970), **92**, 4992.

[4]  a) J. Chandrasekhar, E.-U. Würthwein, P.v.R. Schleyer, *Tetrahedron* (1981), **37**, 921.
     b) P.v.R. Schleyer, A.E. Reed, *J.Am.Chem.Soc.* (1988), **110**, 4453.

[5]  a) R. Keese in O. Chizhov, Ed., *Organic Synthesis*: 'Modern Trends (IUPAC)', Blackwell Scientific Publications, 1987, pg. 43.
     b) B.R. Venepalli, W.C. Agosta, *Chem.Rev.* (1987), **87**, 399.
     c) K. Krohn in *Nachr. Chemie, Technik, Laboratorium* (1987), **35**, 264.

[6]  R. Keese in *Nachr. Chemie, Technik, Laboratorium* (1982), **30**, 844.

[7]  a) W. Luef, R. Keese, H.-B. Bürgi, *Helv.Chim.Acta* (1987), **70**, 534.
     b) W. Luef, R. Keese, *Helv.Chim.Acta* (1987), **70**, 543.

[8]  M. Luyten, R. Keese, *Angew.Chem.* (1984), **96**, 358; *Int.Ed.* **23**, 390.

[9]  R. Mitschka, J. Oehldrich, K. Takahashi, U. Weiss, J.V. Silverton, J.M. Cook, *Tetrahedron* (1981), **37**, 4521.

[10] R.E.R. Craig, A.C. Craig, R.D. Larsen, C.N. Caughlan, *J.Org.Chem.* (1977), **42**, 3188.

[11] a) G. Haas, V. Prelog, *Helv.Chim.Acta* (1969), **52**, 1202.
     b) O.S. Mills, unpublished data. We thank Prof. V. Prelog for the X-ray structural data of 5

[12] W. Luef, R. Keese, manuscript in preparation.

[13] K.B. Wiberg, L.K. Olli, N. Golembeski, R.D. Adams, *J.Am.Chem.Soc.* (1980), **102**, 7467.

[14] W.G. Dauben, D.M. Walker, *Tetrahedron Lett.* (1982), **23**, 711.

[15] K.B. Wiberg, J.V. McClusky, *Tetrahedron Lett.* (1987), **45**, 5411.

[16] J.D. Cox, G. Pilcher, *Thermochemistry of Organic and Organometallic Compounds*, Academic Press London (1970), Table 49.

[17] M. Luyten, R. Keese, *Tetrahedron* (1986), **42**, 1687.

[18] J. Mani, R. Keese, *Tetrahedron* (1985), **41**, 5697.

[19] J. Mani, S. Schüttel, C. Zhang, R. Keese, Manuscript in preparation.

[20] P. Bigler, M. Kamber, *Magn.Res. in Chem.* (1986), **24**, 972.

[21] G. Mehta, K.S. Rao, *J.Org.Chem.* (1988), **53**, 425.

[22] a) H.J. Reich, S. Wollowitz, J.E. Trend, F. Chow, D.F. Wendelborn, *J.Org.Chem.* (1978), **43**, 1697.
b) F.A. Davis, O.D. Stringer, J.M. Billmers, *Tetrahedron Lett.* (1983), **24**, 1213.

[23] R.S. Monson, D.N. Priest, *J.Org.Chem.* (1971), **36**, 3826.

[24] M. Schmid, *Dissertation*, University of Berne, 1987.

[25] O. Mitsunobu, *Synthesis* (1981), 1.

[26] P. Gerber, *Diploma Thesis*, University of Berne, 1987.

# STRAIN RELEASE IN AROMATIC MOLECULES: THE $[2_n]$ CYCLOPHANES

Henning Hopf and Claudia Marquard
Institute of Organic Chemistry
University of Braunschweig
Hagenring 30, D-3300 Braunschweig
Federal Republic of Germany

ABSTRACT: The chemical behavior of the $[2_n]$ cyclophanes is determined to a large extent by strain effects. This is shown by a systematic study of isomerisation and addition reactions, respectively. In particular, the paper describes thermal, photochemical and ionic isomerisations of various $[2_n]$ cyclophanes, as well as their behavior in Diels-Alder additions, epoxidations and carbene addition reactions.

Ever since Brown and Farthing prepared [2.2] paracyclophane more or less accidentally in 1949 (1) this compound and numerous of its derivatives and analogs have fascinated scores of chemists (2). This fascination remains unabated until the present day, and a sentence, written by Cram almost 20 years ago, still is true: " The chemistry of these substances will be completed only when chemists tire of tinkering with them" (3).
What has been accomplished during the last four decades is summarized in fig. 1.

1. Preparation of (almost) any type of phane molecule by (in many cases) novel synthetic methods.
2. Investigation of the spectroscopic properties of phanes by NMR, IR/Raman, UV/vis, photo electron spectroscopy.
3. Study of the electronic interactions in layered organic molecules
4. Determination of the molecular structure of the important phane types by x-ray structural analysis.

Fig.1: Forty years of cyclophane chemistry



A. de Meijere and S. Blechert (eds.), Strain and Its Implications in Organic Chemistry, 297–332.

What remains to be done ? What could be the future of phane chemistry ? Some answers are given in fig.2.

1. Investigation of the chemical properties of phanes on a broad scale; recognition of reactivity patterns.

2. Determination of the thermochemical data (heats of hydrogenation and combustion; experimental strain energies).

3. Exploitation of the planar chirality of many phanes for synthetic purposes ("planar chiral auxiliaries").

4. Incorporation of phane structures into more complex molecular frameworks - heteroorganics, pharmaceuticals, dyestuffs, polymers, inter alia.

Fig. 2: Cyclophane chemistry: Possible future developments.

Although general reactivity patterns are only emerging, it is already clear that the chemistry of the phanes, and in particular of the $[2_n]$-phanes, is strongly influenced and sometimes even controlled completely by strain effects. Many reactions of the phanes are driven by strain release.
Of course, the chemical properties of the $[2_n]$ phanes have been investigated earlier by the pioneers in the field - Cram, Boekelheide, Staab, Misumi, Vögtle and many others (2). However, most of their work was dedicated to the preparation of particular target molecules, highly symmetrical hydrocarbons in many cases. The investigation of reactivity patterns was of secondary importance. But if these molecules are chemically so "interesting" as is often maintained what is this chemistry ?
As already mentioned strain release provides the driving force for many cyclophane reactions.
Unfortunately, in most cases the strain energy of the substrate molecule is unknown. In the $[2_n]$ cyclophane series the only experimentally determined strain energies are those shown in fig.3 .

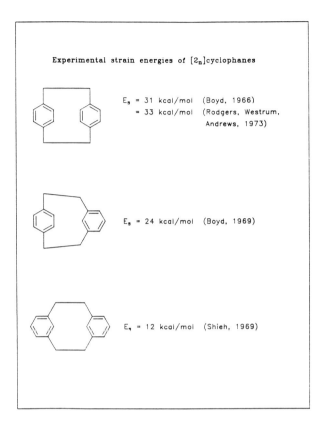

Fig.3: The experimentally determined strain energies of the [2.2] cyclophanes.

However, altogether there are not three but six [2.2] cyclophanes, as illustrated in fig. 4 which also compares their relative strain energies as calculated by MMP2 calculations (4).
As expected the point of attachment of the ethano bridge to the aromatic nucleus exerts a strong influence on the strain energy of a [2.2] cyclophane. The reduction of strain on going from para,para to para,meta to meta,meta will be of particular importance in various rearrangement reactions to be discussed below.

Are there other ways of increasing the strain ? The most obvious one

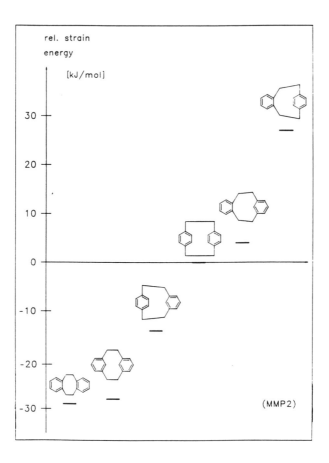

Fig. 4: Relative strain energies of the [2.2]cyclophanes as determined by MMP2 calculations.

involves the introduction of additional bridges. The corresponding force field calculations have been carried out (5) and are reproduced in fig. 5.
In all these cases the bridges are anchored in identical positions of the two aromatic halves; we have called this arrangement parallel in contrast to the skew pattern in which unlike carbon atoms of the two decks function as bridgeheads. The simplest skew-phanes are the meta,para, meta,ortho etc. which we have just seen. Obviously, there are more complex combinations, and some of them will be presented later.
The influence of simple substituents on the strain energy of [2.2]-

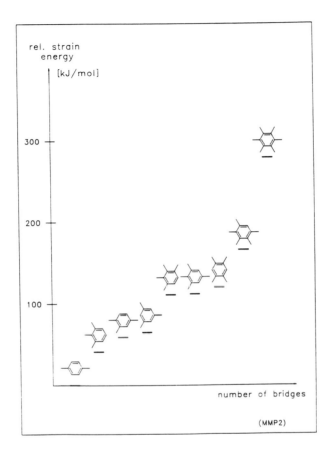

Fig.5: Relative strain energies of the $[2_n]$ cyclophanes (MMP2 calculation). The hydrocarbons are shown in a head-on projection.

paracyclophane was investigated next (4). When one or two methyl groups are incorporated into this hydrocarbon the pattern shown in fig. 6 results.

The least favorable dimethyl derivatives are those carrying the two substituents at one and the same aromatic nucleus.

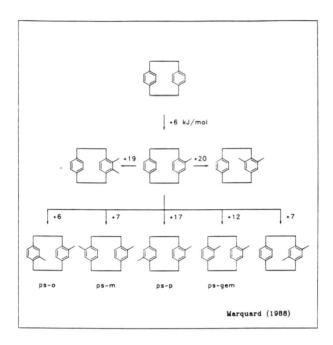

Fig.6: Relative strain energies of the dimethyl-[2.2]paracyclophanes.

A very similar trend is noted (4) for still higher methylated [2.2] paracyclophanes as shown in fig.7

Rearrangements between many of these isomers can be induced by treatment with catalytic amounts of hydrochloric acid in the presence of Lewis acid catalysts as will also be discussed below. In all these reactions the least-strained isomers are produced preferentially or exclusively.

To summarize this first section: The strain energy of the $[2_n]$ cyclophanes depends on a) the number of bridges, b) the position of these bridges, and c) the number and arrangements of the substituents. In the latter case the increase in strain is particularly pronounced when the substituents are in an ortho- or a meta-arrangement at one and the same nucleus (4).

Turning to the reactions of the $[2_n]$cyclophanes, a good point to start is their behavior towards heat, since the corresponding reactions belong to the most simple ones, and sometimes do not even require a further reagent.

Furthermore, it is known from many other classes of hydrocarbons – small-ring systems, highly crowded acyclic molecules (6) – that the

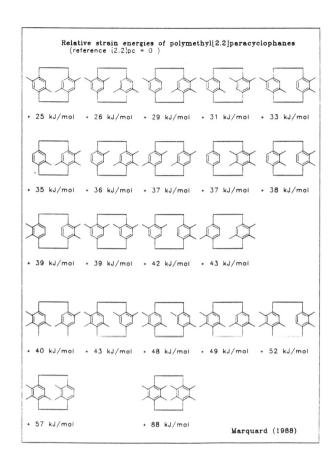

Fig.7: Relative strain energies of polymethylated [2.2]paracyclophanes.

higher the strain energy of the substrate the more readily C-C-bond homolysis will occur. And quite often it is the longest bond of the educt that is broken in the rate-determining step. Clearly, in [2.2]-paracyclophane and its simple derivatives it is the benzyl-benzyl bond, the central bridge bond, that is predicted to be split during thermolysis (fig. 8).
Pyrolysis of [2.2]paracyclophane at temperatures above 200 °C in the presence of hydrogen donors (thiophenol or p- di-isopropylbenzene ) provides 4,4'-dimethyl-bibenzyl in yields up to 74% (7). It is reasonable to assume that the reaction involves the diradical shown in fig.8, even more so since this species may be trapped by either dimethyl maleate or fumarate (200°C, 40h, 60%). That the stereochemi-

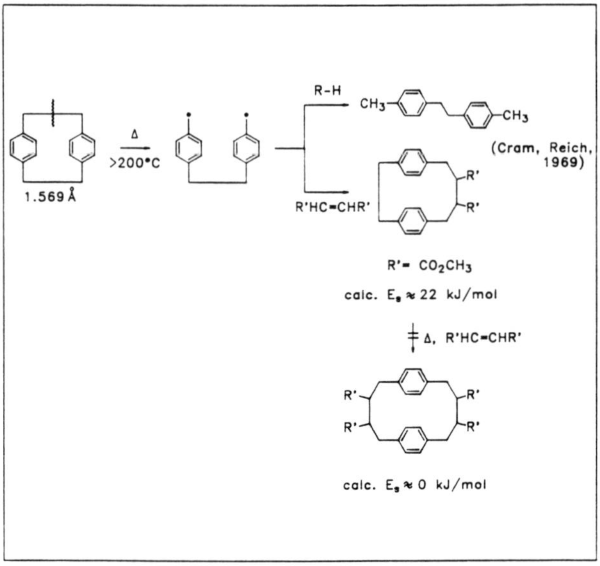

Fig.8: Thermolysis of [2.2] paracyclophane

cal information carried by the trapping reagent is lost during the ring enlargement process is a further support for the diradical pathway (7). To see whether the [2.4] paracyclophane still contains enough "steric driving force" for another ring enlargement step, we have recently repeated Cram and Reich's experiment. Although we looked for a 2:1-addition product by mass spectrometry not even traces of the [4.4] - cyclophane could be detected (8).
Further evidence for the diradical route is presented in fig. 9. Here thermal equlibrations between various cyclophane isomers were carried out. In all cases the material balance was quantitative, and no leakage took place between the different isomers (8,9).

When additional bridges are introduced into the phane system it is of interest to find out which bond is broken preferentially under what conditions.
The first multibridged phane studied was $[2_3]$ (1,2,4)cyclophane (fig. 10).

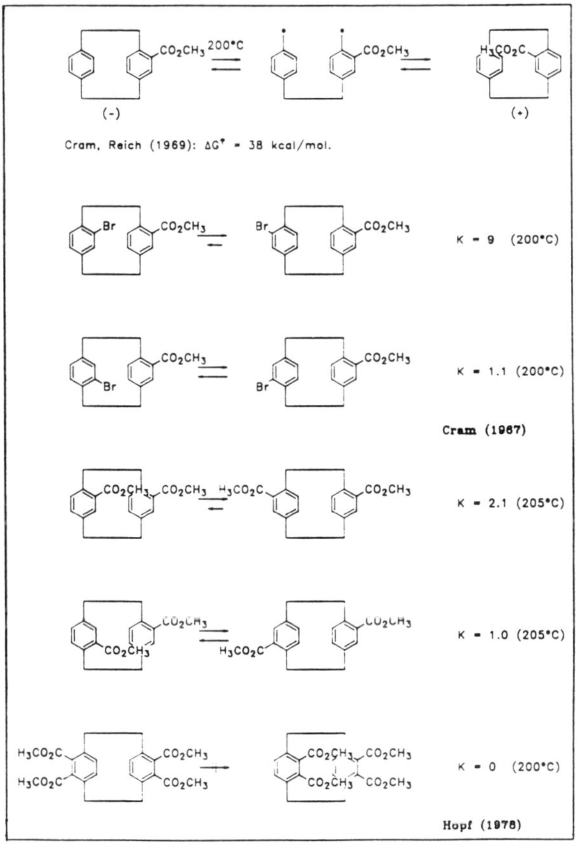

Fig.9: Thermal equilibration of substituted [2.2] - paracyclophanes.

As can be seen from the scheme (fig.10) it is the isolated bridge that is broken exclusively. Had the other bridges been broken, more strained diradical intermediates, derived from [2.2] para- and [2.2] meta-cyclophane, respectively, would have been formed. With an unsaturated trapping reagent a mixture of ring-enlarged products is formed again, indicative of the radical nature of the reaction (10).

$[2_4]$ (1,2,3,5)cyclophane shows an analogous behavior at 290°C (fig.11). However, if the temperature is raised for another 20° a novel strain-reduction process takes over. The second para-bridge is lost as ethylene, and the resulting dihydropyrene ultimately looses hydrogen to yield

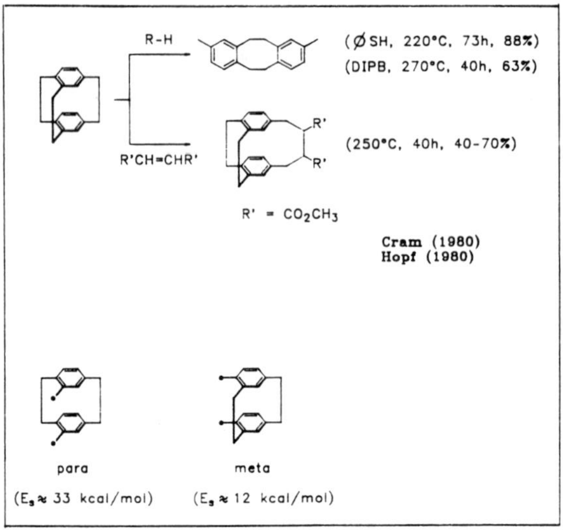

Fig.10: Thermolysis of $[2_3]$(1,2,4)cyclophane.

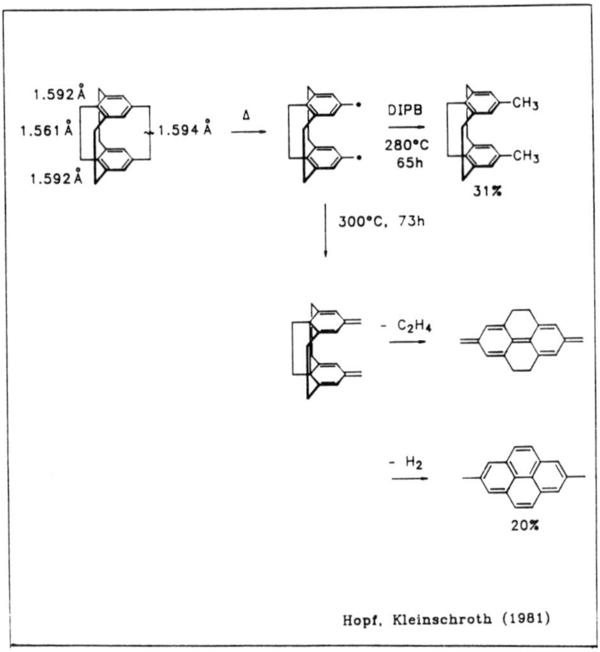

Fig.11: Thermolysis of $[2_4]$(1,2,3,5)cyclophane.

a non-bridged aromatic, 2,7-dimethyl-pyrene (11).

All other polybridged cyclophanes are extremely stable up to very high temperatures (fig.12) (11-13).

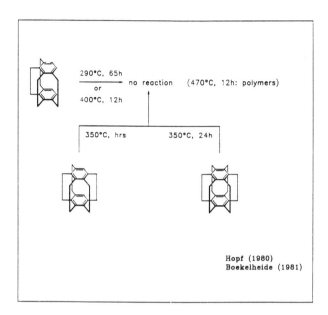

Fig.12: High-temperature behavior of multibridged cyclophanes.

Although the total strain energy of the [$2_n$] cyclophanes increases with the number of bridges (vide supra) the thermal stabilty of these hydrocarbons also increases with the number of bridges.
This may seem paradox on first sight but becomes plausible if one considers that it becomes increasingly difficult to separate the radical centers to the critical distance that allows either hydrogen abstraction or olefin addition when the number of bridges is increased. It is not one bond that eventually must be broken but the whole framework must be stretched almost to the breaking point of one, the reacting bond.
Some of the diradical species postulated above have been observed directly by UV/vis and fluorescence spectroscopy by Kaupp (14) a few years ago (fig.13).
Compared to the photochemistry of classical aromatic molecules, i.e. benzene derivatives, the photochemical behavior of the [$2_n$] cyclophanes remains a largely unexplored area. One of the few published experiments (15) is again due to Cram (fig. 14).

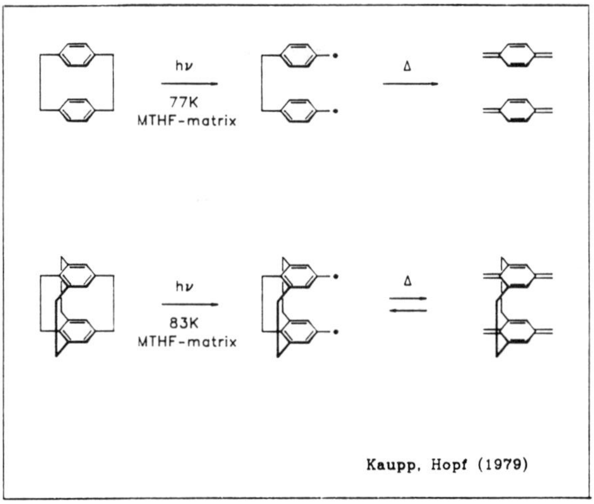

Fig.13: Direct observation of diradical intermediates

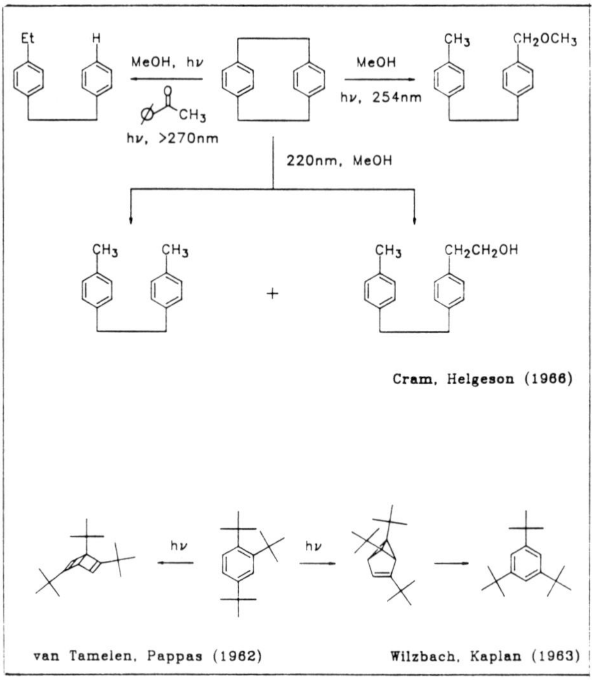

Fig.14: Photochemistry of [2.2]paracyclophane and of crowded benzene derivatives.

Under the conditions shown it is again the bridge which is destroyed.

There could have been other ways, though, of reducing the strain of the starting material since it is well known from the work of van Tamelen, Viehe, Wilzbach and others (16) that sterically crowded and strained benzene derivatives may rearrange to their prismane, benzvalene and Dewar benzene isomers when irradiated (fig.14).

Since the "natural" bending of the benzene rings in [2.2]paracyclophane evidently did not suffice to allow a photochemical synthesis of a prismaphane, benzvalenophane etc. we tried to increase the "warping" by the introduction of tert.-butyl groups into the parent hydrocarbon, hoping that these bulky substituents would push the molecule towards the prismane etc. direction.
Since tert.-butyl-cyclophanes were unknown prior to our studies we had to prepare them (fig.15).

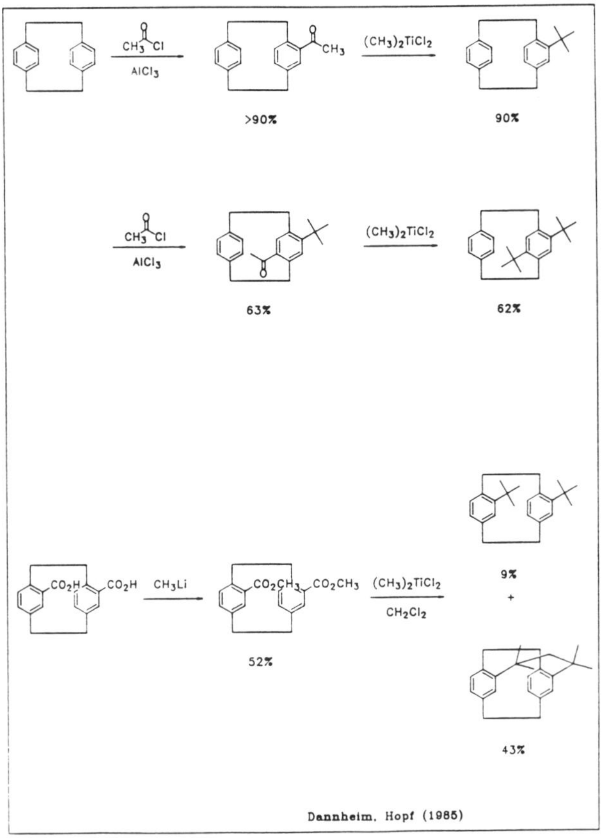

Fig.15: Preparation of tert.-butyl[2.2]cyclophanes.

After several unsuccessful attempts employing classical tert.-butylation reagents (isobutene/acid; tert.-butylchloride/aluminum trichloride etc.) the approach developed by Reetz (17) turned out to be the method of choice, and the three tert.-butyl derivatives shown in fig. 15 were prepared (18).

When these hydrocarbons were irradiated in n-hexane with 254nm light no bridge cleavage at all took place.

After 18h the 4-tert.-butylcyclophane had isomerized to two para, meta and to two meta,meta-cyclophanes (fig. 16), the latter evidently being secondary products. This second isomerisation step had already been observed in one example by Cram in 1972 (19).

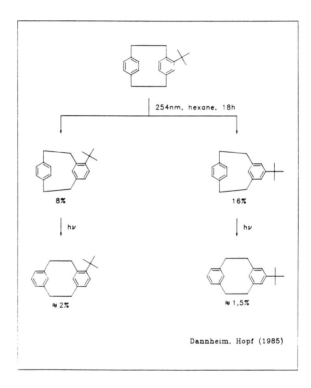

Fig.16: Photoisomerisation of 4-tert.-butyl [2.2] paracyclophane

The pseudo-geminally substituted cyclophane turned out to be much more reactive (fig. 17).

After one hour already the isomeric hydrocarbons shown in fig. 17 had been formed. Again, the mixture consists out of para,meta- and meta, meta-isomers only, and it is a reasonable assumption that it is the strain release which drives the process.

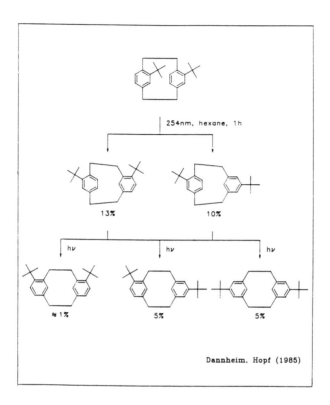

Fig.17: Photoisomerisation of 4,13-di-tert.-butyl[2.2] para-cyclophane

The 4,7-di-substituted isomer provided very similar results- but also a surprise (fig. 18). Besides the now expected para,meta-products and traces of a meta,meta-compound two intensely yellow hydrocarbons were isolated (18) and characterized by spectroscopic methods as the two fulvenophanes shown in the scheme, the first representatives of a hitherto unknown class of cyclophanes. All of these experiments are product studies, and the following mechanistic speculations (fig. 19) may not be more than an educated guess.

Still, there are literature analogies. It is known from benzene photochemistry that this hydrocarbon on irradiation with 254nm light is

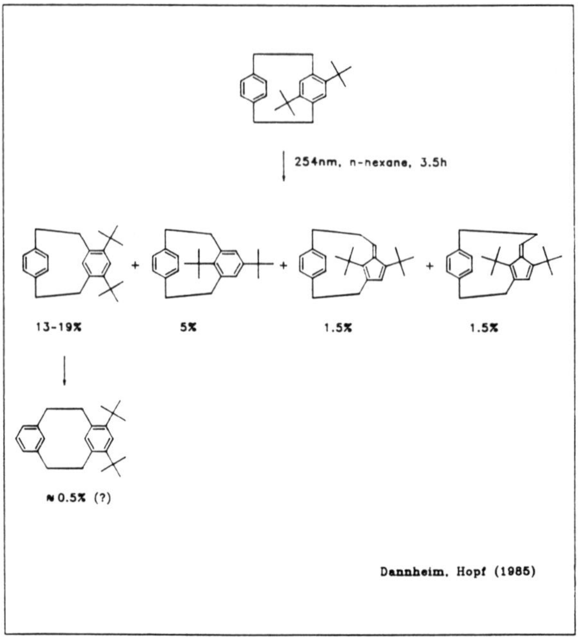

Fig. 18: Photoisomerisation of 4,7-di-tert.-butyl[2.2]para-cyclophane.

excited to a singlet state which subsequently is converted to pre-fulvene, a diradicaloid intermediate (20). This may undergo further isomerisation or be trapped by suitable reagents.

In the 4,7-disubstituted case two isomeric pre-fulvenes may be formed. Shown in fig.19 is the less strained one, which may be obtained from the substrate by a least-motion process. The three-membered ring of this intermediate may be opened in three ways: back to the educt, on to a carbene that cannot insert because of a blocking substituent, and finally to an isomeric carbene with an α-hydrogen atom and there-fore the possibility to react. The two diastereomeric fulvenophanes are the result.
The main route involves 1,2-carbon migration, and since there are two possibilities to rearrange two para,meta-phanes should be pro-duced. As fig.18 shows this is indeed the case.

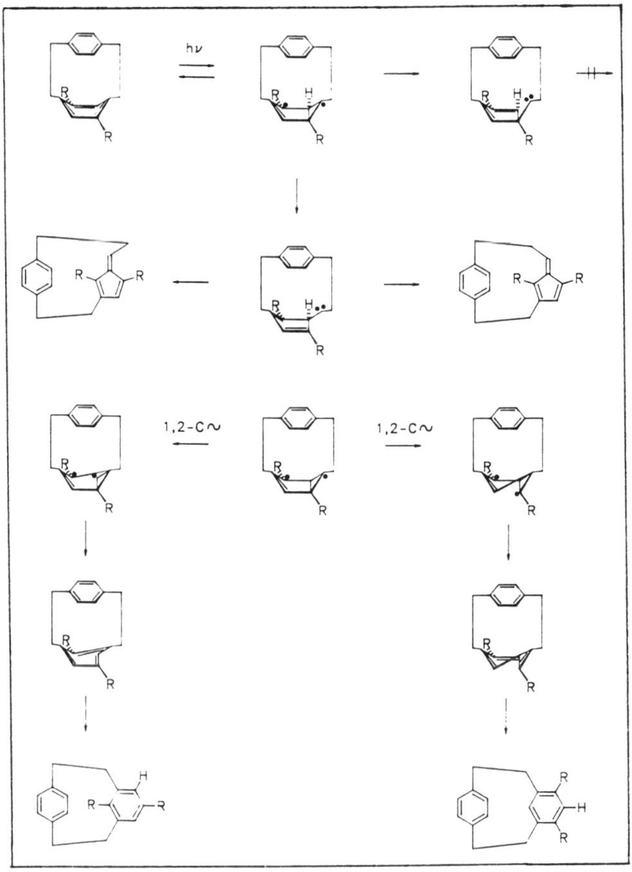

Fig. 19: Mechanism of the photochemical isomerisation of sterically hindered cyclophanes.

The same reaction cascade - para,para ⟶ para,meta ⟶ meta,meta - may also be initiated by treatment of [2.2]paracyclophane with strong acids; again, the basic observation is due to Cram (21). A NMR-spectrum of the protonated [2.2]paracyclophane in $SO_2ClF-HSbF_5$-solution could be obtained; the deep red solution of the conjugated acid is stable to about -10°C (fig.20).

We have repeated and extended this work recently, and have definitely shown that it is the bridgehead protonated carbocation that is formed. Furthermore, in $FSO_3H-SbF_5/SO_2ClF$ at -100°C a pale-yellow dication could be obtained. From the 250 MHz proton NMR spectrum

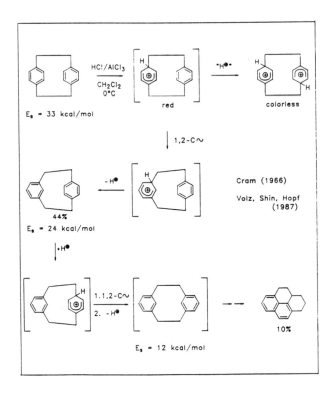

Fig.20: Acid-catalyzed isomerisation of [2.2] paracyclophane.

it follows that the two opposing bridgeheads have been protonated (22). Together with the ensuing 1,2-carbon shift the reaction may be regarded as an ipso-substitution. In classical aromatic chemistry ipso-substitution is often associated with strain release. As will be shown below this reaction type is common in cyclophane chemistry.

The generality of the above reaction is demonstrated by the results summarized in figs. 21 and 22. During all these isomerisations (23) a para,meta-rearrangement takes place, and again it may be attributed to a reduction in strain energy. Actually these acid/Lewis acid initiated processes are quite complex. Whenever the isolated yields are low, sizeable amounts of HCl-addition products are produced -again a sign of the increased reactivity of the cyclophanes. The mostly oily mixtures thus obtained have only rarely been identified or even been qualitatively investigated. A notable exception, showing that this work may turn

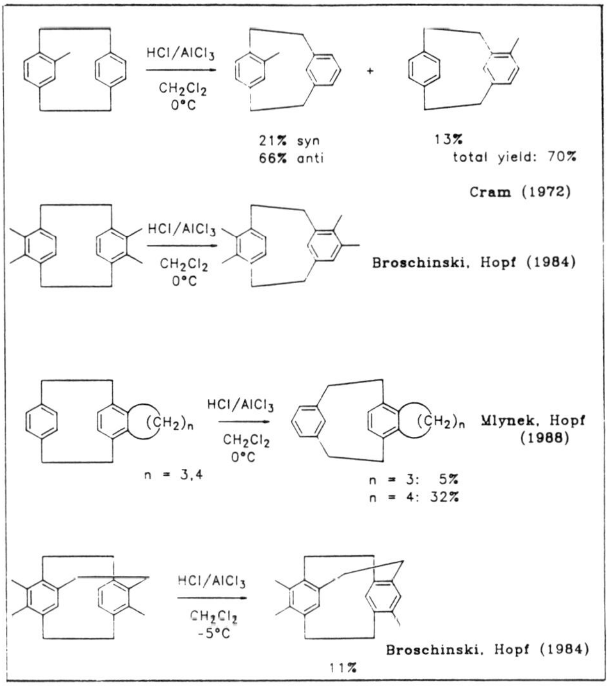

Fig.21: Acid-catalyzed rearrangement of paracyclophanes.

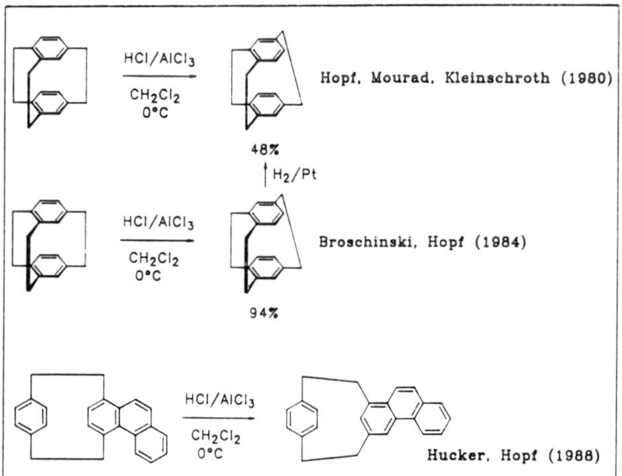

Fig.22: Further acid-catalyzed rearrangements.

out to be quite rewarding, is due to Boekelheide (24) who obtained an interesting tris-adduct from $[2_3]$(1,3,5)cyclophane (fig.23). Here the aromatic subsystems break down completely, and the result is a unique cage hydrocarbon related to hexaprismane.

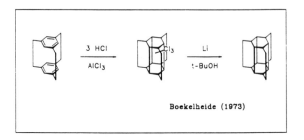

Fig.23: HCl-addition to $[2_3]$(1,3,5)cyclophane.

In another series of experiments we have shown that the nature of the

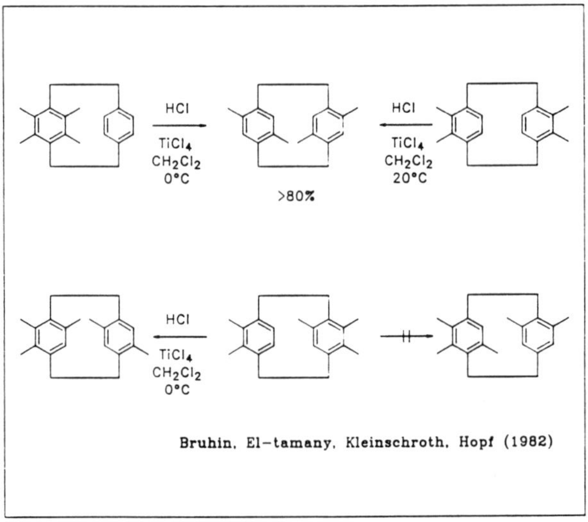

Fig.24: Methyl migrations in $[2.2]$ paracyclophanes.

Lewis acid employed in these isomerisations plays a fundamental role, too (fig. 24). Whereas none of the para,meta-processes described above takes place when the aluminum trichloride is replaced by titanium tetrachloride, certain polymethylated phanes show a very specific intramolecular methyl migration (25) in the presence of this normally less reactive catalyst. In all cases it is the least strained isomer (see MMP2 calculations described above) that is formed. The tendency is always towards a minimization of ortho-methyl group interactions. Some of these hydrocarbons are useful starting materials for the preparation of multilayered cyclophanes.

ipso-Substitutions are not restricted to the [2.2] paracyclophanes. An example illustrating the principle of strain reduction by this process particularly well (11) is shown in the last scheme of this series of experiments (fig. 25).

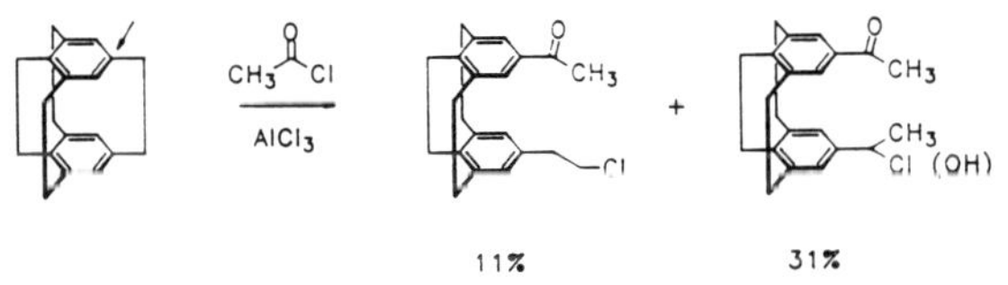

Fig. 25: Bridge cleavage by ipso-substitution of a multibridged cyclophane.

Excessive strain in organic molecules does not only reveal itself in an increased tendency to isomerize but also in an enhanced activity in addition reactions, as has been shown with many polycyclic hydrocarbons, small-ring olefins or alkynes (6).

How do the $[2_n]$cyclophanes behave in this respect?

Although the experimental material is growing, there are again very few general and systematic studies in this area, and some types of addition reactions - nitrene additions or 1,3-dipolar additions for example - have not been carried out at all with the $[2_n]$phanes.

Only three different types of addition reactions of the multibridged cyclophanes will be discussed here: [2+4] cycloadditions, epoxidations, and -briefly- carbene additions. Hydrogenations, whether of the catalytic or Birch type are known, but will not be considered here (11, 26).

The first Diels-Alder addition of a cyclophane was performed by Ciganek at Dupont (27) in 1967 (fig.26):

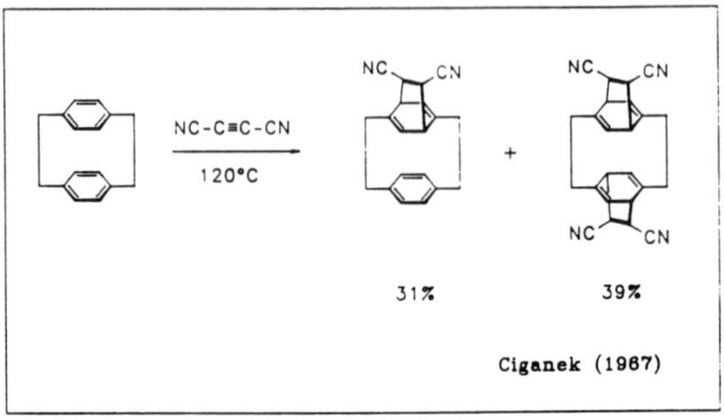

Fig.26: Addition of dicyanoacetylene to [2.2] paracyclophane.

In contrast to aromatics like benzene and the methylbenzenes which add dicyanoacetylene only under very harsh conditions and in the presence of aluminum trichloride, [2.2] paracyclophane reacts at 120°C already, providing a 1:1- and a 1:2-adduct, respectively. Whereas the structure assignment of the former is straightforward, Ciganek wrote the structure shown for the bis-adduct without explaining the reasons for his preference. Obviously, besides this parallel alignment of the two newly formed bridges there can exist a cross-wise arrangement also. A decision between the two alternatives on spectroscopic evidence alone is difficult, and an x-ray structural investigation was not undertaken. In the light of more recent work on another 1:2-adduct (vide infra) it seems more likely that Ciganek's bis-product had the crossed structure. Clearly, these addition products -the second one is a double-barrelene and its parent system a $C_{20}H_{20}$ isomer - are interesting polycyclic materials, and we are presently investigating their chemistry (28).

Other dienophiles were also reacted with [2.2] paracyclophane (fig.27). Again, no chemical studies have been performed with the adducts.

When additional bridges are incorporated, i.e. the strain of the substrate molecule is increased, a drastic increase in reactivity is observed (fig.28).

Interestingly, only those cyclophanes participate in Diels-Alder additions which possess a pair of para-hydrogen substituents (29). When-

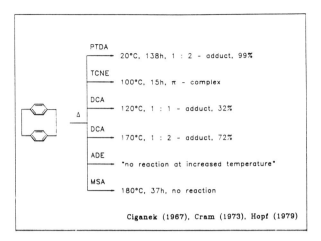

Fig.27: Diels-Alder additions with [2.2] paracyclophane.

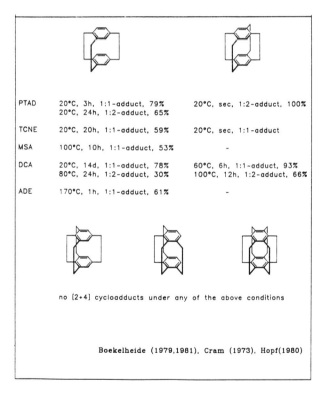

Fig.28: Diels-Alder additions of multibridged cyclophanes.

ever these positions are occupied by one or even two ethano bridges no [2+4] cycloaddition occurs.
Even for [2.2]paracyclophane itself two addition modes exist, and force field calculations (4) show that again the one taking place at the unsubstituted positions should be favored (fig.29).

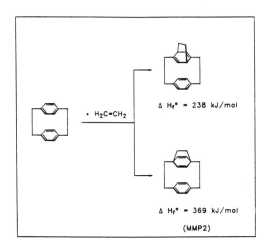

Fig.29: Diels-Alder addition modes to [2.2]paracyclophane.

Returning to the stereochemical problem of the 2:1-adducts a recent report by Matsumoto and co-workers (30) is of importance (fig.30). When N-phenylmaleimide is added to [2.2] paracyclophane at high pressure a bis-adduct is obtained, whose structure was determined by x-ray analysis. It is "crossed", and the two heterocyclic rings point into the same direction.

When we began our epoxidation studies in 1980 we were mostly interested in finding a way to novel syn-benzene-tris-epoxides, which -we hoped - would ring open to new macrocyclic polyethers (fig.31). As it turned out, these plans could not be realized by us (see below). However, in the course of these epoxidation studies some interesting and novel observations about the chemical reactivity of phane systems were made.

With p-nitro-perbenzoic acid 4-hydroxy-[2.2] paracyclophane as well as its p-nitro-benzoate were obtained from [2.2] paracyclophane. With the p-trifluoromethyl derivative again a small amount of the phenol was formed and a linearly conjugated cyclohexadienone. m-Chloro-perbenzoic acid brought us closest to our target molecule (fig.32). One of the oxidized products clearly is an addition product of p-chloro-benzoic

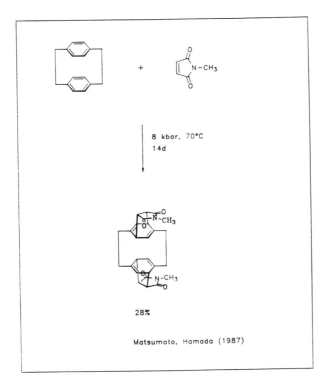

Fig.30: Addition of N-phenylmaleimide to [2.2] paracyclophane under high pressure conditions.

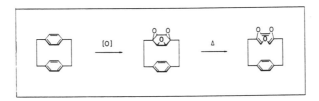

Fig.31: From cyclophanes to crown ethers ?

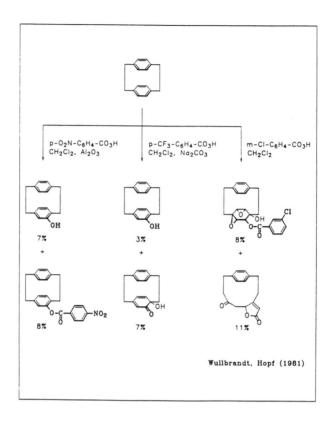

Fig.32: Epoxidation of [2.2] paracyclophane with various peracids.

acid to the anticipated tris-epoxide (fig.32). The other oxidation product has a very unusual structure which could only be established by an x-ray analysis. In this product one of the benzene rings has been destroyed completely. In all cases reaction times are long (sometimes days), und under the same conditions model compounds like p-xylene are not attacked (31).
All products may be rationalized by the mechanism shown in fig.33. The oxidation begins with the formation of an oxepin. This can either add an acid or a peracid molecule, and in both cases two regioisomers may be formed. From the former addition products elimination by either a water or an acid molecule would lead to the ester or the phenol. The peracid addition products could suffer either elimination or ring cleavage, opening-up the routes to the other products isolated. Since the ring-opened cyclophane is an aldehyde it could be oxidized again, and

since the resulting acid is unsaturated it could undergo intramolecular addition providing the unsaturated lactone.

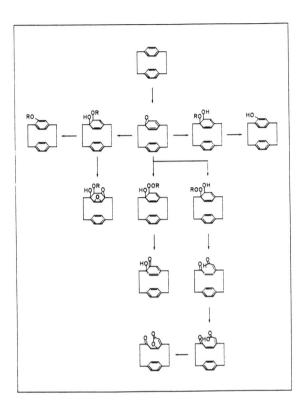

Fig.33: Epoxidation of cyclophanes: reaction mechanism.

Since substituents are expected to have an influence on the course of these epoxidation reactions we next turned to various cyclophanes carrying either electron withdrawing or donating groups.
The tetrakis-methoxycarbonyl paracyclophane (fig.34) could be oxidized to a bis-epoxide (room temperature, 9 days), which was again characterized by x-ray structural analysis (31). The pseudo-para diester behaved similarly, and again yielded a curious oxidation/addition product whose formation may be rationalized along similar lines as discussed above. In both cases no tris- or tetra-epoxides could be obtained under any conditions.

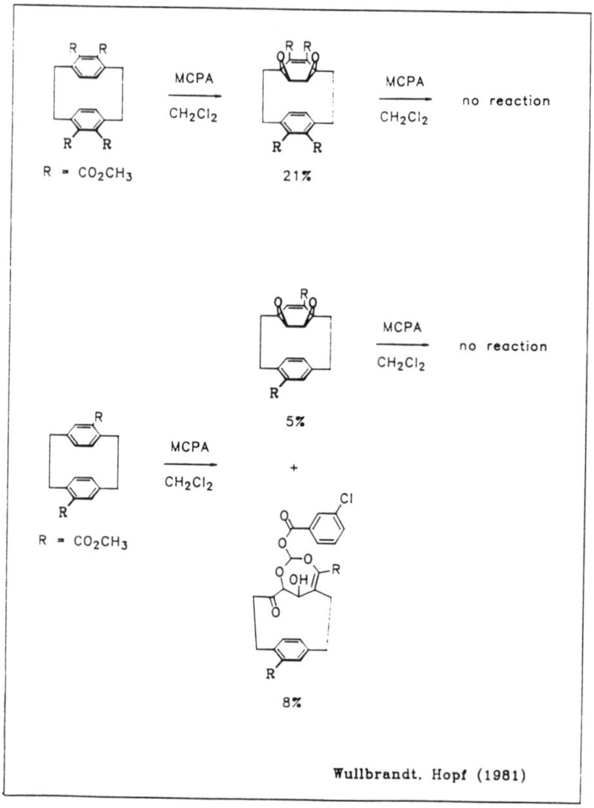

Fig.34: Epoxidation of [2.2] paracyclophane esters.

When the four ester groups are replaced by methyl substituents a drastic increase in reactivity is oberved (fig. 35). In fact, the primary bis-epoxide could not be isolated (31). Even under the extremely mild reaction conditions shown in the scheme this primary adduct undergoes an intra-annular Diels-Alder addition, with the "bottom" benzene ring functioning as the diene component. Note also that for this derivative the epoxidation has taken place right next to the substituents whereas in the di- and tetraester case (fig.34) the unsubstituted double bonds were attacked.
When the polycyclic bis-epoxide was subjected to a peracid again the tris-oxirane shown was formed (31).

While carrying out these oxidations we heard of a related effort in

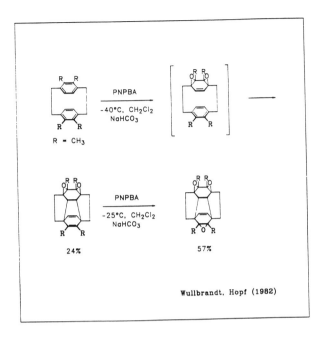

Fig.35: Epoxidation of 4,5,12,13-tetramethyl-[2.2]paracyclophane.

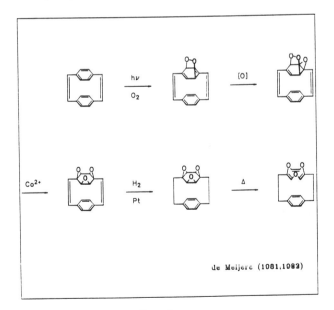

Fig.36: Oxidation of [2.2]paracyclophandiene.

Prof. de Meijere's group (32). Contrary to us, the Hamburg chemists were successful, though, in preparing a syn-benzene-tris-epoxide derivative (fig.36).
They used not only a more reactive phane but also incorporated the oxygen in a different manner. The initially prepared endo-peroxide was epoxidized to a tris-oxygen compound which on isomerisation provided the desired benzene-tris-oxide. Now the activating double bonds were removed by hydrogenation and the unsaturated tris-ether was obtained by thermal cycloreversion (33).

The enhanced propensity of the cyclophanes to participate in addition reaction is also shown by their behavior towards carbenes as will be discussed in the last part of the article.

The motivation to study these reactions derived from classical aromatic chemistry. It is textbook knowledge that homobenzenes cannot be prepared from benzene by methylenation. Furthermore, at the time we started our investigations, no cis-bis-homobenzene derivative was known, and cis-tris-homobenzene, i.e. the parent molecule, has remained elusive until the present day.
Nearly simultaneously three studies concerning the cyclopropanation of [2.2] paracyclophanes were published in 1976. Whereas Misumi (34) only treated the parent hydrocarbon with an excess of diazomethane in the presence of cuprous chloride, de Meijere (35) studied the bis-olefin as well (fig.37).

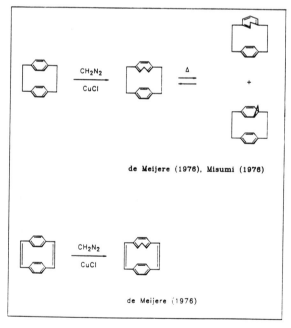

Fig. 37: Cyclopropanation of [2.2] paracyclophane and its diene.

The parent molecule yields two isomeric cycloheptatrienophanes which may be interconverted thermally; the unsymmetrical isomer also closes to a stable norcaradiene.
The diene reacts faster than its saturated analog, and this has been attributed to the higher strain energy of the former (35). Even more interestingly, the benzene rings are attacked more readily than the double bonds. All of the products are oily, and were hence not amenable to an x-ray analysis.

This is not the case for the methylenation products of various [2.2] - paracyclophane derivatives, one early example being the tetrakis-methoxycarbonyl compound (fig. 38).

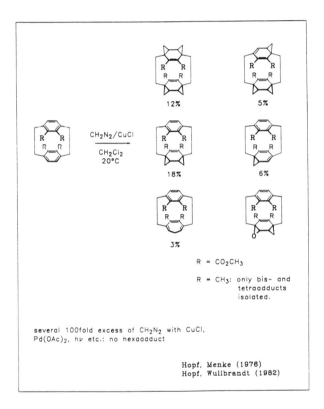

Fig.38: Cyclopropanation of [2.2] paracyclophane derivatives.

With diazomethane/cuprous chloride a complex product mixture was obtained, but all adducts could be separated and characterized (36). The epoxide shown is an artefact produced by air oxidation during work-up as shown by control experiments.

When the corresponding tetramethyl cyclophane is cyclopropanated (31) only bis- and tetra-adducts could be isolated. Under no conditions whatsoever a penta- or a hexa-adduct could be obtained. The particular geometry of the phane molecule prevents <u>anti</u>-attack , but as soon as two three-membered rings have been formed their <u>endo</u>-hydrogen atoms prevent the approach of the third carbene species.

In earlier studies aimed at synthesizing <u>cis</u>-bis-homobenzenes it was noted that these compounds must be very unstable and rearrange to their eight-membered ring isomers. A case in point is provided by an experiment of <u>Whitlock</u> and <u>Schatz</u> (37) who tried to prepare a <u>cis</u>-bis-homobenzene by an intramolecular reductive bisalkylation (fig. 39).

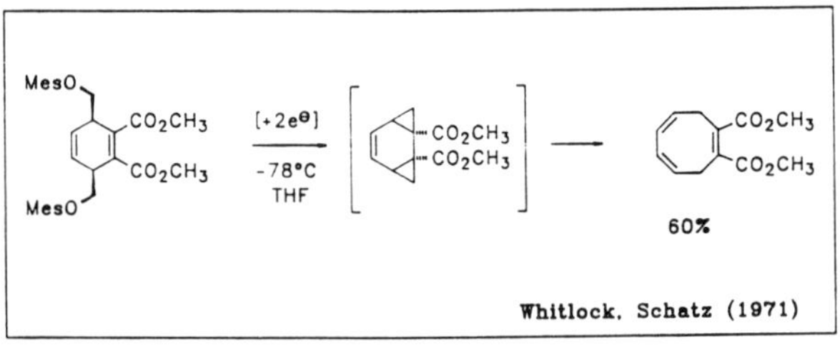

Fig.39: Attempted preparation of a <u>cis</u>-bis-homobenzene.

Although the reaction conditions are extremely mild the expected ring-closed product could not be isolated. Rather, the cycloreversion product was obtained in 60% yield (37). When the less substituted double bond is missing the substrate the corresponding tricyclic intermediate is stable.

The incorporation of the <u>cis</u>-bis-homobenzene moiety into the phane framework leads to a drastic increase in its stability on the other hand (38). In fact, the tetra-adduct of fig. 38 (R=ester groups) only starts to rearrange at 250°C (fig.40).
Unfortunately, the energy available under these conditions is so large that the expected target molecules - possible precursors for [2.2] - cyclooctatetraeneophanes - have only a fleeting existence and are destroyed by a complex process involving hydrogen shifts, electrocyclization and finally fragmentation (fig.40).

If this mechanism is plausible at all, then the bis-homobenzene adduct should lead to one fragmentation product only. This is indeed the case (fig.41): the now expected pyrolysis product being formed in 40% yield (38).

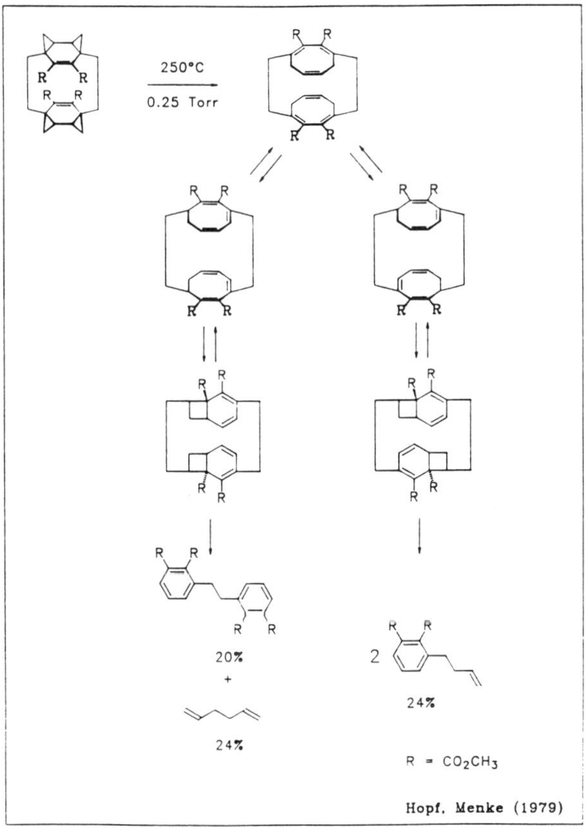

Fig.40: Pyrolysis of a tetra-homo- [2.2] paracyclophane.

Although this last example shows that bridged aromatic molecules may be extremely stable, the overwhelming evidence - presented above - demonstrates that the incorporation of molecular bridges increases the reactivity of a benzene ring. And sometimes this rate enhancement becomes so large that the typical difference between olefinic and aromatic behavior begins to vanish.

On the other hand it would be naive to assume that all reactions of the cyclophanes are dominated by strain effects. There is quite a number of phane reactions in which electronic control dominates. A well studied and understood example being the directional effects of electron-withdrawing groups which are often responsible for the introduction of electrophiles into the pseudo-geminal position of cyclophanes (3).

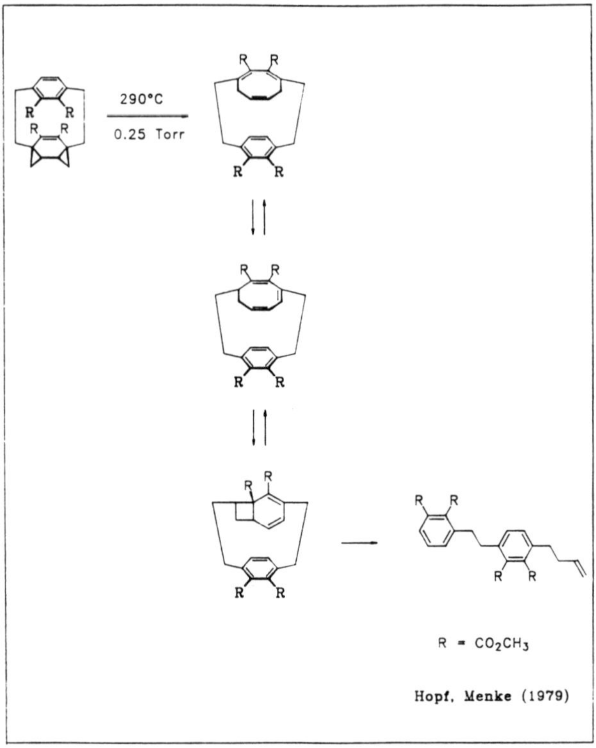

Fig.41: Pyrolysis of a bis-homo- [2.2] paracyclophane.


Acknowledgements: For a continuous support of our work on bridged aromatic molecules we thank the Deutsche Forschungsgemeinschaft, the Fonds der Chemischen Industrie and the Stiftung Volkswagenwerk. The BASF, Bayer, Degussa and Hoechst companies helped with generous supplies of numerous chemicals.

References:

1) C.J.Brown, A.C.Farthing, Nature (London),164 (1949) 915.
2) For recent reviews of cyclophane chemistry see:
   a) V.Boekelheide, Acc.Chem.Res., 13 (1980) 65.
   b) H. Hopf, Naturwiss., 70 (1983) 349.
   c) P.M.Keehn, S.M.Rosenfeld (eds.), The Cyclophanes, vol.
      I and II, Academic Press, New York, 1983.
3) D.J.Cram, J.M.Cram, Acc.Chem.Res., 4 (1971) 204.
4) C.Marquard, H.Hopf, unpublished results.
5) H.J.Lindner, Tetrahedron, 32 (1976) 753.
6) A.Greenberg, J.F.Liebman, Strained Organic Molecules, Academic Press, New York, 1978.
7) D.J.Cram, H.J.Reich, J.Am.Chem.Soc., 91 (1969) 3517.
8) C.Mlynek, H.Hopf, unpublished work.
9) H.Hopf, F.T.Lenich, Chem.Ber., 107 (1974) 1891.
10) H.Hopf, A.E.Mourad, Chem.Ber.,113 (1980) 2358,cf. D.J.Cram, E.A.Truesdale, J.Org.Chem.,45 (1980) 3974.
11) H.Hopf,J.Kleinschroth, Angew.Chem.,94 (1982) 485; Angew.Chem. Int.Ed.Engl.,21 (1982) 469.
12) V.Boekelheide, P.F.T.Schirch, J.Am.Chem.Soc.,103 (1981) 6873.
13) V.Boekelheide, Y.Sekine, J.Am.Chem.Soc., 103 (1981) 1777.
14) G.Kaupp, E. Teufel, H.Hopf, Angew.Chem.,91 (1979) 232; Angew. Chem.Int.Ed.Engl., 18 (1979) 215.
15) R.C.Helgeson, D.J.Cram, J.Am.Chem.Soc., 88 (1966) 509.
16) a) H.G.Viehe, Angew.Chem., 77 (1965) 768; Angew.Chem.Int.Ed. Engl., 4 (1965) 746.
    b) E.E.van Tamelen, Angew.Chem., 77 (1965) 759; Angew.Chem. Int.Ed.Engl., 4 (1965) 738.
    c) Y.Kobayashi, T.Kumadaki, Topics Curr. Chem., 123 (1984) 103.
17) M.T.Reetz, S.-H. Kyung, J.Westermann, Chem.Ber., 118 (1985) 1050.
18) J.Dannheim, H.Hopf, unpublished results, cf. J.Dannheim, Diploma thesis, Braunschweig 1985.
19) R.E.Gilman, M.H.Delton, D.J.Cram, J.Am.Chem.Soc., 94 (1972) 2478.
20) D.Bryce-Smith, A.Gilbert, Tetrahedron, 32 (1976) 1309.
21) D.J.Cram, R.C.Helgeson,D.Loch,L.A.Singer, J.Am.Chem.Soc., 88 (1966) 1324.
22) H.Hopf, J.-H. Shin, H. Volz, Angew.Chem., 99 (1987) 594; Angew.Chem.Int.Ed.Engl.,26 (1987) 564.
23) H.Hopf, C.Mlynek, K.Broschinski, unpublished work, cf.K. Broschinski, Ph.d. dissertation, Braunschweig 1984.
24) V.Boekelheide, R.A.Hollins, J.Am.Chem.Soc., 95 (1973) 3201.
25) J.Kleinschroth, S.El-tamany,H.Hopf, J.Bruhin, Tetrahedron Lett., (1982) 3345.
26) V.Boekelheide, Topics Curr. Chem., 113 (1983) 87.
27) E.Ciganek, Tetrahedron Lett., (1967) 3321.
28) H.Hopf, S.Ehrhardt, unpublished work.
29) H.Hopf, J.Kleinschroth, A.E.Mourad, Angew.Chem., 92 (1980) 388; Angew.Chem.Int.Ed.Engl., 19 (1980) 389.
30) K.Matsumota, T.Okamoto, A.Sera, K.Itoh, K.Hamada, Chem. Lett., (1987) 895.
31) D.Wullbrandt, Ph.d. dissertation, Braunschweig 1982. We thank

Prof.Dr. R. Allmann, University of Marburg, for the x-ray analyses of the epoxidation and carbene addition products.

32) I.Erden,P.Gölitz,R.Näder,A.deMeijere, Angew.Chem.,**93** (1981) 605; Angew.Chem.Int.Ed.Engl., **20** (1981) 583.
33) M.Stobbe, U.Behrens, G.Adiwidjaja, P.Gölitz, A. deMeijere, Angew. Chem., **95** (1980) 904; Angew. Chem.Int.Ed.Engl., **22** (1980) 867.
34) H.Horita, T.Otsubo, Y.Sakata, S.Misumi, Tetrahedron Lett., (1976) 3899.
35) R.Näder, A.deMeijere, Angew.Chem., **88** (1976) 153; Angew.Chem. Int.Ed.Engl., **15** (1976) 166.
36) H.Hopf, K.Menke, Angew.Chem., **88** (1976) 152; Angew.Chem.Int. Ed.Engl., **15** (1976) 165.
37) H.W.Whitlock, P.F.Schatz, J.Am.Chem.Soc., **93** (1971) 3831.
38) K.Menke, Ph.d. dissertation, Karlsruhe 1979.
39) D.J.Cram, R.B. Hornby, E.A. Truesdale, H.J.Reich, M.H.Delton, J.M.Cram, Tetrahedron, **30** (1974) 1757.

# TRIP AROUND THE THREE-MEMBERED CYCLES SYNTHESES

A. KRIEF, D. SURLERAUX, W. DUMONT, P. PASAU, PH. LECOMTE, PH. BARBEAUX
Laboratoire de Chimie Organique
Facultés Universitaires Notre-Dame de la Paix
61, rue de Bruxelles
B-5000 NAMUR (Belgium)

ABSTRACT. We report novel stereo- and enantioselective routes to vinyl cyclopropane carboxylic esters and an original synthesis of aryl cyclopropanes.

S-bioallethrin[1a,b] 1b, deltamethrin[1a-d] 1c and tefluthrin[1c,f] 1d are valuable insecticides, commercially available for domestic and agricultural uses against flying and soil insects respectively. These are unnatural esters of 2,2-dimethyl-3-vinyl cyclopropane carboxylic acids 2 which all belong to the (1R) series. But whereas the former derivative possesses the trans stereochemistry found in the natural pyrethrin I 1a, the others possess a cis relationship. The structure of 2d is clearly the most complex since it requires in addition to the synthesis of a highly substituted cyclopropane ring with the cis stereochemistry, the stereoselective construction of a highly functionalized (E)-carbon-carbon double bond and in fact there does not exist an enantio- or a stereoselective synthesis of such a compound[1e].

Scheme 1

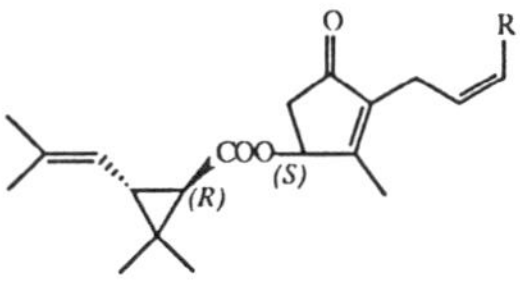


1a R = vinyl
Pyrethrin 1
1b R = H
S-Bioallethrin (RU)

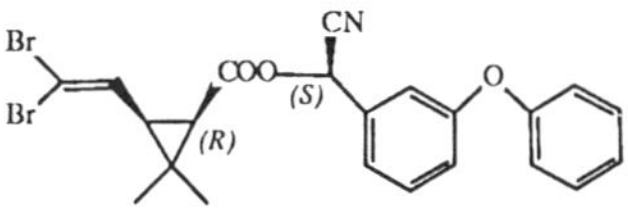


1c
Deltamethrin (RU)

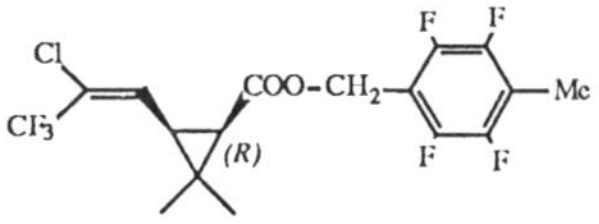


1d
Tefluthrin (ICI)

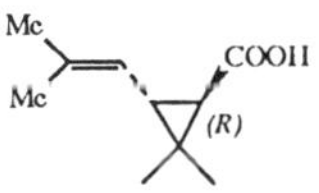


2a'
Chrysanthemic acid

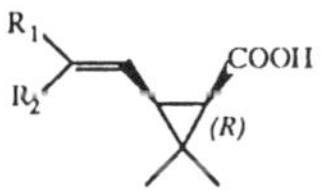


2a $R_1, R_2$ = Me
2c $R_1, R_2$ = Br
2d $R_1$ = Cl, $R_2$ = $CF_3$

We have designed as a model a novel stereoselective synthesis of (d,l) cis-chrysanthemic acid 2a which uses cheap reagents compatible with industrial requirements



A. de Meijere and S. Blechert (eds.), Strain and Its Implications in Organic Chemistry, 333–348.

and which we hope can be further applied to the enantio- and stereoselective synthesis of 1(R) 2d.

2,2,5,5-tetramethyl cyclohexa-1,3-dione [2] 4, readily prepared by di-methylation of dimedone 3 (4 equiv. $K_2CO_3$, EtOH-$H_2O$, 2.5 equiv. MeI, 70°C, 4h, 63 % yield), was chosen as the starting material since it possesses not only the same formula as chrysanthemic acid 2a ($C_{10}H_{16}O_2$) but also the carbon framework and the functionalities placed in suitable positions to allow in a minimum of steps the functional group modifications required for the desired transformation. The key steps of this process are without doubt (i) the cyclopropanation reaction which produces the bicyclo [3.1.0] hexa-1,3-dione [3] 7 and (ii) the Grob fragmentation [4] which was achieved on the sulfonates 9 resulting from its reduction-sulfonation. The stereochemical outcome of each individual step proved as expected particularly important for the success of the whole process shown in the Scheme 2.

Scheme 2

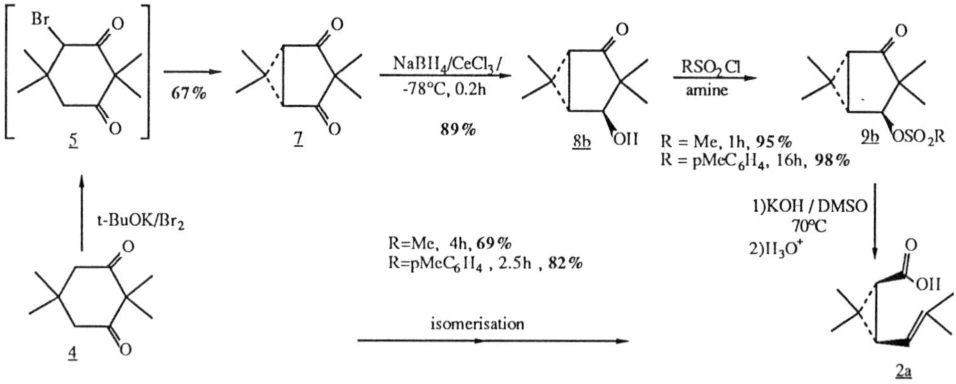

The synthesis of the bicyclo [3.1.0] hexa-1,3-dione 7 was efficiently achieved (67 % yield) from 2,2,5,5-tetramethyl cyclohexa-1,3-dione on sequential reaction with potassium tert-butoxide (2.2 equiv. in THF from -78°C to + 40°C) and bromine (1.6 equiv. in pentane, +40°C, 1h). This one pot transformation offers an original solution to our problem since the 6-bromo-4-potassio-2,2,5,5-tetramethylcyclohexa-1,3-dione 6 intermediary formed is immediately cyclized under these conditions rather than to further react with bromine. This avoids the difficulties we encountered when the transformation was performed stepwise [3].

Although chemoselective mono reduction of the bicyclic dione 7 was achieved by a large array of reducing agents [5] even when used in a fourth fold excess, the stereoselective synthesis of the exo alcohol 8b was tremendously more complex. This stereochemical outcome is very important since the exo alcohol is the only stereoisomer which lead to a sulfonate 9b but also since the later should possess the antiperiplanar arrangement required for the Grob fragmentation [4]. Lithium aluminum hydride in THF, sodium borohydride in methanol and lithium borohydride in DMF lead predominantly to the endo alcohol 8a resulting from the attack of the diketone 7 by the less hindered convex face (Scheme 3). This one can even be exclusively obtained if lithium triethylborohydride is instead used. The approach of the reducing agent by the concave face seems to be particularly difficult due to the concomitant presence of two methyl groups adjacent to the carbonyl group and of the endo methyl group attached to the cyclopropane ring. Lithium borohydride in THF or in ether lead to increased amounts of the desired alcohol 8b, however, all the trials involving a Lewis acid which would activate the carbonyl group allowing thus the production of the more stable exo alcohol 8b by favouring a late transition state were unsuccessful [6]. Performing the reduction with sodium borohydride in ethanol at +20°C as before but in the presence of cerium trichloride [7,8] dramatically increases the amount of the desired stereoisomers 8b. The later reaction proved particularly sensitive to the temperature and 8b was exclusively produced by performing the reaction at -78°C instead at 20°C (Scheme 3).

Scheme 3

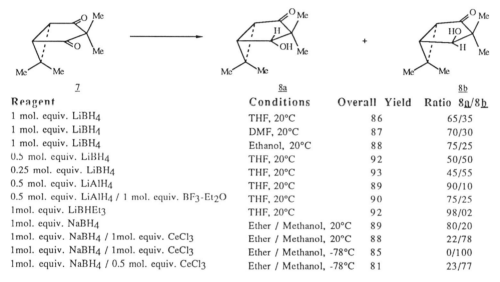

| Reagent | Conditions | Overall Yield | Ratio 8a/8b |
|---|---|---|---|
| 1 mol. equiv. LiBH$_4$ | THF, 20°C | 86 | 65/35 |
| 1 mol. equiv. LiBH$_4$ | DMF, 20°C | 87 | 70/30 |
| 1 mol. equiv. LiBH$_4$ | Ethanol, 20°C | 88 | 75/25 |
| 0.5 mol. equiv. LiBH$_4$ | THF, 20°C | 92 | 50/50 |
| 0.25 mol. equiv. LiBH$_4$ | THF, 20°C | 93 | 45/55 |
| 0.5 mol. equiv. LiAlH$_4$ | THF, 20°C | 89 | 90/10 |
| 0.5 mol. equiv. LiAlH$_4$ / 1 mol. equiv. BF$_3$-Et$_2$O | THF, 20°C | 90 | 75/25 |
| 1mol. equiv. LiBHEt$_3$ | THF, 20°C | 92 | 98/02 |
| 1mol. equiv. NaBH$_4$ | Ether / Methanol, 20°C | 89 | 80/20 |
| 1mol. equiv. NaBH$_4$ / 1mol. equiv. CeCl$_3$ | Ether / Methanol, 20°C | 88 | 22/78 |
| 1mol. equiv. NaBH$_4$ / 1mol. equiv. CeCl$_3$ | Ether / Methanol, -78°C | 85 | 0/100 |
| 1mol. equiv. NaBH$_4$ / 0.5 mol. equiv. CeCl$_3$ | Ether / Methanol, -78°C | 81 | 23/77 |

Further reaction of 8b with mesylchloride-triethylamine leads to the formation of the exo mesylate 9b1 (1.1 equiv. MsCl, 1.5 equiv. NEt$_3$, CH$_2$Cl$_2$, 20°C, 1.5 h , 88% yield) (Scheme 2). Alternatively the synthesis of the corresponding tosylate has been also performed. Although the tosylation is much slower than the mesylation reaction when performed under closely related conditions (1.1 equiv. TsCl., 0.2 equiv. 4-dimethylamino-pyridine, 4 equiv. pyridine, CH$_2$Cl$_2$, 20°C, 16h), it offers the advantage to produce a crystalline material ( 9b2 mp :106°C, ether, 93% yield).

Finally, the synthesis of the cis-chrysanthemic acid 2a from the exo mesylate 9b1 or tosylate 9b2 has been achieved on reaction of the later derivatives with potassium hydroxide in DMSO (6 equiv. KOH, DMSO / H$_2$O, 70°C; 4h, 69% yield from 9b1; 2.5 h, 82% yield from 9b2). Under these conditions, the original cis-stereochemistry is completely retained (Scheme 2).

The synthetic approach which is described in the scheme 2 requires further comments. To our knowledge, cis-chrysanthemic acid is not part of a biologically active compound. It can be however easily transformed to its cis-dibromovinyl analog [9] 2c or isomerized [10] to its trans stereoisomer 2a' whose suitable esters are among the most potent insecticides actually available [1].

We have been also interested in proposing original stereo and enantioselective routes to trans- chrysanthemic acid 2a' and its cis-dibromovinyl analogue 2c from reagents able to transfer an isopropylidene moiety into suitably functionalized α,β-unsaturated esters [11-14]. Isopropylidenetriphenylphosphorane [11,12] 10, isopropylidenediphenylsulfurane [14] 11, 2-metallo-2- nitropropanes[13a-c] 12 and 2-metallo-2-sulfonyl propanes[13d] 13 are among those reagents which successfully allow the desired cyclopropanation reaction. The most straightforward route to methyl (d,l) trans-chrysanthemate involves without doubt isopropylidenetriphenylphosphorane (2.2 equiv.) and methyl γ-oxobutenoate. [11a]

γ-Oxobutenoates derived from chiral alcohols also produce on reaction with isopropylidenetriphenylphosphorane the corresponding trans-chrysanthemates but the asymmetric induction was rather low and does not reach the actual standard. [12c] We found clear cut differences of reactivity between isopropylidenetriphenylphosphorane and diphenylsulfurane. The former reacts with (Z) and (E)-α,β unsaturated esters and leads exclusively to trans-cyclopropane carboxylic ester. It enolises γ-butyrolactone 14 and 4-methoxy γ-butyrolactone 15 and affords dimenthyl caronate 17 in high yield with reasonably high asymmetric induction from dimenthyl fumarate 16 [12b].

Isopropylidenediphenylsulfurane 11 however reacts stereospecifically with the α,β unsaturated esters leading to trans- cyclopropyl esters from (E) isomers and to cis-cyclopropyl derivatives from their (Z) isomers [14a]. It provides tricyclo [3.1.0] hexane derivatives on reaction with γ-butyrolactones but affords[12b] dimenthyl caronate 17 in high yield but with dramatically low asymmetric induction from dimenthyl fumarate 16.

As a continuation of this work, we became interested in performing the above mentioned cyclopropanation reaction on the diester 18 (Scheme 4).

Scheme 4

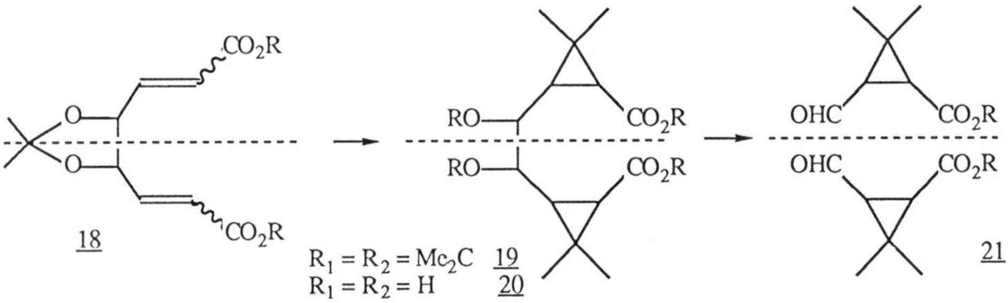

$R_1 = R_2 = Me_2C$  19
$R_1 = R_2 = H$  20

This yet unknown building block should possess exceptional features due not only to the presence in double of the γ-alkoxy α,β-unsaturated ester pattern (analogous to masked γ-oxobutenoates) but also due to the chiral nature of some of its forms susceptible to allow asymmetric inductions (inter alia in the double cyclopropanation reaction envisaged). Last

4h, 60 % yield of the (E,E) stereoisomer 18a containing 4% of the (E,Z) isomer 18c] or with methyl(triphenylphosphoranilidene)acetate [16b] [2.5 equiv. MeOH, -78°C to +20°C, 3h; 44 % yield of the (Z,Z) isomer 18b].

Scheme 7

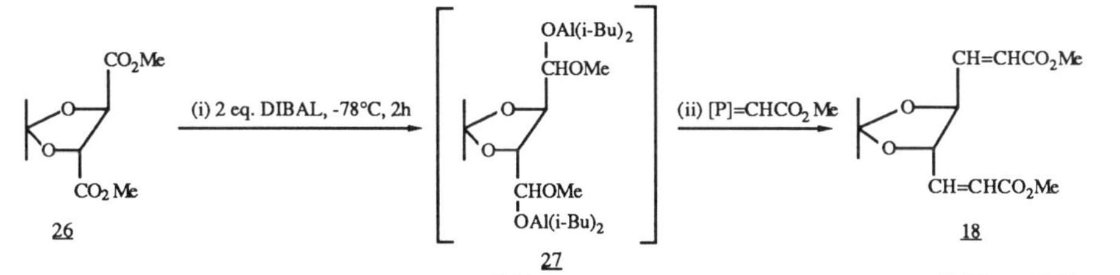

|  | E, E | Z, Z | Z, E |
|---|---|---|---|
| 2.5 equiv. (EtO)$_2$P(O)CH(Na)CO$_2$Me/DME, -78°C, 0.1h, 20°C, 4h | 60% | 96 | 4 |
| 2.5 equiv. Ph$_3$PCHCO$_2$Me/MeOH, -78°C to 20°C, 3h | 83% | 2 | 60 | 38 |

Remarkably, these Wittig-Horner or Wittig reactions presumably take place on the dialuminate 27 rather than on the corresponding dialdehyde 25 and therefore this approach avoids the tedious isolation of such highly water soluble and enolisable compound. It should also allow the synthesis of the enantiomers 18, already synthesized from D-mannitol, which are now readily available from the still cheap unnatural tartaric acid.

Cyclopropanation of the chiral diester 18a has been achieved with 2.5 equiv. of isopropylidenetriphenylphosphorane 10 (generated from isopropyl triphenylphosphonium iodide and n-butyl lithium in THF-hexane),the reaction being performed at 0°C for 1h then at 20°C for one more hour (Scheme 8). Typically, the diadduct 19 is obtained in 80 % yield as a 87/13 mixture of the 19a / 19a' stereoisomers from which the major one ( 19a, mp 123°C, cyclohexane) has been isolated in 50 % overall yield after one crystallization. The diadduct 19a was in turn transformed to (1R) - trans-, hemicaronaldehyde [9] 21a after hydrolysis of the dioxolane moiety (2N aq. HClO$_4$, THF, 20°C, 6h, 98 % yield) and cleavage of the resulting diol (1.5 equiv. NaIO$_4$, MeOH, phosphate buffer pH 7.2, 20°C, 1h, 68 % yield, ee 98%). The synthesis of the natural methyl trans-chrysanthemate 2a' from 21a is straightforward and has already been described[1] (Scheme 8).

Scheme 8

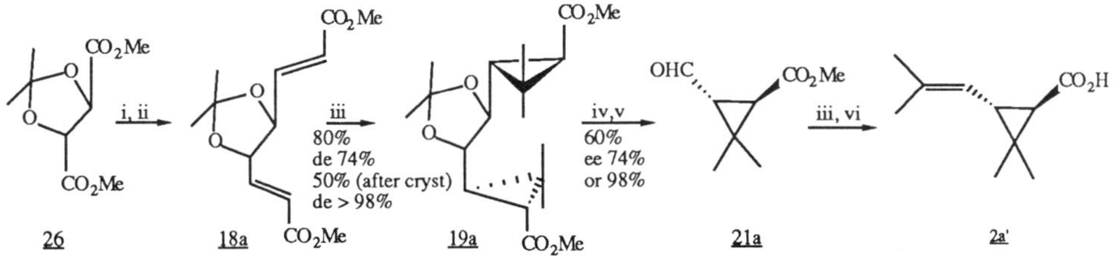

(i) 2 equiv. DIBAH, toluene, -78°C, 2h - (ii) 2.5 equiv. (EtO)$_2$P(O)CH(Na)CO$_2$Me, DME, -78°C, 0.1h then 20°C, 4h - (iii) 2.5 equiv. Ph$_3$P=C(Me)$_2$, LiI, THF, 0°C, 1h then 20°C, 1h - (iv) 4 equiv. 2N aq. HClO$_4$, THF, 20°C, 6h - (v) 1.5 equiv. NaIO$_4$, MeOH, Phosphate buffer pH 7.2, 20°C, 1h - (vi) KOH, MeOH then aq. HCl.

but not least it was discovered that both the (E,E) 18a and the (Z,Z) 18b stereoisomers in each series [(4R, 5R) and (4S, 5S)] are available from natural mannitol 22 and from natural tartaric acid 23 respectively via the acetonides (Scheme 5).

Scheme 5

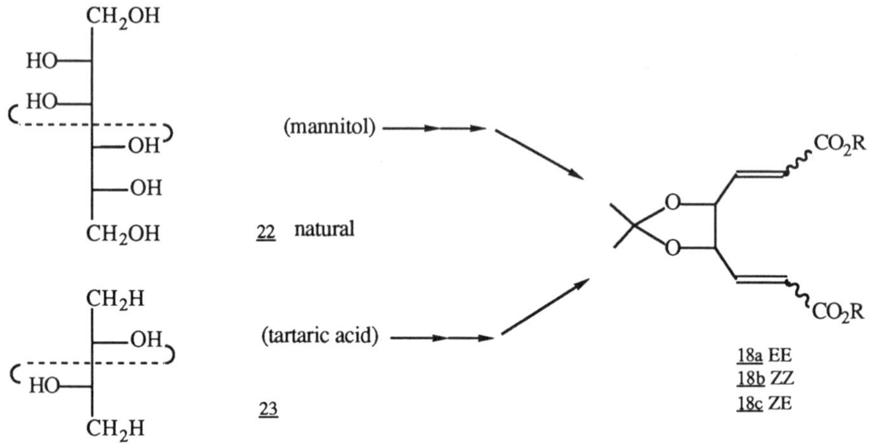

The cleavage of the tetrol [15] 24 derived from D-mannitol with sodium periodate (3 equiv. NaIO$_4$, MeOH, phosphate buffer pH : 7.2, 0°C, 0.5 h, 66 % yield) leads after removal of the solvent and extraction of the organic phase, to 2,3-isopropylidene tartraldehyde 25' as an amorphous powder (Scheme 6).

Scheme 6

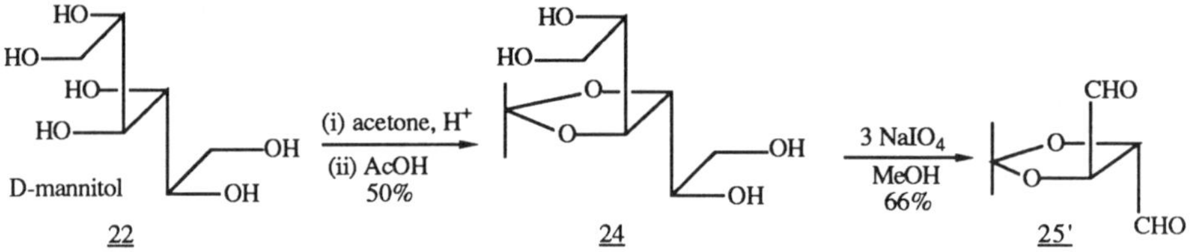

This on further reaction with sodio carbomethoxymethyldiethyl phosphonate 16a produces the (E,E) - diester 18a' contaminated with small amounts of the (Z,E) stereoisomer 18c' (E,E / Z,E : 90/10) in 35 % yield. If the reaction is instead performed with methyl(triphenylphosphoranilidene)acetate 16b in methanol, a mixture of all the three stereoisomers (Z,Z, E,E, Z,E) is produced in high yield. The Z,Z stereoisomer 18b' which predominates can be easily purified by chromatography on silicagel and has been obtained, free from the others, in 25 % yield.

The synthesis of their enantiomers 18a and 18b proved even more expeditious since we found (scheme 7) that they can be produced in one pot from the diester 26 derived from natural (2R,3R) - (+) - tartaric acid 23 by sequential treatment with diisobutyl aluminum hydride (DIBAL, 2 equiv. of a 1.5N solution in toluene, -78°C, 2h) followed by reaction with sodio carbomethoxymethyl diethylphosphonate 16a [2.5 equiv., DME, -78°C to 20°C,

The conditions described above for the cyclopropanation reaction are crucial for its success since for example a mixture of monoadducts and diadducts, in which the 19a / 19a' ratio decreases, is produced if the cyclopropanation reaction is carried out between -78°C and 20°C instead of between 0°C and 20°C.

The same set of reactions applied to 18b, the (Z, Z) stereoisomer of 18a, produces 19b in 65% yield. This on further reactions finally led to trans-hemicaronaldehyde 21 whose composition is close to a racemate. In order to have a more detailed insight into the intimate mechanism of this process, we have reacted 18b with only one equivalent of isopropylidenetriphenylphosphorane (Scheme 9). We surprisingly found that the monoadduct 28b, obtained beside unreacted 18b and the diadduct 19b ( 18b / 19b / 28b ratio : 1 / 1 / 1), possesses an (E) instead of the expected (Z) carbon - carbon double bond.

Scheme 9

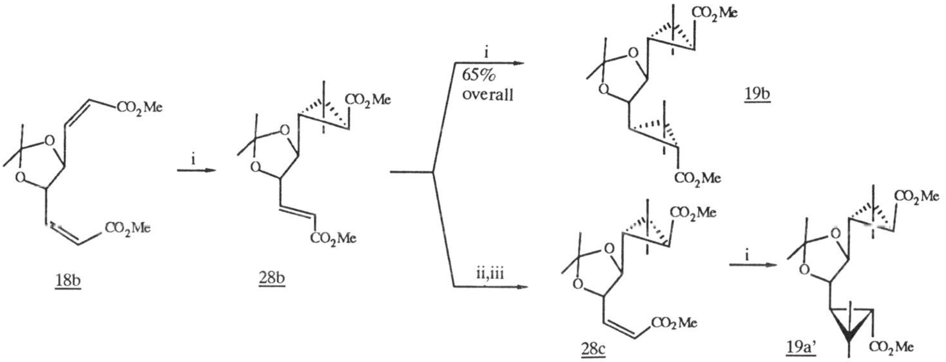

(i) Ph$_3$P=C(Me)$_2$, LiI, THF, 0°C, 1h then 20°C, 1h - (ii) O$_3$, CH$_2$Cl$_2$, -78°C then Me$_2$S, -78°C to +20°C (iii) Ph$_3$PCHCO$_2$Me, MeOH, 0°C, 2h then 20°C 1h

This mono adduct 28b has been transformed to its stereoisomer 28c possessing now a (Z) carbon-carbon double bond by sequential ozonolysis and reaction of the resulting aldehyde with methyl (triphenylphosphoranilidene)acetate[16b]. This, on cyclopropanation (1.5 equiv. of isopropylidenetriphenylphosphorane, 0°C, 1h then 20°C, 1h) gave the new diadduct 19a' (Scheme 9) which when subjected to the sequence of reactions described in scheme 8 led to (1S) - trans-, hemicaronaldehyde 21a' enantiomer of the one directly obtained from the (E, E) diester 18a (Scheme 8).

Rationalization of these results led us to conclude that : (i) The relation between the substituents on the cyclopropane is always trans whether a (Z) or a (E) unsaturated ester is used. (ii) The stereochemistry of the β carbon on the adducts 19 is dependent upon the stereochemistry of the carbon - carbon double bonds in 18. Therefore the transformation of 18b to 19b implies the following sequence of reactions : (a) The addition of the ylide across one of the (Z) carbon-carbon double bond leading to a betaine. (b) The stereoselective isomerisation (Z to E) of the remaining olefinic linkage. (c) Cyclization of the betaine to 28b. (d) Cyclopropanation of the second carbon-carbon double bond.

Diastereoselective addition to γ-alkoxy-α,β-unsaturated carbonyl compounds is well documented [17]. It has been used inter alia for the enantioselective synthesis of βlactame thienamycin[17a], brefeldin [17b], santalene [17c], prostanoids[17c], galactitol [17d] and (1R) - trans-chrysanthemic acid [17e] **2a'** (Scheme 10).

Scheme 10

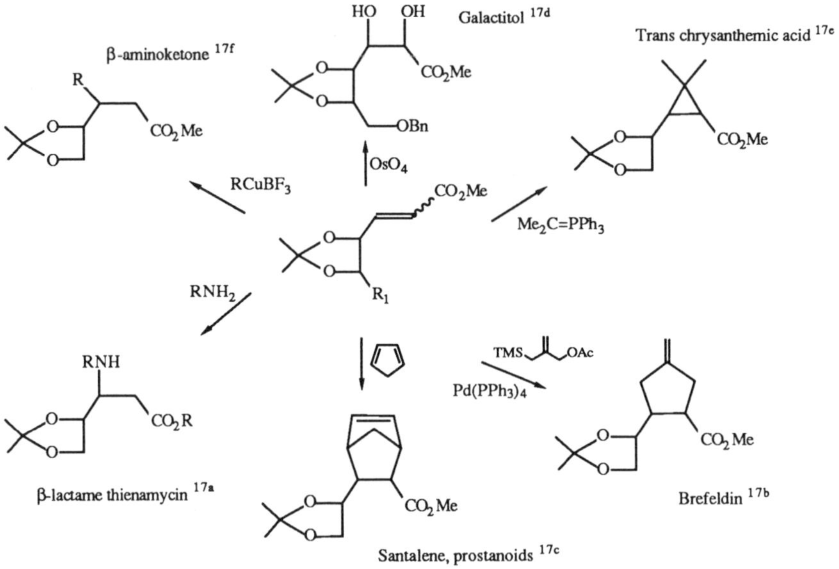

In most of the cases the stereochemistry of the β carbon on the adduct proved independent of the stereochemistry of the carbon-carbon double bond [17a-c] (Scheme 11). In few cases however the stereochemistry of the β carbon on the adduct is directly related to the stereochemistry of the starting material [17f,h] (Scheme 11), and interestingly the stereochemical outcome of the reactions on the β carbon of the adduct is identical in both series when the (Z) stereoisomer is involved. The models used to rationalize these results however remain unsatisfactory (Scheme 11).

Scheme 11

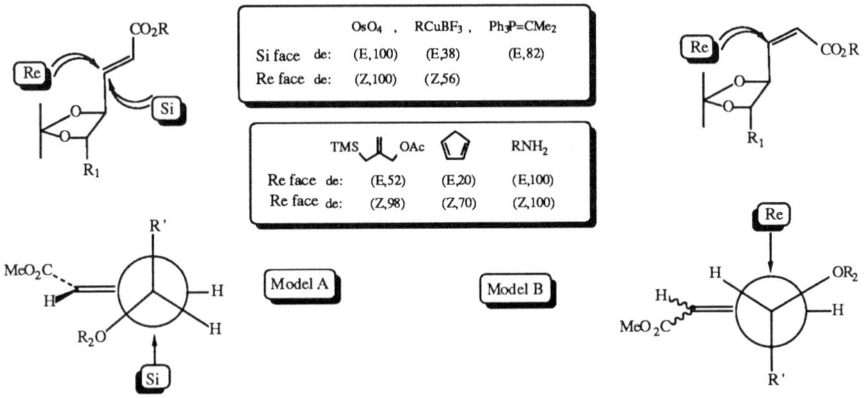

In order to generalize our observations, we investigated the more simple case of unsaturated esters 29a and 29b derived from D-glyceraldehyde and possessing respectively the (E) and (Z) stereochemistry [16] (Scheme 12). When reacted with isopropylidenetriphenylphosphorane, the former product leads, as previously described [17c] to the methyl trans- cyclopropane carboxylate 30 as a 91/9 mixture of diastereoisomers in which 30a predominates, whereas the other diastereoisomer 30a' is almost exclusively formed (30a' / 30a : 97 / 3) starting from the (Z) unsaturated ester 29b (Scheme 12).

Scheme 12

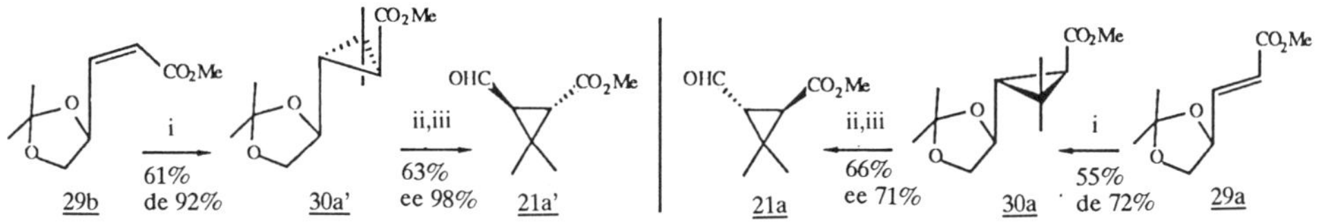

(i) 1.5 equiv. Ph$_3$P=C(Me)$_2$, LiI, THF, 0°C, 1h then 20°C, 1 h - (ii) 4 equiv. 2N aq. HClO$_4$, THF, 20°C, 6h - (iii) 1.5 equiv. NaIO$_4$, MeOH, Phosphate buffer pH 7.2, 20°C, 1h.

These results firmly support our proposal and clearly show that the (1R) - cis- series precursor of deltamethrin 1c is not directly accessible from isopropylidenetriphenylphosphorane.

It was therefore interesting to test another reagent able to transfer an isopropylidene moiety with conservation of the stereochemistry of the starting α,β-unsaturated ester. On the basis of Corey's original work [14a] and of the results already obtained in our laboratory,[11b] we expected that isopropylidenediphenylsulfurane could fulfil these requirements.

Thus isopropylidenediphenylsulfurane 11 (3 mol. equiv.) was reacted in place of its phosphorus analogue with the (Z,Z) diester 18b (DME,-78°C, 0.3 h then -60°C to -50°C, 0.7 h, then -50°C to 20°C), the diadduct 19c possessing two cis- disubstituted cyclopropane moieties is produced in 70% overall yield almost as a single diastereoisomer (de >92%) (Scheme 13). This adduct has been in turn transformed to the desired (1R) - cis- hemicaronaldehyde 21c precursor of deltamethrin 1 after hydrolysis of the dioxolane moiety (2N aq. HClO$_4$ THF, 20°C, 6h, 98% yield) and cleavage of the resulting diol (1.5 equiv. NaIO$_4$, MeOH, phosphate buffer pH 7.2, 20°C, 1h, 62% yield, ee: 92 %).

The same set of reactions applied to the E,E diester 18a leads (scheme 13) to the adduct 19a' (82% yield, de >94%) which on further reactions produces (1S) - trans- hemicaronaldehyde 21a' (70% yield, ee 93.7% ).

These results are remarkable since they involve two cyclopropanation reactions, each one occurring in particularly high yield (>85 %) and high stereochemical control (de > 96%).

Scheme 13

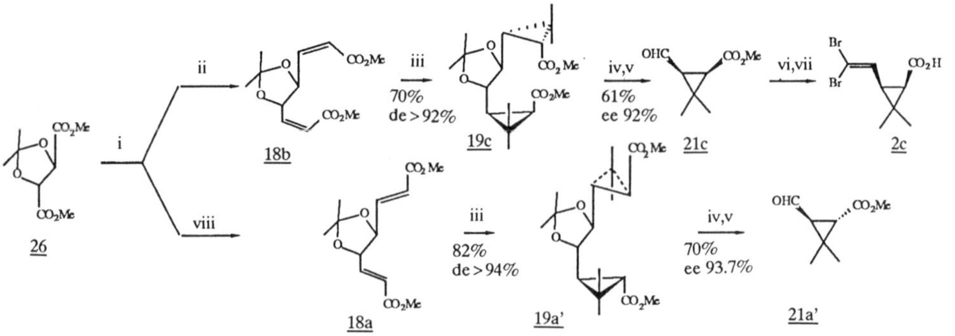

(i) 2 equiv. DIBAH, toluene, -78°C, 2h - (ii) 2.5 equiv.Ph₃PCHCO₂Me, MeOH, -78°C then -78°C to +20°C, 3h - (iii) 3 equiv. Ph₂S=C(Me)₂, DME, -78°C, 0.3 h then -60°C to -50°C, 0.7 h then -50°C to 20°C - (iv) 6 equiv. 2N aq. HClO₄, THF, 20°C, 6h - (v) 1.5 equiv. NaIO₄, MeOH, Phosphate buffer pH 7.2, 20°C, 1h - (vi) 4 CBr₄, PPh₃, CH₂Cl₂ - (vii) KOH, MeOH then aq. HCl - (viii) 2.5 equiv. (EtO)₂P(O)CHNaCO₂Me, DME, -78°C then -78°C to +20°C, 4h.

The stereochemical observations reported above have been confirmed on the more simple case of the unsaturated esters 29b and 29a derived from D-glyceraldehyde and possessing the Z and E stereochemistry (Scheme 14) respectively . These, on reaction with isopropylidenediphenylsulfurane (1.5 equiv., same conditions as above) produce respectively the cyclopropyl esters cis- 30c (84% yield, de >96%) and trans- 30a' (92% yield, de >98%) in very high yield and with high diastereoselection, and later on lead to hemicaronaldehyde (1R) - cis- 21c (60% yield, ee 96% ) and (1S) - trans- 21a' (63% yield, ee 98%) respectively .

Scheme 14

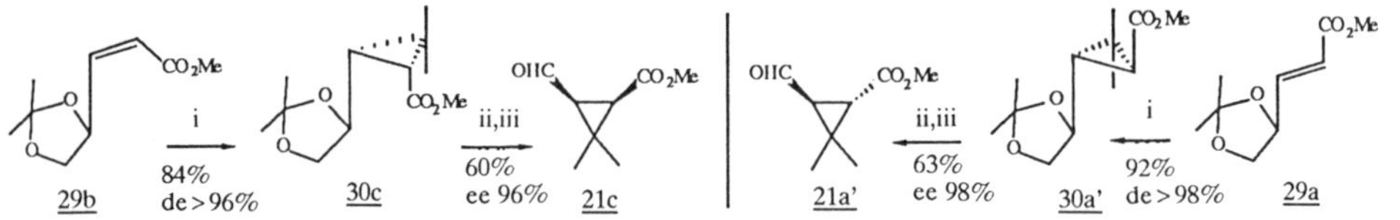

(i) 1.5 equiv. Ph₂S=C(Me)₂, DME, -78°C, 0.2 h  then -78°C to -50°C, 0.7 h  then -50°C to  20°C, 0.3 h  (ii) 4 equiv. 2N aq. HClO₄, THF, 20°C, 6h  - (iii) 1.5 equiv. NaIO₄, MeOH, Phosphate buffer pH 7.2, 20°C, 1h.

These results are in sharp contrast with those obtained with isopropylidenetriphenylphosphorane reported above. These differences are particularly evident on the more simple unsaturated ester 29 derived from D-glyceraldehyde. (i) The reaction with isopropylidenediphenylsulfurane is almost completely stereospecific as far as the relative stereochemistry is concerned (trans- 30a' from (E)-29a ; cis- 30c from Z - 29b) whereas it is almost completely stereoselective with its phosphorus analogue (trans- 30a and 30a' from (E) - 29a and (Z) - 29b respectively) . (ii) The stereochemistry of the β carbon of the adduct 30 is independent of the stereochemistry of the carbon-carbon double bond of 29 when the sulfur ylide is used (always (S)) whereas it is completely dependent when its phosphorus analogue is involved ((R) from (E) - 29a and (S) from (Z) - 29b ).

These unexpected results are not yet rationalized, although they fit in the series of results already published [17] and collected in the scheme 11, but as far as the synthetic aspect is concerned, it is now possible to device straightforward enantioselective routes to natural pyrethrins or to deltamethrin from natural tartaric acid or from natural D-glyceraldehyde by the proper choice of the isopropylidene transferring reagent (phosphorous or sulfur ylides respectively).

α-Selenoalkyllithiums[18] 32, α-selenoxyalkyllithiums[18b,19] 34 and α-metalloselenones[20] 36 also belong to the same family of α-heterosubstituted organometallics. The former derivatives are easily available from selenoacetals 31 and alkyllithiums whereas the two others are produced on reaction of the corresponding selenoxides 33 and selenones 35 with lithium diisopropylamide and potassium t-butoxide respectively (Scheme 15).

Scheme 15

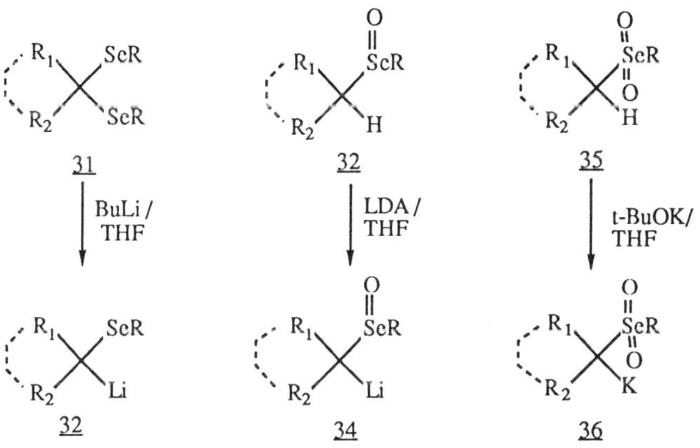

Although their reactivity is related to the one of their thio analogues, the seleno moiety in all the compounds cited proved by far a better leaving group. For example the selenoxide elimination leading to olefins takes place at much lower temperature than the related sulfoxide elimination [18,21].

It was therefore expected that α-metalloselenones would have a higher tendency than α-metallosulfones to produce cyclopropylesters from α,β-unsaturated esters. Only very few reports described the synthesis of selenones [22] when we started to work in that field and α-metalloselenones were unknown. The straightforward synthesis of selenones involve the oxidation of selenides. Although a large array of oxidants such as hydrogen peroxide, organic hydroperoxides, periodates, peracids and permanganates smoothly oxidize selenides to selenoxides [18,21] only the two later are able to overoxidize them to selenones. We have also found that the second oxidation is much more difficult than the first one and is dramatically slowered by steric hindrance around the active site. However the success of the whole process requires that the second oxidation step is rapid enough so that the elimination of the selenoxide leading instead to olefins cannot compete. Otherwise the seleninate moiety proved to be an exceptionally good leaving group, much better than the iodide[20] and therefore we experienced great difficulty in trying to synthesize for example benzylselenones which seem to be unstable and lead instead to products resulting from the substitution of the seleno moiety. Thus dimethyl, arylmethyl, primary alkylphenyl and methylselenides and secondary alkylmethylselenides have been transformed [22a,b] in reasonable to good yield to the corresponding selenones by potassium permanganate or with peracids (Scheme 16). Even the simplest procedure involves pereseleninic acids prepared in situ from seleninic acid and hydrogen peroxide as the effective oxidant [22c] (Scheme 16).

Secondary alkylphenylselenides did not produce the corresponding selenones probably due to the higher propensity of the selenoxide intermediate to decompose. Cyclopropyl phenylselenides whose selenoxide is much less prone to produce the strained cyclopropenes are however cleanly transformed [22b,c] to cyclopropyl phenylselenones. Substitution of the cyclopropane ring by methyl groups greatly lowers the rate of the second step and for example, we have been unable [22b] to synthesize 2,2,3,3-(tetramethyl) cyclopropyl phenylselenone since the corresponding selenoxide is produced in almost quantitative yield instead and cannot be further oxidized [22c].

Scheme 16

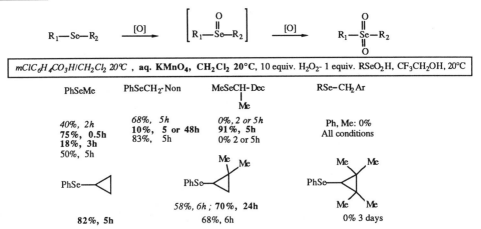

Metallation of phenylselenones can be carried out with potassium tert-butoxide alone or with potassium diisopropylamide (KDA). If the reaction is performed in the presence of alkylidene malonates or of α,β unsaturated esters the corresponding cyclopropyl diesters or esters are formed [22c] in 40-70% yield and are easily separated from the potassium phenylseleninate concomitantly produced (Scheme 17).

The reaction proceeds under mild conditions in THF and allows the transfer of a methylene, an 1-alkylidene as well as of a cyclopropylidene moiety. In that respect the reaction clearly differs from the one of the corresponding alkyl phenylsulfones[13d, 22c] which produces cyclopropane derivatives only (i) when they bear a sec-alkyl group (ii) when they are reacted with an alkylidene malonate and (iii) with the use of DMSO as solvent.

Scheme 17.

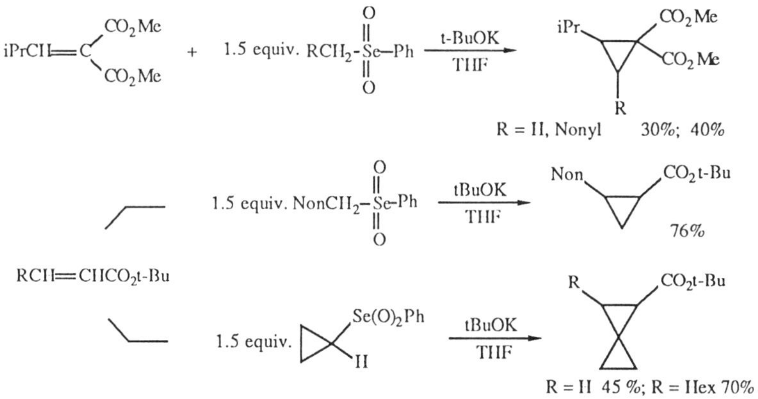

[Scheme 17 depicting reactions of iPrCH=C(CO$_2$Me)$_2$ with 1.5 equiv. RCH$_2$-Se(O)-Ph, t-BuOK/THF giving cyclopropane with iPr, CO$_2$Me, CO$_2$Me, R substituents; R = H, Nonyl, 30%; 40%.

RCH=CHCO$_2$t-Bu with 1.5 equiv. NonCH$_2$-Se(O)-Ph, tBuOK/THF giving cyclopropane with Non, CO$_2$t-Bu, 76%.

1.5 equiv. of compound II with Se(O)$_2$Ph, tBuOK/THF giving cyclopropane with R, CO$_2$t-Bu; R = H 45%; R = Hex 70%.]

The phenylselenyl moiety is by far a less potent leaving group, for example α-selenoalkyllithiums in THF-HMPT [23] or in the presence of trimethylsilylchloride [23b] and α-selenobenzyllithiums [23b] react with α,β unsaturated carbonyl compounds and lead to γ-selenocarbonyl compounds after hydrolysis rather than to cyclopropylcarbonyl compounds. We recently however found [24] that α-selenobenzyllithiums add accross the carbon-carbon double bond of ethylene, styrene, butadiene, vinylsilane and vinylsulfide around 0°C and immediately lead to the corresponding cyclopropane derivatives in moderate to good yield (Scheme 18)

Scheme 18.

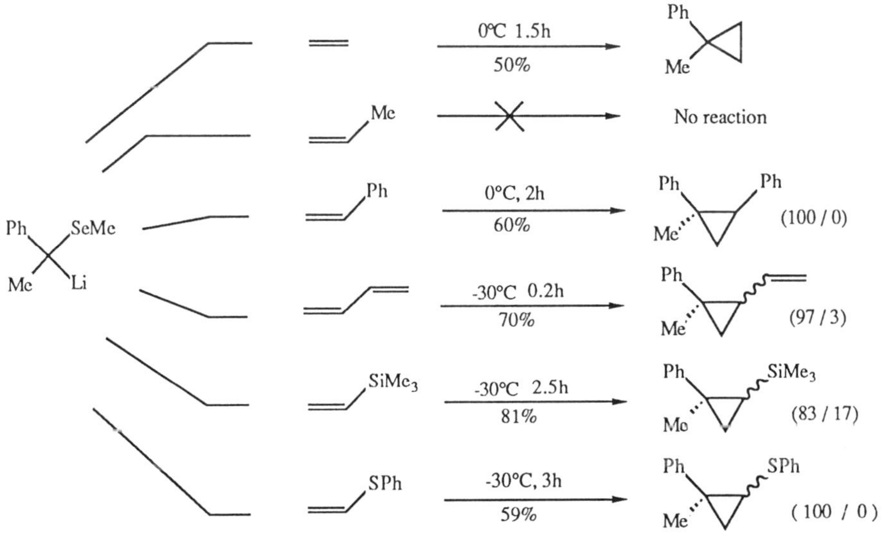

[Scheme 18: Ph(Me)(SeMe)C-Li reacting with various alkenes:
- ethylene, 0°C 1.5h, 50% → cyclopropane with Ph, Me
- propene (Me-substituted), no reaction
- styrene (Ph), 0°C, 2h, 60% → Ph, Me / Ph cyclopropane (100/0)
- butadiene, -30°C 0.2h, 70% → Ph, Me / vinyl cyclopropane (97/3)
- vinylsilane (SiMe$_3$), -30°C 2.5h, 81% → Ph, Me / SiMe$_3$ cyclopropane (83/17)
- vinylsulfide (SPh), -30°C, 3h, 59% → Ph, Me / SPh cyclopropane (100/0)]

The same reaction takes place with α-selenobenzyllithiums bearing a built in carbon-carbon double bond in a suitable position to allow the synthesis of a bicyclo [3.1.0] hexane or [4.1.0] heptane ring (Scheme 19). It does not seem that these reactions proceed via a carbene since more nucleophilic olefins such as for example 2-propene are inert under similar conditions.

Scheme 19

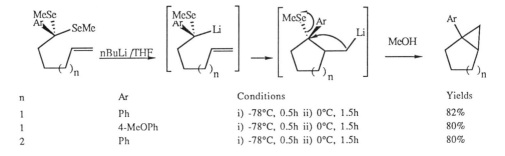

| n | Ar | Conditions | Yields |
|---|---|---|---|
| 1 | Ph | i) -78°C, 0.5h ii) 0°C, 1.5h | 82% |
| 1 | 4-MeOPh | i) -78°C, 0.5h ii) 0°C, 1.5h | 80% |
| 2 | Ph | i) -78°C, 0.5h ii) 0°C, 1.5h | 80% |

In conclusion, various organometallics bearing a heteroatomic moiety on the carbanionic center are able to perform the cyclopropanation of carbon-carbon double bond. However, subtle differences have been found which still remain unexplained between closely related reagents. Thus, although both the phosphorous and sulfur ylide react with α,β unsaturated esters to provide cyclopropyl esters, the stereochemistry and also the diastereoselection in case of γ-alkoxyesters dramatically differ. Nevertheless, we have taken advantage of these reactions to synthesize some valuable pyrethroid insecticides.

α-Metalloselenones proved more suitable than their sulfur analogues to perform the cyclopropanation of α,β unsaturated esters and finally α-selenobenzyllithiums add on ethylene and heterosubstituted olefins and directly produce the cyclopropane ring. Since α-selenobenzyllithiums are available from arylketones, the whole process allows the two-pot geminal dialkylation of the carbonyl group of aromatic ketones.

REFERENCES

1.  (a) Elliot M. and Janes N.F., Chem. Soc. Rev. **7**, 473 (1978). (b) Arlt D., Jautelat M. and Lantzsch R., Angew. Chem. int. ed., **20**, 703 (1981). (c) Naumann K., Chemie der Pflanzenschutz- und Schädlingsbekämpfungsmittel, by Springer-Verlag, Ed., Heidelberg, 1981, ISBN 3-540-10452-6. (d) Deltaméthrine, Roussel-Uclaf, 1982, ISBN 2-904125-00-0. (e) A. R . Justum, R.F.S. Gordon and C.N.E. Ruscoe, Proceedings 1986, British Corp. Protection Conference, **1**, 97 (1986). (f) E. Mc Donald, N. Punja and A.R. Justum, Proceedings 1986, British Corp. Protection Conference, **1**, 199 (1986).
2.  Kiwus R., Schwarz W., Rosnagel I. and Musso H., Chem. Ber. **120**, 435 (1987).
3.  (a) This compound has been already synthesized by Nozaki [3b]. However the problem lies in the preparation of 3-bromo -2,2,5,5-tetramethylcyclohexa-1,3-dione **5** which we were unable to perform selectively under various experimental conditions. Under the conditions cited by Nozaki [3b] (ref. to Stetter's work) and using NBS, the bromination is very slow and provides a mixture of products from which the desired pure product **5** can be isolated in low yield by crystallization. (b) Okada T., Kamogawa K., Kawanisi M. and Nozaki H. Bull. Chem. Soc. Jpn., **43**, 2908 (1970).
4.  (a) Grob C.A., Angew. Chem. Int. Ed. **8**, 535 (1969) and references cited. (b) Carruthers W., Some Modern Methods of Organic Chemistry Synthesis, Cambridge University Press, Cambridge, 107 (1971).
5.  House H.O., Modern Synthetic Reactions, W.A. Benjamin Inc. Menlo Park., 2nd Ed., (1972).
6.  (a) Eliel E.L. and Senda Y., Tetrahedron **26**, 2411 (1970), Report 61. (b) Wigfield D.C., Tetrahedron **35**, 449 (1979).
7.  (a) Luche J-L., J. Am. Chem. Soc., **100**, 2226 (1978). (b) Luche J-L., J. Am. Chem. Soc., **103**, 5454 (1981) and references cited.
8.  (a) Wender P.A., Heterocycles, **25**, 263 (1987). (b) Declercq P., University of Ghent, private communication.
9.  Martel J., Roussel Uclaf, Fr. Appl. 159,066 (1968), Ger. Offen. 1,966,839 (1984), Chem. Abst. **81**, 135530 (1974).
10. (a) Suzukamo G. and Yasuda M., Bunri-Gijutsu **16**, 345 (1986) (b) Sakita Y.and Suzukamo G., Chem. Lett. 621 (1986) and references cited.
11. (a) Devos M-J., Denis J.N., Krief A., Tetrahedron Lett., 1847 (1978). (b) Devos M-J. and Krief A., Tetrahedron Lett., 1891 (1979).
12. (a) Devos M-J., Hevesi L., Bayet P., Krief A., Tetrahedron Lett., 3911 (1976). (b) Devos M-J. and Krief A., Tetrahedron Lett. **24**, 103 (1983) and references cited. (c) Devos M-J. PhD Thesis Namur 1981 and unpublished results from our laboratory.
13. (a) Devos M-J., Hevesi L., Mathy P., Sevrin M., Chaboteaux G. and Krief A., Organic Synthesis : an interdisciplinary challenge, Blackwell, 123 (1985). (b) Chaboteaux G. and Krief A., Bull. Soc. Chim. Belg., **94**, 495 (1985). (c) Krief A. and Devos M-J., Tetrahedron Lett., **26**, 6115 (1985). (d) Krief A., Hevesi L., Chaboteaux G., Mathy P., Sevrin M. and Devos M-J., J. Chem. Soc. Chem. Commun.,1693 (1985).

14. (a) Corey E.J. and Jautelat M., J. Amer. Chem. Soc. **89**, 3913 (1967). (b) Sevrin M., Hevesi L., Bayet P. and Krief A., Tetrahedron Lett., 3915 (1976).
15. (a) Debost J.L., Gelas J. and Horton D., J. Org. Chem. **48**, 1381(1983). (b) Carmack M. and Kelly O.J., J. Org. Chem. **33**, 2971 (1968).
16. (a) Wadsdworth W.S., Organic Reactions **25**, 73 (1977). (b) Minami N., Ko S.S., Kishi Y., J. Am. Chem.Soc., **104**, (1982). (c) Jurczak J., Pikul S. and Bauer T., Tetrahedron, **42**, 447 (1986).
17. (a) Matsunaga H., Sakamaki T., Nagoka H. and Yamada Y., Tetrahedron Lett., **23**, 3009 (1983). (b) Trost B.M. and Mignani S.M., Tetrahedron Lett. **35**, 4137 (1986). (c) Mulzer J. and Kappert M., Tetrahedron Lett. **26**, 1631 (1985). (d) Katsuki T., Lee A.W.M., Ma P., Machin V.S., Masamune S., Sharpless K.B., Tuddenham D. and Walker F.J., J. Org. Chem. **47**, 1378 (1982). (e) Mulzer J. and Kappert M., Angew. Chem. Int. Ed. **22**, 63 (1983). (f) Yamamoto Y., Nishii S. and Ibuka T., J.C.S. Chem. Commun. 464 (1987). (g) Roush W.R. and Lesur B.M., Tetrahedron Lett. **23**, 2231 (1983). (h) Stork G. and Kahn M., Tetrahedron Lett. **24**, 3951 (1983).
18. (a) Krief A., Tetrahedron **36**, 2531 (1980) report 94. (b) Krief A., "The chemistry of organoselenium and tellurium compounds", Patai S. ed. John Wiley, Chichester, **vol 2**, chapter 17, 675 (1988).
19. Reich H.J. and Shah S.K., J. Amer. Chem. Soc **97**, 3250 (1975).
20. Krief A., Dumont W. and Denis J-N., J. C. S. Chem Commun. 570 (1985).
21. (a) Clive D.L.J., Tetrahedron **34**, 1049 (1978). (b) Reich H.J., Acc. Chem. Res. **22** (1979).
22. (a) Krief A., Dumont W., Denis J-N., Evrard G. and Norberg B., J.C.S. Chem Commun. 569 (1985). (b) Krief A., Dumont W. and J.L. Laboureur Tetrahedron Lett. **29** 3265 (1988).(c) Krief A., Dumont W. and A.F. De Mahieu Tetrahedron Lett. **29**, 3269
23. (a) Wartski L., El Bouz M.E., Seyden-Penne J., Dumont W. and Krief A., Tetrahedron Lett. 1801 (1979). (b) Unpublished results from our laboratory.
24. Barbeaux P., unpublished results from our laboratory.

# EXPERIMENTAL AND THEORETICAL STUDIES OF BRIDGED CYCLOPROPENES

Kenneth B. Wiberg and Dean R. Artis
Department of Chemistry
Yale University
New Haven, Connecticut 06511
U. S. A.

ABSTRACT: The formation of cyclopropenes with small bridges has been studied via the dehalogenation of the corresponding dihalides. In solution, the reaction occurs and gives the cyclopropene which may subsequently be trapped as a Diels-Alder Adduct. In the gas phase, however, the main reaction leads to ring cleavage giving a methylenecycloalkene. The compounds have been studied theoretically, and at the 6-31G* level, bicyclo[2.1.0]pent-1(4)-ene is a saddle point and opens to methylenecyclobutylidene. The properties of these compounds have been studied via numerical integration of their charge densities.

There has been much interest in the effects of bond angle distortion on the energies, structures and properties of alkenes. One mode of distortion is pyramidalization, and some examples of such alkenes include

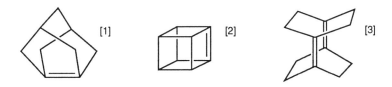

In these cases, the pyramidalization is enforced by the geometrical constraints. There are other groups of alkenes in which pyramidalization occurs without these constraints. One such group is the bridged cyclopropenes.

In the first detailed theoretical examination of the compounds 1-4, Wagner et al[4] found that the preferred geometries were non-planar, and that the pyramidalization decreased as the size of the bridging ring increased. Our later theoretical study using a larger basis set confirmed these conclusions,[5] and allowed estimates of the enthalpies of formation and of the olefinic strain[6] (the difference in strain energy between the cycloalkene and the corresponding cycloalkane). These data are summarized in Table I.

1　　　　　2　　　　　3　　　　　4

Table I
Calculated Energies of Bridged Cyclopropenes

| Compound | Energy, 6-31G* | ΔH$_f$ | SE | OS |
|---|---|---|---|---|
| Cyclopropene | -115.82305 | 69 | 55 | 27 |
| 1,2-Dimethylcyclopropene | -193.91214 | 4 | 45 | 23 |
| Bicyclo[1.1.0]butene, planar | -153.55949 | 158 | 143 | |
| Bicyclo[1.1.0]butene, bent | -153.58020 | 145 | 130 | 66 |
| Bicyclo[2.1.0]pentene, planar | -192.59656 | 152 | 142 | |
| Bicyclo[2.1.0]pentene, bent | -192.62153 | 136 | 126 | 71 |
| Bicyclo[3.1.0]hexene, bent | -231.72314 | 89 | 84 | 53 |
| Bicyclo[4.1.0]heptene | -270.79159 | 63 | 63 | 36 |

We wished to try to prepare some of these compounds because of our long-standing interest in cyclopropenes, and a desire to compare their properties with that of bicyclo[2.2.0]hex-1(4)ene (5), which is one of the most reactive of the strained alkenes which can be directly observed at room temperature.[7]

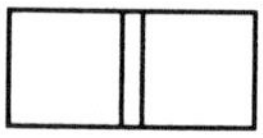

5

Closs and Böll have prepared some bridged cyclopropenes via photolysis of pyrazolines:[8]

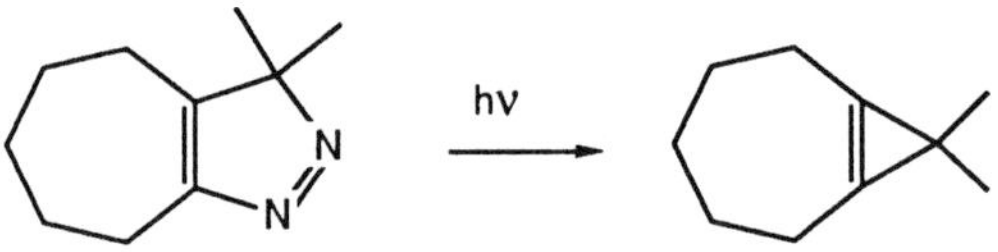

6　(1840 cm$^{-1}$)

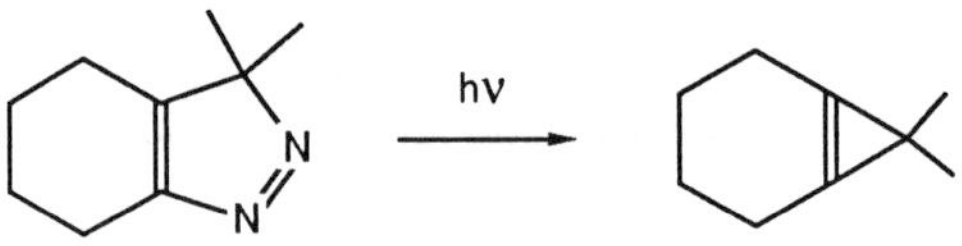

7　(1740 cm$^{-1}$)

The alkene, 6, was quite stable and could be heated to 100° in solution without change. However, 7 could be observed only at low temperature, and on warming underwent a rearrangement:

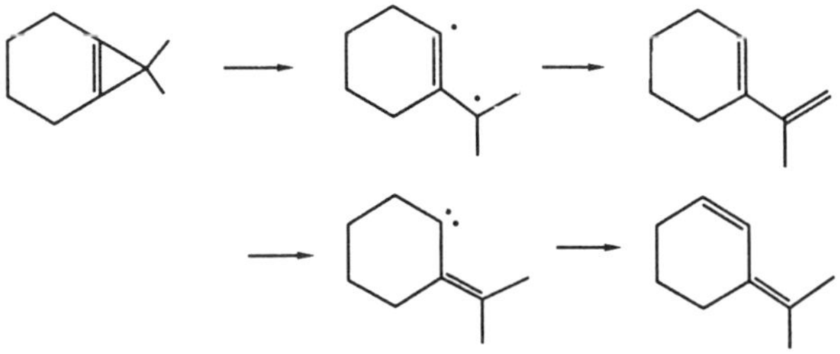

Szeimies has provided strong evidence for compounds related to **1** and **2** as intermediates in reactions.[9] They could only be identified via the Diels-Alder adducts.

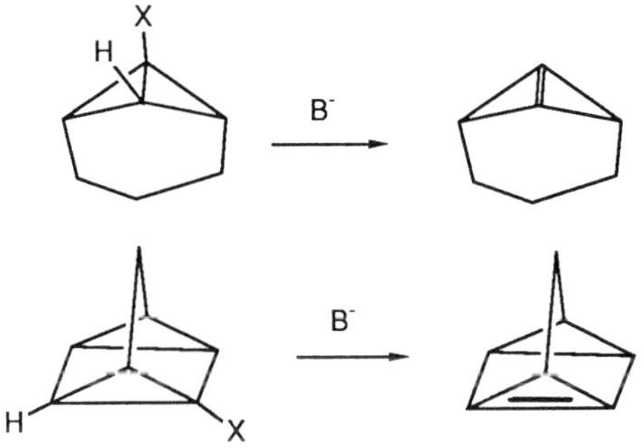

Neither procedure appeared very useful for the formation of **3**, and in the case of **4**, the Closs procedure is successful only with the gem-dimethyl group which prevents bond migration. The dehalogenation of dihalides appeared to be a more generally useful procedure. As a test, 1,2-dibromo-1,2-dimethylcyclopropane was treated with t-butyllithium, and 1,2-dimethylcyclopropene was found as the product:

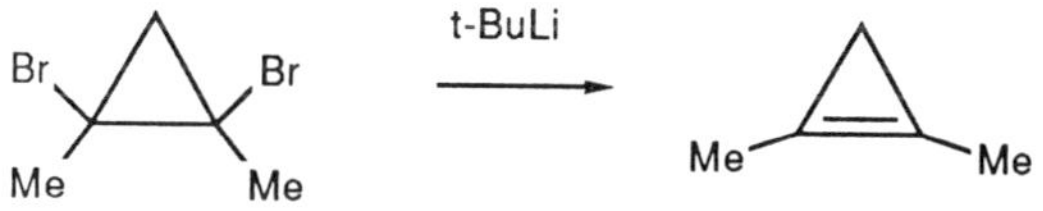

1,5-Dibromobicyclo[3.1.0]hexane (**8**) and 1,6-dibromobicyclo[4.1.0]heptane (**9**) were prepared via brominative decarboxylation of the corresponding diacids. Treatment with t-butyllithium in pentane gave the following results:[10]

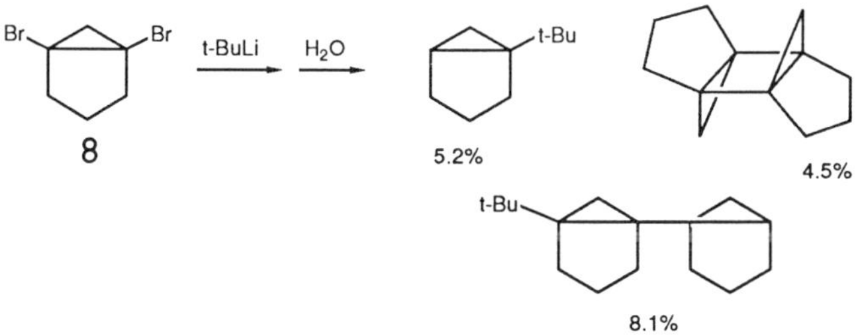

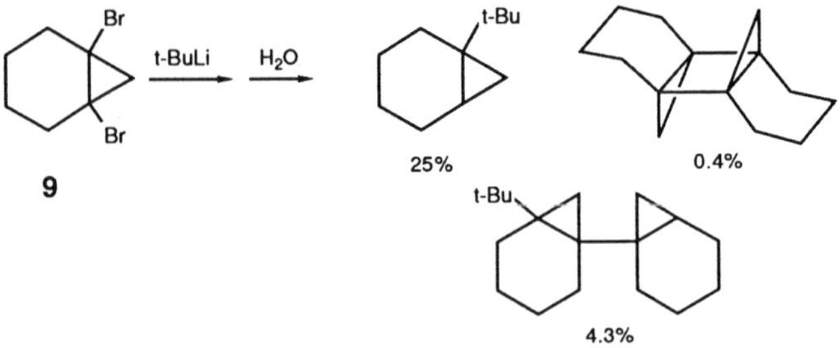

[Scheme: 9 + t-BuLi, H₂O → t-Bu-substituted bicyclic (25%), dimer (0.4%), and bis-t-Bu coupled product (4.3%)]

Although the products were formed in low yield, they suggested the formation of the corresponding cyclopropene as an intermediate. Better evidence was obtained by carrying out the reaction in the presence of diphenylisobenzofuran as a trapping agent. In this case, the Diels-Alder adduct was formed in good yield.

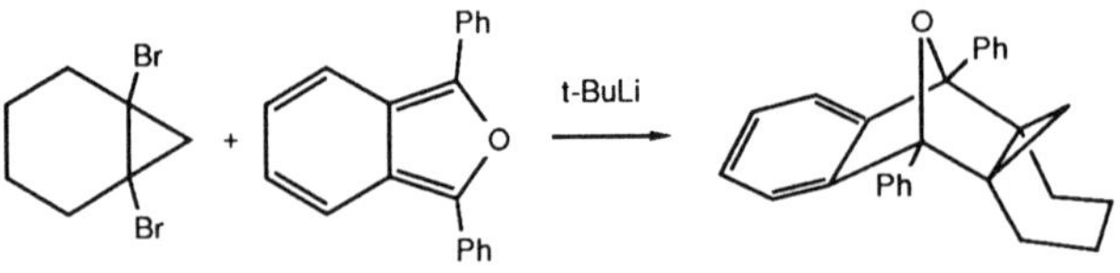

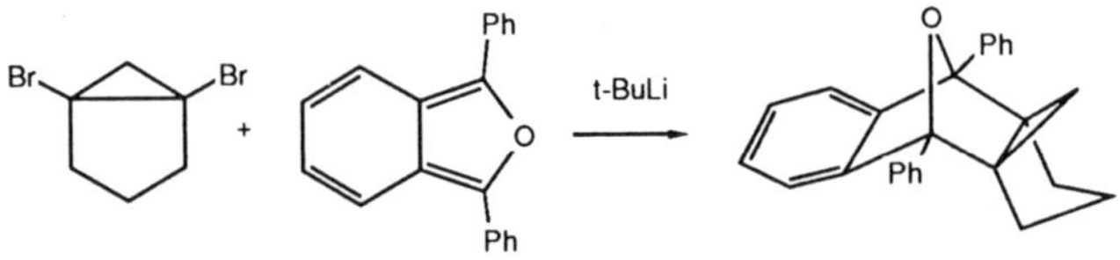

[Scheme: dibromobicyclic + diphenylisobenzofuran, t-BuLi → Diels-Alder adduct]

[Scheme: isomeric dibromobicyclic + diphenylisobenzofuran, t-BuLi → Diels-Alder adduct]

Despite the success in forming the DA adducts, the question still remains, was the adduct formed from the free alkene, or from an intermediate organolithium species? In order to explore this question, we have carried out a number of experiments designed to trap the alkene, if formed, in an argon matrix at 20K.

The success we have had in forming strained propellanes via gas phase dehalogenation of dihalides with potassium atoms[11] led us to try this procedure for the preparation of the cyclopropenes. Again, 1,2-dibromo-1,2-dimethylcyclopropane gave dimethylcylopropene. However, the reactions of 8-11 led to ring opened dienes as the main products

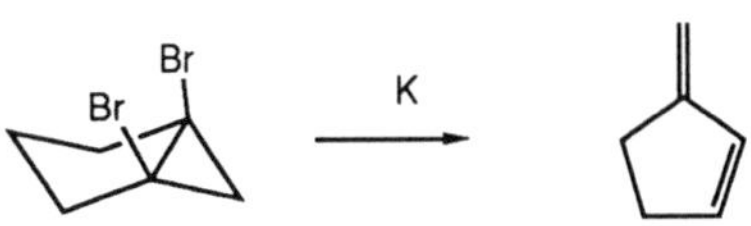

[Schemes showing:
8 (dibromo) + K → methylenecyclopentene
10 (diiodo) + K → methylenecyclopentene
10 + MeLi → methylenecyclopentene
11 + MeLi → methylenecyclohexene]

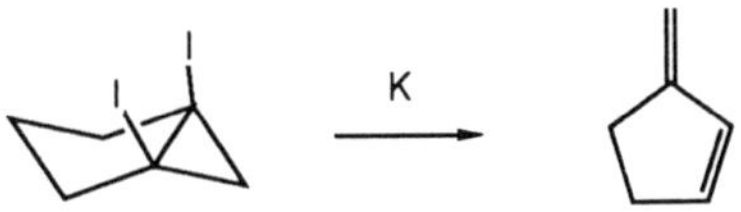

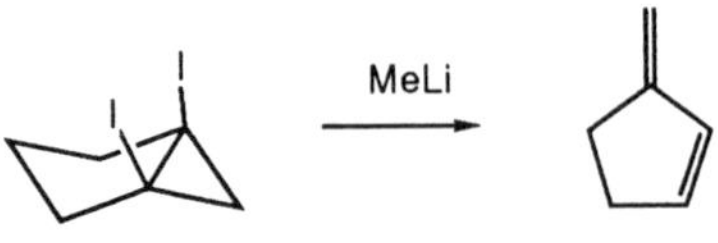

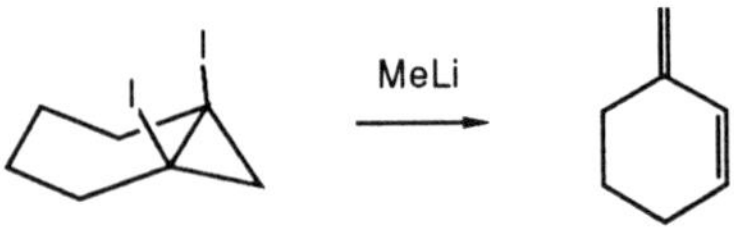

The difference in reaction between the gas phase and solution led us to further study the latter. In order to minimize or avoid addition to any alkene formed, methyllithium was chosen as the reducing agent and the diiodides were used in order to increase reactivity at low temperature. The reaction of 1,6-diiodobicyclo[4.1.0]heptane with methyllithium in ether was carried out at -100° where the reaction is rather slow. After several hours, precooled methanol was added to quench the alkyllithium, and after five minutes diphenylisobenzofuran was added. The main component of the mixture thus formed was unreacted diiodide. No monoiodide could be detected. The other products were the Diels-Alder adduct previously obtained and a tetramer. No dimeric or trimeric products were found.

The isolation of the Diels Alder adduct under these conditions demonstrates that bicyclo[4.1.0]hept-1(6)ene was formed in the reaction, and the preferred mode of reaction is tetramer formation. This suggests a rapid ene-type reaction to form dimer, followed by a rapid coupling of dimers giving the tetramer.

Since we now have evidence for 4 as a discrete species in the solution phase studies, and Billups has demonstrated that 4 can be trapped as a Diels-Alder adduct from the gas phase fluoride-catalyzed elimination from 6-chloro-1-trimethyl-siloxybicyclo[4.1.0]heptane,[12] we wished to know if some of the alkene was formed in our gas-phase elimination reactions. We use infrared spectroscopy as the analytical tool, and therefore we wished to have an estimate of the infrared spectrum of 4, as well as those of 1-3.

It is now possible to make good estimates of such spectra via ab initio molecular orbital calculations of force constants for molecular vibrations.[13] These calculations were carried out for each of the alkenes, 1-4, and some of the results are shown in Table II. The calculated double bond stretching frequencies for both cyclopropene and 1,2-dimethylcyclopropene were well reproduced by the calculations after applying the usual scaling factor of 0.88.[14]

Table II
Calculated Vibrational Frequencies for Bridged Cyclopropenes

| Compound | ν(C=C) | ν(ring open.) |
|---|---|---|
| Cyclopropene | 1638 (1653) | 617 (596) |
| 1,2-Dimethylcyclopropene | 1895 (1880) | |
| Bicyclo[1.1.0]butene (1) | 1391 | 356 |
| Bicyclo[2.1.0]pentene (2) | 1530 | -130 |
| Bicyclo[3.1.0]hexene (3) | 1703 | 315 |
| Bicyclo[4.1.0]heptene (4) | 1854 | 378 |

Values in parentheses are observed frequencies.

The calculated vibrational frequency for 4 is considerably higher than that reported by Closs and Boll[8] for their gem-dimethyl substituted derivative. In view of the generally good agreement between calculated and experimental frequencies, it appears that the 1730 cm$^{-1}$ band observed by them was due to some other species in the complex mixture which was formed on photolysis of the diazo precursor.

The matrix isolated spectrum derived from the dehalogenation of 1,5-diiodobicyco[3.1.0]hexane was examined for the bands calculated for 3. Although the presence of the cycloalkene could not be ruled out, its concentration was at best very low. The same is true for the product of dehalogenation of 1,6-diiodobicyclo[4.1.0]heptane.

One of the most interesting of the computational results was the finding that 2 had an negative eigenvalue, leading to an imaginary frequency. This shows that at the level of theory used, 2 is a transition state rather than a stable species. The course of the process going downhill from this structure was followed computationally by mode-walking along the vibrational coordinate corresponding to the imaginary frequency until the negative eigenvalue disappeared, and then the geometry was varied in a steepest descent mode. The force constants were calculated analytically at each step. The course of the reaction is shown in Figure 1.

The reaction starts by lengthening one of the outer cyclopropane C-C bonds, which is followed by an upwards motion and rotation of the methylene group. The final structure achieved is that of methylenecyclobutylidene, which would be expected to undergo a hydrogen migration giving methylenecyclobutene as the final product.

This reaction course appears to be followed by both 3 and 4 as long as bimolecular reactions can be avoided. However, when their concentrations are higher, as in the solution dehalogenation, the bimolecular ene reaction becomes predominant.

A study of the wave functions derived from the theoretical calculations provide further information concerning these cycloalkenes. Here, we make use of Bader's theory of atoms in molecules.[15] One begins by locating the bond paths which are the paths of maximum charge density between a pair of bonded atoms. Along this path there is a point known as the bond critical point at which the charge density is a minimum along the path, but a maximum in

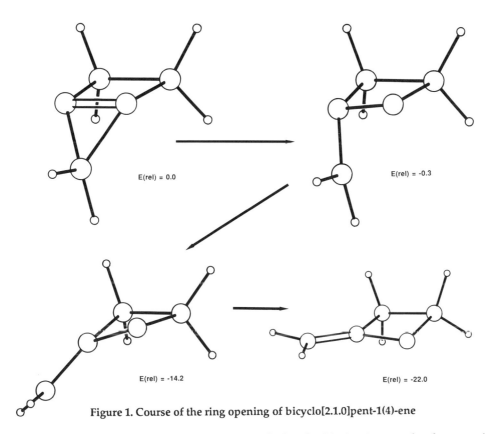

Figure 1. Course of the ring opening of bicyclo[2.1.0]pent-1(4)-ene

directions perpendicular to the path. Starting at the bond critical point, one develops a series of rays corresponding to paths of maximum decrease in charge density. The set of rays serves to define a surface which separates a given pair of atoms. The set of these surfaces divides the molecule into atomic subspaces, each of which obeys the usual quantum mechanical rules such as the virial theorem.

The electron population for each of these atomic subspaces has been obtained by numerical integration of the charge density. The populations have been converted to atomic charges by subtracting the nuclear charge, and are shown in Figure 2.

The olefinic carbons in cyclopropene are more negative than the methylene carbon because of their increased s-character. As a result, they have a negative charge. The relatively large negative charge at the methylene protons, as compared to the hydrogens in cyclopropane, is associated with the reversed sign of the dipole moment of cyclopropene as compared to other alkenes.

On going from cyclopropene to the increasingly more strained cycloalkenes: bent **2**, planar **2**, bent **1** and planar **1**, the charge at the bridgehead carbon becomes progressively more

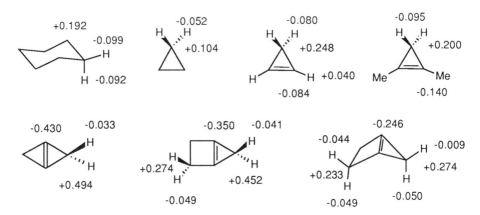

Figure 2. Atomic charges for cyclopropene derivatives

negative, indicating increased electronegativity. This suggests a simple model for the hybridization and for the preference for non-planar structures. Consider the following model for planar **1**:

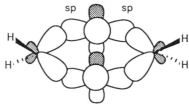

The bridgehead carbon can at best use a *p*-orbital to form the central C-C bond, and one to form the π-bond. This leaves a pair of sp orbitals to form the bonds to the methylene group. The high s-character leads to increased electronegativity and a large negative charge. When the structure is bent, a much improved hybridization scheme is possible. It may be visualized by starting with bicyclo[1.1.0]butane and removing the two bridgehead hydrogens, The π-bond would be formed from the two carbon orbitals which had been bonded to the hydrogens. This would imply a relatively weak π-bond, and is in accord with the very low calculated double bond vibrational frequency (Table I).

Further information may be gained by examining the angles between bond paths at a nucleus. With most cyclopropane derivatives, the bond path angles are about 18° larger than the conventional bond angles, and correspond to the bent bonds in these molecules. The angles found for some of the compounds in this study are shown in Figure 3. With cyclopropene, the bond path angles are again greater than the conventional angles (given in parentheses). However, with bicyclo[1.1.0]but-1(3)-ene, a different pattern is found. The bond path angle at the methylene group is now only 37.7° as compared to the conventional angle of 54.0°. Both the bent and planar structures show these unusual angles. Bicyclo[2.1.0]pent-1(4)-ene has more normal angles, with only one small bond path angle, and **3** and **4** have quite normal angles. The peculiar angles for bicyclobutene are readily seen in projection density plots as well as total density contour plots. It is clear that the bonding in bicyclobutene is different than that for most cyclopropenes.

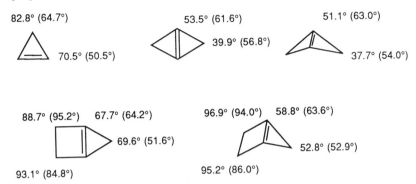

Figure 3. Bond path angles (conventional angles in parentheses)

The kinetic energies of the electrons associated with a given atom may also be obtained by numerical integrations of the appropriate function. The virial theorem indicates that the total energy of an atom is given by the negative of its kinetic energy. We have found that these energies are transferable among groups in identical environments.[16] An analysis of the changes in energy from those of "standard" group (such as a central $CH_2$ group of a long hydrocarbon chain) allows us to identify the source of strain in a molecule. For example, in the case of cyclopropane, the atom energies derived via numerical integration were -37.7118 for the carbon and -0.6554 for the hydrogen. In the case of the standard methylene group, the energies were -37.6898 for the carbon and -0.6741 for the hydrogen. The carbon of cyclopropane is seen to be 13.8 kcal/mol more stable than the standard methylene, and the hydrogen is 11.7 kcal/mol less stable.

The decrease in energy of the cyclopropane carbon is a result of its increased electronegativity, which leads to a larger electron population and an increased kinetic energy. Conversely, the cyclopropane hydrogen has a decreased population, and an increased energy. The energy of each methylene group changes by 9.7 kcal/mol, and the strain energy should be three times this value, or 29 kcal/mol. This compares well with the experimental value of 28 kcal/mol. This analysis indicates the the origin of the strain in cyclopropane is found in the increase in energy of the hydrogens.

We may now examine the bridged cyclopropenes in the same fashion. Using -39.0379 as the energy of a standard methylene group (6-31G**), -39.6191 as the energy of a standard methyl group, and -37.8545 as the energy of a standard olefinic carbon, the changes in energy with respect to these energies have been obtained and are shown in Figure 4. With 1,2-dimethylcyclopropene, the olefinic carbon has become more stable by 39 kcal/mol, whereas the methylene and methyl groups have become less stable by 62 and 34 kcal/mol respectively.

The increased strain at the bridgehead carbons of bicyclo[3.1.0]hex-1(5)-ene makes the carbon more electronegative, gives it a greater electron population, and a reduced energy. The cyclopropene methylene group is affected by both bridgehead carbons, and suffers the greatest increase in energy. The other methylene groups attached to the bridgehead carbons have an energy increase about half as large. The remote methylene is relatively little affected. The sum of the energy changes corresponds to the calculated strain energy.

Bicyclo[2.1.0]but-1(4)ene has significantly different energy changes in the bent and planar forms. Its planar form, and that of bicyclo[1.1.0]but-1(3)ene have very large energy decreases at the bridgehead carbons, and corresponding large increases in energy at the methylene carbons. The details of these energy changes will require further study.

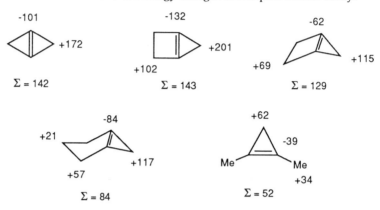

Figure 4. Changes in atom or group energies from the standard values.

**Conclusions:** The nature of the bonds in bicyclo[1.1.0]but-1(3)ene (**1**) is different from that of other cyclopropene derivatives. Bicyclo[2.1.0]pent-1(4)-ene (**2**) is calculated to be a transition state, and it undergoes ring opening to methylenecyclobutylidene. The compounds with larger bridges, **3** and **4**, are predicted and found to be stable species with low activation energies for ring opening to give the corresponding methylenecycloalkylidene. They also undergo the ene reaction with a very low activation energy.

**References:**

1. G. E. Renzoni, J. K.Yin and W. T. Borden, *J. Am. Chem. Soc.* **108**, 7121 (1986)

2. P. E. Eaton and M. Maggini, *J. Am. Chem. Soc.* **110**, 0000 (1988).

3. K. B. Wiberg, M. G. Matturro, P. J. Okarma and M. E. Jason, *J. Am. Chem. Soc.* **106**, 2194 (1984).

4. H. U. Wagner, G. Szeimies, J. Chandrasekhar, P. v. R. Schleyer, J. A. Pople and J. S. Binkley, *J. Am. Chem. Soc.* **100**, 1210 (1978).

5. K. B. Wiberg, G. Bonneville and R. Dempsey, *Isr. J. Chem.* **23**, 85 (1983).

6. W. F. Maier and P. v. R. Schleyer, *J. Am. Chem. Soc.* **103**, 1891 (1981).

7. K. B. Wiberg, M. G. Matturro, P. J. Okarma, M. E. Jason, W. P. Dailey, G. J. Burgmaier, W. F. Bailey and P. Warner, *Tetrahedron*, **42**, 1895 (1986).

8. G. Closs and W. Böll, *Angew. Chem. Int. Ed.* **2**, 399 (1963). G. Closs and W. Böll, *J. Am Chem. Soc.* **85**, 3094 (1963). G. Closs, W. Böll, H. Heyn and V. Dev, *J. Am. Chem. Soc.* **90**, 173 (1968).

9. G. Szeimies, J. Harnisch and O. Baumgärtl, *J. Am. Chem. Soc.* **99**, 5183 (1977). U. Szeimies-Seebach, A. Schöffer, R. Römer and G. Szeimies, *Chem. Ber.* **114**, 1767 (1981). J. Harnisch, O. Baumgärtel, G. Szeimies, M. Van Meerssche, G. Germain and J.-P. Declercq, *J. Am. Chem. Soc.* **101**, 3370 (1979).

10. K. B. Wiberg and G. Bonneville, *Tetrahedron Lett.* 5385 (1982).

11. F. H. Walker, K. B. Wiberg and J. Michl, *J. Am. Chem. Soc.* **104**, 2056 (1982). K. B. Wiberg, F. H. Walker, W. E. Pratt and J. Michl, *J. Am. Chem. Soc.* **105**, 3638 (1983).

12. W. E. Billups and L.-J. Lin, *Tetrahedron* **42**, 1575 (1986).

13. B. A. Hess, Jr., L. J. Schaad, P. Carsky and R. Zahradnik, *Chem. Rev.* **86**, 709 (1986).

14. K. B. Wiberg, V. A. Walters and W. P. Dailey, *J. Am. Chem. Soc.* **107**, 4860 (1985).

15. R. F. W. Bader, *Accts. Chem. Res.* **9**, 18 (1985).

16. K. B. Wiberg, R. F. W. Bader and C. D. H. Lau, *J. Am. Chem. Soc.* **1987**, *109*, 985, 1001.

# FROM BICYCLO[1.1.0]BUTANES TO [n.1.1]PROPELLANES

G. SZEIMIES
Institut für Organische Chemie der Universität München
Karlstraße 23, D-8000 München 2
Germany

ABSTRACT. In this article, three different synthetic methods are presented dealing with the conversion of a bicyclo[1.1.0]butane into the corresponding [n.1.1]propellane. The first method starts from 1-halobicyclo[1.1.0]butanes which, on treatment with strong bases, eliminate hydrogen halide to form short-lived bicyclo[1.1.0]but-1(3)-ene derivatives. These elusive intermediates can be trapped by cyclic 1,3-dienes in a Diels-Alder reaction to afford [4.1.1]- or [3.1.1]propellanes. The second method is a "classical" cyclization with a rather broad application leading to [n.1.1]propellanes with n = 1 to 4. In the third method, the bicyclo[1.1.0]butane central bond is cleaved after appropriate substitution of the bridgehead positions to give rise to 1-halo- or 1,1-dihalo-3-methylenecyclobutanes which are converted on a carbenoid route to [1.1.1]propellanes. Some aspects of the diverse chemistry of the propellanes are also discussed.

## 1. INTRODUCTION

Within the group of small-ring propellanes,[1] the [n.1.1]propellanes have doubtlessly attracted most attention.[2a,b] The syntheses of these compounds followed two lines: **1**,[3] **2**[4] and **3**[5] were obtained by reduction of the dihalides of type **4**. **5** was synthesized by intramolecular cycloaddition of type **6** carbene.[6] Majerski has used this method to prepare some [3.1.1]propellanes derived from adamantane and, respectively, from nortricyclane.[7] In all these reactions the propellane central CC-bond is formed in the last step of the synthesis.

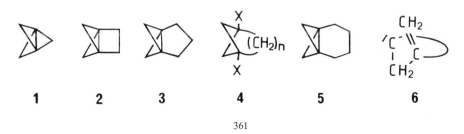

1     2     3     4     5     6

Our interest in the synthesis of [n.1.1]propellanes started ten years ago when we found that treatment of 1-halobicyclo[1.1.0]butanes with strong bases generated bicyclo[1.1.0]but-1(3)-ene derivatives, which could be trapped as Diels-Alder adducts containing [4.1.1]- or [3.1.1]propellane structural subunits. An overview of this field of our research is given in section 2. Cyclization of 1-lithio-3-chloro-alkylbicyclo[1.1.0]butanes has proved to be an efficient route to [n.1.1]propellanes. In section 3 an account of our experience with this reaction is presented. In section 4 we report on the carbenoid route to [1.1.1]propellanes.

## 2. [4.1.1]- AND [3.1.1]PROPELLANES BY DIELS-ALDER REACTION OF BICYCLO-[1.1.0]BUT-1(3)-ENE DERIVATIVES

### 2.1. 1-HALOBICYCLO[1.1.0]BUTANES

1-Halobicyclo[1.1.0]butanes (Hal = Cl, Br, I) were prepared from the corresponding bicyclo[1.1.0]butanes by metalation of the bridgehead position of the hydrocarbon with n-butyllithium (BuLi) in ether, followed by chlorination or bromination of the 1-lithiobicyclo[1.1.0]butane with tosyl chloride or bromide. Elemental iodine was used for the synthesis of the corresponding 1-iodobicyclo[1.1.0]butanes. Typical results are summarized in Table 1.

Table 1. Yields of 1-Halobicyclo[1.1.0]butanes from the Hydrocarbons via the Bridgehead Lithiated Intermediates

| Hydrocarbon | (X = H) | Halide X | % Yield | Reference |
|---|---|---|---|---|
|  [structure] | (7a) | Cl (7b) | 54 | 8) |
|  |  | Br (7c) | 47 | 8) |
|  |  | I (7d) | 38 | 8) |
| 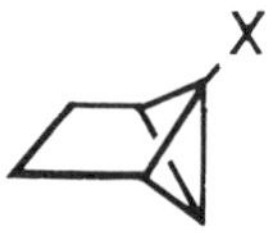 [structure] | (8a) | Cl (8b) | 48 | 9) |
| 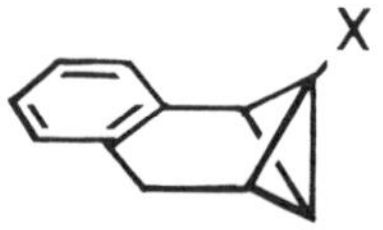 [structure] | (9a) | Cl (9b) | 51 | 9) |
| 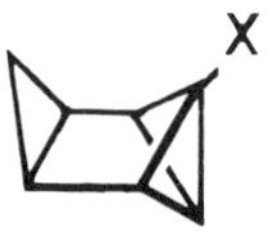 [structure] | (10a) | Cl (10b) | 64 | 9) |
|  |  | 3:1 mixture of *anti/syn* isomers |  |  |

Skattebøl and Baird have reported an alternative method for the synthesis of 1-bromobicyclo[1.1.0]butanes, which is depicted in Scheme 1.[10] Starting materials are allyl halides, which are converted into the corresponding 1,1-dibromo-2-halomethylcyclopropanes by dibromocarbene addition. Treatment of the adduct with methyllithium at low temperature generates a carbenoid, which cyclizes to give the corresponding 1-bromobicyclo[1.1.0]butane.

Scheme 1:

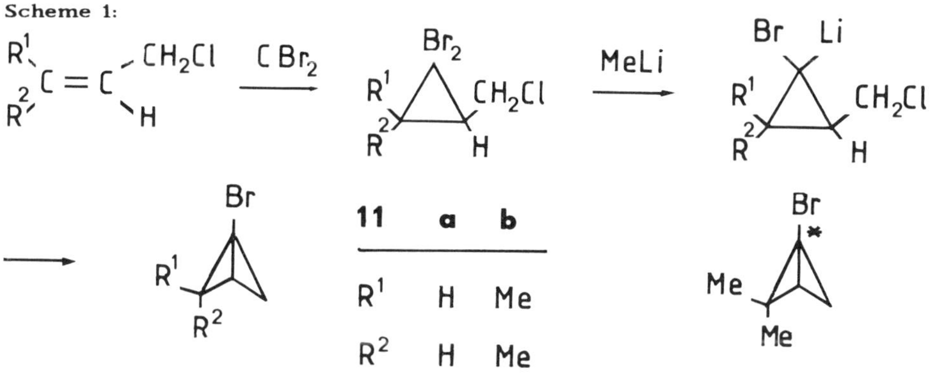

|     | 11 | a | b |
|-----|----|---|---|
| $R^1$ |   | H | Me |
| $R^2$ |   | H | Me |

Besides the preparation of **11a** and **b**,[10] this method allowed the synthesis of bridgehead $^{12}$C-labelled **11b*** in 34% yield[11]. Likewise, $^{12}$C-labelled **7c*** was obtained from **12** in 42 % yield by reaction with methyllithium in ether at -78 °C.[12] 1-Bromotricyclo[5.1.0.0$^{2,8}$]octane (**13**), the higher homologue of **7c**, was formed in 58% yield from **14** by the same procedure.[12] It is worth noting that this reaction is the most convenient entry to the tricyclo[5.1.0.0$^{2,8}$]octane system.[13]

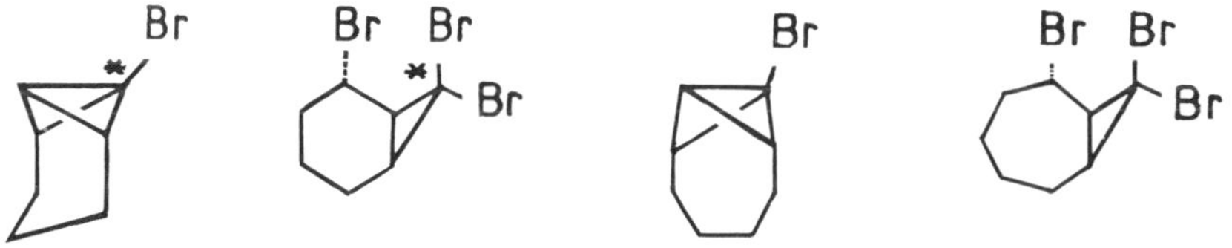

### 2.2. PROPELLANE FORMATION

When 1-halobicyclo[1.1.0]butanes carrying a bridge between C-2 and C-4 were treated with a strong, non-nucleophilic base (lithium diisopropylamide, LDA, or lithium 2,2,6,6-tetramethylpiperidide, LTMP) in the presence of cyclic 1,3-dienes, like anthracene and substituted anthracenes, furan, 2-methylfuran, 2,5-dimethylfuran, diphenylisobenzofuran, trimethylisoindole, or spiro[2.4]hepta-4,6-diene in tetrahydrofuran at about -40 °C, Diels-Alder adducts of the corresponding bicyclo[1.1.0]but-1(3)-ene derivatives were formed in variable yields. The 1-chlorotricyclo[4.1.0.0$^{2,7}$]heptane **7b** was used most often as starting material. Propellane formation is believed to proceed as shown in Scheme 2. The results are summarized in Table 2.

Scheme 2:

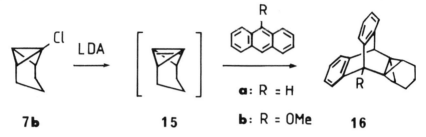

Table 2. [4.1.1]- and [3.1.1]Propellanes by Diels-Alder Reaction of Bicyclo[1.1.0]-but-1(3)-ene Derivatives

| Halobicyclo-[1.1.0]butane | 1,3-Diene | % Yield (isolated) of Propellane | Reference |
|---|---|---|---|
| 7b | anthracene | 31 | 14) |
| 7b | 9-methoxyanthracene | 34 | 14) |
| 7b | 9-methylanthracene | 44 | 14) |
| 7b | 9-chloroanthracene | 14 | 14) |
| 7b | 9-bromoanthracene | 5 | 14) |
| 7b | anthracene-9-carbonitrile | 0.7 | 14) |
| 7b | 9,10-dimethoxyanthracene | 43 | 15) |
| 7b | 9,10-dimethylanthracene | 26 | 16) |
| 7b | furan | 47 | 14) |
| 7b | 2-methylfuran | 32 | 15) |
| 7b | 2,5-dimethylfuran | 53 | 14) |
| 7b | diphenylisobenzofuran | 42 | 14) |
| 7b | spiro[2.4]hept-4,6-diene | 38 | 14) |
| 7b | N-methylisoindole | 25 | 17) |
| 7b | 1,2-dimethylisoindole | 61 | 17) |
| 7b | 1,2,3-trimethylisoindole | 53 | 17) |
| 8b | anthracene | 36 | 9) |
| 8b | 9-methoxyanthracene | 44 | 9) |
| 8b | 2,5-dimethylfuran | 18 | 9) |
| 8b | diphenylisobenzofuran | 20 | 9) |
| 8b | 1,2,3-trimethylisoindole | 52 | 18) |
| 9b | diphenylisobenzofuran | 39 [a] | 9) |
| 9b | 1,2,3-trimethylisoindole | 35 [a] | 9) |
| 13 | 2,5-dimethylfuran | 11 | 12) |

a) Yield by $^1$H NMR spectroscopy; mixture of stereoisomers.

Structure proof of the propellanes rested heavily on NMR and mass spectroscopic data. For **16a**,[19] **17**,[20] **18**,[21] and **19**[18] the X-ray structure has been determined. A common feature is the "inverted" geometry of the propellane central carbon atoms.[2] The distance between these atoms ranges from 1.54 Å in **18** to 1.57 Å in **19** and is only slightly longer than the CC bond in ethane.

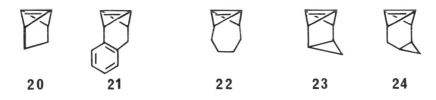

**17**   **18**   **19**

### 2.3. ON THE INTERMEDIACY OF BICYCLO[1.1.0]BUT-1(3)-ENES

If Diels-Alder adduct formation is accepted as evidence for the intermediacy of the corresponding bicyclo[1.1.0]but-1(3)-ene derivative, then our results indicate the existence of the pyramidalized bridgehead olefins **15**, **20**, **21**, and **22** as reactive, short-lived species. In addition, trapping experiments with nucleophiles furnished evidence for the formation of **23**[9] and **24**.[22]

**20**   **21**   **22**   **23**   **24**

Propellane formation was used as a basis to further establish the intermediacy of tricyclo[4.1.0.0$^{2,7}$]hept-1(7)-ene (**15**). In a competition experiment, the chloride **7b**, the bromide **7c** and the iodide **7d** were treated with LDA or with potassium *tert*-butoxide in THF at -20 °C in the presence of a mixture of diphenylisobenzofuran and 9-methoxyanthracene. By carefully determining the yields of propellane **17** and **16b** the competition constant K for propellane formation could be obtained. The results are given in Table 3.[14]

Table 3. Competition Constant K for the Reaction of **15**, generated from **7b**, **7c** and **7d** with Diphenylisobenzofuran and 9-Methoxyanthracene at -20 °C

| 7 | Base | K(17/16b) |
|---|------|-----------|
| b | LDA | 7.7 ± 0.2 |
| c | LDA | 7.5 ± 0.4 |
| d | LDA | 7.8 ± 0.4 |
| b | KO-*t*-Bu | 8.0 ± 0.2 |

According to Table 3 the competition constant K of the formation of the propellanes **17** and **16b** is independent of the starting halide **7** and of the base. This result is consistent with tricyclo[4.1.0.0$^{2,7}$]hept-1(7)-ene (**15**) as an intermediate.

This was also deduced from a second experiment: When **7b** was treated with lithium thiophenolate in THF no reaction occurred. However, when a strong but non-nucleophilic base was added (LDA, LTMP), tricyclo[4.1.0.0$^{2,7}$]hept-1-ylphenyl thioether (**25**) was isolated in yields of 70 - 85 %.[12,23] When the same reaction was carried out with **7c*** instead of **7b**, the thioethers **25a*** and **25b*** were obtained in a 56:44 ratio. The $^{12}$C label distribution was determined by $^{13}$C NMR spectroscopy, using the gated decoupling technique.[12,24]

|  Br  |  Br  |     |     | SPh | SPh | SPh |
| :--: | :--: | :-: | :-: | :-: | :-: | :-: |
| **7c\*-** | **7c\*-** | **15\*-** | **15\*-** | **25** | **25\*a** | **25\*b** |
| syn | anti | syn | anti |  |  |  |

Although the scrambling of the label in **7c*** on its way to the products is excessive, and thus in accordance with the proposed mechanism, **25a*** and **25b*** are not formed in the expected ratio of 1:1. The reason for this phenomenon is at present not quite clear. It should, however, be realized that the three-methylene bridge in **7c*** can take two boat-like conformations (**7c*-syn** and **7c*-anti**) which might be populated unevenly. This could lead to a **15*-syn** and **15*-anti** mixture deviating from 1:1. If the rate of equilibration of the two conformers of **15*** is not faster than the reaction of **15** with thiophenolate, one could not expect a 1:1 ratio of the labelled thioethers.

## 2.4. AN ALTERNATIVE ROUTE TO TRICYCLO[4.1.0.0$^{2,7}$]HEPT-1(7)-ENE AND ITS ISOMERIZATION TO 1,2,3-CYCLOHEPTATRIENE

Besides **7b - d**, the bromosilane **26** served as precursor for the generation of **15** by reaction with alkali fluorides in dimethyl sulfoxide. Using potassium fluoride, the temperature for disappearance of **26** within reasonable reaction times (24 hours) was at least 55 °C. Formation of **15** could be demonstrated by trapping experiments with cyclic 1,3-dienes of Table 2. However, in the presence of anthracene, not only **16a** was obtained but also the isomeric adduct **27a**.[25] It could be shown conclusively that **27a** was not formed by isomerization of **16a**. The **16a/27a** ratio decreased at higher reaction temperatures and rised at higher concentrations of 1,3-diene. These observations and the results of further mechanistic investigations are consistent with a thermal isomerization of **15** to 1,2,3-cycloheptatriene (**28**) as depicted in Scheme 3.[25] The double bond at C-2-C-3 is pyramidalized and highly reactive, prohibiting the isolation of **28** and making it a short-

Scheme 3.

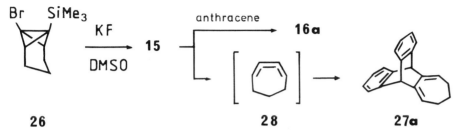

lived intermediate. Under selected reaction conditions (80 °C, one equivalent of 1,3-diene), the sequence **26** → **15** → **28** could be used for the preparation of

Table 4. Diels-Alder Reactions of 1,2,3-Cycloheptatriene (**28**) at 80 °C

| 1,3-Diene | Adduct | % Yield |
|---|---|---|
| diphenylisobenzofuran | | 69 |
| 2,5-dimethylfuran | | 75 |
| tetraphenylcyclopentadienone | | 30 |
| 1,3-cyclohexadiene | | 64 |
| 2,3-dimethyl-1,3-butadiene | | 78 |

Diels-Alder adducts of **28** which were free from or accompanied only by traces of the corresponding propellanes. Typical results are given in Table 4.

Besides 1,3-dienes, 1,3-dipoles like organic azides, ethyl diazoacetate, or nitrones were also successfully used as trapping reagents for **28**.[25]

In addition to the conversion of **15** → **28**, a related isomerization has been observed for **21** → **29**[9], and for **24** → **30**[22]. However, all efforts to effect the rearrangement of **20** → **31** were in vain.[9]

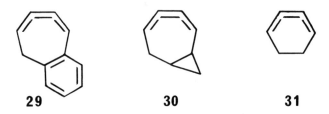

          **29**                    **30**                    **31**

### 2.5. "UNBRIDGED" 1-BROMOBICYCLO[1.1.0]BUTANES AND LDA

If orbital symmetry restrictions are responsible for the slow isomerization of **15**, **21** and **24** to **28**, **29** or **30**, bicyclo[1.1.0]but-1(3)-enes, **32a** and **b**, generated from **11a** or **b**, could rearrange much faster to the corresponding 1,2,3-butatrienes **33a** and **b**. Thus, although thorough theoretical investigations have confirmed that **32a** is a local minimum on the $C_4H_4$ energy hypersurface,[26] isomerization of **32a** to **33a** could be faster than trapping reactions. Results of all our trapping efforts starting from **11a** and **b** and LDA in the presence of lithium thiophenolate or several cyclic 1,3-dienes at -40 °C did not lead to the isolation of the expected bicyclo[1.1.0]butene adducts. Instead, the alkynes **34a** and **b** were formed as sole products.[11] It was shown conclusively that the alkynes were generated by base isomerization of the 1,2,3-butatrienes **33a** and **b**.[11]

$$R_2C=C=C=CH_2 \qquad R_2C=CH-C\equiv CH$$

**a**: R = H
**b**: R = Me

    **32**                    **33**                  **34**

The obvious assumption that butatriene formation of **11a** and **b** proceeded via **32a** and **b** turned out to be wrong. When $^{12}$C-labelled **11b**\* was treated with LDA, $^{13}$C NMR analysis of alkyne **34b**\* revealed that the CH signal of **34b** at δ 105.4 was completely absent.[11] All the label of **34b**\* is retained at C-3. This result eliminates **32b** as an intermediate, which would lead to a 1:1 mixture of [2-$^{12}$C]**34b** and [3-$^{12}$C]**34b**. The outcome of this reaction asks for a concerted (but probably not synchronous) elimination of lithium bromide and the selective cleavage of the bonds C-1-C-4 and C-2-C-3 of **35**, as depicted in Scheme 4.

Scheme 4.

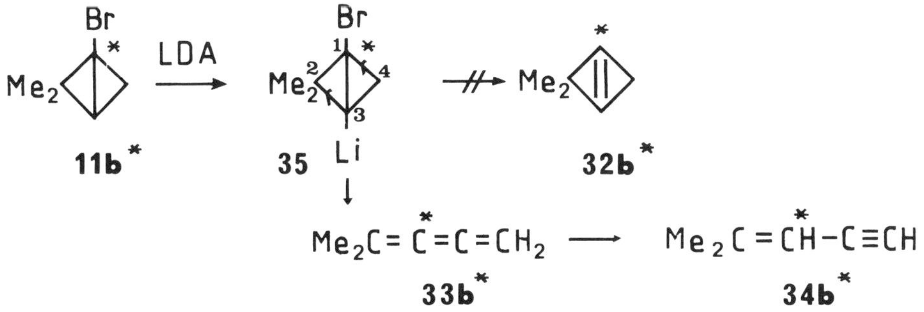

$$11b^* \xrightarrow{LDA} 35 \xrightarrow{\,\,\,/\!/\,\,\,} 32b^*$$

$$\downarrow$$

$$Me_2C=\overset{*}{C}=C=CH_2 \longrightarrow Me_2C=\overset{*}{C}H-C\equiv CH$$
$$33b^* \qquad\qquad\qquad 34b^*$$

Summing up this section we conclude that 1-halobicyclo[1.1.0]butanes on treatment with strong bases are converted into 1,2,3-butatrienes circumventing the high energy bicyclo[1.1.0]but-1(3)-ene intermediates. The parent bicyclo[1.1.0]but-1(3)-ene (**32a**) is still a challenge for the experimentalist. If the 1-halobicyclobutane carries a two-, three-, or four-atom bridge connected at C-2 and C-4, strong bases will generate the corresponding bicyclo[1.1.0]but-1(3)-ene derivative as a reactive, short-lived species.

## 3. "CLASSICAL" CYCLIZATION REACTIONS OF BICYCLO[1.1.0]BUTANES TO [n.1.1]PROPELLANES

### 3.1. PRELUDE

The facile preparation of **35a** and its easy conversion into **35b**[27] led to the synthesis of **36a-i**.[27, 28] In a formal sense, these compounds contain the [4.1.1]propellane framework as a structural subunit.

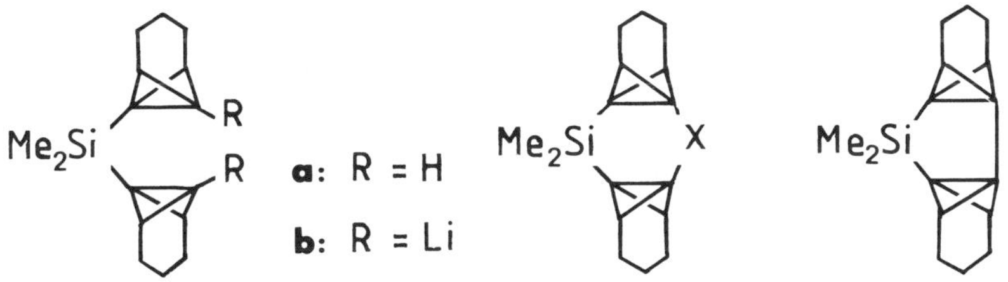

**a**: R = H
**b**: R = Li

35          36          37

| 36 | a | b | c | d | e | f | g | i |
|---|---|---|---|---|---|---|---|---|
| X | SiMe$_2$ | GeMe$_2$ | SnMe$_2$ | S | SO$_2$ | CO | CHOH | TiCp$_2$ |

Attempts at contracting the six-membered ring of **36e** and, respectively, of **36f** by thermolysis or photolysis were not successful.[28] However, on irradiation of **36i** with a 150 W mercury high-pressure lamp the [3.1.1]propellane **37** was generated.[28]

## 3.2. [4.1.1]- AND [3.1.1]PROPELLANES

Connecting the C-1 and C-7 bridgehead positions of tricyclo[4.1.0.0$^{2,7}$]heptane by a four- or a three-methylene bridge was achieved via the dihalides **42a** and **b**. The synthesis of these compounds proceeded as indicated in Scheme 5.[29,30]

Scheme 5.

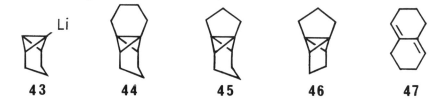

The crucial step was the preparation of the alcohols **40a** and **b**, which were obtained in yields of 70% and 58% by a cuprate-catalyzed cross-coupling reaction of two equivalents of tricyclo[4.1.0.0$^{2,7}$]hept-1-ylmagnesium bromide (**38**) with the corresponding bromoalcohols **39**. It is worth noting that the reaction of tricyclo[4.1.0.0$^{2,7}$]hept-1-yllithium (**43**) and oxetane led also to the formation of **40b**, but only in 14% yield. The bromination of **40a** and **b** at the bridgehead position followed standard procedures as described in section 2.1. The bromoalcohols **41a** and **b** were converted into the dihalides **42a** and **b** by means of triphenylphosphine and CCl$_4$.[29]

Ring closure of **42a** and **b** to the propellanes **44** and **45** was achieved by reaction with butyllithium in ether at -40 °C in yields of 96% and 51%. Both compounds were characterized by their NMR and mass spectra. Starting from tricyclo-[3.1.0.0$^{2,6}$]hexane (**8a**), the propellane **46** was also synthesized by the same procedure.

Whereas **44** could be stored in the refrigerator, **45** isomerized within a few hours at 0 °C to the diene **47**, probably catalyzed by electrophilic reagents.[30]

## 3.3. THE ELUSIVE [2.1.1]PROPELLANES

Wiberg et al. have shown convincingly that [2.1.1]propellane **2** exists in a nitrogen matrix at about 30 K.[4] However, the lifetime of **2** at room temperature under laboratory conditions seems too short for its isolation.[4] Our own results are in accordance with these observations.
2-(Tricyclo[4.1.0.0$^{2,7}$]hept-1-yl)ethanol (**48a**) was prepared in 77% yield from **43** and ethylene oxide in ether.[30,31] The bromination of the 7-position of **48a** affording the bromoalcohol **48b** was achieved in 46% yield following the procedure of Scheme 5, and the yield of **48c** was 76%.[29,30]

| 48 | a | b | c | d | e |
|---|---|---|---|---|---|
| R | H | Br | Br | H | Li |
| X | OH | OH | Cl | Cl | Cl |

When **48c** was treated with n-butyllithium in ether/pentane (1:3) at -40 °C, aqueous workup gave **48d**. Obviously, the bromine lithium exchange had occurred leading to **48e**. However, at -40 °C the cyclization of **48e** was rather slow. When **48e** was generated at 20 °C, the "ether adduct" **49a** was isolated in 57% yield along with **50** (2%), **51a** (2%) and **51b** (1%). In THF as solvent, the mixture of **48c** and butyllithium produced a 50% yield of **52**.[29,30] The formation of the solvent adducts **49a** and **52** is consistent with the intermediacy of the [2.1.1]propellane **53a**, if a radical chain addition reaction with the radicals **53c** and **53d** as transient species is assumed.

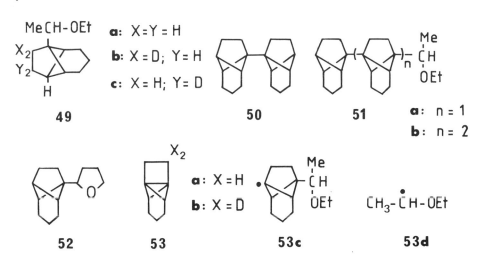

Further evidence for **53b** as an intermediate was obtained, when **48f** was reacted with BuLi in ether.[29,30] NMR analysis of the ether adduct showed that a 1:1 mixture of **49b** and **49c** had been produced. This result implies that an intermediate with equivalent bridgehead positions is involved in the formation of **49**. Neglecting secondary isotope effects, the [2.1.1]propellane **53b** fulfils this requirement.

### 3.4. [1.1.1]PROPELLANES

*3.4.1. [1.1.1]Propellanes with Additional Bridges.* "Classical" ring closure reactions as described in sections 3.1. - 3.3. were successfully applied for the synthesis of the [1.1.1]propellanes **54a-d** and **55-57**, which could be isolated in pure form and which were fully characterized.[31-33] The results are summarized in Table 5.

Table 5. Isolated Yields of [1.1.1]Propellanes **54a - d** and **55 - 57**

| [1.1.1]Propellane | | % Yield | Reference |
|---|---|---|---|
|  **54** | **a**: R = H | 71 | 31, 32) |
| | **b**: R = Me | 66 | 33) |
| | **c**: R = i-Pr | 60 | 33) |
| | **d**: R = Cyclohexyl | 61 | 33) |
| **55** | | 51 | 32) |
| **56** | | 10 a) | 32) |
| **57** | | 67 | 32) |

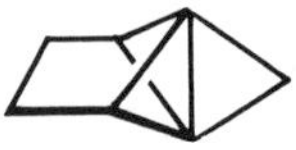

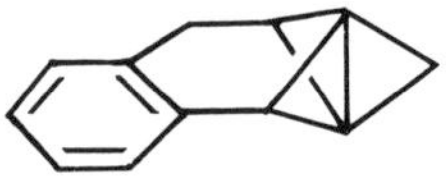

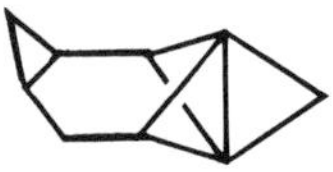

a) Isolated yield after distillation, which proceeded under heavy loss of material, and after crystallisation.

In all examples of Table 5, the corresponding bicyclo[1.1.0]butanes served as starting materials. The sequence leading to the preparation of the [1.1.1]propellane **54** is depicted in Scheme 6.

Scheme 6.

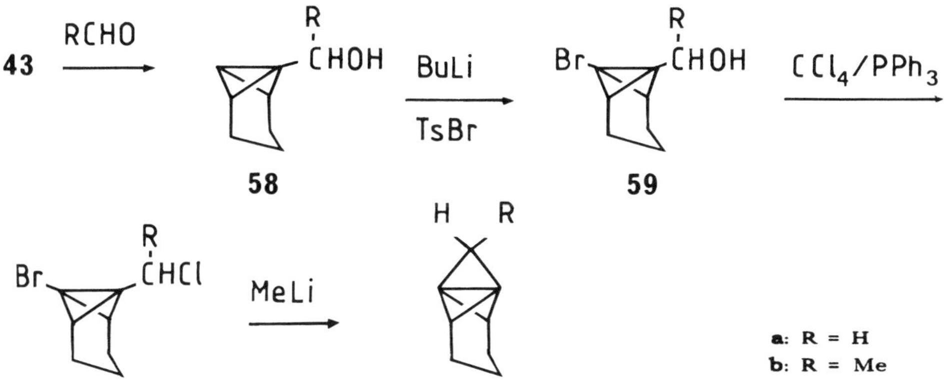

a: R = H
b: R = Me
c: R = i-Pr
d: R = Cylohexyl

A crucial step in this Scheme was the conversion of the bromocarbinols **59** to the dihalides **60**. Triphenylphosphine/$CCl_4$ was the reagent of choice for the synthesis of **60a** (73%)[32] and **b** (70%).[33] This method, however, was insufficient for the preparation of **60c** and **d**. These dihalides were obtained from **59c** and **d** by reaction with the immonium chloride **61**[34] in yields between 50-60%.[33] The secondary chlorides **60b-d** were considerably less stable than **60a**, showing an enhanced tendency towards a bicyclo[1.1.0]butyl/3-methylenecyclobutyl rearrangement, affording the 6,6-dihalo-7-alkylidenenorpinanes **62**. In section 4 it will be shown that these compounds are valuable precursors for [1.1.1]propellanes.

The propellanes **55** and **56** could be prepared from the corresponding doubly lithiated bicyclo[1.1.0]butanes (**63** and **64**) in a very simple way. Reaction with chloroiodomethane afforded a 50% yield of **55** and, respectively, a 10% yield of **56**.[32] This elegant method is unfortunately confined to bicyclo[1.1.0]butanes, which allow bridgehead dilithiation under mild conditions with two equivalents of BuLi.[35]

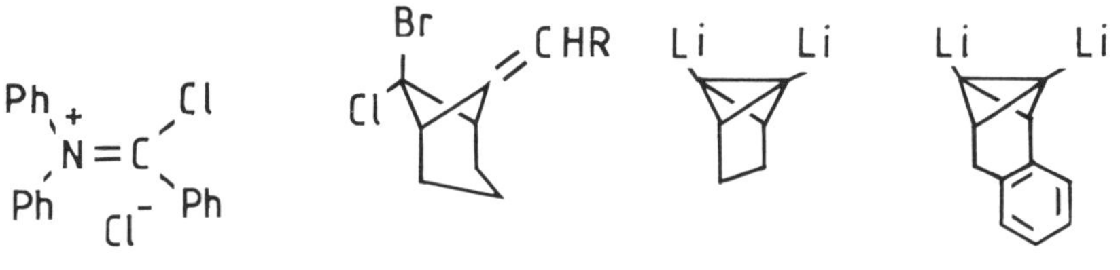

61    62    63    64

Propellanes **54-57** were stored without visible decomposition at dry-ice temperature under nitrogen. At room temperature, even in the absence of oxygen, polymerization was fast. At -78 °C, **54a** and **55** are crystalline solids. X-ray structures of both models have been determined by Dunitz and Seiler.[36]

3.4.2. *The Parent Hydrocarbon.* Besides **55** and **56**, the parent hydrocarbon **1** could be prepared rather efficiently. Commercially available dichloride **65** was converted into the dibromocarbene adduct **66** in 45% yield.[31,37] Reaction of **66** with two equivalents of methyllithium gave rise to a 70% yield of **1** in ether.[31,32] If one equivalent of methyllithium was used, 1-bromo-3-chloromethylbicyclo[1.1.0]butane (**67**) was isolated as the main product.[38] **68** and **69** are probable intermediates on the way from **66** to **1**.

$H_2C = C(CH_2Cl)_2$

**65**

Br Br
△ $CH_2Cl$
**66** $CH_2Cl$

Br
△
**67** $CH_2Cl$

Br Li
△ $CH_2Cl$
**68** $CH_2Cl$

Li
△
**69** $CH_2Cl$

Removal of ether from **1** by fractional distillation proved difficult and was accompanied by extensive polymerization. Solvent-free [1.1.1]propellane **1** was obtained in 70% yield from **66** by reduction with a dispersion of lithium in triglyme at about 70 °C.[32,39,40] This procedure is another convenient route to **1**.

## 4. [1.1.1]PROPELLANES BY CARBENOID CYCLIZATION

### 4.1. INTRODUCTION

Our first observation, which led to the idea that [1.1.1]propellane formation could be achieved by carbenoid cyclization, was obtained when the diiodide **70** was treated with methyllithium in ether at -30 °C, which afforded propellane **54a** in 43% yield.[41] As **70** was the product of iodine addition to **54a**, the usefulness of this route to [1.1.1]propellanes had still to be established.

**70**

Br
□ $=C{\overset{H}{\underset{H}{}}}$
Br
**71**

$Me_2$
$N_2=$ □ $=CH_2$
$Me_2$
**72**

At the beginning of this project it seemed difficult to synthesize precursors of type **71** efficiently. **71** was prepared from **67** by reaction with magnesium bromide in the presence of lithium bromide in ether. When treated with methyllithium, **71** afforded a 76% yield of **1**.[38] Decomposition of diazoalkanes of type **72**, however, does not seem to lead to an efficient synthesis of [1.1.1]propellanes.[42]

## 4.2. RESULTS

**4.2.1. Synthesis of Precursors.** Precursors for [1.1.1]propellane formation were obtained via the bicyclo[1.1.0]butylcarbinyl cation rearrangement starting either from bicyclo[1.1.0]butylcarbinols of type **73a**, from bromobicyclo[1.1.0]butylcarbinols of type **73b** or from bromo-vinylbicyclo[1.1.0]butanes of type **75**. Alcohols of type **73a** and **b** were converted into halides of type **74a** or dihalides of type **74b** by reaction with thionyl chloride in pyridine, with triphenylphosphine and carbon tetrachloride or with triphenylphosphine and 1,2-dibromo-tetrachloroethane.[43] Addition of iodine to **75** afforded **76a**, the allylic iodide of which could be exchanged against phenolate (**76b**) or against hydrogen (**76c**) by reduction with $LiAlH_4$.[44]

**a:** X = H
**b:** X = Br

**73**  **74**

**a:** X = I
**b:** X = OPh
**c:** X = H

**75**  **76**

**4.2.1. Propellane Formation.** The preparation of [1.1.1]propellanes from dihalides of type **74b** or **76** was achieved by treatment with methyllithium in ether at -78 °C. In most cases the yields were high and side products were not observed. To our surprise, monohalides **74a** could also be converted into [1.1.1]propellanes by reaction with lithium diisopropylamide (LDA) in ether at 20 °C.[45] Typical reaction times were 48 to 72 h and the yields were only moderate. However, as the bromination step **73a** → **73b** is omitted, this sequence is still an efficient route to [1.1.1]propellanes. The results are summarized in Table 6.

We assume that propellane formation in both reactions proceeds via carbenoids, which are generated either by halogen-lithium exchange or by metalation. At present it is unclear if the cyclization reaction is due to the carbenoids, or if free carbenes are involved.

Table 6. [1.1.1]Propellanes by Carbenoid Cyclization

| Propellane | | | | % Yield | Ref. |
|---|---|---|---|---|---|

A. From Type **74b** Dihalides and Methyllithium

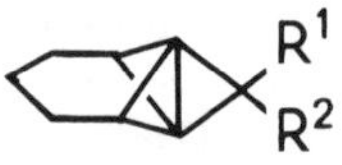

| | $R^1$ | $R^2$ | | |
|---|---|---|---|---|
| | **e:** Ph | H | 35 | 33) |
| | **f:** Me | Me | 57 | 44) |
| | **g:** Me | $CH_2OPh$ | 95 | 44) |
| **54** | | | | |

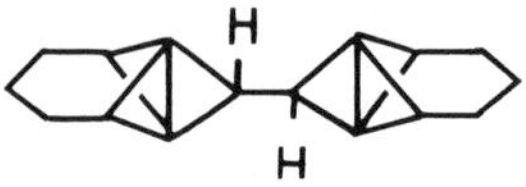

| | | | | |
|---|---|---|---|---|
| **77** | | | 94 | 44) |

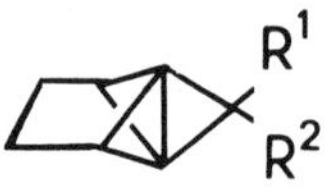

| | $R^1$ | $R^2$ | | |
|---|---|---|---|---|
| | **b:** H | 4-MeO-$C_6H_4$ | 81 | 33) |
| | **c:** H | α-Naphthyl | 70 | 33) |
| | **d:** Ph | Ph | 87 | 33) |
| | **e:** Ph | 2-Br-$C_6H_4$ | 76 | 33) |
| **55** | | | | |

B. From Type **74a** Monohalides and LDA

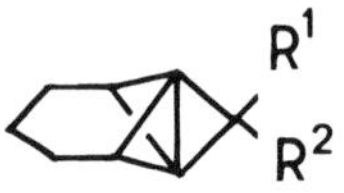

| | $R^1$ | $R^2$ | | |
|---|---|---|---|---|
| | **a:** H | H | 67 | 45) |
| | **c:** H | i-Pr | 58 | 45) |
| | **h:** H | 2-Propenyl | 55 | 45) |
| | **i:** | $(CH_2)_4$ | 43 | 45) |
| | **j:** | $(CH_2)_5$ | 50 | 45) |
| **54** | | | | |

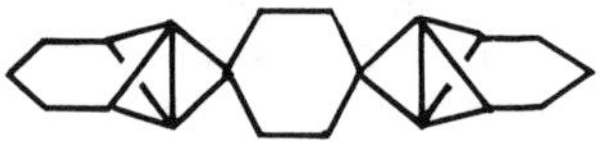

| | | | | |
|---|---|---|---|---|
| **78** | | | 18 | 45) |

The X-ray structures of **77**,[44] **55d**[46] and **78**[47] have been determined. The propellane central CC bonds were found as follows (in Å): 1.577(1) for **77**, 1.586(4) for **55d**, and 1.58(1) for **78**.

## 5. SELECTED PROPERTIES OF [1.1.1]PROPELLANES

[n.1.1]Propellanes are reactive species which show a diverse chemistry. It is at this time not intended to present a review on the emerging material. Rather, a few topics of the chemistry of [1.1.1]propellanes will be discussed, including thermal behavior and addition of Grignard reagents and disulfides.

### 5.1. THERMAL BEHAVIOR OF SOME [1.1.1]PROPELLANES

[1.1.1]Propellanes are sensitive towards traces of electrophilic reagents, which will isomerize the propellane framework to give olefinic compounds, or will initiate polymerization. Reproducible results were obtained when the propellanes were distilled in vacuo through a thoroughly cleaned glass or quartz tube heated by an electrical furnace. In most cases, the condensed material contained variable amounts of starting propellane and the procedure had to be repeated to gain full conversion. Occasionally, the product was further isomerized by further heating. Selected examples are depicted in Scheme 7:

Scheme 7.

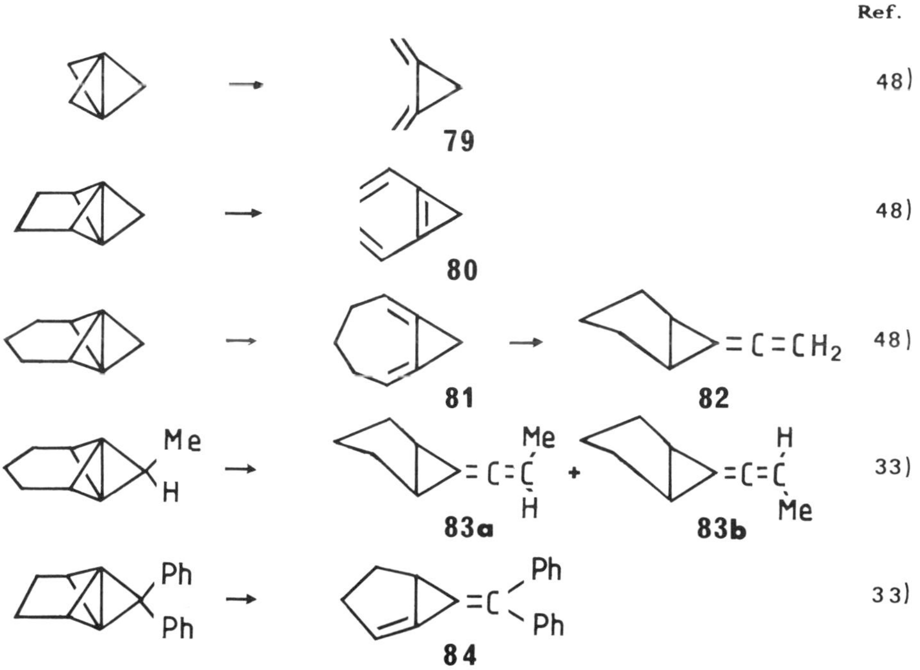

It is interesting to notice that in all cases the propellane central bond is retained in the products. As all starting propellanes of Scheme 7 can be prepared without difficulty, thermal isomerization of these compounds is an efficient route to products **79 - 84**.

## 5.2. ADDITION OF GRIGNARD REAGENTS

The [1.1.1]propellane skeleton is capable of reacting with electrophiles,[3a,49] radicals[41,50] and with organolithium compounds, which initiate polymerization.[51] We have recently found that Grignard reagents can be added to **54a** in boiling ether in a slow reaction to produce substituted bicyclo[1.1.1]pentyl derivatives of type **85**, which, after hydrolysis, afford the hydrocarbons **86** in acceptable yields. Workup with electrophiles like methyl chloroformate, benzaldehyde, allyl bromide, and tert-butyl perbenzoate led to the corresponding products **87**.[52] A remarkable feature of these reactions is that the intermediate Grignard reagent **85** adds much more slowly to the central CC bond of **54a**, keeping dimer and oligomer formation low.

| 86 | a | b | c | d | e | f |
|---|---|---|---|---|---|---|
| R | Et | i-Pr | t-Bu | Ph | Ph-CH$_2$ | Allyl |
| % Yield | 21 | 46 | 32 | 67 | 68 | 67 |

| 87 | a | b | c | d |
|---|---|---|---|---|
| R$^1$ | i-Pr | Et | Allyl | PhCh$_2$ |
| R$^2$ | CO$_2$Me | CHOHPh | Allyl | O-t-Bu |
| % Yield | 50 | 40 | 68 | 37 |

Similar results were obtained with **1**.[52]

## 5.3. ADDITION OF ORGANIC DISULFIDES

Organic disulfides could be added to the central bond of bicyclo[1.1.0]butanes via a radical chain process.[53] [1.1.1]Propellanes behave similarly giving adducts of type **88**. However, reaction of radical **89** with **1** leading to **90** and **91** competes with formation of **88**. Furthermore, **90** can react with **1** and produce coupled trimers of type **92**(n=1), and higher oligomers **92**(n>1) are also observed. Results of reactions of **1** with dimethyl disulfide and diphenyl disulfide are given in Table 7.[54] It is worth noting that the coupled trimer **92a**(n=1) and the coupled tetramer **92a**(n=2) could be characterized spectroscopically. The hydrocarbon **91c** was obtained from **91b** in 21% yield by reduction with lithium in ethylamine. Oxidation of **91a** by hydrogen peroxide in acetone afforded a 76% yield of sulfone **91d**.[54]

**88 – 92:**

| | a | b | c | d |
|---|---|---|---|---|
| R: | SMe | SPh | H | SO$_2$Me |

Table 7. Disulfide Addition to **1**; Product Yields (%)[54]

| R | Ratio **1** : Disulfide | **88** | **91** | **92**(n=1) | **92**(n=2) |
|---|---|---|---|---|---|
| SMe | 1 : 1 | 38 | 6.5 | -- | -- |
| SMe | 2 : 1 | 40 | 12 | 0.4 | 0.1 |
| SMe | 3 : 1 | 49 | 21 | 7 | 4 |
| SPh | 3 : 1 | 63 | 27 | -- | -- |

The X-ray structure of **91a** and **d** showed a short bond (1.480(3) and 1.469(6) Å) between the bicyclo[1.1.1]pentyl units.[54]

## ACKNOWLEDGEMENT

*It is a pleasure to acknowledge the invaluable help of the many collaborators cited in the references, whose effort and dedication brought this work to fruition. I am also grateful to the* DEUTSCHE FORSCHUNGSGEMEINSCHAFT *and to the* FONDS DER CHEMISCHEN INDUSTRIE *for financial support of this research.*

## REFERENCES

1) For definition see Ginsburg, D., *Propellanes: Structure and Reactions*, Verlag Chemie, Weinheim 1975.
2) For recent reviews containing the chemistry of [n.1.1]propellanes, see 2a) Wiberg, K. B. *Acc. Chem. Res.* **1984**, *17*, 379. 2b) Ginsburg, D. *[m.n.1]Propellanes*, in *The Chemistry of the Cyclopropyl Group, Part II*, Rappoport, Z. Ed.; J. Wiley & Sons; Chichester, New York, Brisbane, Toronto, Singapore, 1987, p. 1193.
3) 3a) Wiberg, K. B.; Walker, F. H. *J. Am. Chem. Soc.* **1982**, *104*, 5239. 3b) Wiberg, K. B.; Dailey, W. P.; Walker, F. H.; Waddell, S. T.; Crocker, L. S.; Newton, M. *J. Am. Chem. Soc.* **1985**, *107*, 7247.
4) Wiberg, K.B.; Walker, F. H.; Pratt, W. E.; Michl, J. *J. Am. Chem. Soc.* **1983**, *105*, 3638.
5) Gassman, P. G.; Proehl; G. S. *J. Am. Chem. Soc.* **1980**, *102*, 6862.
6) Hamon, D. P. G.; Trenerry, V. C. *J. Am. Chem. Soc.* **1981**, *103*, 4962.
7) 7a) Mlinarić-Majerski, K.; Majerski, Z. *J. Am. Chem. Soc.* **1983**, *105*, 7389. 7b) Vinković, V.; Majerski, Z. *J. Am. Chem. Soc.* **1982**, *104*, 4027.
8) Szeimies, G.; Philipp, F.; Baumgärtel, O.; Harnisch, J. *Tetrahedron Lett.* **1977**, 2135.
9) Schlüter, A.-D.; Harnisch, H.; Harnisch, J.; Szeimies-Seebach, U.; Szeimies, G. *Chem. Ber.* **1985**, *118*, 2883.
10) Nilsen, N. O.; Skattebøl, L.; Baird, M. S.; Buxton, S. R.; Slowey, P. *Tetrahedron Lett.* **1984**, 2887.
11) Düker, A.; Szeimies, G. *Tetrahedron Lett.* **1985**, 3555.
12) A. Düker, Thesis, University of Munich, **1986**.
13) Christl, M.; Herzog, C.; Kemmer, P. *Chem. Ber.* **1986**, *119*, 3045.
14) Szeimies-Seebach, U.; Schöffer, A.; Römer, R.; Szeimies, G. *Chem. Ber.* **1981**, *114*, 1767.
15) Baumgart, K.-D. Thesis, University of Munich, **1983**.
16) Baumgart, K.-D.; Harnisch, H.; Szeimies-Seebach, U.; Szeimies, G. *Chem. Ber.* **1985**, *118*, 2883.
17) Zoch, H.-G.; Schlüter, A.-D.; Szeimies, G. *Tetrahedron Lett.* **1981**, 3835.
18) Chakrabarti, P.; Seiler, P.; Dunitz, J. D.; Schlüter, A.-D.; Szeimies, G. *J. Am. Chem. Soc.* **1981**, *103*, 7378.
19) Declercq, J.-P.; Germain, G.; Van Meerssche, M. *Acta Crystallogr. Sect. B*, **1978**, *34*, 3472.
20) Szeimies-Seebach, U.; Szeimies, G.; Van Meerssche, M.; Germain, G.; Declercq, J.-P., *Nouv. J. Chim.* **1979**, *3*, 357.
21) Szeimies-Seebach, U.; Harnisch, J.; Szeimies, G.; Van Meerssche, M.; Germain, G.; Declercq, J.-P. *Angew. Chem.* **1978**, *90*, 904; *Angew. Chem., Int. Ed. Engl.* **1978**, *17*, 848.
22) Römer, R. Thesis, University of Munich, **1983**.
23) Harnisch, J.; Legner, H.; Szeimies-Seebach, U.; Szeimies, G. *Tetrahedron Lett.* **1978**, 3683.
24) Freeman, R.; Hill, H. W. D.; Kaptein, R. *J. Magn. Resonance* **1972**, *7*, 327.
25) Zoch, H.-G.; Szeimies, G.; Römer, R.; Germain, G.; Declercq, J.-P. *Chem. Ber.* **1983**, *116*, 2285.
26) Hess, B. A., Jr.; Allen, W. D.; Michalska, D.; Schaad, L. J.; Schaefer, III, H. F. *J. Am. Chem. Soc.* **1987**, *109*, 1615. See also Hess, B. A., Jr.; Michalska, D.; Schaad, L. J. *J. Am. Chem. Soc.* **1987**, *109*, 7546.

27) Zoch, H.-G.; Szeimies, G.; Butkowskyj, T.; Van Meerssche, M.; Germain, G.; Declercq, J.-P. *Chem. Ber.* **1981**, *114*, 3896.
28) Butkowskyj-Walkiw, T.; Szeimies, G. *Tetrahedron* **1986**, *42*, 1845.
29) Morf, J.; Szeimies, G. *Tetrahedron Lett.* **1986**, 5363.
30) Morf, J. Thesis, University of Munich, **1988**.
31) Semmler, K.; Szeimies, G.; Belzner, J. *J. Am. Chem. Soc.* **1985**, *107*, 6410.
32) Belzner, J. Thesis, University of Munich, **1988**.
33) Gareiß, B. Thesis, University of Munich, **1988**.
34) Fujisawa, T.; Iida, S.; Sato, T. *Chem. Lett.* **1984**, 1173.
35) Schlüter, A.-D.; Huber, H.; Szeimies, G. *Angew. Chem.* **1985**, *97*, 406; *Angew. Chem., Int. Ed. Engl.* **1985**, *24*, 404.
36) Private communication of Prof. J. D. Dunitz and P. Seiler, ETH Zürich.
37) Semmler, K. Thesis, University of Munich, **1986**.
38) Fuchs, J.; Szeimies, G. unpublished.
39) Schlüter, A.-D.; Opitz, K. unpublished.
40) Bunz, U.; Szeimies, G. unpublished.
41) Belzner, J.; Szeimies, G. *Tetrahedron Lett.* **1987**, 3099.
42) Brinker, U. H.; Erdle, W. *Angew. Chem.* **1987**, *99*, 1290; *Angew. Chem., Int. Ed. Engl.* **1987**, *26*, 1260.
43) Bringmann, G.; Schneider, S. *Synthesis* **1983**, 139.
44) Kottirsch, G.; Polborn, K.; Szeimies, G. *J. Am. Chem. Soc.* **1988**, *110*, 0000.
45) Schmid, W.; Szeimies, G. unpublished.
46) Polborn, K.; Gareiß, B.; Szeimies, G., unpublished. See also ref. 33).
47) Polborn, K.; Schmid, W.; Szeimies, G., unpublished.
48) Belzner, J.; Szeimies, G. *Tetrahedron Lett.* **1986**, 5839.
49) Wiberg, K. B.; Waddell, S. T. *Tetrahedron Lett.* **1987**, 151.
50) Wiberg, K. B.; Waddell, S. T.; Laidig, K. *Tetrahedron Lett.* **1986**, 1553.
51) Schlüter, A.-D. *Angew. Chem.* **1988**, *100*, 283; *Angew. Chem., Int. Ed. Engl.* **1988**, *27*, 296.
52) Godt, A.; Szeimies, G. unpublished.
53) Dietz, P.; Szeimies, G. *Chem. Ber.* **1978**, *111*, 1938.
54) Bunz, U.; Polborn, K.; Wagner, H.-U., Szeimies, G. *Chem. Ber.* **1988**, *121*, 0000.

NEW ASPECTS OF HIGHLY STRAINED RING CHEMISTRY

Z. Yoshida
Department of Synthetic Chemistry, Kyoto University
Yoshida, Kyoto, 606
Japan

ABSTRACT. Electronic features of small ring systems are described. Control of stability and reactivity of highly strained systems has been achieved thru modification with substituents. Thus, the diaminochlorocyclopropenium ion, its Grignard reagent or lithium carbenoid are shown to be useful building blocks for the introduction of a diamino-cyclopropenylidene moiety. The trithiocyclopropenium ion is a versatile building block in the synthesis of heterocycles. A theoretically interesting 16$\pi$ aromatic hydrocarbon, "cyclic bicalicene" was synthesized on the basis of the dithiodichlorocyclopropene strategy. A new type of molecular energy storage system involving a photovalence isomerization to a highly strained system is described for a donor-acceptor substituted norbornadiene going to the corresponding quadricyclane ("DONAC"), as well as a bulkily substituted polyacene going to the corresponding valene isomer ("AROVA"). The concept and experimental data for DONAC as an excellent solar energy storage system are shown in detail. Both systems, in particular the former will be candidates for an information storage system.

1. INTRODUCTION

Among strained systems those small ring systems, whose strain energy (SE) is bigger than 10 kcal per ring carbon, may be expected to have interesting physical and chemical properties compared with cycloalkanes containing three- and four-membered rings where SE is 9 and 6.5 kcal per ring carbon, respectively. The remarkable electronic features of such small ring systems are described below. Each ring carbon of these systems has nonequivalent $\sigma$ orbitals (for example, one sp and two sp$^3$ orbitals for the cyclopropenium ring carbon) in contrast to the non-strained organic molecules in which the hybridization of $\sigma$ orbitals is equivalent (for instance, methane carbon has four equivalent sp$^3$ orbitals). The nonquivalent $\sigma$ orbitals between the ring carbons take part in forming the bent $\sigma$ bonds which have electron-donating character. Due to the peculiar electronic nature of small ring systems, electronic effects of substituents in such systems are not always normal. For example, the trimethylcyclopropenium ion is more stable than the triphenylcyclopropenium ion. This is in sharp contrast to a usual carbenium ion which is stabilized by phenyl groups much more than by methyl groups. This nonequivalence in hybridization of $\sigma$ orbitals of the ring carbons is considered to be responsible for the strain. Generally ring systems become unstable with increase of the molecular strain engergy. However, if we can control the stability and reactivity of the unstable strained ring systems by suitable ways (for example, introduction of appropriate substituents) it may be possible to find potential applications of such systems (1) to develop new useful reactions for organic synthesis, (2) to create novel electron systems, or (3) to create novel functions leading to high technology applications. The first two aspects will be discussed for the examples of the amino- and alkylthio-substituted cyclopropenium systems, while the photovalence isomerization involving highly strained rings will be used to highlight the third.

## 2. HETEROATOM-SUBSTITUTED CYCLOPROPENIUM SYSTEMS

The parent cyclopropenium ion is stabilized through its $2\pi$ electron aromatic character but destabilized through the strain of the $\sigma$ framework (its strain energy is estimated to be *ca.* 23 kcal/ring carbon). Breslow[1] reported that the salt of the parent cyclopropenium ion is so unstable that it decomposes within a week. Introduction of alkyl or aryl groups on the cyclopropenium ring ($C_3$ ring) increases their stability so that they become stable even at room temperature. However, we have been interested in heteroatom-substituted cyclopropenium systems as useful building blocks in view of the potential flexibility of reaction behaviors. At present, the triaminocyclopropenium salt and the diaminochlorocyclopropenium salt are very easy to prepare on a large scale according to our reaction[2] shown below.

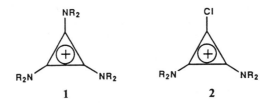

XH=aliphatic or aromatic secondary amine

The reaction of tetrachlorocyclopropene with an excess of secondary amine in methylene chloride smoothly proceeds to exclusively afford the triaminocyclopropenium salt (**1**) in almost quantitative yield. This is the most direct and simplest method for synthesis of **1**. Instead of tetrachlorocyclopropene, pentachlorocyclopropane obtained by the reaction of trichloroethylene with dichlorocarbene can be used in the presence of base. If the reaction temperature is kept low (for example -70°C) the diaminochlorocyclopropenium salt (**2**) is produced.

Since this reaction is so rapid its kinetic data have not yet been determined. From a detailed study, however, it may be deduced that the reaction of tetrachlorocyclopropene with a secondary amine leading to triamino- and diaminochlorocyclopropenium ions follows the pathway depicted in Scheme 1.

Scheme 1

It should be emphasized that the reaction may proceed by a repetitive concerted process ($S_N2'$ type). If the reaction stops at 1-amino-2-chloro-3-amino-3-chlorocyclopropene (III) due to a decreased reactivity of its cyclopropene double-bond carbon toward the secondary amine, the diaminochlorocyclopropenium perchlorate can be obtained. Amino-cyclopropenium salts, in particular triaminocyclopropenium salts ($pK_R+ \sim 13$) are stable in contrast to the trichloro- and triphenylcyclopropenium salts. The force constants and bond distances of tris(dimethylamino)cyclopropenium ion obtained by normal coordinate analysis and X-ray diffraction indicate a strong delocalization of the nitrogen lone-pair electrons into the $C_3$ ring as well as an increase in the C-C $\sigma$ bond order in the ring by the amino groups.[3] Our INDO calculation suggests that this stabilization arises from an increase in the overlap population of the C-C bonds in the ring by the electron-donating groups as well as from the delocalization of positive charge into the electron-donating groups as exemplified below.

INDO overlap population

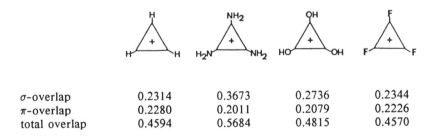

|  |  |  |  |  |
|---|---|---|---|---|
| $\sigma$-overlap | 0.2314 | 0.3673 | 0.2736 | 0.2344 |
| $\pi$-overlap | 0.2280 | 0.2011 | 0.2079 | 0.2226 |
| total overlap | 0.4594 | 0.5684 | 0.4815 | 0.4570 |

The INDO orbital energies of the lowest unoccupied molecular orbitals (LUMO) and the highest occupied molecular orbitals (HOMO) are given in TABLE I.

TABLE I. Orbital energies (eV) for trisubstituted cyclopropenium ions

|  | H | $CH_3$ | OH | $NH_2$ |
|---|---|---|---|---|
| LUMO ($\pi^*$) | -6.125 | -5.206 | -5.224 | -3.738 |
| HOMO ($\pi$) | -21.902 | -19.912 | -17.277 | -14.849 |

Electronic effects of the substituent on the ring carbon are indicated in TABLE II.

TABLE II. Electronic effects of substituents on the ring carbons

|  | $CH_3$ | OH | $NH_2$ |
|---|---|---|---|
| $\pi$-Conjugative | +0.005 | +0.115 | +0.190 |
| $\sigma$-Inductive | +0.152 | -0.012 | -0.050 |

It should be noted that for the $CH_3$ group the $\sigma$-inductive effect is larger than the $\pi$-conjugative effect, in contrast to that in the benzene system in which $\sigma_R > \sigma_I$. This novel substituent effect explains why the trialkylcyclopropenium ion is more stable than the triphenylcyclopropenium ion. On the other hand, the aminogroup has a large electron-donating $\pi$-conjugative effect as compared with its electron-withdrawing $\sigma$-inductive effect, which is in accord with its stability.

The general reaction behavior of heteroatom-substituted cyclopropenium ions toward nucleophiles (Nu) is shown in eqs (2) and (3).

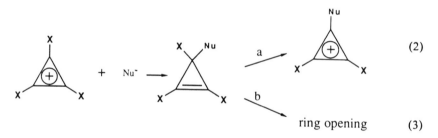

From our detailed investigation, reaction of triamino- and diaminochlorocyclopropenium ions toward various nucleophiles have been disclosed to proceed in the direction (2). This is due to the strong ring stabilizing effect of the amino groups. Along this route various diaminocyclopropenyl systems substituted with Nu can be synthesized. Some examples are given in Scheme 2.

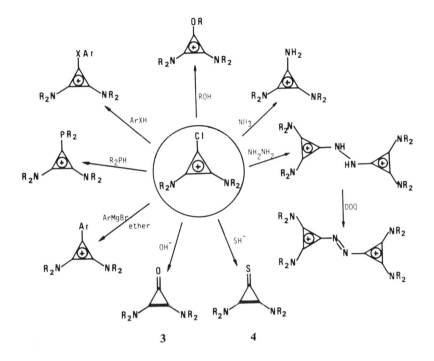

Scheme 2

It is noteworthy that diaminocyclopropenone (3) and triaminocyclopropenium ion (1) closely resemble the biological substances urea and protonated guanidine $((H_2N)_3C^+)$ respectively, in their chemical properties.[4] Very interestingly diaminocyclopropenethione (4) reacts with $HNO_3$ to provide the diaminocyclopropenium ion (5) from which the stable diaminocyclopropenylidene (6) can be prepared in the form of its lithium carbenoid.[5] The diaminochlorocyclopropenium ion (2) reacts with magnesium to produce 7 as a new type of Grignard reagent.

[Scheme 3 diagram]

Scheme 3

The diaminochlorocyclopropenium ion (2), diaminocyclopropenethione (4), and Li (6) and Mg (7) complexes of diaminocyclopropenylidene are useful reagents to introduce a diaminocyclopropenylidene moiety. For example, by using these building blocks triafulvene (8) and various triafulvalenes (9) (n = 2,3,4) can be synthesized.[6,7]

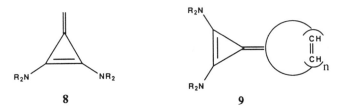

These are nice models for examining interactions of bent $\sigma$ electrons in the $C_3$ ring with $\pi$ electrons in the opposite ring as well as $\pi$ interactions between both rings. Cyclopropenylidenes are essentially undissociated when they form complexes with certain transition metals. Thus, the Li complex of diaminocyclopropenylidene 6 has a merit to give transition metal complexes of low symmetry[8] as shown in Scheme 4.

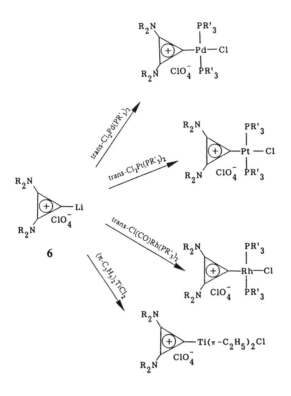

Scheme 4

Another type of reaction is the reduction (electron transfer reaction) of **1** which is reported to afford hexaaminobenzene.[9]

If the secondary amine in reaction (1) is replaced by mercaptan (RSH), the trithiocyclopropenium salt (**10**), which is stable at room temperature, is exclusively produced in good yields. **10** has $D_{3h}$ symmetry with respect to $(C_3S_3)^+$ and values within 1245-1260 cm$^{-1}$ for the asymmetric ring deformation mode (E') which correlate to the value of $K_{C-C}$ (force constant for the C-C bond of the $C^3$ core).[10] The E' band for trithiocyclopropenium ion is

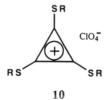

10

at lower wavenumbers than those of other trisubstituted cyclopropenium ions (e.g. 1553 for $Me_2N$-, 1446 for $CH_3$-, 1411 for Ph-, 1321 for Cl-), indicating that the C-C bond of **10** should be weak compared with that of **1**. As CNDO calculations[10] suggest, this bond weakening is due to the interaction between sulfur 3d orbitals and bent σ bonds of the $C_3$ ring. Reflecting this feature of bonding in the tri-thiocyclopropenium ion (**10**), its reaction with nucleophiles usually proceeds in the direction (3) to provide useful products. For example, the reaction of trithiocyclopropenium salt with a primary or secondary amine in the presence of t-BuOK in DMF produces 2,3-dithiopyrrole.[11]

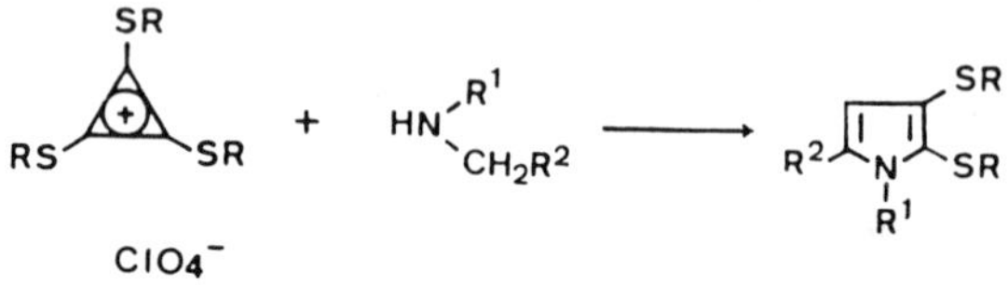

This one pot synthesis of pyrroles probably follows the pathway depicted in Scheme 5.

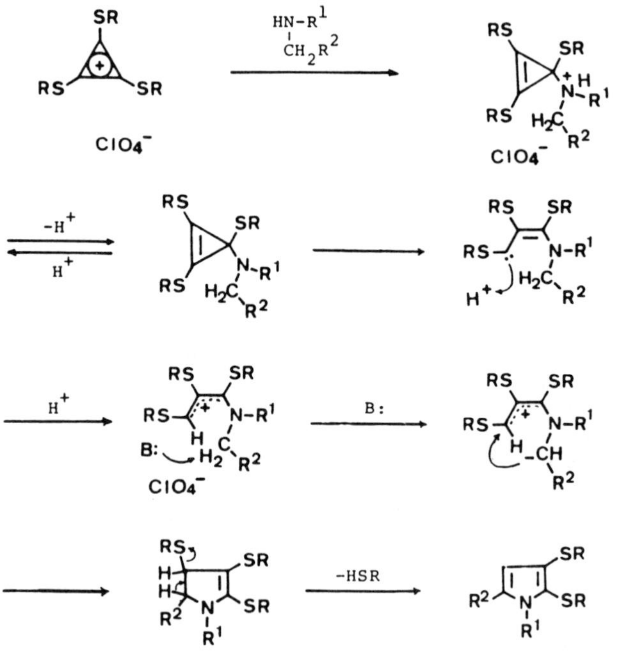

Scheme 5

In a detailed study[12] pertaining to the reaction of trithiocyclopropenium ion with various nucleophiles we succeeded in synthesizing about twenty heterocycles, the formation of which can be formally divided into three types (Figure 1).

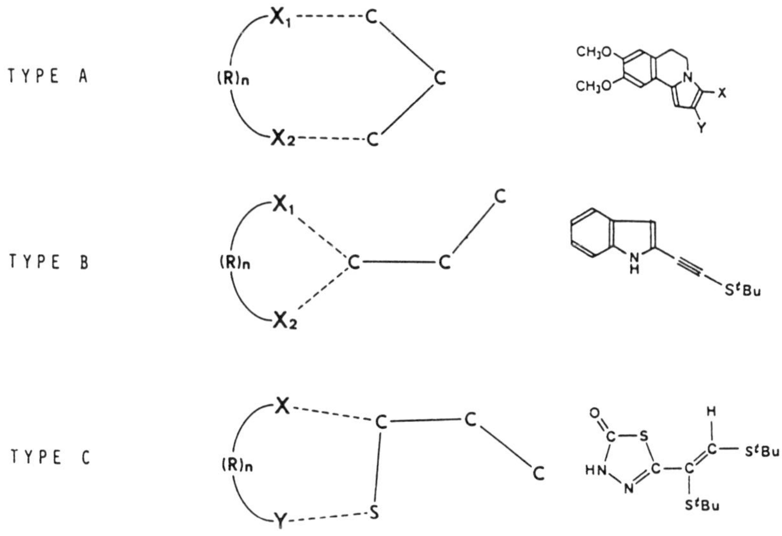

Fig. 1. Three types of heterocycle formation from trithiocyclopropenium ions **10**

In type A, all of the ring carbons of **10** end up in the heterocyclic ring. In type B, only one carbon of the $C_3$ ring is involved in the heterocyclic ring formation and the other two carbons of the $C_3$ ring are left as a vinyl functionality. Very recently we have found the formation of the 1,3,4-thiadiazolin-5-one derivative **11** upon acid treatment of the product from the reaction of **10** with semicarbazide.

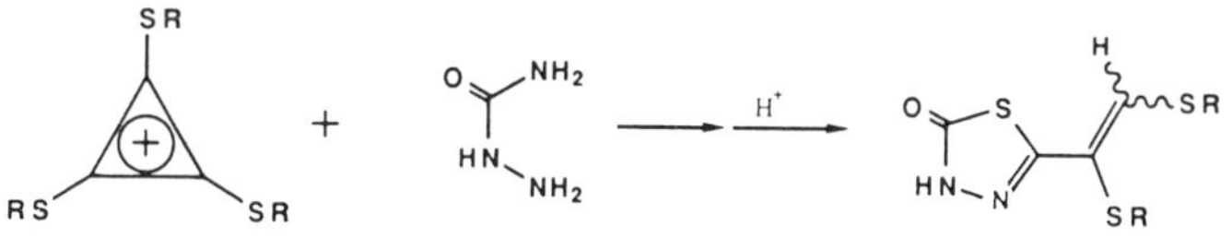

$$\text{(5)}$$

**11**

In this type C, the sulfur atom of SR and one carbon of the $C_3$ ring are incorporated into the heterocyclic ring. Formations of pyrroles (Nu: amines), furans (Nu: alcohols), thiophenes (Nu: thiols), pyrazoles (Nu: hydroxylamine), pyridines (Nu: β-amino esters or nitriles), pyrimidines (Nu: amidines) and benzodiazepines (Nu: o-phenylenediamines) belong to type A. Formations of s-triazines (Nu: biguanide), oxazoles (Nu: benzamidoxime), thiadiazoles (Nu: thiosemicarbazides), benzimidazoles (Nu: o-phenylenediamine), benzoxazoles (Nu: o-amiophenol), and benzothiazoles (Nu: o-aminothiophenol) belong to type B. The types A and B are summarized[13] in Figure 2.

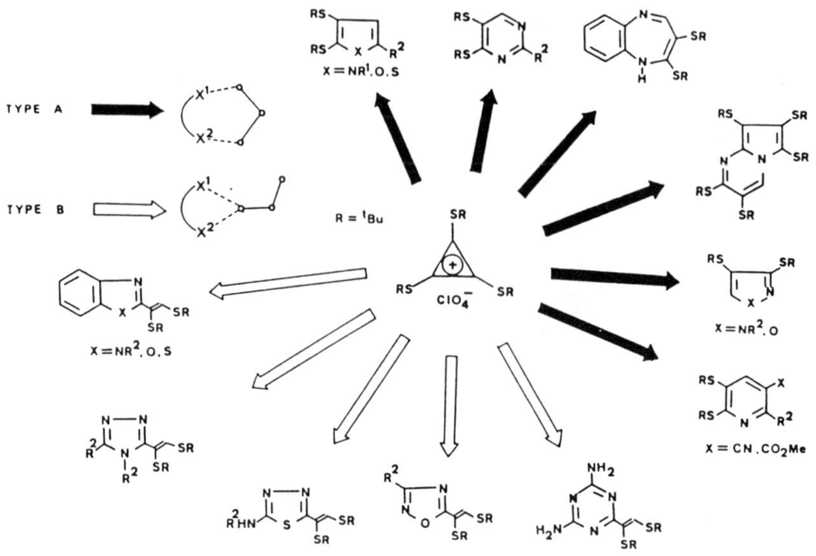

Fig. 2. One pot synthesis of various heterocycles from trithiocyclopropenium salts **10**

By applying the Type A strategy to tetrahydroisoquinolines, pyrroloisoquinoline derivatives (**12**) can be prepared in one step.

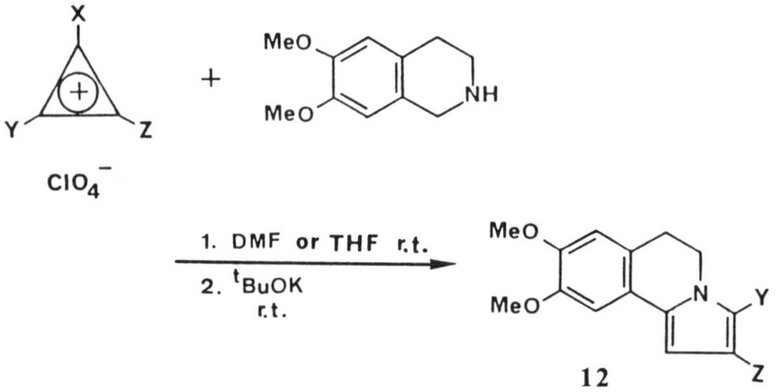

(6)

(X=SBu$^t$; Y=SBu$^t$, SMe, Ph; Z=SBu$^t$, SMe, Ph)

Indole derivatives (**13**) can also be prepared in a similar way as shown in reaction (7).

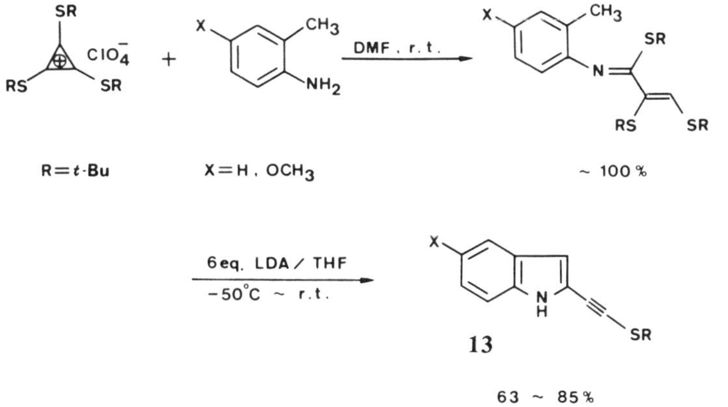

(7)

Upon treatment of dichlorodithiocyclopropene (14) with lithium cyclopentadienid, the theoretically very interesting "4nπ aromatic" hydrocarbon, "cyclic bicalicene" 16 is formed (Scheme 6).

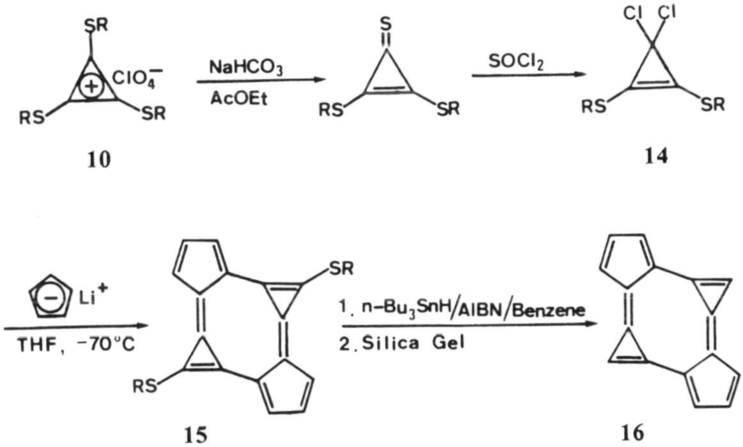

Scheme 6

Cyclic bicalicene 16 with a 16π electron perimeter is stable and has a completely planar structure.[14,15] The observed anisotropic magnetic susceptibility indicates that cyclic bicalicene 16 is aromatic. As derived from X-ray and NMR data, 16 must be regarded as a π system with an alternating tetrapolar structure 17.

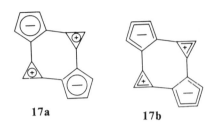

17a        17b

## 3. PHOTOCHEMICAL VALENCE ISOMERIZATIONS INVOLVING HIGHLY STRAINED SYSTEMS — THE CREATION OF NOVEL FUNCTIONS

So far organic chemistry, in particular synthetic organic chemistry has been directed toward the synthesis of natural products and theoretically interesting systems, where "structure synthesis" is the target. On the other hand "function synthesis" has recently been recognized as a very important new field in chemistry (chemical science). This chemistry is directed toward the creation of organic compounds with novel functions. Here we would like to introduce a new molecular energy storage system for solar energy and an information storage system which make use of photochemically interesting functions.

A molecular energy storage system is composed of a cycle of reactions including an endergonic process (8) leading to a highly strained molecule and the reverse exergonic process (9) initiated with a catalyst:

$$A \xrightarrow[(100\%)]{\text{solar irradiation}} B \quad (8)$$

$$B \xrightarrow{\text{catalyst}} A + \text{heat} \quad (9)$$

To establish a molecular energy storage system for sunlight the following seven conditions should be similtaneously satisfied.

- a) Reactant A must absorb sunlight in the UV and the visible region. The absorption band of A should be broad towards the visible region and should have a low absorption coefficient ($\epsilon \approx 10^2$).
- b) The reaction A→B should be photochromic, and the photoproduct (valence isomer B) must not absorb sunlight.
- c) The quantum yield for the reaction A→B should be approximately unity even in the presence of oxygen.
- d) The reaction A→B should have a large positive ground state enthalpy (high energy storage). The energy stored should be larger than 100 kcal kg$^{-1}$.
- e) The photoproduct B must be stable and must withstand long-term storage at ambient temperatures.
- f) The reaction B→A must be controlled with a heterogeneous catalyst.
- g) Reactions A→B and B→A must proceed quantitatively.

It seemed to be quite impossible to obtain compounds A and B which meet all seven conditions. Although various photovalence isomerization systems (A/B) are known, absorbtion only occurs upon direct UV (200-300 nm) irradiation (which does not satisfy condition (a)) or upon sensitized irradiation. It might be possible to absorb a limited part of the UV ($\lambda > 300$ nm) and the visible region of sunlight through sensitized irradiation by choosing a suitable sensitizer. However, the reaction system could become contaminated on repetition of processes (8) and (9) because of side reactions (e.g. the generation of singlet oxygen and/or the reaction of A with the sensitizer). One such example is the benzophenone photosensitized isomerization of norbornadiene (SE=29 kcal/mol) to quadricyclane (SE=95 kcal/mol). The introduction of chromophoric groups gives rise to a bathochromic shift, but it reduces the quantum yield and decreases the amount of energy stored because of the increase in molecular weight. In order to design a functional molecule A, which satisfies condition (a) without any sacrifice on the other conditions, it is necessary to introduce a new concept of color development, which is not accompanied by a large increase in molecular weight. We succeeded in solving this extremely difficult problem using a quite new concept (i.e. control of the distance and angle between the donor olefin and acceptor olefin parts of molecule A, which are connected with each other in a non-conjugated manner, and a change in the nature of the donor and acceptor groups). Then, we investigated the problem with respect to condition (b), i.e. that A absorbs the UV ($\lambda > 300$ nm) and visible regions of sunlight and B does not. Eventually, we found a solution to this problem by the skeletal reorganization of colored molecule A ($\pi$ system) to colorless molecule B ($\sigma$ system). As exemplified above, the quantum yields for the usual organic photochemical reactions are very low (about $10^{-3} - 10^{-2}$) even in the absence of molecular oxygen, which is responsible for a further decrease in the quantum yield. Therefore it should be extremely difficult to solve the problem for condition (c). We succeeded in solving this problem by designing a rigid structure (e.g. a cage structure) for A so that an interaction of $\pi$ electrons should be possible between both the donor olefin part and the acceptor olefin part of molecule A. Then the solution of the problem for condition (d) was examined. The energy stored ES (kcal mol$^{-1}$) in solution can be calculated according to equation (10).

$$ES = -(\Sigma^B \Delta h_i - \Sigma^A \Delta h_j) + (SE^B - SE^A) - (RE^B - \Sigma E^A) - (LE^B - LE^A) \quad (10)$$

In this equation, $\Sigma^A \Delta h_i$ and $\Sigma^B \Delta h_j$, are the sum of the energies of disappearing bonds and the sum of the energies of newly forming bonds upon photovalence isomerization respectively, $SE^B$ and $SE^A$ are the strain energies of B and A respectively, $RE^B$ and $RE^A$ are the resonance energies of B and A respectively, and $LE^B$ and $LE^A$ are the solvation energies for B and A, respectively. The first term of the right-hand side in eqn. (10) turns out always negative (exothermic) but constant for a given valence isomerization pair (A/B). Also the difference in solvation energies should be small in most cases. Therefore the amount of energy stored is largely governed by both the strain energy difference (the second term) and the resonance energy difference (the third term). In practice, the amount of energy stored in kilocalories per kilogram is more important than that quoted in kilocalories per mole. In the molecular design of the A/B pair we took these points into account. In relation to the problem for condition (e), it is necessary that compounds A and B are stable enough to sunlight, oxygen and heat (at least at about 100 - 150°C). In addition, it is essential that the A/B pair has an appropriate energy barrier in its potential energy surface, as is shown in Figure 3, to prevent thermal cycloreversion (B A) both during storage and during the energy-storing process (8).

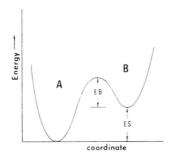

Fig. 3. Potential energy diagram for the A/B valence isomer pair. EB: energy barrier; ES: energy stored.

We also succeeded in solving the problems concerning conditions (e)-(g).
According to the principle of function synthesis we repeated (1) the better molecular design for solar energy storage systems, (2) its synthesis and (3) a check of the above seven requirements in order to obtain compounds A and B which satisfy all the conditions. Finally we made the discovery that the donor-acceptor (DA) norbornadienes (A) and the corresponding quadricyclanes (B) formed an extremely good molecular storage system for solar energy.[16-19] The simplest model which represents the present author's new concept for the molecular design of A is shown in Figure 4.

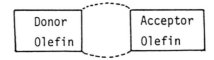

Fig. 4. Non-conjugatively connected donor olefin and acceptor olefin (----, plural bonds).

This sort of system is generally colorless if both olefin moieties are colorless. However, we thought that colored molecules could be made by controlling the distance and the angle between the olefins in the non-conjugated system. We proved this concept to be correct by examining the absorptions of DA-bicyclo[2.2.$n$]alkadienes (18) and DA-cyclohexa-1,4-diene (19).

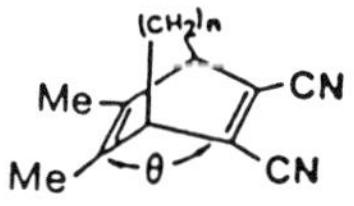

**18** 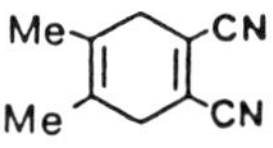 **19**

TABLE III. Additional absorption bands which appear in the non-conjugatively connected donor olefin and acceptor olefin

| System | | | New absorption band | | |
|---|---|---|---|---|---|
| | $n$ | Trend of increasing $\theta$ | $\lambda_{max}$ (nm) | Trend of increasing $\lambda$ | $\epsilon_{max}$ |
| 19 | Non-bridged | ↑ | 283 | | 496 |
| 18 | 3 | | 305 | ↓ | 178 |
| | 2 | | 306 | | 162 |
| | 1 | | 338 | | 200 |

Some examples are shown in TABLE III. Molecular orbital calculations[20,21] ascribe the absorption band to the charge transfer from the donor olefin (highest occupied molecular orbital (HOMO)) to the acceptor olefin (lowest unoccupied molecular orbital (LUMO)). The data in TABLE III show that the bathochromic shift of the new band increases with decreasing $n$ or decreasing dihedral angle $\theta$ and reaches a maximum in the DA-norbornadiene with $n = 1$.

The color of DA-norbornadienes can be shifted to longer wavelengths by increasing the electron-donating power of the donor olefin part and the electron-withdrawing power of the acceptor olefin part, as depicted in Figure 5. By mixing such DA-norbornadienes, the absorption spectrum of the mixture can be brought close to the solar spectrum.

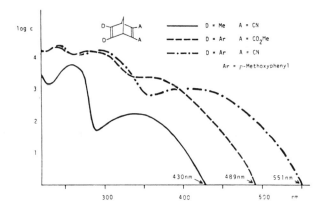

Fig. 5. The characteristic charge transfer spectra of DA-norbornadienes in MeCN.

We observed that solar irradiation of DA-norbornadienes **20** afforded quantitatively the corresponding DA-quadricyclanes **21**, which have no absorptions in the wavelength region above 300 nm (i.e. **21** is colorless).

$$20 \xrightarrow{h\nu, 100\%} 21 \quad (11)$$

The quantum yields ($\Phi$ for 313 nm) for this energy-storing process are given in TABLE IV.

TABLE IV. Quantum yields for the photoisomerization of norbornadienes to quadricyclanes

| Norbornadienes | Φ(degassed) | Φ(non-degassed) |
|---|---|---|
| (CO₂Me, CO₂Me) | 0.45 | |
| (CN, CN) | 0.59 | 0.52 |
| (Me-substituted, CO₂Me, CO₂Me) | 0.87 | 0.85 |
| (Me-substituted, CN, CN) | 0.96 | 1.01 |
| (Ph, Ph, CO₂Me, CO₂Me) | 1.06 | 0.97 |
| (Ph, Ph, CN, CN) | 0.95 | 0.97 |

As can be seen from TABLE IV the quantum yields are close to unity for DA-norbornadienes even in the presence of oxygen . These quantum yields remained the same when the wavelength of the light was changed (from 313 nm to 334 nm and 366 nm). In contrast, the quantum yield for the dicyano-norbornadiene was 0.59 in the absence of oxygen. It is noteworthy that photochemical valence isomerization for DA-norbornadienes by lower energy photons occurs with a higher quantum yield (close to unity). The solvent polarity does not substantially affect the quantum yield for DA-norbornadienes.

DA-quadricyclanes are stable in the solid state. Their stability in solution varies with the substituents as shown in TABLE V. Using an aryl group as a donor decreases the thermal stability of DA-quadricyclanes, which could be improved by introducing some substituents at the bridgehead positions.

The values of $t_{1/2}$ at 33.5°C indicate the usefulness of these DA-quadricyclanes as solar energy storage systems. These DA-quadricyclanes are not decomposed by sunlight, molecular oxygen or heat (between 100-150°C). The amount of energy stored is estimated from the heat of reversion for DA-quadricyclanes, as shown in TABLE VI. It is of interest that the amount ($-\Delta H$) of energy stored in these quadricyclanes is almost independent of the substituents. Thus we measured for the DA-quadricyclanes an energy storage value of about 120-130 kcal kg$^{-1}$.

TABLE V. Half-life times $t_{1/2}$ for the thermal reversion of donor-acceptor substituted quadricyclanes

| Quadricyclane | Solvent | $t_{1/2}$ at 33.5°C |
|---|---|---|
| 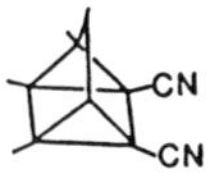 CN, CN | Xylene | 124 days |
| 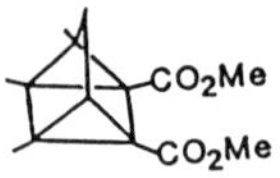 CO$_2$Me, CO$_2$Me | Xylene | 4.3 years |
| 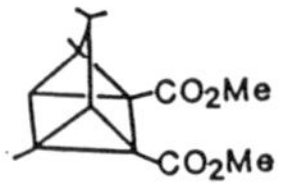 CO$_2$Me, CO$_2$Me | Xylene | 38 years |

TABLE VI. Heat of reversion for donor-acceptor quadricyclanes

| Donor | Acceptor | Solvent | Catalyst | Reaction temp.(°C) | $-\Delta H$ (kcal mol$^{-1}$) | $-\Delta H$ (kcal(kg) |
|---|---|---|---|---|---|---|
| Me | CO$_2$Me | Neat | None | DSC | 19[a] | 90 |
| Me | CN | MeCN | AgO$_3$SCF$_3$ | 30 | 21[b] | 120 |
| | | C$_6$H$_5$Cl | None | DSC | 21[a] | |
| H | H | C$_6$H$_6$ | Rh$_2$(NOR)$_2$Cl$_2$ | 30 | 22[b] | 250 |

DSC, differential scanning calorimetry.- [a]Determined by differential scanning calorimetry. [b]Determined on an LKB microcalorimetry batch system.

We also tested different catalysts for the cycloreversion process of DA-quadricyclanes to DA-norbornadienes (eqn. (12)).

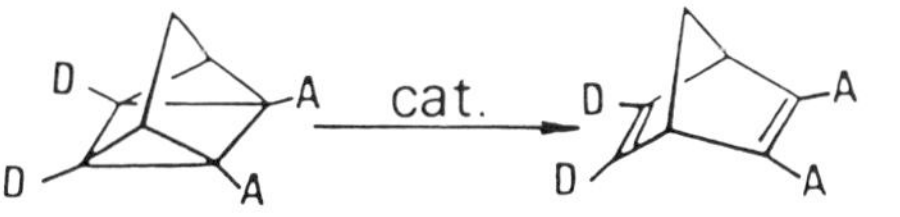

$$\underset{21}{\text{[structure]}} \xrightarrow{\text{cat.}} \underset{20}{\text{[structure]}} + \text{heat} \quad (12)$$

We observed that Lewis acids (e.g. $BF_3O(C_2H_5)_2$) and Brønsted acids (e.g. $CF_3SO_3H$, $H_2SO_4$, 2,4-dinitrobenzenesulfonic acid, m-nitrobenzenesulfonic acid and p-toluenesulfonic acid) catalyze the cycloreversion of 1,2,3-trimethyl-5,6-dicyanoquadricyclane, but they are not good catalysts because of side reactions (e.g. cationic oligomerization). Although various transition metal complexes are reported to have catalytic activity for the isomerization of quadricyclane to norbornadiene, most of them have only poor catalytic activity for quadricyclanes with high ionization potentials such as DA-quadricyclanes. We found that not only silver but also some transition metal complexes with planar ligands (e.g. porphyrin, salen and salphen) have catalytic effects on reaction (9). In particular Co(II)salphen 22 and its immobilized catalysts were excellent.[22,23]

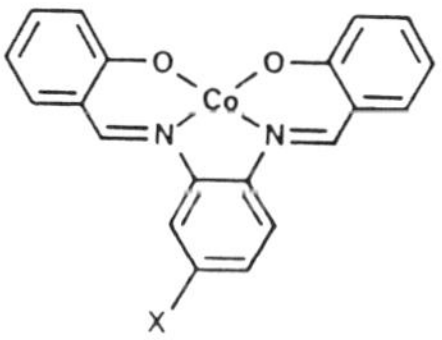

22a X=H
22b X=COOMe
22c X=COOH

In conclusion, the DA-norbornadiene/DA-quadricyclane system (photochromic system) satisfies all seven conditions required for a molecular energy storage system for solar energy. The author refers to this system as DONAC ($D_1,D_2$ : donor; $A_1,A_2$ : acceptor; $R_1,R_2,R_3,R_4$ : H or alkyl etc.):

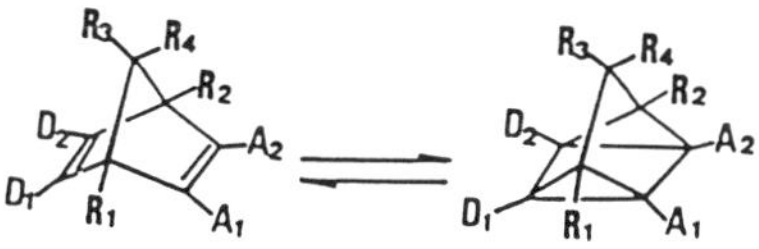

$$\text{[structure]} \rightleftarrows \text{[structure]} \quad (13)$$

Another molecular energy storage system was discovered in the following t-butyl substituted naphthalene/t-butyl substituted naphthvalene system[24]:

$$\text{(diagram of 23} \rightleftarrows \text{24)} \quad (14)$$

| | | |
|---|---|---|
| 23a | $R_1=R_2=R_3=t\text{-Bu}$ | $R_4=H$ |
| 23b | $R_1=R_2=R_4=t\text{-Bu}$ | $R_3=H$ |
| 23c | $R_2=R_3=t\text{-Bu}$, | $R_1=R_4=H$ |

Although naphthalene itself does not undergo this sort of photovalence isomerization, we found that 23a-23c cleanly photoisomerize to the corresponding naphthvalenes ($\Phi=0.01$-0.2) containing a highly strained bicyclobutane moiety. However, the naphthvalenes 24a-24e obtained are stable and colorless. One of the major problems of this system is the failure of 23 to absorb in the visible region of the solar spectrum.

$$\text{(diagram of 25} \rightleftarrows \text{26)} \quad (15)$$

By replacing the naphthalene ring with polyacene ring as shown in 25, the absorbtion shifts into the visible region without decreasing the quantum yield. However, this sort of structure change reduces the energy storage (kcal/kg). We call this A/B system AROVA. Preliminary examination of the DONAC system and the AROVA system have suggested that they may be candidates for an information storage system (as a photoresponse system).

4. CONCLUSION

The author has succeeded in controlling the stability and reactivity of highly strained small ring systems, and thereby in developing new types of reactions leading to various heterocycles, in creating novel $\pi$ systems including a $16\pi$ aromatic hydrocarbon and in creating a novel system (so-called "DONAC") with photochemically interesting functions. These are really new aspects of highly strained ring systems. Extrapolating from these remarkable results, the author feels the necessity that many organic chemists in the world recognize the importance and usefulness of highly strained systems in theoretical, physical and synthetic organic chemistry, and participate in this field.

## ACKNOWLEDGMENTS

The author wishes to acknowledge the important contributions of many collabborators (in particular, Prof. H. Ogoshi, late Prof. S. Yoneda, Prof. N. Kasai, Dr. T. Sugimoto, Dr. S. Miki, Dr. T. Maruyama, Dr. T. Kobayashi, Dr. Y. Kai, Dr. H. Konishi, Dr. S. Araki, Dr. Y. Kamitori, Dr. H. Hirai, Dr. M. Shibata, Dr. T. Ohno, Dr. Y. Asako, N. Nakayama, S. Kida, K. Yamada, Y. Kato) who participated in this work. The work was supported by the Ministry of Education, Science and Culture, Japan.

## REFERENCES

1. R. Breslow, *J. Am. Chem. Soc.*, 92, 984 (1970).
2. Z. Yoshida, Y. Tawara, *J. Am. Chem. Soc.*, 93, 2573 (1971).
3. Z. Yoshida, *Topics in Current Chem.*, 40, 47-72 (1972).
4. Z. Yoshida, H. Konishi, H. Ogoshi, *J.Am. Chem. Soc.*, 95, 3043 (1973).
5. Z. Yoshida, *Pure & Appl. Chem.*, 54, 1059 (1982).
6. H. Konishi, *Ph. D. Thesis of Kyoto Univ.* (1976).
7. S. Araki, *Ph. D. Thesis of Kyoto Univ.* (1978).
8. Y. Kamitori, *Ph. D. Thesis of Kyoto Univ.* (1981).
9. R. Breslow, *J. Am. Chem. Soc.*, 110, 3970 (1988)
10. S. Miki, *Ph. D. Thesis of Kyoto Univ.* (1978).
11. S. Yoneda, H. Hirai, Z. Yoshida, *Heterocycles*, 15, 865 (1981).
12. H. Hirai, *Ph. D. Thesis of Kyoto Univ.* (1980).
13. Z. Yoshida, Partly presented at *International Symposium on Chemistry of Carbocations*, 7-11 September, 1981, Bangor, UK.
14. S. Yoneda, M. Shibata, Z. Yoshida, Y. Kai, N. Kasai, *Angew. Chem. Int. Ed. Engl.*, 23, 63 (1984).
15. T. Sugimoto, M. Shibata, S. Yoneda, Z. Yoshida, *J. Am. Chem. Soc.*, 108, 7032 (1986).
16. Z. Yoshida, T. Hijiya, Y. Umehara and S. Miki, *40th Natl. Meet. Chemical Society of Japan, Fukuoka, 1979*, Abstract, p. 558.
17. Z. Yoshida, T. Hijiya and S. Miki, *41st Natl. Meet. Chemical Society of Japan, Osaka, 1980*, Abstract, p. 1023 .
18. Z. Yoshida, T. Hijiya and S. Miki, *43rd Natl. Meet. Chemical Society of Japan, Tokyo, 1981*, Abstract, p. 848.
19. Partly reproduced from Z. Yoshida, "New Molecular Energy Storage Systems", *J. Photochem.* 29, 27-40 (1985) under the permission of the Publisher (Elsevier Sequoia S.A.), for which the author thanks.
20. Y. Asako, *Ph. D. Thesis of Kyoto Univ.* (1987).
21. T. Kobayashi, Z. Yoshida, Y. Asako, S. Miki, *J. Am. Chem. Soc.*, 109, 5103 (1987).

22. S. Miki, T. Maruyama, T. Ohno, Z. Yoshida, (a) *Chem. Lett. 1988*, 861, (b) *Bull. Chem. Soc. Jpn.* 61, 973 (1988).
23. T. Ohno, *Ph. D. Thesis of Kyoto Univ.* (1988).
24. Z. Yoshida, *General Research on Solar Energy Conversion by Means of Photosynthesis*, Ministry of Education, Science and Culture, Tokyo, 1987, p. 49.

# STRAINED CAGE SYSTEMS
## Synthetic and structural implications

B. ZWANENBURG AND A.J.H. KLUNDER
*Department of Organic Chemistry*
*University of Nijmegen*
*Toernooiveld*
*6525 ED, Nijmegen, The Netherlands*

ABSTRACT. The synthesis and chemical properties of strained polyclic bridgehead alcohols and derivatives are described. Particular attention is given to regio- and stereochemical aspects of their cage opening reactions. In section 2 various synthetic approaches to bridgehead functionalized polycyclic compounds, especially with the cubane, homocubane, 1,3-bishomocubane, basketane and homocuneane skeleton, are described. One bond cleavage reactions governed either by strain effects or by an appropriate substituent are discussed in section 3. Two bond cleavage reactions are the subject of section 4. In the final section some results with cage closure reactions to oxa-polycyclic compounds are described.

## 1. Introduction

Strained polycyclic cage compounds appeal to the imagination of many chemists because of their special structural features, especially the deformation of the ideal carbon-carbon bond angle, the inherent ring strain, their novel and distinctive architecture and their synthetic challenge. Some typical examples of such structures are depicted in Fig. 1, together with their relative strain energies.

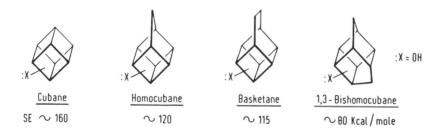

Figure 1

Since the first synthesis of cubane by Eaton and Cole[1] in 1964 many chemists felt encouraged to investigate the synthesis and chemical behaviour of strained polycyclic compounds. Cubane appeared to be surprisingly thermally stable in spite of earlier theoretical predictions[2]. This thermal stability of cubane and its congeners is merely caused by the inability of the cage compound to undergo a concerted bond reorganization reaction in the ground state[3]. When there is the possibility of a non-concerted conversion, cage degradation or fragmentation reactions may take place, even at low temperatures.

In this account special attention is devoted to the synthesis of strained cage compounds functionalized at the bridgehead positions and the study of their chemical behaviour in relation to their cage strain energy. In this respect bridgehead cage alcohols are appropriate substrates since the hydroxyl function can interact electronically with the electron deficient cage moiety and can initiate a non-concerted process.

## 2. Synthesis

The key step[4] in the synthesis of bridgehead substituted cage compounds is the intramolecular $[\pi^2+\pi^2]$ photocyclization of an appropriate tricyclodecadienone. In a subsequent step further modification of the cage is then accomplished either by a ring contraction or a ring expansion reaction. In many cases the bridgehead functionality has to be adjusted. An obvious but essential step in the sequence is the preparation of a suitable photoprecursor. A typical example of a synthesis of a cage compound is depicted in Scheme 1. The photoprecursor is obtained through a Diels-Alder type

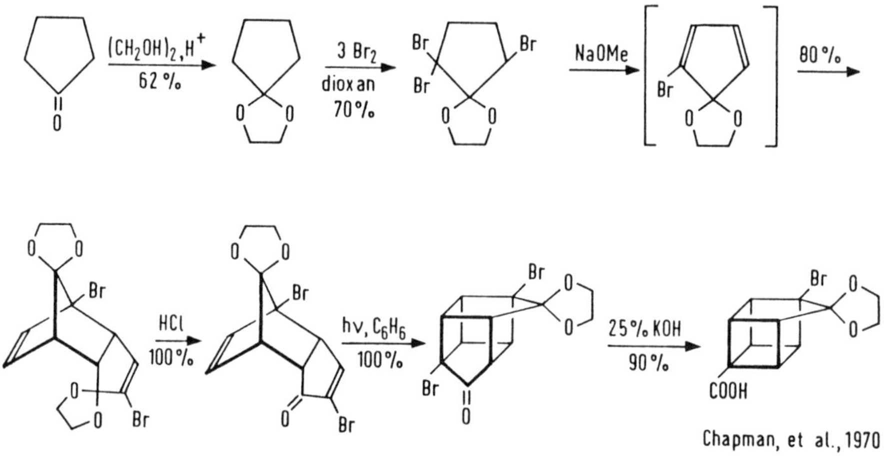

Scheme 1

dimerization of bromocyclopentadienone acetal followed by a selective hydrolysis of one of the acetal functions. The 1,3-bishomocubane system obtained upon irradiation is subjected to a Favorskii-type cage contraction reaction. Further contraction can be accomplished by the same type of reaction, as is shown in Scheme 2. This synthesis of

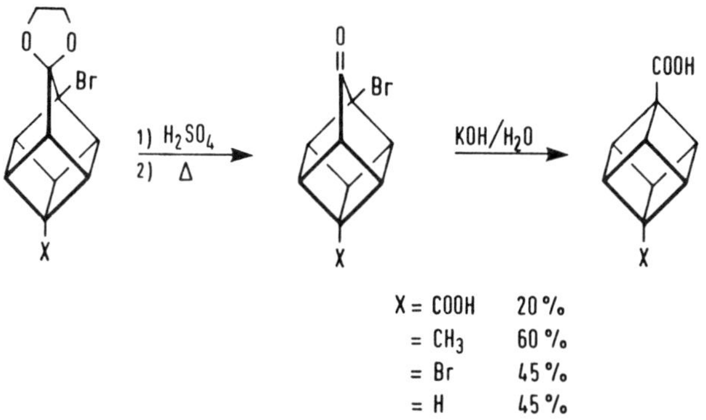

Scheme 2

cubane derivatives[5], as depicted in the Schemes 1 and 2, is actually an improved modification[6] of Eaton and Cole's original cubane synthesis.[1]

A cage expansion reaction of a homocubanone leading to the basketane system[7] is shown in Scheme 3. This expansion reaction is a regiospecific process giving the

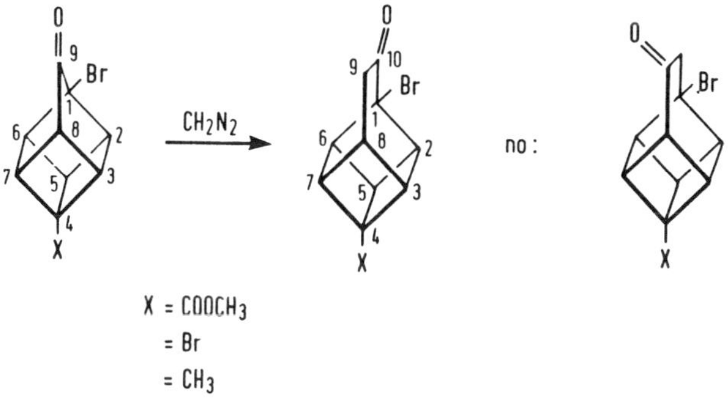

Scheme 3

10-ketone exclusively. Migration of the more electron rich $C_8$-$C_9$ bond is favoured over the migration of the $C_1$-$C_9$ bond, in full accord with the accepted mechanism for this conversion.

Another useful cage expansion reaction (Scheme 4) is the regiospecific cationic rearrangement of homocubyl carbinols producing, after basic methanolysis, 1,3-bishomocubane bridgehead alcohols in good yields.[8] The driving force in this regiospecific rearrangement reaction is the relief of ring strain of about 40 kcal/mol. Force field calculations[9] reveal that migration of the $C_3$-$C_4$ bond (or the equivalent $C_4$-$C_7$ bond) releases more strain energy than migration of the central $C_4$-$C_5$ bond, which would have led to a 1,4-bishomocubane.

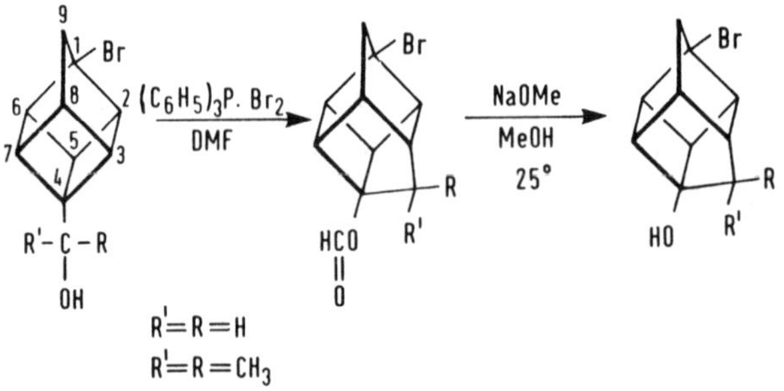

Scheme 4

The bridgehead cage alcohols were, in most cases, obtained from the corresponding bridgehead carboxylic acid by a Curtius rearrangement followed by a diazotation of the bridgehead amine thus obtained in acetic acid; careful ethanolysis of the bridgehead acetates under acidic conditions then produces the alcohols (Schema 5).[5,7] As an

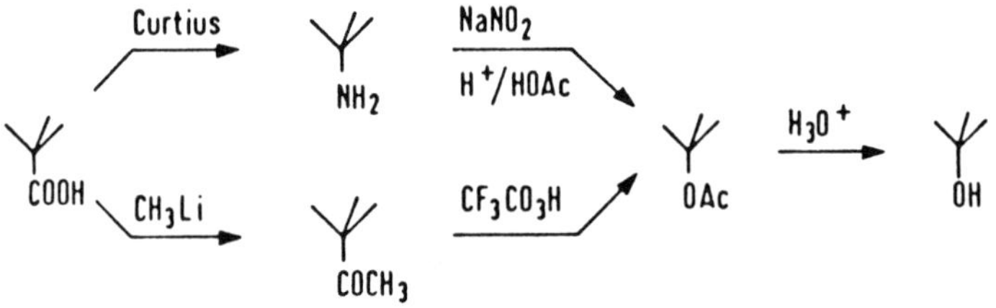

Scheme 5

alternative, a Baeyer-Villiger oxidation of methyl ketones can be used (Scheme 5). The preparation[5,7,10] of some representative bridgehead alcohols and their acetates are depicted in Scheme 6.

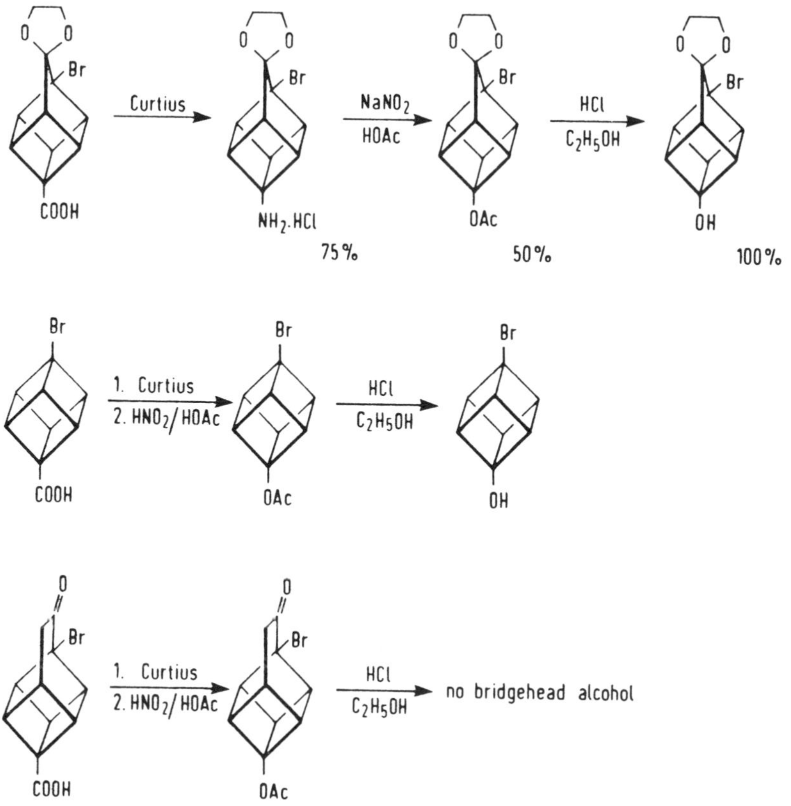

Scheme 6

The silver ion catalyzed transformation of a homocubane[11] is an effective method to prepare a homocuneane bridgehead acetate,[12,13] as is shown in Scheme 7. It should be

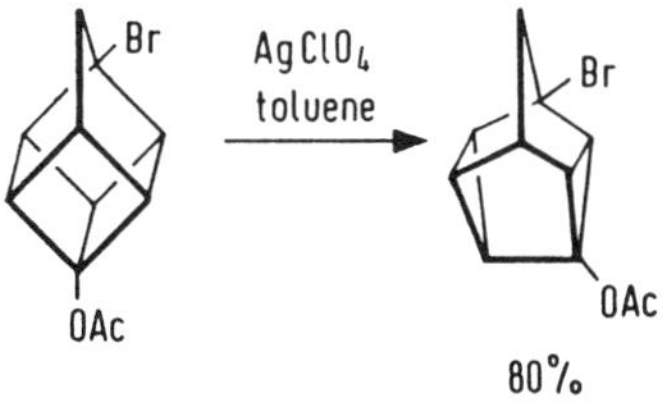

Scheme 7

noted that this silver ion catalyzed skeletal rearrangement cannot be accomplished with homocubane-4-carboxylic acid. This problem can be circumvented by first reducing the carboxylic acid to the corresponding homocubylmethanol, then performing the silver ion catalyzed rearrangement with silver nitrate in aqueous methanol and subsequently reoxidizing the -CH$_2$OH group to the carboxylic acid.[13]

It should be emphasized here that the order of events in the synthetic sequence to bridgehead cage alcohols or acetates needs careful consideration. It is advisable to plan the transformation of a bridgehead substituent, e.g. the carboxyl function to the alcohol or acetate as the last step, in view of the high reactivity of the bridgehead alcohols and acetates under various conditions.

## 3. Cage Opening Reactions

### 3.1. STRAIN CONTROLLED ONE BOND CLEAVAGE REACTIONS
(Base-induced homoketonizations)

In studying the chemical behaviour of highly strained cage systems the intriguing question arises whether these compounds will undergo reactions with a high degree of regio- and stereospecificity or relief of strain would lead to uncontrolled degradation reactions. It is of interest that in an attempt to prepare homocubane bridgehead alcohols by a base-catalyzed alcoholysis of the corresponding acetates an exclusive one bond cleavage reaction was observed.[14,15] It was found that the seco-cage ketone is the result of a regiospecific scission of the C$_3$-C$_4$ (or the equivalent C$_4$-C$_7$) bond (Scheme 8); cleavage of the central C$_4$-C$_5$ bond does not occur at all. Analogous cage opening reactions were encountered for the bridgehead cubane,[15] 1,3-bishomocubane[15] and basketane alcohols[16] (or acetates) as depicted in Scheme 8. In the case of cubanol the seco-cubanone could not be isolated, but its formation was deduced from the further degradation products.[15] The half-cage ketone derived from the basketane bridgehead acetate could be isolated only when short reaction times were applied.[16] Prolonged base treatment induced further break down to a bicyclo-octene ester.[16] It is noteworthy that such extended degradation was not observed in the case of seco-homocubanone.[15]

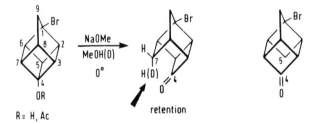

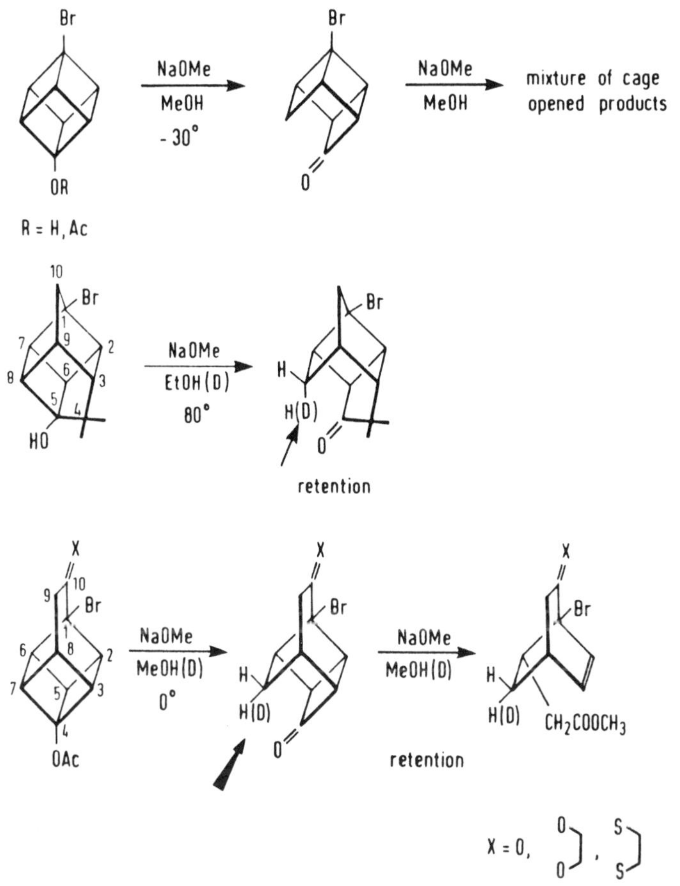

Scheme 8

The one bond cleavage reactions of bridgehead cage alcohols shown in Scheme 8 can, in general terms, be formulated as homoketonizations (Scheme 9).[17] The cyclanol constrained in the polycyclic structure is actually a homoenol. Whether the homoenolate is in equilibrium with the ketocarbanion in the cases under consideration will be discussed later.

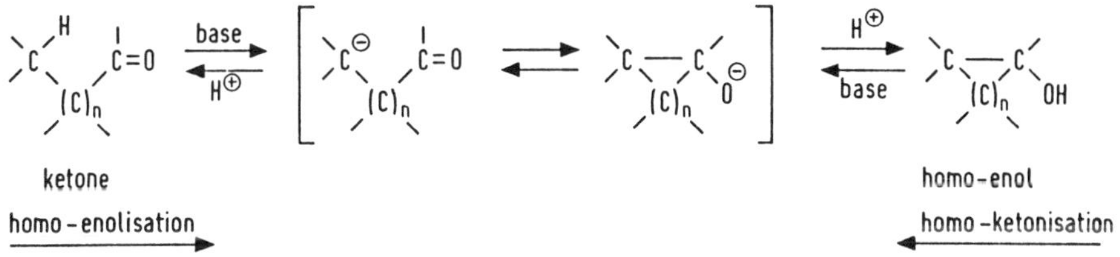

Scheme 9

The regiochemistry of the one bond cleavage reaction of the cage bridgehead alcohols (or acetates) shown in Scheme 8 is primarily governed by relief of cage strain, resulting in the exclusive formation of the thermodynamically most stable half cage ketones. This is demonstrated by the enthalpy data[9] for the conceivable half-cage ketones from homocubanol and 1,3-bishomocubanol, respectively, given in Fig. 2.

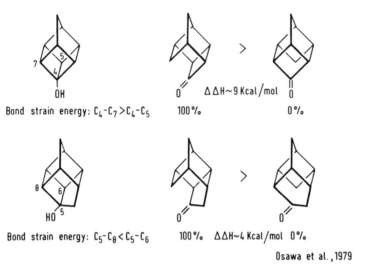

Figure 2

An alternative explanation for the regiospecificity of the cage opening reaction would be that one of the bonds of the original cage compound is significantly more strained than the others, already in the ground state, and therefore react preferentially. Calculations revealed,[9] that in the homocubane system the bond strain energy of the $C_4$-$C_7$ bond is indeed higher than that of the $C_4$-$C_5$ bond, but that the reverse is true in the 1,3-bishomocubane system in which the central $C_5$-$C_6$ bond appears to be somewhat more strained than the $C_5$-$C_8$ bond (Fig. 2). Hence, this explanation seems inconsistent with the experimental findings. Consequently, these homoketonization reactions can probably be best understood by assuming that the product developing stage is located rather late along the reaction coordinate and are accordingly influenced strongly by the thermodynamic stability of the product.[9]

The reactivity of the cage alcohols (or acetates) in this homoketonization process *grosso modo* parallels the total cage strain energy: cubane > basketane > homocubane > 1,3-bishomocubane. The basketane system behaves somewhat unexpectedly since basketyl acetate homoketonizes much faster than the more strained homocubyl acetate. This increased reactivity is probably attributable[16] to the outbending effect of the ethylene bridge which increases the constraint around the $C_4$ and $C_5$ atoms in basketane relative to homocubane. The structural features of homocubane and basketane derivatives, determined by X-ray diffraction analyses, show clearly that the C-C bonds around $C_4$ and $C_5$ in basketane are in fact somewhat compressed compared with those in the homocubane system.[16]

This phenomenon is also nicely demonstrated by the difference in reactivity of the 4- and 1-homocubanols shown in Fig. 3. The 4-substituted compound reacts smoothly

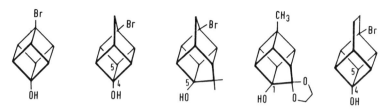

Figure 3

(Scheme 8), whereas the 1-hydroxy derivative does not homoketonize at all, even during prolonged treatment with sodium methoxide in methanol at 100°C. However, the 1,3-bishomocubanol shown, which has about 40 kcal/mol. less strain energy, readily undergoes a cage opening reaction at 80°C (cf. Scheme 8). The extra methylene group present in this system when compared with the 1-homocubanol apparently causes extra compression around C-5 and accordingly, elicits an increased reactivity.

An interesting aspect of the homoketonization reactions of bridgehead cage alcohols is that they invariably take place with complete retention of configuration[14,15,16] (Scheme 10, cf. Scheme 8). This stereochemical course of the cage opening reactions

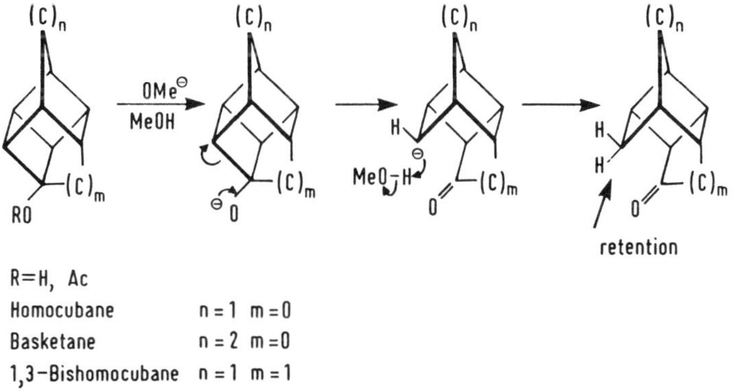

Scheme 10

could readily be established by means of deuterium labeling experiments. Similar stereospecific reactions, i.e. with retention of configuration, were observed for other types of strained polycyclic bridgehead alcohols; some examples[18,19,20] are depicted in Scheme 11. This stereochemical behaviour seems to be typical for polycyclic bridgehead alcohols in which the bridgehead is flanked by four- and/or five- membered rings, with no exception as yet.[17]

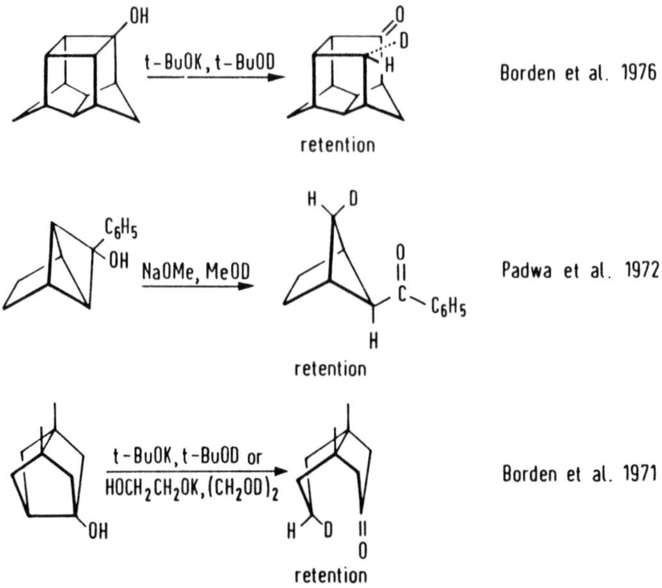

Scheme 11

In contrast, ring opening reactions of cyclopropanols constrained in polycyclic structures show a stereochemical behaviour that strongly depends on the nature of the polycyclic structure. Some typical examples are shown in Scheme 12. For nortricyclanol and related compounds homoketonization proceeds with complete inversion of configuration.[21,17] Triaxane bridgehead acetate[22] (middle line in Scheme 12) also gives a one bond cleavage reaction with inversion of configuration upon treatment with base, whereas the homocuneane bridgehead acetate, shown at the bottom line shows complete retention for the cycloprapanol ring opening reaction.[12,13] The striking difference in stereochemistry of the cyclopropanol cleavage reaction in the triaxane and homocuneane system can only be attributed to the difference in total strain energy of the two structures. The extra three-membered ring present in homocuneane when compared with triaxane, is responsible for a considerably higher strain energy in homocuneane as is evidenced by inspection of molecular models.[13] The results depicted in the Schemes 10, 11 and 12 strongly suggest that retention of configuration is the stereochemical outcome for homoketonization reactions of highly strained polycyclic bridgehead alcohols, whereas inversion of configuration will result for cyclanol cleavage reactions of weakly strained compounds (so far observed only for cyclopropanol containing compounds). The borderline between these stereochemical pathways has not been established as yet. An elegant demonstration of the influence of strain on the stereochemistry of the cyclopropanol opening was described by Miller and Dolce[23] (Scheme 13). The homocuneane diol shown undergoes a double homoketonization reaction, the first of which proceeds with retention and the second one with inversion of configuration. This is in full accordance with the established behaviour of the highly strained homocuneane and the considerably less strained triaxane, respectively.

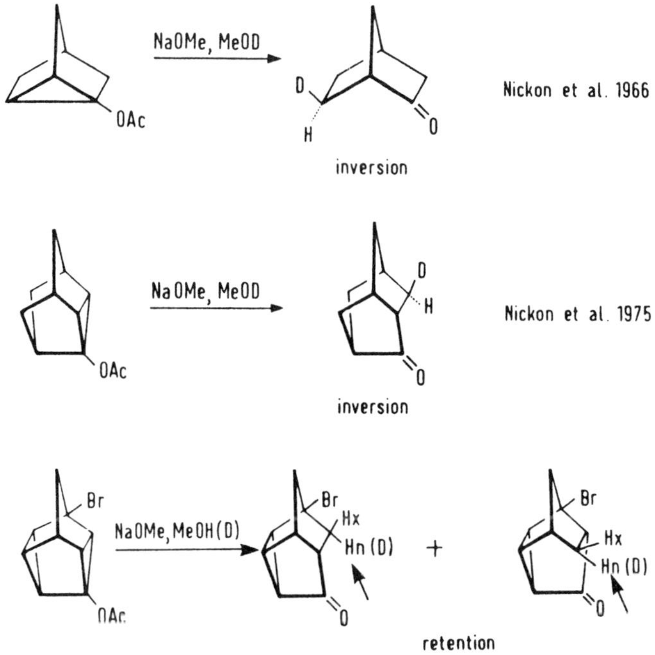

Scheme 12

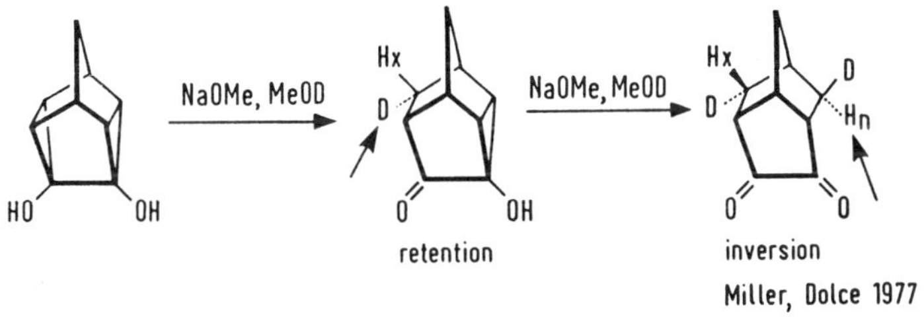

Scheme 13

In order to rationalize the divergency in the stereochemistry of the respective homoketonization processes, it is assumed that the first step involves the formation of a carbanionic species. It may be expected that in highly strained structures the carbanionic center will rapidly move away from the developing carbonyl function with a concomitant release of a considerable amount of strain energy. The carbanionic center as a consequence, will hardly be shielded by this departing carbonyl function and will be rapidly protonated on its open face by solvent molecules which are favourably disposed around the polar carbonyl group.[13] The stereochemical result is retention of configuration.

In less strained systems the strain is not large enough to enforce the carbanionic center to separate completely from the electrophilic carbonyl function, leaving a substantial homoconjugative stabilization of the incipient carbanionic species. Subsequent protonation of this carbanionic species in which the *endo*-side is homoconjugatively shielded by the carbonyl group will preferentially take place from the *exo*-side and accordingly will result in inversion of configuration. Both mechanisms are depicted in Scheme 14.

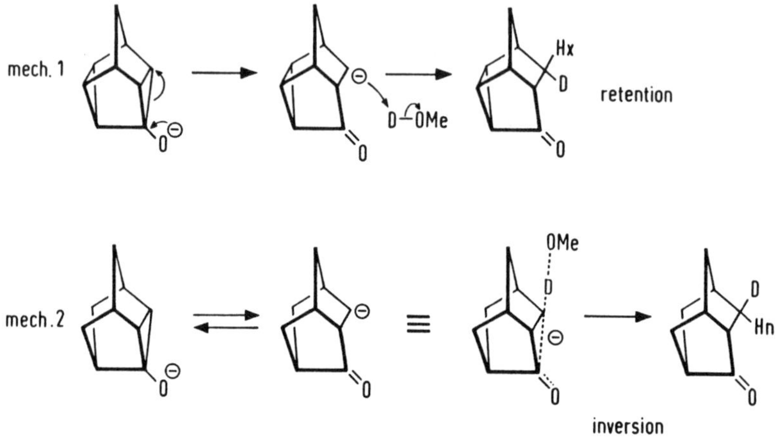

Scheme 14

It is suggested that the stereochemistry of the base induced homoketonization of strained polycyclic alcohols, in general, is highly dependent on the possibility of homoconjugative stabilization during the C-C bond cleavage of the homoenolate anion. Effective homoconjugative stabilization can only be envisaged in those cases where the strain energy associated with the C-C bond to be cleaved in the homoketonization process, permits effective orbital overlap.

The cyclopropanol ring opening reactions that take place with inversion have as a characteristic feature that the homoketonization is reversible. This was demonstrated by deuterium exchange experiments of appropriate ketones, e.g. *exo*- and *endo*-5,5,6-trimethylbicyclo[2.2.1]heptan-2-one (*exo*- and *endo*-isocamphanone), camphor and fenchone.[17] The reversibility of the homoketonization has never been observed for cyclanols that ring open with retention of configuration, and, in fact, would be highly unlikely in view of the high strain energy involved.

The exact nature of the proposed intermediate carbanionic species in homoketonization reactions cannot be deduced from the experimental data available so far. Whether or not a proton is incorporated into the transition state of the one bond cleavage reaction, and if so, to what extent, is a question that cannot be answered, since no kinetic isotope effect or any other kinetic parameters have been determined.

Attempts to prepare the putative carbanion intermediate in the base-induced homoketonization of 3,7 dimethyltricyclo[3.3.0.0$^{3,7}$]octan-1-ol (see reaction at the bottom line of Scheme 11) by an alternative route,[24] did not provide conclusive evidence about the nature of the carbanionic species since the effect of strain could not be properly taken into account.

## 3.2. SUBSTITUENT DIRECTED ONE BOND CLEAVAGE REACTIONS

The homoketonization reactions of bridgehead cage alcohols can also be viewed as nucleophilic eliminative ring fissions in which a carbon oxygen double bond is formed by elimination of a carbon leaving group.[25,26] The occurrence of unactivated carbon leaving groups in acyclic systems is hardly known; in alicyclic compounds, particularly in small ring systems, an increasing number of examples have become available.[25] In the case of highly strained molecules the release of strain energy during the bond fission process will compensate for the high activation energy required for the expulsion of a non-activated nucleofugal carbon.[25,26] As discussed in the preceding section the regiochemistry of the base-induced cage opening reaction of highly strained bridgehead polycyclanols is primarily determined by the thermodynamic stability of the conceivable half-cage structures. Electronic factors do not play a role, since in none of the three possible bond cleavages the developing carbanion is particularly stabilized.

The intriguing question now is whether attachment of a carbanion stabilizing group at a nucleofugal carbon will alter the regiochemistry of the homoketonization process.[26,27,28] The 1,3-bishomocubane ketone acetate shown in Fig. 4 is an appropriate substrate to investigate such a directive effect of a substituent. The strain energies for the three conceivable diketones (see Fig. 4), as calculated by MM2 molecular mechanics are 56.75 kcal/mol for the $C_5$-$C_6$, 49.80 kcal/mol for the $C_6$-$C_7$ and 52.90 kcal/mol for the $C_6$-$C_2$ cleavage product.[29] The strain data for the corresponding acetal ketones (Fig 4) are 58.67, 48.02 and 54.00 kcal/mol, respectively.[29]

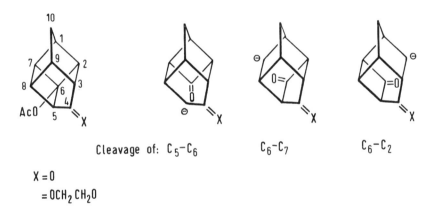

Cleavage of: $C_5$-$C_6$     $C_6$-$C_7$     $C_6$-$C_2$

X = O
 = OCH$_2$CH$_2$O

Figure 4

The ketone acetate shown in Fig 4 could be readily prepared[26,27,28] from cyclopentenedione, as depicted in Scheme 15, following the principles outlined in section 2. Acetalization of the ketone function could be accomplished in the usual manner, i.e. treatment with glycol and a trace of p-toluenesulfonic acid (yield 87%) in spite of the instability of the ketone acetate towards both acidic and basic reagents.

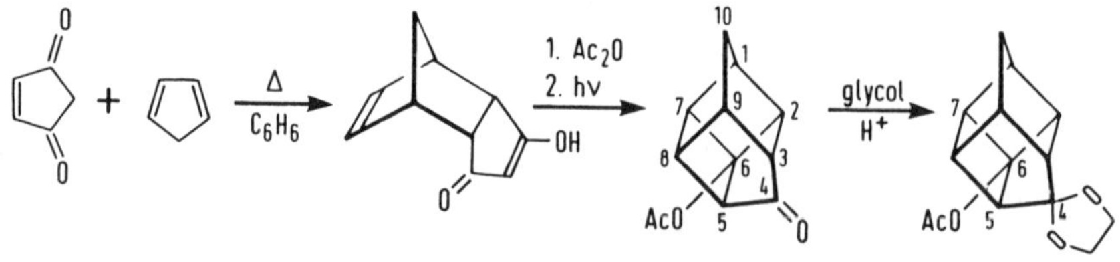

Scheme 15

Mild treatment of the ketone acetate with sodium methoxide in methanol at 0°C gave an almost instantaneous cage opening reaction to furnish the $C_5$-$C_6$ cleavage product in quantitative yield[26,27,28] (Scheme 16, top line). The acetal acetate, (Scheme 16 bottom line), being the substrate with the protected carbonyl function, underwent base-induced homoketonization to produce the $C_6$-$C_7$ cleavage product.[26,27] In both cases the cage opening reactions proceed with retention of configuration.[26,27]

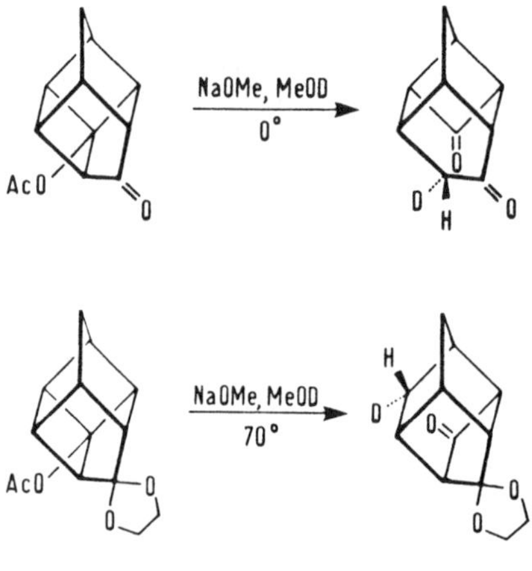

Scheme 16

The observed divergency in regiochemistry of these two base-induced homoketonization reactions clearly demonstrates the influence of the cage substituent. The regiospecific formation of the $C_6$-$C_7$ cleavage product from the acetal acetate (bottom line Scheme 16) conforms entirely with the general pattern observed for the non-activated nucleophilic eliminative ring fission of strained cage bridgehead acetates, (section 3.1), i.e. producing the thermodynamically most stable half-cage ketone of the three possible structures (cf. strain energies given above). The one bond cleavage product from the ketone acetate (Scheme 16 top line) is, according to the calculated strain energies for the three conceivable diketones, the most strained one.[26,29] The formation of this 'contra thermodynamic product' shows convincingly that the conjugative stabilization of the nucleofugal carbanion is sufficient to overrule the aforementioned thermodynamic control.[26]

A 'contra-thermodynamic' cage opening reaction may also be enforced by a 1,3-through-cage *elimination* reaction in a bridgehead cage acetate (or alcohol) appropriately substituted with a leaving group at the β-position with respect to the acetate (or alcohol).[30,26] The 1,3-bishomocubane ketone acetate shown in Fig. 4 was converted into the β-substituted mesylate depicted in Scheme 17 by a stereoselective reduction with lithium aluminumtri(t-butoxy)hydride in ether followed by mesylation

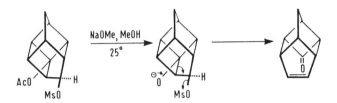

Scheme 17

in the usual manner. An X-ray diffraction analysis unequivocally revealed that the mesylate predominantly obtained (in an excess of 80%) has the anti-configuration (see scheme 17).[26] Treatment of this anti-mesylate with sodium methoxide in methanol at room temperature for one hour gave the interesting olefine ketone shown in Scheme 17 in an excellent yield.[30,26] It is noteworthy that the syn-mesylate acetate did not show cage cleavage at all during a treatment with sodium methoxide in methanol. There was only conversion to the corresponding syn-mesylate alcohol. This difference in behaviour of the syn- and anti- compound strongly suggests that the 1,3-through-cage elimination process is subject to a strict stereoelectronic control, resembling a Grob-type elimination reaction with the leaving group and the bond to be cleaved in a trans-antiparellel orientation.[26] The alkenone obtained is an interesting structure since it contains two isolated orthogonal π-electron systems in close spatial promixity. Some chemical aspects of these alkenones will be discussed in Section 5.

In the more strained homocubane systems the difference in thermodynamic stability between the two possible seco-cage ketones for the homoketonization of 4-homocubanol amounts to ca. 9 kcal/mol (Fig 2). Hence, in comparison with the 1,3-bishomocubane system discussed hitherto, it may be much more difficult to overrule the thermodynamic control of the cage cleavage reaction by means of an appropriate substituent placed at C-5. The original plan to use 5-acetyl or 5-benzoyl homocubyl-4-acetates for this purpose failed due to the impossibility to prepare these compounds.[28] Therefore, 5-bromo-homocubyl-4-acetate was considered as a candidate for a $C_4$-$C_5$-bond cleavage reaction.[28] The preparation of this compound is outlined in Scheme 18 and involves bromination of the cycloadduct of cyclopentadiene and cyclopentenedione (Scheme 15) with bromine in acetic acid, followed sequentially by conversion of the enol into the vinyl bromide, photocyclization, a Favorskii cage contraction, a transformation of the carboxyl function into a methyl ketone and finally by a Baeyer-Villiger oxidation.[28] Careful experimentation ensured that these successive steps could be performed in good yields. Homoketonization of 5-bromo-homocubyl-4-acetate under the usual conditions with sodium methoxide in methanol gave the, at first sight unexpected, acetal product shown in Scheme 19. The rationale for the formation of this acetal clearly demonstrates that the bromine substituent has the ability to direct the base-induced homoketonization in a contrathermodynamic fashion. The bromoketone formed initially by cleavage of the $C_4$-$C_5$ bond reacts with methanol from the *exo*-side giving a hemi-acetal that will intramolecularly expel the bromine at C-5 and result in the acetal shown. In this reaction there was no indication whatsoever of fission of the alternative $C_4$-$C_3$ (or equivalent $C_4C_7$) bond.[28]

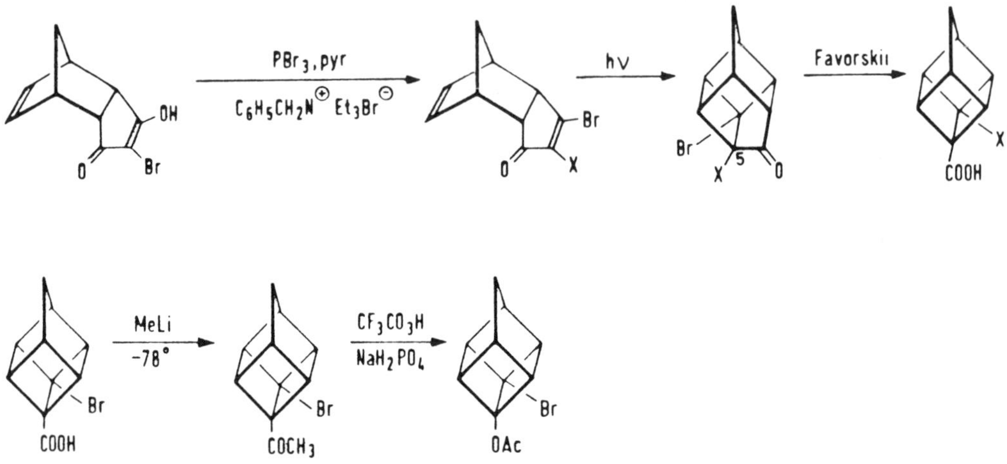

Scheme 18

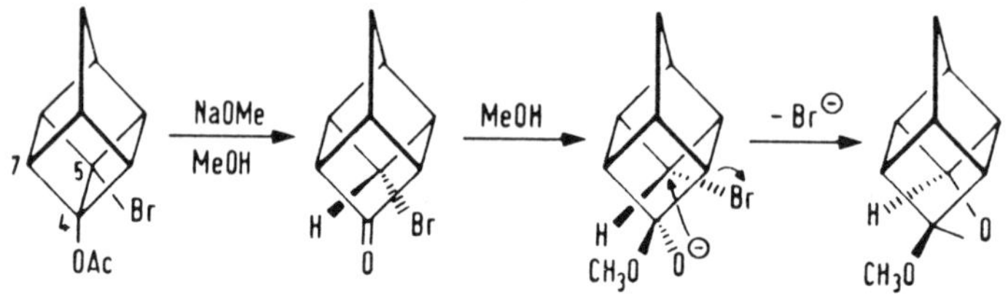

Scheme 19

The directive effect exerted by the relatively poor carbanion stabilizing bromine substituent illustrates the subtle balance between thermodynamic and electronic parameters determining the regiochemistry of the base-induced bond cleavage reaction in strained polycyclanols.

An attempt[31] was made to divert the ring opening reaction of a homocuneane bridgehead alcohol by means of an anion stabilizing substituent at the alternative nucleofugal carbon C-5. It was found, however, that in 5-benzoylhomocuneyl-4-acetate the direction of the bond cleavage is not affected by the carbonyl containing function at C-5, since the cyclopropanol ring opening reaction observed was the same as is shown in Scheme 12. Apparently, the thermodynamic control cannot be overruled in this case.

## 4. Two Bond Cleavage Reactions

### 4.1. BASED INDUCED REACTIONS INVOLVING TWO BONDS

In the preceding section the cage opening reactions were all performed in aprotic media in which incipient carbanionic species are protonated very rapidly. In the absence of protons, i.e. in aprotic media, homocubane bridgehead alcohols undergo a two bond cleavage reaction[15] as depicted in Scheme 20. It is suggested that the formation of a stabilized carbanion, i.e. a lithium enolate, is the driving force of this reaction. In Scheme 20 the reaction is pictured, for reasons of clarity, as a two step process involving a free carbanion after the first bond cleavage. It seems likely however that the two bonds cleave in a concerted fashion. It is remarkable, that the homoketonization reaction with sodium methoxide in dioxane or acetonitrile containing a trace of methanol[15] stops at the stage of one bond fission giving the seco-homocubanone shown in Scheme 8. Attempts to accomplish a two bond cleavage reaction of 1,3-bishomocubane bridgehead alcohols of the type shown in Scheme 8, were unsuccessful; no reaction was observed not even under more drastic conditions.[15]

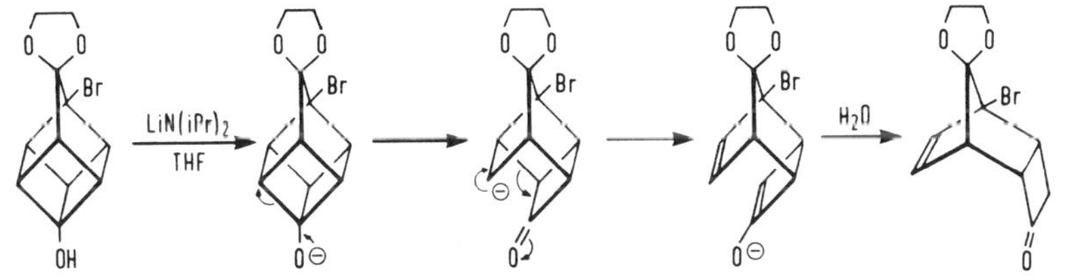

Scheme 20

A two bond fission reaction similar to that of 4-homocubanol was observed for cubane and homocubane derivatives having an active methylene group at the 4-bridgehead position (Scheme 21).[32,33] These cage degradation reactions are designated as homoallylic rearrangements. Such reactions could not be accomplished

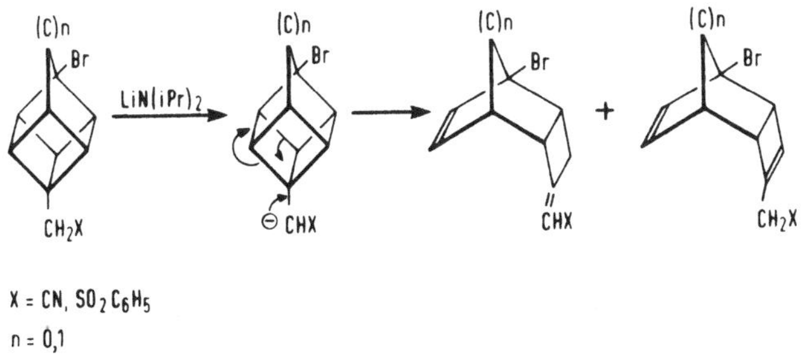

X = CN, $SO_2C_6H_5$
n = 0,1

Scheme 21

under protic conditions. When the activating substituent is an ethoxycarbonyl group, treatment with lithium diisopropylamide does not cause any bond cleavage.[33] Apparently, the ester enolate anion in question is less prone to undergo a homoallylic rearrangement than the anions shown in Scheme 21.

## 4.2. THERMAL AND ACID CATALYZED TWO BOND CLEAVAGE REACTIONS

Heating a 4-homocubanol in benzene brings about a two bond cleavage reaction to give the product shown in Scheme 22 in nearly quantitative yield.[34] The same product was

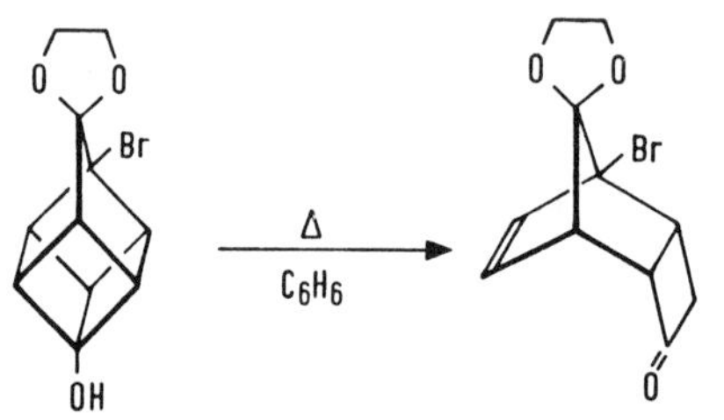

Scheme 22

obtained by treatment of the homocubanol with aqueous hydrochloric acid in methanol at room temperature in a rather slow reaction[34] (Scheme 23). At first glance, an acceptable pathway for this double cage opening reaction would involve the intermediacy of a seco-homocubanone produced by an initially formed proton-homocubane sigma complex, as shown in Scheme 23. However, when this supposed seco-homocubanone was treated with acid (HCl or HBr) under the same conditions, entirely different products were obtained (Scheme 24).[34] The mechanistic

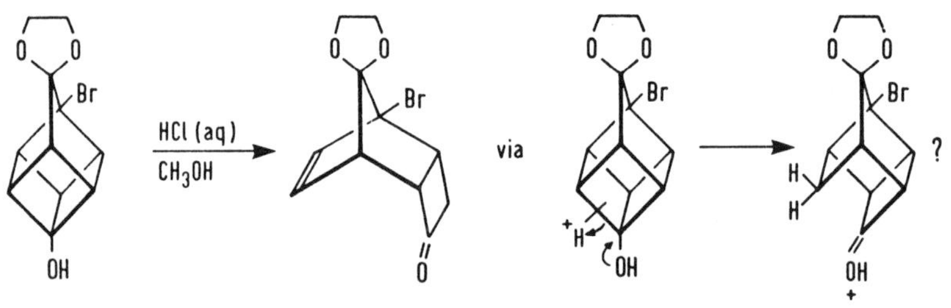

Scheme 23

rationale for their formation is also shown in this Scheme. Protonation of the cyclobutanone carbonyl group induces a series of consecutive cyclobutyl-cyclopropylcarbinyl cation rearrangements ultimately leading to the most stable cyclopropyl carbinyl cation. Subsequent opening of the cyclopropane ring by nucleophilic attack of the halide ion in a regio- and stereo-controlled manner furnishes

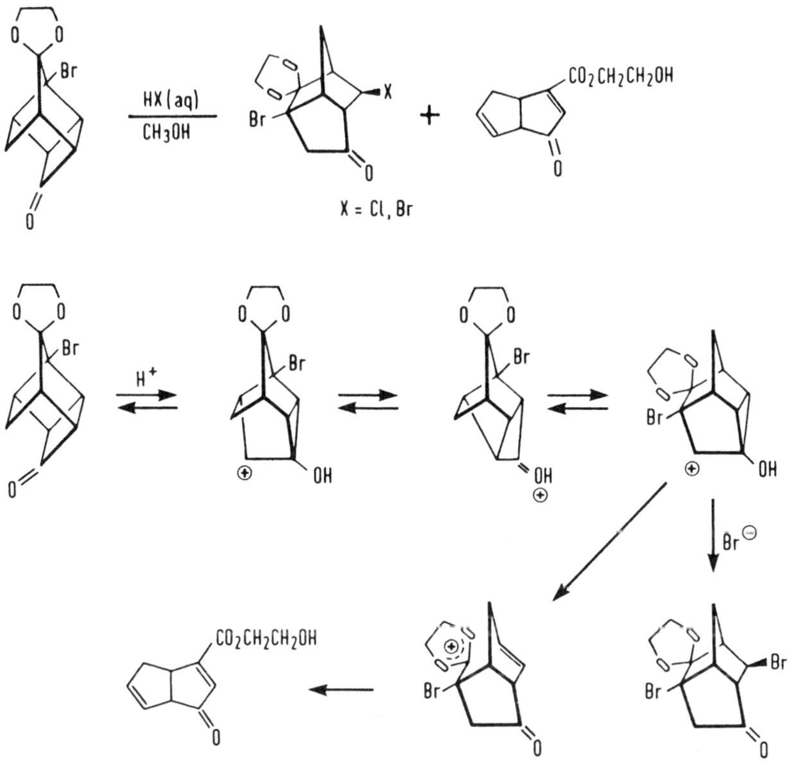

Scheme 24

the brendanone product. (The alternative nortwistanone is about 11 kcal/mol less stable). Intramolecular cyclopropane ring opening with involvement of the acetal function explains the formation of the pentalenone.[34] The cyclic acetal function has a profound influence on the cationic rearrangement, because, when this acetal is absent, a different type of product is isolated upon treatment with hydrobromic acid, viz. 2,7-dibromobrendane-4-one.[35] The pentalenone is the sole product when the seco-homocubanone with an acetal group (structure, see Scheme 24) is treated with a non-nucleophilic acid such as aqueous perchloric acid.[34] This reaction is of synthetic interest since it provides an annelated cylopentenone in high yield.

Methoxy substituted 1,3-bishomocubanone shown in Scheme 25 gave a smooth thermal cycloreversion[28] which was most efficiently carried out using the technique of flash vacuum thermolysis (400°C/0.35 Torr). The initially formed enolether gave upon hydrolysis the syn doubly annelated cyclopentane shown. This thermal cage cleavage probably proceeds by a radical pathway, involving the initial formation of a 1-methoxy-1,4-diradical by cleavage of the $C_4$-$C_5$ bond, followed by further bond scission.[28]

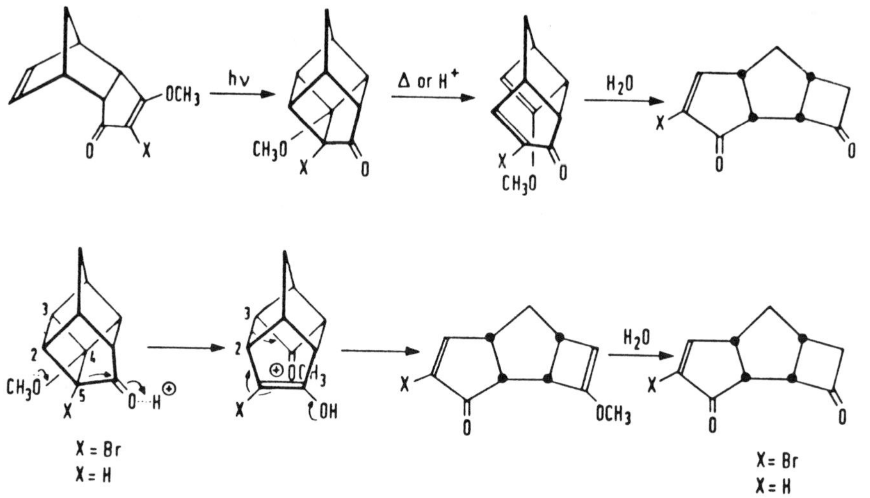

Scheme 25

The same product was obtained when the methoxy-1-3-bishomocubanone was treated with acid in toluene solution (Scheme 25).[36,28] A mechanistic explanation of this cage unfolding process is depicted at the bottom line of Scheme 25. Carbonyl protonation initiates the scission of the $C_4$-$C_5$ bond, which is facilitated by the electron donating methoxy substituent. Subsequent further cleavage, as indicated, results in the vinyl ether which readily undergoes hydrolysis. It is of interest to note that the sequence shown in the top line of Scheme 25 can be carried out in one operation, simply by performing the irradiation in toluene containing a small amount of acid.[28] The starting material is easy to prepare from cyclopentadiene and cyclopentenedione in analogy with the bridgehead acetate shown in Scheme 15 by methylation instead of acetylation and therefore, the preparation of the syn doubly annelated cyclopentane may have considerable synthetic merit. The steric course of this cage unfolding reaction is an illustration of *cage directed stereochemistry*.

## 5. Oxa Cage Compounds by Cage Closure Reactions

The polycyclic compound containing two π-systems in orthogonal proximity, the preparation of which is shown in Scheme 17, may exhibit interesting transannular effects. Its ultraviolet spectrum shows a maximum absorption at 204 nm in hexane (ε 3200) which may be indicative of an orbital-orbital interaction between the two π-systems.[26,30] This absorption was absent in the compound in which the olefinic bond was reduced.[37]

The alkenone, which can be named 2,9-carbonyl-brendene, shows a great propensity to undergo cationic rearrangement reactions. For example, acetalization with ethylene glycol under the usual conditions only produces the acetal of the rearranged ketone, viz. 2,4-carbonyl-brexene (Scheme 26).[37] Even a brief treatment with Grignard reagent causes this rearrangement (Scheme 26, bottom line).[37] This conversion can be readily explained by invoking a cyclobutyl-cyclopropylcarbinyl cation rearrangement. The driving force is clearly relief of strain. MM2 calculations give a strain energy of 67,7 kcal/mol for the brendene and of 65,7 kcal/mol for the rearranged brexene structure.[37]

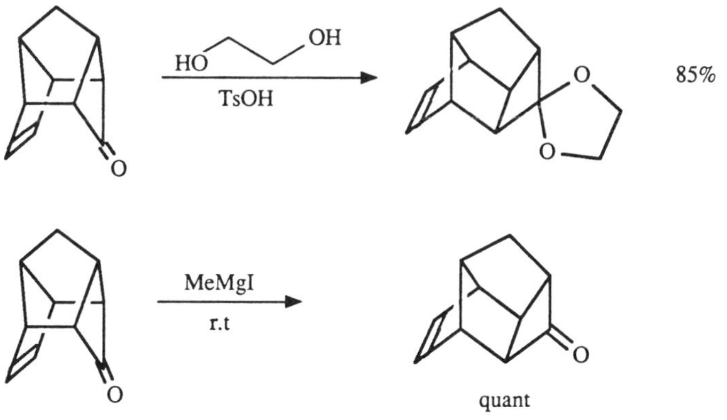

Scheme 26

Catalytic reduction of the olefinic bond can only be accomplished under slightly basic conditions (Scheme 27), otherwise rearrangement does take place.[37] Methylidenation with methylene triphenylphosphorane leads to a structure having two olefinic bonds in orthogonal proximity (Scheme 27).[37] The properties of this interesting molecule have not yet been explored. Reduction with lithium aluminumhydride gives the *endo*-alcohol[37,38] shown in Scheme 28. Methyllithium (not methylmagnesium iodide, see Scheme 26) similarly, produces an *endo* alcohol (Scheme 28). These reactions also were carried out for the compound having an ethylene (n=2) instead of a methylene bridge (n=1). The preparation of this homologue follows the same sequence of events as shown for n=1 (Scheme 15 +17), in this case starting with the cycloadduct of cylopentenedione and cyclohexa-1,3-diene.[37]

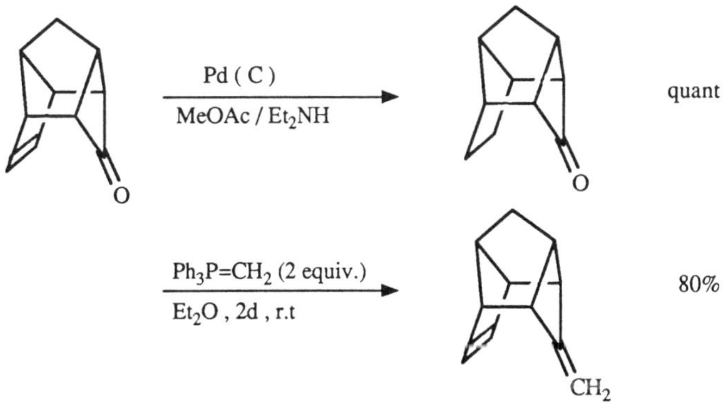

Scheme 27

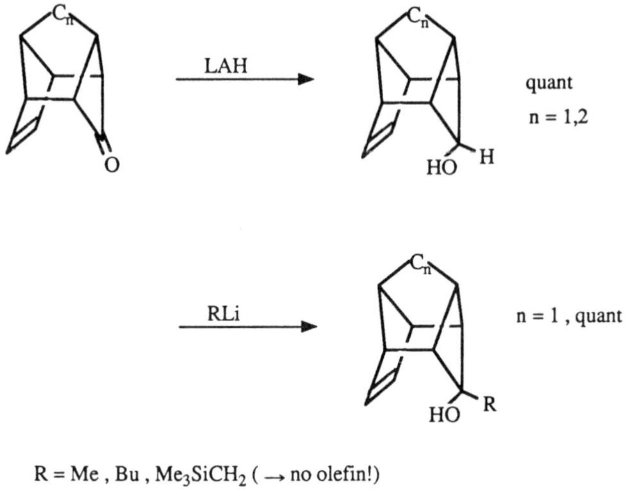

R = Me , Bu , Me$_3$SiCH$_2$ ( → no olefin!)

Scheme 28

An interesting question is whether an intramolecular addition of the alcohol function to the olefinic bond can be accomplished. Alkenols that have some structural resemblance indeed show such an intramolecular reaction, as was described by Ganter et al.[39] (Scheme 29) It should be stressed however, that these compounds are considerably less strained than their counterparts in Scheme 28, according to MM2 calculations.[37] Several attempts were made to accomplish a base-catalyzed cage closure reaction for the polycyclenol shown in the top line of Scheme 30, however, all of them were in vain. With potassium tert-butylate in tert-butyl alcohol a cage opening reaction (a homoketonization) was observed in addition to an entirely unexpected oxidation, probably by atmospheric oxygen, to the corresponding 2,9-carbonyl-brendene.

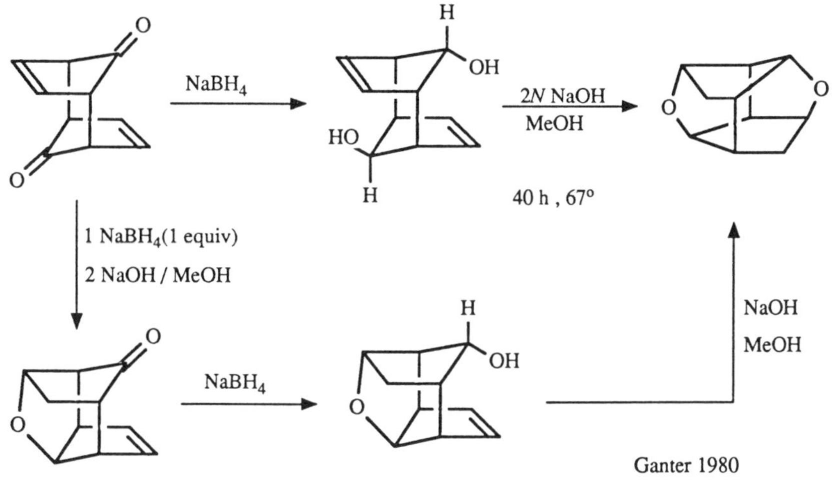

Scheme 29

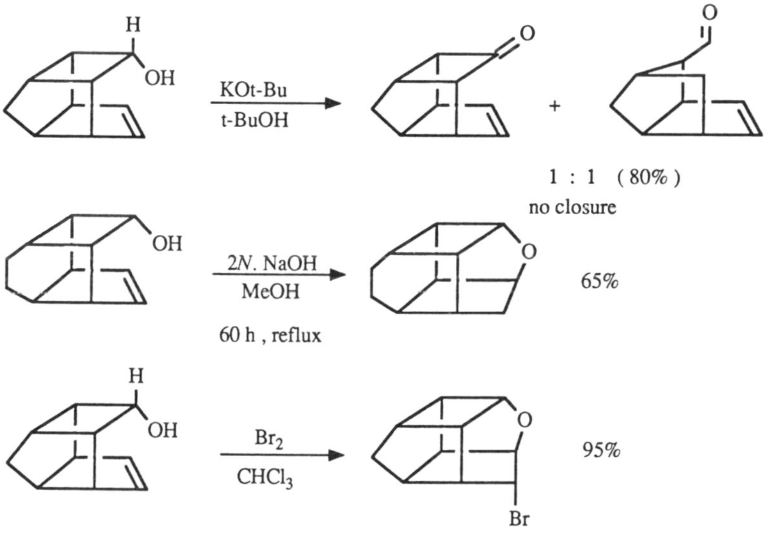

Scheme 30

Inspection of molecular models reveals that the hydroxyl function has a very unfavourable spatial position with respect to the olefinic carbon atoms. Following the principles underlying the Baldwin rules a reaction of the symmetrically placed hydroxyl function with an olefinic carbon atom would be virtually impossible. In the case of the alkenols studied by Ganter et al.[39] (Scheme 29) the polycyclic skeleton is more flexible and less strained. As a consequence the hydroxyl function can reach a position that allows a reaction with one of the olefinic carbon atoms.

The behaviour of the *endo*-alcohol derived from the polycyclenone with n=2 towards treatment with base was entirely expected, viz. cyclization to a oxa-cage compound took place in an acceptable yield (Scheme 30, middle line). The polycyclenol with an ethylene bridge is more strained (a strain energy of 70.7 kcal/mol was calculated by MM2) than that with a methylene bridge (67,7 kcal/mol). This strain effect brings the hydroxyl function and the olefinic bond in close proximity, the spatial position however, is still highly unfavourable. Further experimentation is necessary to understand this cage closure reaction.

An alternative method to accomplish a cage closure reaction is presented at the bottom line of Scheme 30. Bromination of the olefinic bond, probably to a bromonium ion, followed by an intramolecular displacement reaction results in a facile cage closure. The steric constraints for this reaction are apparently less severe than those in the direct nucleophilic addition to the olefinic bond. The electrophilic triggering of the olefinic bond in polycyclenols to affect a cage closure will be the subject of future research.

**Acknowledgement**

Part of the work described in this account was supported by the Netherlands Foundation for Chemical Research (SON) with financial aid from the Netherlands Organization for the advancement of Pure Research (ZWO). The authors are indebted to their co-workers for their enthusiastic and skilful contributions.

References
1. P.E. Eaton and T.W. Cole, *J. Amer. Chem. Soc.* **86**, 962 (1964); P.E. Eaton and T.W. Cole, *J. Amer. Chem. Soc.* **86**, 3157 (1964).
2. W. Weltner, Jr., *J. Amer. Chem. Soc.* **75**, 4224 (1953).
3. R.B. Woodward and R. Hoffmann, *The Conservation of Orbital Symmetry*, Verlag Chemie, Weinheim/Germany, 1970.
4. R.C. Cookson, E. Crundwell, R.R. Hill and J. Hudec, *J. Chem. Soc.*, 3062 (1964).
5. A.J.H. Klunder and B. Zwanenburg, *Tetrahedron* **28**, 4131 (1972).
6. N.B. Chapman, J.M. Key and K.J. Toyne, *J. Org. Chem.* **35**, 3860 (1970).
7. A.J.C. van Seters, M. Buza, A.J.H. Klunder and B. Zwanenburg, *Tetrahedron* **37**, 1027 (1981).
8. A.J.H. Klunder and B . Zwanenburg, *Tetrahedron* **29**, 161 (1973).
9. E. Osawa, K. Aigami and Y. Inamoto, *J. Chem. Soc. Perkin II*, 181 (1979).
10. B. Zwanenburg and A.J.H. Klunder, *Tetrahedron Lett.*, 1717 (1971).
11. L.A. Paquette, *Accounts Chem. Res.* **4**, 280 (1971); L.A. Paquette, *Synthesis*, 347 (1975).
12. N.B.M. Arts, A.J.H. KLunder and B. Zwanenburg, *Tetrahedron Lett.*, 2359 (1976).
13. N.B.M. Arts, A.J.H. Klunder and B. Zwanenburg,*Tetrahedron* **34**, 1271 (1978).
14. A.J.H. Klunder and B. Zwanenburg, *Tetrahedron Lett.*, 1721 (1971).
15. A.J.H. Klunder and B. Zwanenburg, *Tetrahedron* **29**, 1683 (1973).
16. A.J.H. Klunder, A.J.C. van Seters, M. Buza and B. Zwanenburg, *Tetrahedron* **37**, 1601 (1981).
17. For an extended review on homoenolization and homoketonization, see: N.H.

Werstiuk, *Tetrahedron* **39**, 205 (1983).
18. R. Howe and S. Winstein, *J. Amer. Chem. Soc.* **87**, 915 (1965); T. Fukunaga, *J. Amer. Chem. Soc.* **87**, 916 (1965); A.B. Crow and W.T. Borden, *Tetrahedron Lett.* 1967 (1976).
19. A. Padwa and W. Eisenberg, *J. Amer. Chem. Soc.* **94**, 5852 (1972).
20. W.T. Borden, V. Varma, M. Cabell and T. Ravindranathan, *J. Amer. Chem. Soc.* **93**, 3800 (1971).
21. A. Nickon, J.L. Lambert, R.O. Williams and N.H. Werstiuk, *J. Amer. Chem. Soc.* **88**, 3354 (1966).
22. A. Nickon, D.F. Coven, G.D. Pandit and J.J. Frank, *Tetrahedron Lett.*, 3681 (1975).
23. R.D. Miller and D.L. Dolce, *Tetrahedron Lett.*, 1023 (1977).
24. A.B. Crow and W.T. Borden, *J. Amer. Chem. Soc.* **101**, 6666 (1979).
25. For an excellent review on nucleophilic eliminative ring fission, see: C.J.M. Stirling, *Chem. Rev.*, 517 (1978).
26. A.J.H. Klunder, W.C.G.M. de Valk, J.M.J. Verlaak, J.W.M. Schellekens, J.H. Noordik, V. Parthasarathi and B. Zwanenburg, *Tetrahedron* **41**, 963 (1985).
27. W.C.G.M. de Valk, A.J.H. Klunder and B. Zwanenburg, *Tetrahedron Lett.*, 971 (1980).
28. A.J.H. Klunder, G.J.A. Ariaans, E.A.R.M. v.d. Loop and B. Zwanenburg, *Tetrahedron* **42**, 1903 (1986).
29. P.M. Ivanov, E. Osawa, A.J.H. Klunder and B. Zwanenburg, *Tetrahedron* **41**, 975 (1985).
30. A.J.H. Klunder, J.W.M. Schellekens and B. Zwanenburg, *Tetrahedron Lett.*, 2807 (1982).
31. N.B.M. Arts, H. Weenen, A.J.H. Klunder and B. Zwanenburg, *Tetrahedron* **39**, 2825 (1983).
32. A.J.H. Klunder and B. Zwanenburg, *Tetrahedron Lett.*, 2383 (1972).
33. A.J.H. Klunder and B. Zwanenburg, *Tetrahedron* **31**, 1419 (1975).
34. P.J.D. Sakkers, J.M.J. Vankan, A.J.H. Klunder and B. Zwanenburg, *Tetrahedron Lett., 897 (1979)*.
35. J.M.J. Vankan, A.J.H. Klunder, J.H. Noordik and B. Zwanenburg, *Recl. Trav. Chim. Pays-Bas* **99**, 213 (1980).
36. A.J.H. Klunder, G.J.A. Ariaans and B. Zwanenburg, *Tetrahedron Lett.*, 5457 (1984).
37. H.L.E. Depré, A.J.H. Klunder and B. Zwanenburg, *to be published.*
38. T. Visser, J.H. van der Maas, H.L.E. Depré, R.C.W. Zwanenburg, A.J.H. Klunder and B. Zwanenburg, *Tetrahedron* **44**, 1413 (1988).
39. W. Ammann, F.J. Jäggi and C. Ganter, *Helv. Chim. Acta* **63**, 2029 (1980); R.A. Pfund and C. Ganter, *Helv. Chim. Acta*, **62**, 228 (1979); W.B. Schweizer, J.D. Dünitz, R.A. Pfund, G.M. Ramos Tombo and C. Ganter, *Helv. Chim. Acta*, **64** 2738 (1981).

# STUDIES ON THE SYNTHESIS OF CYCLOPENTANOIC SESQUITERPENES VIA REARRANGEMENT ROUTES: (±)MODHEPHENE AND (±)ISOCOMENE [1]

Lutz Fitjer[*], Andreas Kanschik and Marita Majewski
Institut für Organische Chemie der Universität Göttingen
Tammannstraße 2, D-3400 Göttingen, FRG

**Abstract:** Based on model studies with dispiroundecane **9**, dispiroundecane **1** has been synthesized and rearranged to both (±)modhephene **6** and (±)isocomene **7**.

## Introduction

The cyclopentanoic sesquiterpenes (±)modhephene **6** and (±)isocomene **7** have been the focus of considerable interest during the past decade [2]. A large number of ingenious syntheses have emerged [2], but none of them took advantage of the fact that **6** and **7** may formally be derived from each other by three consecutive 1,2-shifts (**6-2-3-4-5-7**). We have now found that cascade rearrangement of dispiroundecane **1** is not only a well suited method for a synthesis of **6**, but also gives access to **7**, via **3** and **4** [3].

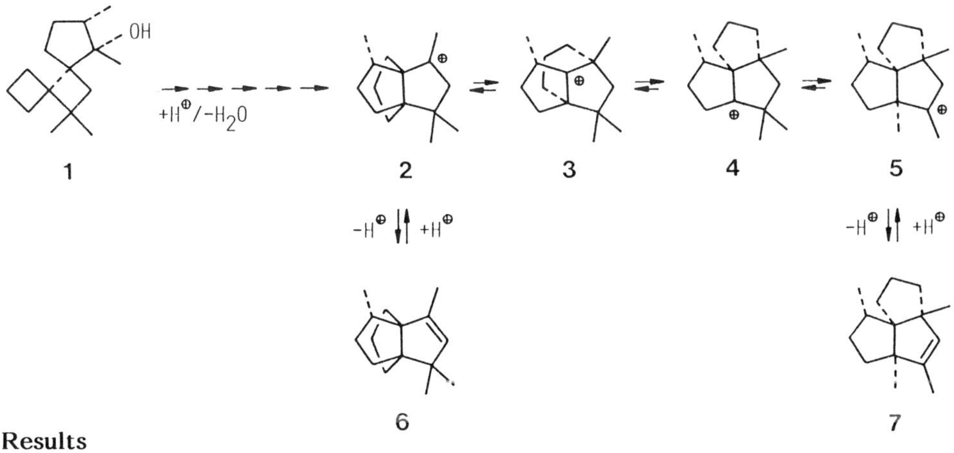

## Results

Based on the fact that on treatment with acids [7,7-$D_2$]-**8** rearranges regiospecifically and without loss of deuterium to [8,8-$D_2$]-**10** we have recently recognized [4] dispiroundecane **1** as a promising candidate for a direct conversion to (±)modhephene **6**. Assuming that the rearrangement of **1** proceeds analogously to the rearrangement of [7,7-D$_2$]-**8**, the methyl group at C-7 was thought to preserve its stereochemistry and to end up exclusively at C-8 of **6**. However, as an analogous rearrangement seemed endangered by the presence of the geminal dimethyl group at C-11, we first studied the rearrangement of dispirane **9** as a model until we rearranged dispirane **1**.

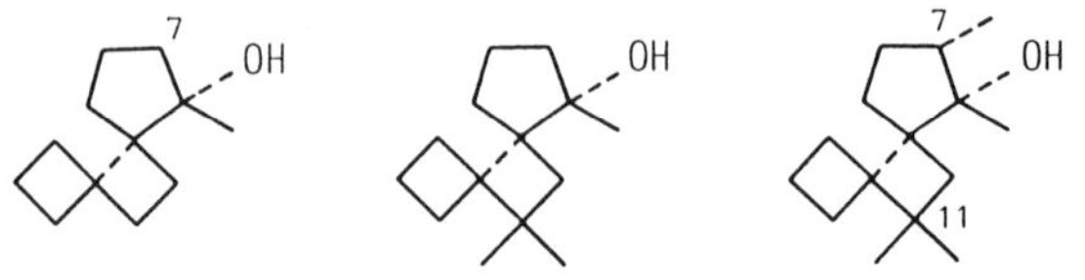

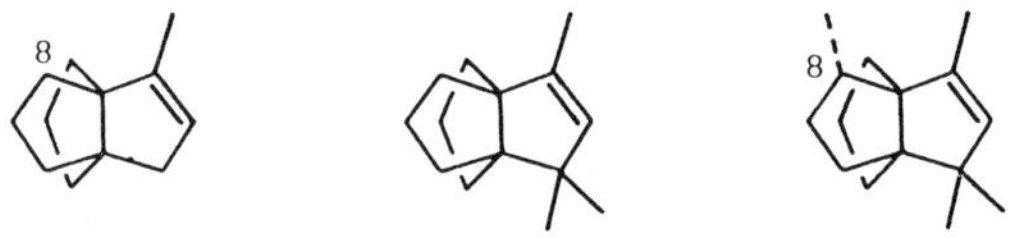

Dispiranes **9** and **1** were obtained as follows: addition of dichloroketene [5] to isopropylidenecyclobutane **12** and subsequent dechlorination [6] yielded spiroketone **13** [7] which was cyclobutylidenated to **14** [7], and then epoxidized and rearranged to dispiroketone **15** [7]. The synthesis of **16** [7] was achieved by a sequence of methylation, enolization and reprotonation [8], and stereoselective addition of methyllithium to **15** and **16** then yielded **9** [7] and **1** [7], respectively.

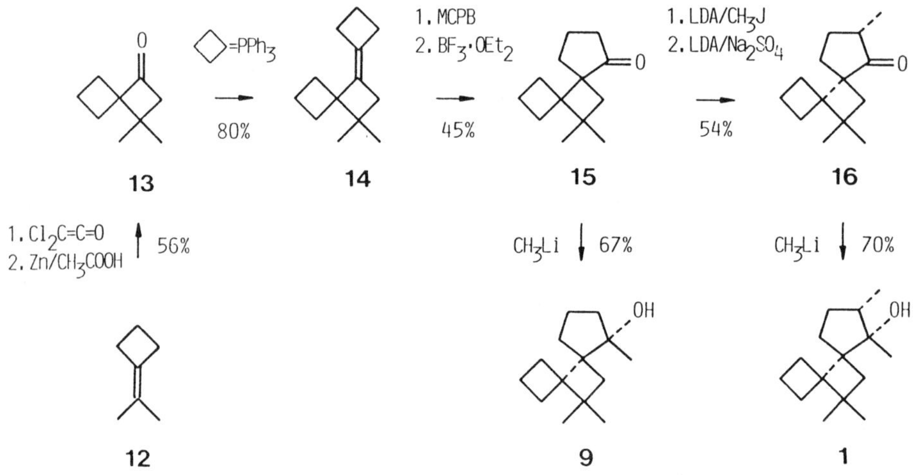

Rearrangement of **9** could be accomplished by heating 0.50 molar solutions in benzene with equivalent amounts (w/w) of Nafion-H [9]. Complete conversion to a 85:15 mixture of propellane **11** [7] and triquinane **21** [7] was observed within 2.5 h, at 70°C, but after 18 h at 70°C **21** was the only product. It thus turned out, that the formation of **11** is kinetically controlled, and the formation of **21** is thermodynamically controlled. Interestingly, this result agrees well with the calculated heats of formation [10] of **11, 21, 22** and **23**.

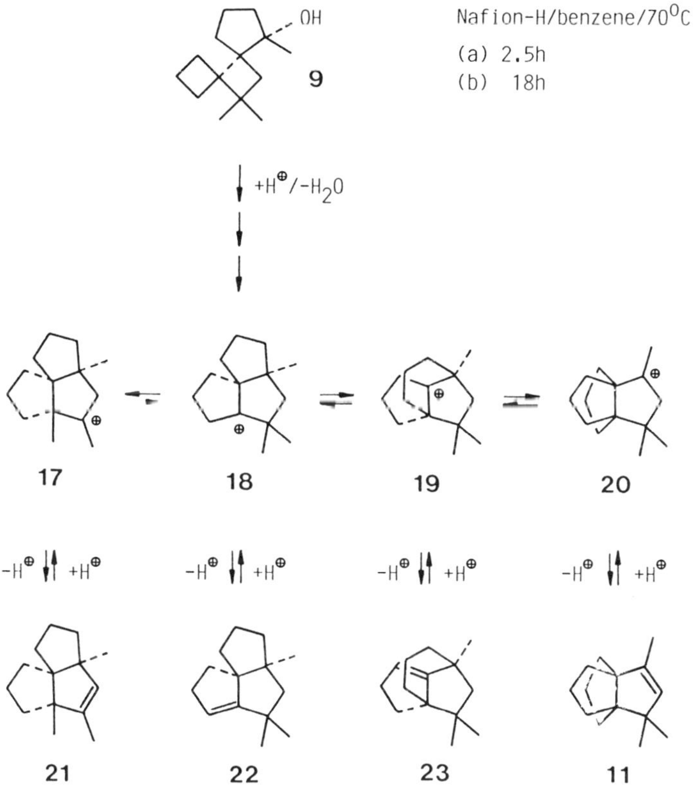

$\Delta H_f$ [kcal/mol]   −21.9        −15.9        −14.5        −19.0

(a) 15%                              (a) 85%
(b) 100%                             (b)  −

We therefore have used the calculated heats of formation of all olefins which could be derived from the two series of diastereoisomeric tertiary carbenium ions (**24-26, 2-5** and **32-38**) in a search for probable rearrangement products of **1**. Since bridgehead olefins (**22,23**) had not been formed during the rearrangement of **9**, ( )modhephene **6** was the candidate of choice for a rearrangement of **1** under kinetic control, but triquinane **27**, ( )isocomene **7** and triquinane **39** were possible candidates for a rearrangement under thermodynamic control.

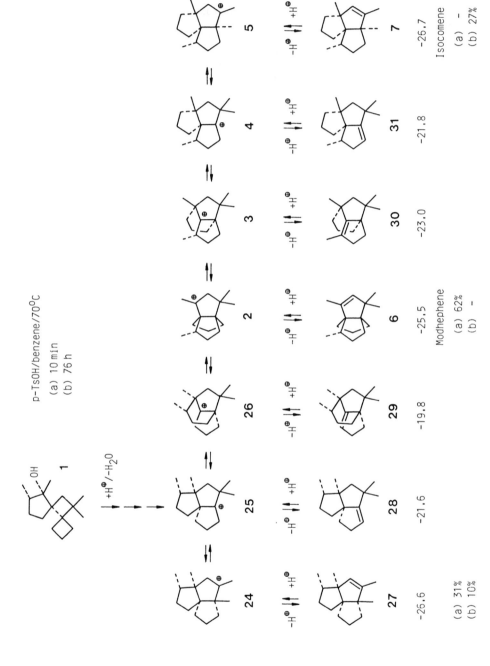

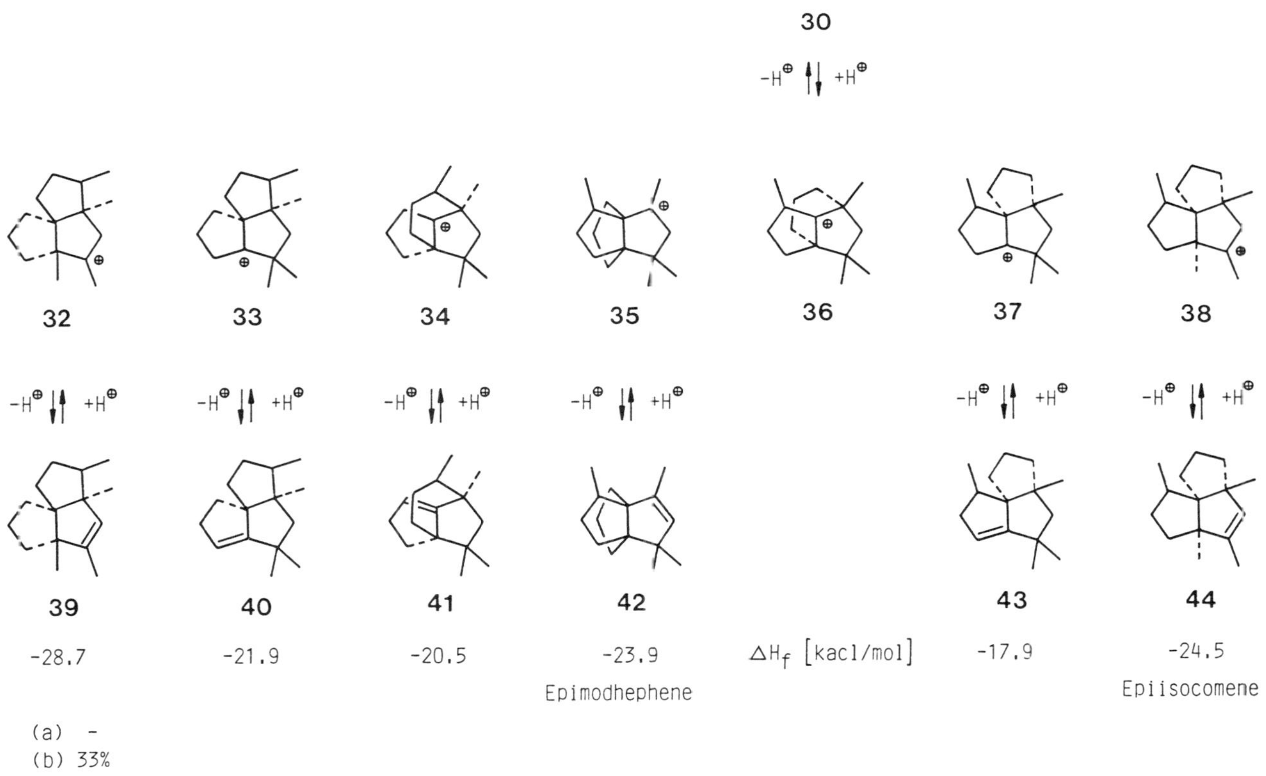

$\Delta H_f$ [kcal/mol]

| | | | | | | |
|---|---|---|---|---|---|---|
| 32 | 33 | 34 | 35 | 36 | 37 | 38 |
| 39 | 40 | 41 | 42 | 43 | 44 | |
| −28.7 | −21.9 | −20.5 | −23.9 | | −17.9 | −24.5 |
| | | | Epimodhephene | | | Epiisocomene |

(a) —
(b) 33%

The experimental verification has been achieved by the rearrangement of **1** with an equimolar amount of a 0.040 molar solution of p-toluenesulfonic acid in benzene-$d_6$. When this solution was heated to $+70°C$ and the reaction progress monitored by $^1$H-NMR spectroscopy [11], the following facts could be observed: after 10 min **1** had been completely consumed and (±)modhephene **6** [12] (62%) and triquinane **27** [7] (31%) had formed instead. However, after 76h (±)modhephene **6** had disappeared, the yield of triquinane **27** (10%) had substantially decreased, and (±)isocomene **7** [12] (27%) and triquinane **39** [7] (33%) were the principal products. This clearly indicates that the rearrangement of **1** to (±)modhephene **6** and triquinane **27** is kinetically controlled, whereas the subsequent rearrangement of both **6** and **27** to (±)isocomene **7** and triquinane **39** is thermodynamically controlled. From the multitude of rearrangement patterns observed one is tempted to speculate that triquinanes **27** and/or **39** may well be naturally occuring sesquiterpenes yet to be discovered.

Interestingly to note, the bridgehead olefin **28** [7], not beeing formed under acidic rearrangement conditions, has been obtained by treatment of **1** with thionyl chloride in pyridine. Under these conditions, reprotonation is impossible, and hence no further rearrangement is observed.

In summary, from a total of thirteen conceivable rearrangement products of **1**, five have been isolated and identified. Moreover, an efficient entry to (±)modhephene **6** has been established, and its rearrangement to (±)isocomene **7** has been demonstrated for the first time.

**Acknowledgements:** Financial support of the Deutsche Forschungsgemeinschaft (project Fi 191/8-2) and the Fonds der Chemischen Industrie is gratefully acknowledged.

## References and notes

[1] Polyspiranes, 19, Cascade Rearrangements, 14; for communications 18 and 13 see L.Fitjer and U.Quabeck, **Angew. Chem. 100** (1988), in press; part of this work has been submitted to **Tetrahedron Lett.** for publication.

[2] Reviews: L.A.Paquette, **Top. Curr. Chem. 79**, 41 (1979); **119**, 1 (1984); L.A.Paquette and A.M.Doherty, **Polyquinane Chemistry**, Springer-Verlag, Berlin Heidelberg 1987.

[3] Carbenium ion **4** has been has been proposed as a biogenetic precursor of both **6** and **7**: L.H.Zalkow, R.N.Harris, III and D. van Derveer, **J. Chem. Soc. Chem. Commun. 1978**, 420.

[4] L.Fitjer, M.Majewski and A.Kanschik, **Tetrahedron Lett. 29**, 1263 (1988)

[5] G.Mehta and H.S.P.Rao, **Synth. Commun. 15**, 991 (1985).

[6] D.A.Bak and W.T.Brady, **J. Org. Chem. 44**, 101 (1979).

[7] The new compounds **1,9,11,13,14,15,16,21,27** and **28** gave correct elemental analyses and/or high resolution mass spectroscopic data. **39** could not be separated from **27** and **6** and was characterized by 1H-

and 13C-NMR spectroscopy only. The IR, 1H-NMR, 13C-NMR and mass spectroscopic data are consistent with the given structures. 13C-NMR data (CDCl3) are as follows: **1**: $\delta$ = 12.20 (Cprim), 17.69 (Csek), 21.34, 26.85 (Cprim), 27.48 (Csek), 27.62 (Cprim), 27.96, 28.25 (Csek), 33.26 (Cquart), 33.69 (Csek), 41.68 (Ctert), 44.19 (Csek), 50.47, 51.75, 83.21 (Cquart); **9**: $\delta$ = 17.47, 18.20 (Csek), 26.71 (Cprim), 27.54 (Csek), 27.61 (Cprim), 28.11 (Csek.), 32.85 (Cquart), 39.05, 39.68, 42.46 (Csek), 49.51, 51.42, 82.79 (Cquart); **11**: $\delta$ = 13.52 (Cprim), 27.20 (Csek), 27.65 (Cprim), 37.45, 38.07 (Csek), 46.21, 64.84, 72.15 (Cquart), 135.15 (Ctert), 140.02 (Cquart); **13**: $\delta$ = 15.52, 24.10, 25.84, 30.51, 56.82, 68.11, 214.69; **14**: $\delta$ = 15.74, 17.95 (Csek), 24.69 (Cprim), 28.27, 29.32, 29.68 (Csek), 35.29 (Cquart), 39.75 (Csek), 54.95, 129.41, 134.93 (Cquart); **15**: $\delta$ = 16.05, 19.86 (Csek.), 25.23 (Cprim), 25.76 (Csek), 26.76 (Cprim), 27.16 (Csek), 33.72 (Cquart), 35.28, 37.61, 39.89 (Csek), 51.45, 54.16, 220.71 (Cquart); **16**: $\delta$ = 14.22 (Cprim), 15.51 (Csek), 25.06 (Cprim), 25.28 (Csek), 26.59 (Cprim), 27.53, 29.57, 33.36 (Csek), 33.51 (Cquart), 39.92 (Csek), 42.78 (Cprim), 51.33, 54.40, 222.14 (Cquart); **21**: $\delta$ = 12.98, 22.43, 23.93 (Cprim), 24.42, 24.63, 37.14, 37.39, 39.93, 42.67 (Csek), 55.78, 59.33, 61.30 (Cquart), 133.46 (Ctert), 142.64 (Cquart); **27**: $\delta$ = 13.00, 15.99, 17.59, 23.24 (Cprim), 24.87, 32.87, 34.97, 38.70, 39.61 (Csek), 45.55 (Ctert), 57.97, 59.49, 63.03 (Cquart), 133.96 (Ctert), 141.67 (Cquart); **28**: $\delta$ = 17.61, 20.65, 31.33, 31.54 (Cprim), 33.34, 35.21 (Csek), 36.10 (Cquart), 36.56, 38.41 (Csek), 44.34 (Ctert), 50.78, 61.44 (Csek), 69.04 (Cquart), 115.61 (Ctert), 167.03 (Cquart); **39**: $\delta$ = 13.17, 14.96, 22.26, 22.44 (Cprim), 24.57, 33.36, 35.30, 37.25, 39.85 (Csek), 47.13 (Ctert), 58.03, 59.56, 62.07 (Cquart), 129.63 (Ctert), 143.67 (Cquart).

[8] T.Kametani, M.Tsubuki, K.Higurashi and T.Honda, **J. Org. Chem. 51**, 2932 (1986).

[9] Review: G.A.Olah, **Synthesis 1986**, 513.

[10] Program MM2: N.L.Allinger, **J. Am. Chem. Soc. 98**, 8127 (1977).

[11] Measurements were performed on a Varian VXR 500 spectrometer operated at 500 MHz and +70°C using the original rearrangement mixtures and $\delta$H(C6HD5) = 7.25 ppm as reference line. The vinylic protons of **6, 7, 39** and **27** showed narrow quartets (J = 1.3 Hz) at $\delta$ = 4.93, 4.95, 5.03 and 5.14 ppm, respectively.

[12] (±)Modhephene **6** [3,13] and (±)isocomene **7** [14] were identified using their known 13C-NMR data. The same technique revealed the absence of (±)epimodhephene **42** [15] and (±)epiisocomene **44** [16] throughout the whole rearrangement.

[13] H.Schostarez and L.A.Paquette, **Tetrahedron 37**, 4431 (1981).

[14] F.Bohlmann, N. Le Van and J.Pichardt, **Chem. Ber. 110**, 3777 (1977).

[15] M.Karpf and A.S.Dreiding, **Helv. Chim. Acta 64**, 1123 (1981).

[16] B.C.Ranu, M.Kavka, L.A.Higgs and T.Hudlicky, **Tetrahedron Lett. 25**, 2447 (1984).

# DIRECTIONALITY IN FORMATION OF SMALL RINGS BY INTRAMOLECULAR NUCLEOPHILIC SUBSTITUTION

C.J.M. Stirling
Department of Chemistry
University College of North Wales
Bangor, Gwynedd LL57 2 UW
United Kingdom

ABSTRACT. The formation of small rings by intramolecular nucleophilic substitution is examined with respect to the activation parameters for such processes set beside calculations of the transition structure dimension and the problem of alignment of nucleophile, electrophile and leaving group. The consequences of restraint on the leaving group trajectory are considered in the light of synthetically useful procedures involving intramolecular attack on strained rings and the often inefficient harnessing of strain in the acceleration of elimination reactions.

Formation of strained rings by intramolecular nucleophilic displacement reactions (Scheme 1) is a familiar process. Until investigations by Dr. A.C. Knipe, working in the author's group in the mid-1960's,[1] it was assumed on the basis of investigations of the cyclisation of ω-halogeno amines and alcohols (Scheme 1 Nu = $Ru_2N$ or O, E = $CH_2$ and LG = halogen) that formation of three-membered and four-membered rings was very slow by comparison with that of 5-membered rings.[2] Knipe[1,3] showed that cyclisations

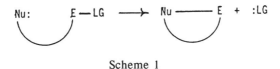

Scheme 1

involving carbon or sulphur nucleophiles gave 3- and 5-rings substantially more rapidly than 4-rings and assumptions that had been previously made about the role of strain in controlling such processes had to be put aside. These results stimulated a substantial amount of work aimed at unravelling the effect of conjugative groups,[4] the ranking of leaving groups in such processes,[5] and the timing of the deprotonation - leaving group expulsion sequence.[6] An important advance in this field came when Benedetti[7] determined activation parameters for the mechanistically unambiguous system of Scheme 2.

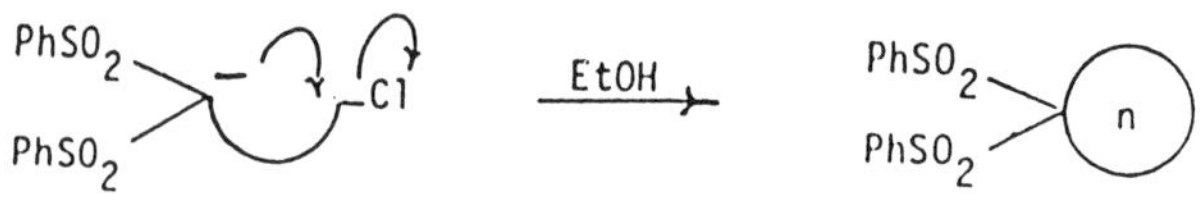

Scheme 2

This system, employing a carbon acid of $pK_a$ 12, allows direct observation of the cyclisation of the carbanion. Rates and activation parameters are in Table 1. The following striking features emerge
- (i) 3-ring formation is much faster than any other ring size and this is because of the positive entropy term.
- (ii) $\Delta H^{\#}$ for 3-ring formation is <u>lower</u> than for the less strained 4-ring
- (iii) $\Delta H^{\#}$ for the 6-ring rises again to almost the value found for the 3-ring and for the whole series, $\Delta H^{\#}$ varies by only 4.2 kcal mol$^{-1}$ against a strain energy differential of 27.4 kcal mol$^{-1}$.

## INTRAMOLECULAR NUCLEOPHILIC DISPLACEMENT

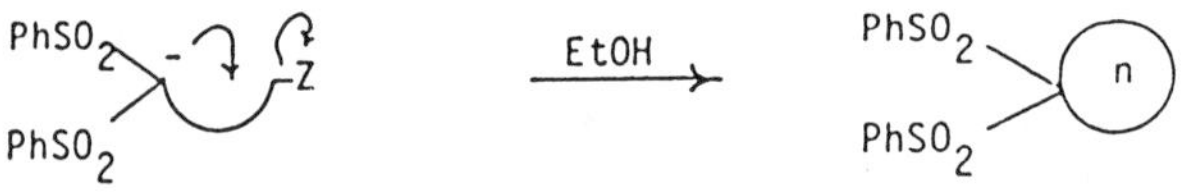

| Ring Size n | Ring Strain | $k_{rel}$ | $\Delta H^{\neq}$ kcal mol$^{-1}$ | $\Delta S^{\neq}$ cal K$^{-1}$ mol$^{-1}$ |
|---|---|---|---|---|
| 3 | 27.4 | 1 | 20.5 | 10 |
| 4 | 26.0 | $6.7 \times 10^{-6}$ | 21.8 | -9 |
| 5 | 6.1 | $1.6 \times 10^{-2}$ | 16.3 | -12 |
| 6 | 0 | $2.8 \times 10^{-6}$ | 19.8 | -18 |

Table 1

An attempt was made to assess the extent of leaving group departure in the transition structure for the system using Hammett leaving groups (Table 2).[8] The system is insentitive but the manifestly greater value of $\rho$ for the 3-ring suggests larger leaving group departure for this ring size and hence, presumably, a greater extent of ring closure.

Calculations, using MNDO, in Verhoeven's[9] laboratory are consistent with the experimental conclusions (Scheme 3). In both unstabilized (first line) and stabilized (second line) carbanion systems, the degree of ring closure and leaving group departure is consistently greater for the 3-ring than for the 4-ring, 5-ring, or intermolecular processes.

# CARBANION CYCLISATION WITH HAMMETT LEAVING GROUPS

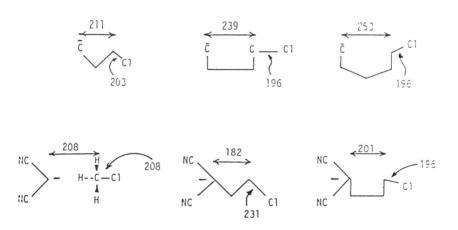

| Ring Size | krel(Ar=Ph) | ρ (m- Me, H, OMe, Cl) |
|---|---|---|
| 3 | 7.8 x 10$^5$ | 1.52 |
| 4 | 1 | 1.30 |
| 5 | 2.8 x 10$^4$ | 1.45 |

Table 2

## CALCULATED TRANSITION STRUCTURE DIMENSIONS (pm)
## FOR CARBANION CYCLISATION

Scheme 3

This clear evidence for a substantially formed ring in the transition structure for three-ring closure raises the question of alignment of nucleophile, electrophile and leaving group in this transition structure. Two, by now classical, investigations bear on the question of alignment in transition structures for displacement reactions. The first showed[10] that *gem*-dimethyl substitution on carbon adjacent to that bearing the leaving group (i.e. a neopentylic structure) reduced reactivity in the intermolecular process by a factor of $10^5$ (Scheme 4). This severe retardation was attributed to obstruction of the nucleophile trajectory by what is effectively an α-t-butyl group. For the intramolecular reaction however (Scheme 4) *gem*-dimethyl substitution is strongly accelerative[5] giving

ALIGNMENT IN SMALL RING FORMATION

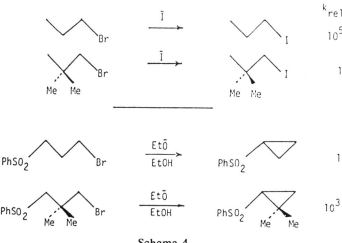

Scheme 4

an out-of-match of $10^8$ between the two processes. In this latter case there is no obstruction of the nucleophile and proximation of nucleophile and electrophile is encouraged both by diminution of the C-$CMe_2$-C angle and by increased population of conformers favourable for cyclisation.
The second is the characteristically elegant and exacting study of Eschenmoser[11] on intramolecular displacement (Scheme 5). The convenient and superficially reasonable assumption that alkylation of the sulphone-stabilized carbanion should occur intramolecularly was clearly disproved with a direct bearing on the trajectory for such a displacement namely that nucleophile, electrophile, and leaving group should be collinear in the transition structure or nearly so.

INTER- VS INTRAMOLECULAR NUCLEOPHILIC SUBSTITUTION

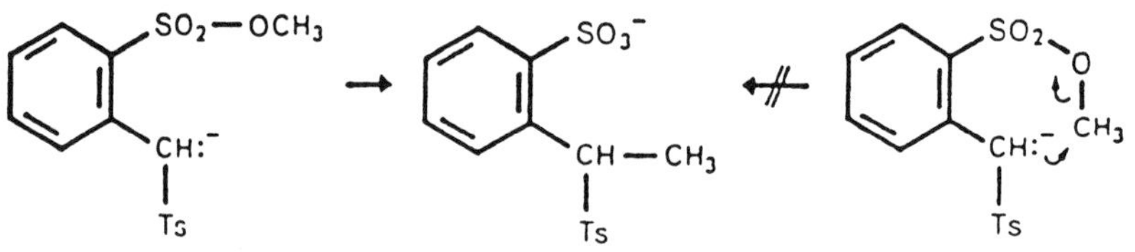

Scheme 5

With this sort of evidence available, consideration should be given as to the directionality of the intramolecular displacements of Scheme 2. The deviation angles for cyclisation to various ring sizes in Scheme 6 are simply derived from measurements on models. The values vary according to whether an $sp^3$ or $sp^2$ hybridised carbanion is the appropriate description. Note that deviations may be positive, as in the case of 3- and 4-membered rings, requiring a decrease in one or more of the C-C-C bond angles intervening between nucleophile and electrophile. For 5- and 6-membered rings, <u>widening</u> of inter-

### DEVIATION ANGLES FROM LINEAR TRAJECTORIES

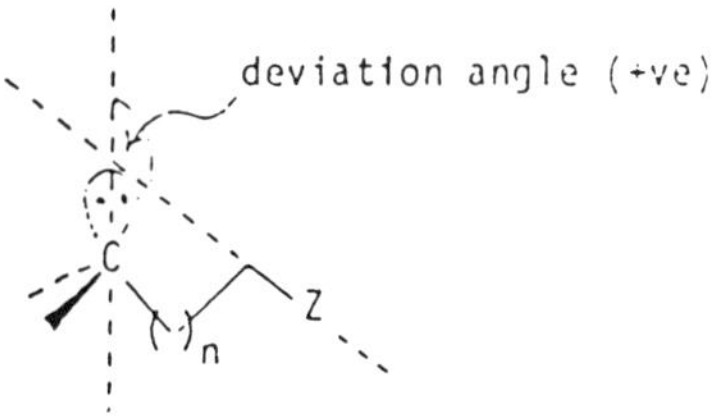

|        | 3-Ring | 4-Ring | 5-Ring | 6-Ring (Chair) |
|--------|--------|--------|--------|----------------|
| $sp^3$ | +112   | +48    | −22    | −67            |
| $sp^2$ | +96    | +30    | −39    | −85            |

Scheme 6

vening C-C-C bond angles is required in the approach to collinearity. The deviations from linearity in the small rings are enormous. What, then, is the enthalpic price of such deviation? An answer is portrayed in Figure 1. Maier and his co-workers[12] have calculated using MNDO, the variation of transition structure energy as a function of

### CALCULATED (MNDO) ENTHALPIC PRICE OF NON-LINEARITY IN $S_N2$ DISPLACEMENTS

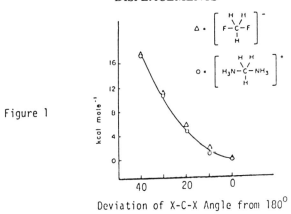

Figure 1

Deviation of X-C-X Angle from $180°$

deviation from linearity for two reactions giving negatively and positively charged structures respectively. It can be seen that energy rises very rapidly as the deviation angle increases and at an angle of 40° is greater than 16 kcal mol$^{-1}$. How then can cyclisation occur in a transition structure whose details have been quite intimately explored? Even if it is allowed that, especially for 3-membered rings, a simple geometric structure is a deficient portrayal of reality,[13] a considerable out-of-match exists. It is interesting that $\Delta H^{\#}$ for the 4-, 5-, and 6-membered rings correlates with the algebraic deviation angle.

For the reactions of Table 1, there is no restriction on the departure trajectory of the leaving group. It is of considerable interest in this general connection, therefore, to know of the effect of such restriction. The matter has been considered[14,15] in respect of the synthetic application of intramolecular displacement on epoxides (Scheme 7). The trajectory was considered to determine which of the alternative ring sizes was produced

### INTRAMOLECULAR NUCLEOPHILIC SUBSTITUTION ON EPOXIDES

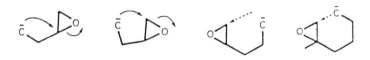

Scheme 7

(although this is an indecisive matter[15]) but until very recently no quantitative data were available. The pertinent results are in Table 3.[16]

### REGIOSPECIFICITY AND DIRECTIONALITY IN INTRAMOLECULAR ATTACK ON EPOXIDES

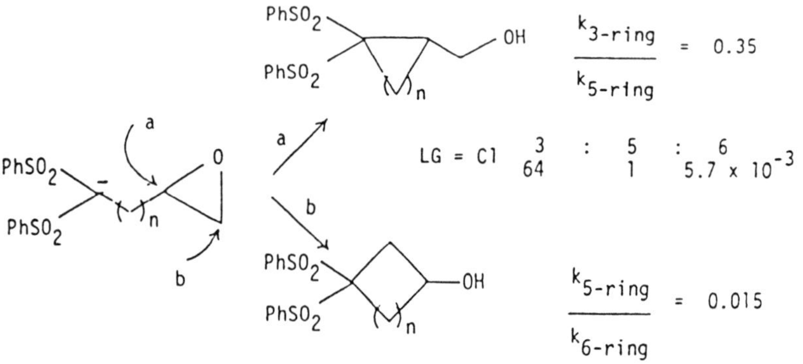

$$\frac{k_{3\text{-ring}}}{k_{5\text{-ring}}} = 0.35$$

$$LG = Cl \quad \begin{matrix} 3 \\ 64 \end{matrix} : \begin{matrix} 5 \\ 1 \end{matrix} : \begin{matrix} 6 \\ 5.7 \times 10^{-3} \end{matrix}$$

$$\frac{k_{5\text{-ring}}}{k_{6\text{-ring}}} = 0.015$$

Table 3

At once, the effect of imposition of an additional trajectory restraint in the 3-ring system is apparent. With Cl as an unrestricted leaving group the 3-ring to 5-ring ratio is 64; with the epoxy leaving group in pathway a, the ratio decreases to 0.35. The other striking trajectory effect is seen in the 5-ring 6-ring comparison for pathway b in which the distal epoxy carbon atom is the electrophile. When there is no leaving group restraint, the 5:6 ratio is 1:5.7 x $10^{-3}$. For pathway b, the 5:6 ratio is 1:66.

In conclusion, therefore, it is clear that directionality effects make themselves apparent, but in rather unpredictable ways. Certainly the gross effects which might be expected in these reactions do not seem to make themselves felt when there is no leaving group restraint. When there is leaving group restraint, the system becomes much more sensitive. It is very noticeable that in our earlier work[17] on the harnessing of strain in the acceleration (Table 4), the proportions of strain energy expressed in acceleration are rather small and this may be due to unfavourable leaving group-nucleophile trajectrories. Further work is in progress to illuminate this problem.

## HARNESSING OF STRAIN

| | | % | Ref |
|---|---|---|---|
| 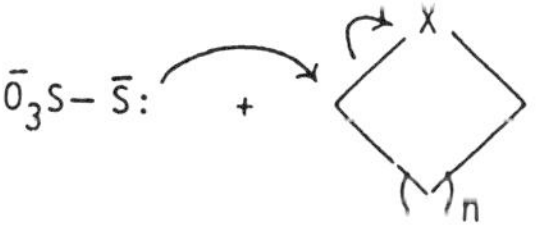 $\bar{O}_3S-\bar{S}:$ + (epoxide/thiirane) | $X = O$    $\frac{3\text{-ring}}{5\text{-ring}} = 10^{11}$ | 60 | 16a |
| | $X = S$    $\frac{3\text{-ring}}{5\text{-ring}} = 10^{9}$ | 85 | |
| 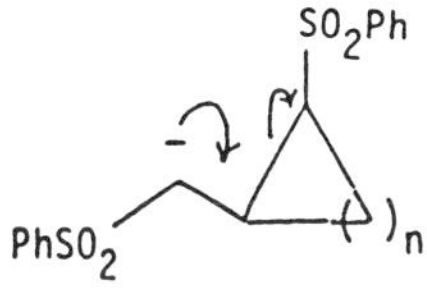 PhSO$_2$ ... SO$_2$Ph | $\frac{3\text{-ring}}{\text{open-chain}} = 10^{11}$ | 45 | 16b |
| | $\frac{4\text{-ring}}{\text{open chain}} = 10^{7}$ | 26 | |
| 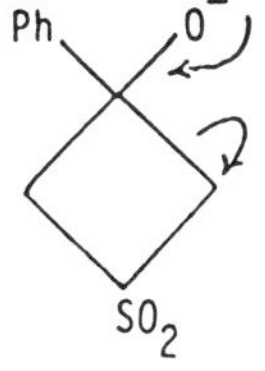 Ph, O$^-$ / SO$_2$ | $\frac{4\text{-ring}}{\text{open chain}} = 10^{7}$ | 33 | 16c |

Table 4

**References and Footnotes**

1.  A.C. Knipe and C.J.M. Stirling, *J. Chem. Soc. (B)*, 1967, 808.
2.  C.H.M. Stirling, *J. Chem. Ed.*, 1973, **50**, 84.
3.  A.C. Knipe and C.J.M. Stirling, *J. Chem. Soc. (B)*, 1968, 67.
4.  R. Bird, A.C. Knipe and C.J.M. Stirling, *J. Chem. Soc., Perkin II*, 1973, 1215.
5.  B. Issari and C.J.M. Stirling, *J. Chem. Soc. Perkin Trans. II*, 1984, 1043.
6.  R. Bird, G. Griffiths, G.F. Griffiths and C.J.M. Stirling, *J. Chem. Soc. Perkin Trans. II*, 1982, 579.
7.  F. Benedetti and C.J.M. Stirling, *J. Chem. Soc. Perkin Trans. II*, 1986. 605.
8.  F. Benedetti, S. Fabrissin, A. Rusconi and C.J.M. Stirling, *Gazz. Chim. Ital.*, 1988, **118**, 233.
9.  S.M. van der Kerk, J.W. Verhoeven and C.J.M. Stirling, *J. Chem. Soc. Perkin Trans. II*, 1985, 1355.
10. N. Isaacs, 'Physical Organic Chemistry', Longmans, London, 1987, p. 381.
11. L. Tenud, S. Farooq, J. Seibl and A. Eschenmoser, *Helv. chim. Acta*, 1970, **53**, 2059.
12. D.H. Heathcock, T.W. von Geldern, C.B. Lebrilla and W.F. Maier, *J. Org. Chem.*, 1985, **50**, 968.
13. L.A. Paquette, Workshop comment.
14. G. Stork, L.D. Cama and D.R. Coulson, *J. Am. Chem. Soc.*, 1974, **96**, 5268; G. Stork and J.F. Cohen, *ibid*, 1974, **96**, 5270.
15. J.Y. Lallemand and M. Onanga, *Tetrahedron Lett.*, 1975, 585.
16. F. Benedetti, S. Fabrissin, T. Gianferrara and A. Risaliti, *J. Chem. Soc. Chem. Commun.*, 1987, 406.
17. (a) J.I. Lynas-Gray and C.J.M. Stirling, unpublished work.
    (b) S. Hughes, G. Griffiths and C.J.M. Stirling, *J. Chem. Soc. Perkin Trans. II*, 1987, 1253.
    (c) D.J. Young and C.J.M. Stirling, *J. Chem. Soc. Chem. Commun.*, 1987, 552.

# SYNTHESIS, CHIROPTICAL PROPERTIES AND SYNTHETIC APPLICATIONS OF PERHYDROTRIQUINACENE-1,4,7-TRIONE

*C. Almansa*[a], *M.L. García*[a], *C. Jaime*[b], *A. Moyano*[a], *M.A. Pericás*[a] *and F. Serratosa*[a,*]

[a]Department of Organic Chemistry. University of Barcelona. E-08028 Barcelona; [b]Department of Chemistry. Autonomous University of Barcelona. E-08138 Barcelona. Spain.

ABSTRACT. The racemic and the (+)-enantiomer of the title compound have been synthesized by an intramolecular Pauson-Khand bis-annulation, and their chiroptical properties and synthetic applications have been studied.

*1.1 Triquinacene as the Target.* $C_{10}H_{10}$ polycyclic hydrocarbons are a class of theoretically important compounds that have been exhaustively investigated in the last 25 years, and include molecules such as bullvalene -an unusual *fluxional*[1]molecule that undergoes *pandegeneration*- and triquinacene that may be regarded as an *homoaromatic* neutral molecule[2], as well as the building block for an "ideal" convergent synthesis of dodecahedrane.[3] Therefore, after our successful synthesis of bullvalene (**4**),[4] an obvious synthetic objective was the synthesis of triquinacene (**2**) starting from the same tris-*alfa*-diazoketone and following a similar synthetic scheme, which involved the synthesis of perhydrotriquinacene-1,4,7-trione (**1**) as the key intermediate. In practice, however, the synthesis failed and the reaction product, formed in more than 70% yield, was in fact a triasterane derivative (**3**) (see Scheme 1).[5]

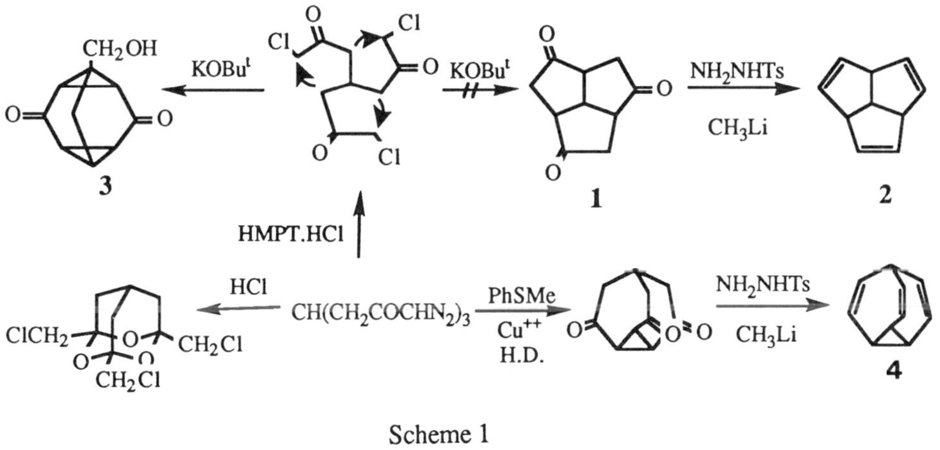


Scheme 1



A. de Meijere and S. Blechert (eds.), Strain and Its Implications in Organic Chemistry, 447–450.

*1.2. New Strategies for a New Target Molecule.* In the meantime, a rational synthetic analysis of dodecahedrane let us to realize that the racemic form of perhydrotriquinacene-1,4,7-trione would be the "ideal" synthon for the long-sought convergent synthesis of this highly symmetrical molecule.[6] Accordingly, new strategies for the synthesis of the triketone were designed.(see Scheme 2) However, since most of these strategies started either from triquinacene itself (**2**) or from some of its simple derivatives or synthetic intermediates, the synthesis of triquinacene turned out to be redundant and all our efforts were then directed to the synthesis of perhydrotriquinacene-1,4,7-trione (**1**) as the starting material for a new synthesis of dodecahedrane (**5**).

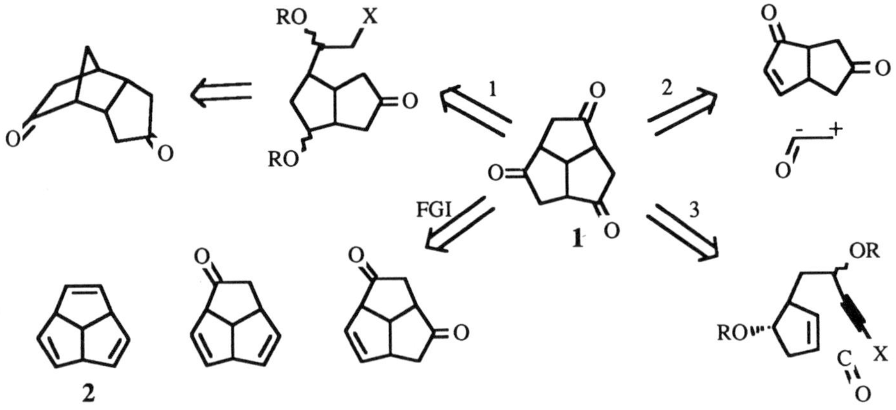

Scheme 2

The strategies for the synthesis of perhydrotriquinacene-1,4,7-trione imply either :
i) a regioselective control in order to create the three 1,4-dissonant relationships between the three carbonyl groups; or, ii) a stereoselective control in order to achieve the *all-cis* configuration characteristic of polyquinanes.

Although successful syntheses were developed following these strategies,[6,7] a new one involving an intramolecular Pauson-Khand bis-annulation was also visualized, which does not require any type of control, provided that the precursor is properly designed.

*1.3 Reconsidering the Objectives once more.* The accomplishment of this synthesis via an intramolecular Pauson-Khand bis-annulation (see Scheme 3)[6,8,] starting from lactol **6** let us to reconsider again our objectives, in special:

i) the *mise au point* of the new synthesis of triquinacene as originally planned from the very beginning (**1** --- **2**),

ii) the synthesis of an enantiomerically pure form of the triketone (+)-**1** starting from an optical active lactone (+)-**7**, commercially available, and study of its chiroptical properties,

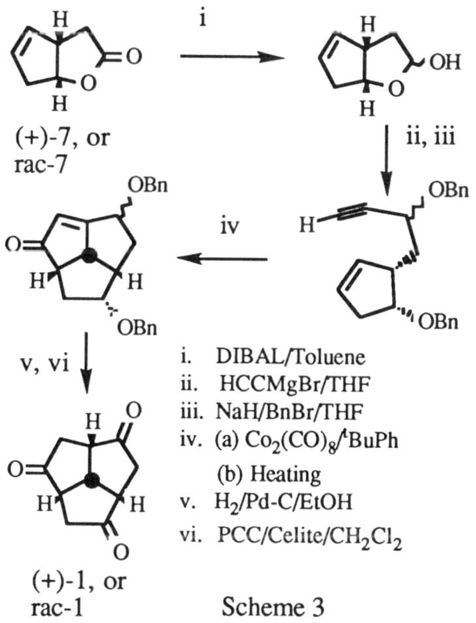
i
ii, iii
iv
v, vi
(+)-7, or
rac-7
(+)-1, or
rac-1
i. DIBAL/Toluene
ii. HCCMgBr/THF
iii. NaH/BnBr/THF
iv. (a) $Co_2(CO)_8$/$^t$BuPh
(b) Heating
v. $H_2$/Pd-C/EtOH
vi. PCC/Celite/$CH_2Cl_2$

Scheme 3

iii) the synthesis of dodecahedrane (**5**) by the "Narcissistic coupling"[9] of the racemic triketone **1**, -a process that involves a series of acid-induced aldol condensations in which all the steps are reversible, except the last one leading to hexahydroxydodecahedrane (**6**), that implies an irreversible transannular cyclization of a "rigid cyclooctanone" in which the interacting carbon atoms are almost at the proper bond distance (see Scheme 4), and finally,

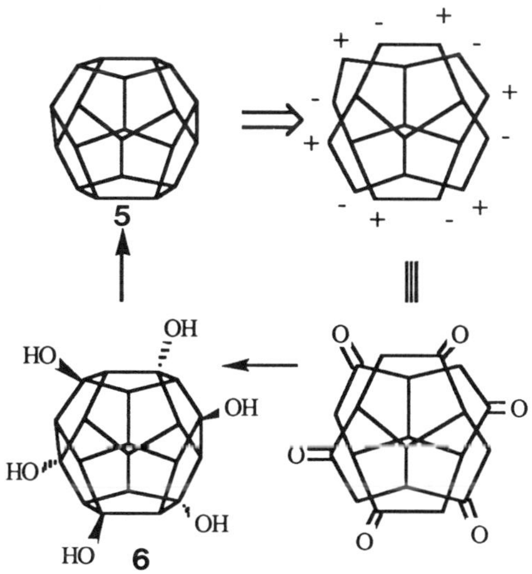
5
6

Scheme 4

iv) the ready accessibility of triketone **1** either in the racemic or the enantiomerically pure forms, together with the symmetrical functionalization around the perimeter of its rigid hemispherical framework, let us to explore its usefulness as a "concave chiral cap" for the synthesis of some model hosts (tripodands and cavitands), which could be of interest either in ion transport or molecular and chiral recognition.[10]

In this short communication a brief account of the work done in oder to achieve all these objectives will be given. Although the "Narcissistic coupling" of the racemic triketone, under a variety of different experimental conditions, did not give the expected hexahydroxydodecahedrane (**6**) -in spite of being detected in some experiments by the CI(NH3)/MS technique-, some empirical molecular mechanics (MM2) and semiempirical (MNDO) molecular orbital calculations were performed in order to evaluate the strain and the heat of formation of all the intermediates involved in the process.


REFERENCES

1. a) Doering, W. von E., and Roth, W.R.: *Angew. Chem. Internat. Edn.*, **1963**, *2*, 115; *Tetrahedron*, **1963**, *19*, 715; b) Schroeder, G: *Chem. Ber.*, **1964**, *97*, 3140.

2. Liebman, J.F., Paquette, L.A.,Peterson, J.R., and Rogers, D.W.; *J. Am. Chem. Soc.*, **1986**, *108*, 8267.

3. Woodward, R.B, Fukunaga, T., and Kelly, R.C.: *J. Am. Chem. Soc.*, **1964**, *86*, 3162.

4. Serratosa, F, Lopez, F., and Font, J.; *Anales de Quim.*, **1974**, *70*, 893.

5. Herranz, E., and Serratosa, F.: *Tetrahedron*, **1977**, *33*, 995.

6.Carceller, E., García, M.L., Moyano, A., Pericás, M.A., and Serratosa,F.;*Tetrahedron*, **1986**, *42*, 1831.

7.a) Almansa, C, Carceller, E., Moyano, A., and Serratosa, F.; *Tetrahedron*, **1986**, *42*, 3637; b) Carceller, García, M.L., Serratosa, F., Font-Altaba, M., and Solans, X; *Tetrahedron*, **1987**, *43*, 2147.

8. Almansa, C, Carceller, E., García, M.L., Torrents, A., and Serratosa, F.; *Synthetic Communications,* **1988**, *18*, 381.

9. *Narcissistic*, from *Narkissos*, youth who fell in love with his reflection in the water of a pond. See *The Oxford English Dictionary*.

10. See, for instance, Voegtle, F., and Weber, E.; *Host Guest Complex Chemistry - Macrocycles,* Verlag-Chemie, Berlin (1985).

REACTIONS OF STRAINED CARBON-CARBON BONDS WITH METAL ATOMS

W. E. Billups
Department of Chemistry
Rice University
P.O. Box 1892
Houston, TX 77251, USA

ABSTRACT. The reactions between atomic iron and cyclopropane have been investigated in cryogenic matrices. Photolysis of the matrix leads to carbon-carbon bond activation of the cyclopropane by photoexcited iron atoms to yield ferracyclobutane. This species is converted rapidly to methyl vinyl iron via an undetected ethylene iron carbene complex. Prolonged photolysis of the matrix yields methane and ethynyl iron hydride. Unligated iron carbene ($Fe{=}CH_2$) can be microsynthesized by cocondensing iron atoms with diazomethane in argon. The chemistry of this species can be investigated by doping the matrix with a third reagent.

## Introduction

The first photoinsertion of a metal atom into a carbon-hydrogen bond was reported in 1980 with the insertion of iron atoms into the C-H bonds of methane.[1] Subsequent work demonstrated the reversibility of

$$CH_4 + Fe \underset{400\ \text{nm}}{\overset{\text{uv}}{\rightleftharpoons}} CH_3FeH$$

the reaction and also showed that other first row transition metals including Cr, Mn, Co, Ni, and Cu insert, in their excited states, into the C-H bonds of methane and other low molecular weight *n*-alkanes.[2-4] Carbon-carbon bond activation could not be detected when ethane or other *n*-alkanes were cocondensed with iron atoms and photolyzed.

## Results and Discussion

The activation of carbon-carbon bonds in cyclopropane has been achieved in cryogenic matrices at 15 K by iron atoms.[4] The products (Scheme I) have been characterized by FTIR spectroscopy using isotopically labeled precursors. No carbon-hydrogen bond activation



*A. de Meijere and S. Blechert (eds.), Strain and Its Implications in Organic Chemistry, 451–456.*

could be detected as shown by the absence of Fe-H stretching frequencies.

The ligand-free metallacycle, ferracyclobutane, is converted rapidly upon photolysis to methyl vinyl iron. This species is

SCHEME I

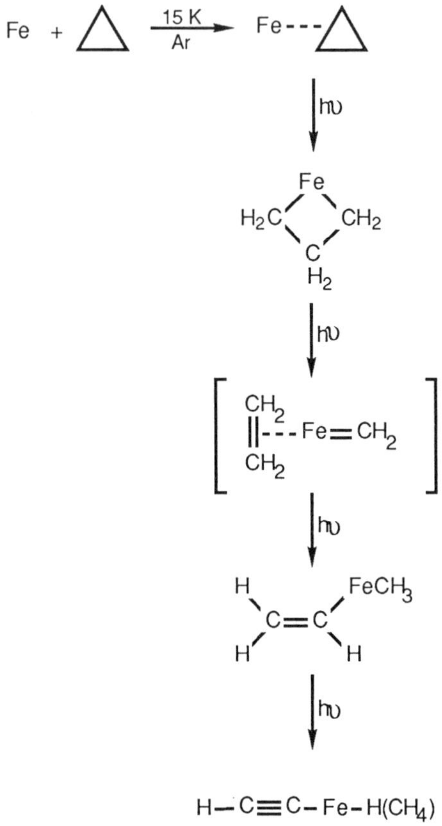

$$Fe + \triangle \xrightarrow[Ar]{15 K} Fe\text{---}\triangle$$

$$\downarrow h\nu$$

$$\begin{array}{c} Fe \\ H_2C \diagup \diagdown CH_2 \\ \diagdown C \diagup \\ H_2 \end{array}$$

$$\downarrow h\nu$$

$$\left[ \begin{array}{c} CH_2 \\ \| \text{---} Fe=CH_2 \\ CH_2 \end{array} \right]$$

$$\downarrow h\nu$$

$$\begin{array}{c} H \diagdown \phantom{xx} \diagup FeCH_3 \\ C=C \\ H \diagup \phantom{xx} \diagdown H \end{array}$$

$$\downarrow h\nu$$

$$H\text{---}C\equiv C\text{-}Fe\text{-}H(CH_4)$$

postulated to arise via the undetected ethylene iron carbene complex shown in the brackets. Prolonged photolysis of the matrix yields methane and ethynyl iron hydride.

Similar observations were made when iron/ethylene oxide/argon matrices were photolyzed. In this case the final product is vinyl iron hydroxide (Scheme II).[5]

Evidence for the simple iron carbene species as postulated in Scheme I can be obtained by cocondensing iron atoms with diazomethane in cryogenic matrices. The infrared spectra of $FeCH_2$, $N_2FeCH_2$, and isotopically labeled species are presented in Figure 1.

The observation that residual dihydrogen reacted with $FeCH_2$ and $N_2FeCH_2$ to yield $CH_3FeH$ and $N_2CH_3FeH$, respectively, was particularly salient, as it suggested that ternary reactions can be investigated readily using matrix isolation spectroscopy. This hypothesis was confirmed by experiments in which excess dihydrogen was cocondensed with the matrix. Surprisingly, $N_2FeCH_2$ is much more reactive than $FeCH_2$ towards hydrogenation, an observation not easily rationalized.

SCHEME II

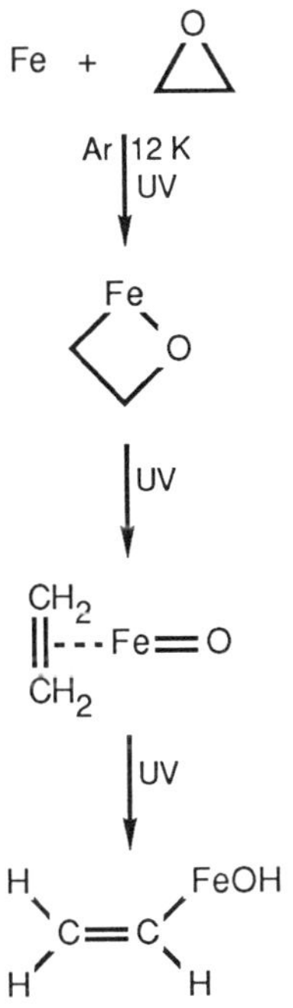

$$Fe + \underset{\triangle}{\overset{O}{\phantom{x}}}$$

$$\downarrow Ar\,|\,12\,K\;\;UV$$

$$\underset{\diamondsuit}{Fe\diagdown O}$$

$$\downarrow UV$$

$$\underset{CH_2}{\overset{CH_2}{\|}}\text{---}Fe{=}O$$

$$\downarrow UV$$

$$\underset{H}{\overset{H}{\diagdown}}C{=}C\underset{H}{\overset{FeOH}{\diagup}}$$

Matrix reactions with water as the ternary reagent led to the formation of $CH_3FeOH$, a product characterized previously when iron/$CH_3OH$/argon matrices were photolyzed.[6]

Photolysis of an iron/diazomethane/argon matrix with $\lambda \geq 500$ nm leads to significant bleaching of the bands assigned to $N_2FeCH_2$;

$$N_2FeCH_2 \xrightarrow{\lambda \geq 500 \text{ nm}} Fe + CH_2N_2$$

however, no new bands associated with an iron/diazomethane reaction product could be detected, suggesting that a reductive elimination reaction may occur with the low energy photolysis.

Photolysis under the same conditions or with $\lambda \geq 400$ nm irradiation showed little effect on the absorptions arising from $FeCH_2$; however, UV photolysis ($360 \geq \lambda \geq 280$ nm) leads to the rapid conversion of $FeCH_2$ to a new species with absorptions at 1681.6, 674.2, and 632.1 cm$^{-1}$. These new absorptions are assigned to $HFe{\equiv}CH$. Thus the band at 1681.6 cm$^{-1}$ can be assigned readily to a Fe-H stretching mode. The remaining two bands at 674.2 and 632.1 cm$^{-1}$ are assigned to a Fe≡C stretching mode and a C-H bending mode, respectively. The C-H stretching band of this species was not observed. Photolysis of the carbyne through a cutoff filter with $\lambda \geq$ 400 nm leads to the reverse process. Although carbene ⟶ carbyne rearrangements are thought to be

$$FeCH_2 \rightleftharpoons HFeCH$$

symmetry forbidden for certain 14 electron complexes, the barrier is apparently lifted in the photoinduced rearrangement observed here.

These iron/diazomethane reactions are summarized in Scheme III.

SCHEME III

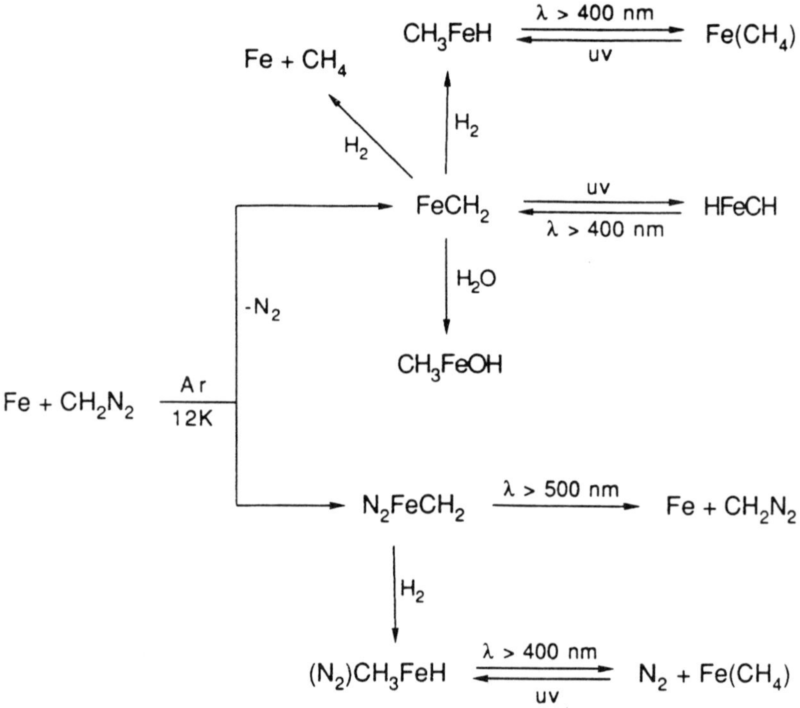

$$Fe + CH_2N_2 \xrightarrow[12K]{Ar} \begin{cases} FeCH_2 \\ N_2FeCH_2 \end{cases}$$

- $Fe + CH_4 \xleftarrow{H_2} CH_3FeH \underset{uv}{\overset{\lambda > 400 \text{ nm}}{\rightleftharpoons}} Fe(CH_4)$
- $FeCH_2 \underset{\lambda > 400 \text{ nm}}{\overset{uv}{\rightleftharpoons}} HFeCH$
- $FeCH_2 \xrightarrow{H_2O} CH_3FeOH$
- $CH_3FeH \xrightarrow{H_2}$ (from FeCH$_2$)
- $N_2FeCH_2 \xrightarrow{\lambda > 500 \text{ nm}} Fe + CH_2N_2$
- $N_2FeCH_2 \xrightarrow{H_2} (N_2)CH_3FeH \underset{uv}{\overset{\lambda > 400 \text{ nm}}{\rightleftharpoons}} N_2 + Fe(CH_4)$
- $FeCH_2$ formed via $-N_2$ from $N_2FeCH_2$

The cocondensation of iron vapor with cyclopropane and other n-alkanes at -196° C yields organometallic powders, with or without photolysis, which result from the reactions of clusters reacting with the organic substrate. These powders can be used to reduce alkenes in the absence of added hydrogen gas.

Efforts are currently under way to characterize the products resulting from the spontaneous reactions of iron atoms with cyclopropene.

### Acknowledgment

I am indebted to my co-workers, whose names appear in the literature cited. This work was supported by The Welch Foundation and the 3M Company.

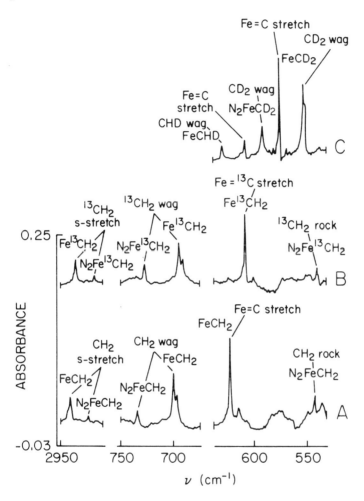

Figure 1. FTIR spectra of selected regions of A, $FeCH_2$ and $N_2FeCH_2$; B, $Fe^{13}CH_2$ and $N_2Fe^{13}CH_2$; C, FeCHD, $FeCD_2$, and $N_2FeCD_2$ in argon matrices.

### References

(1) W. E. Billups, M. M. Konarski, R. H. Hauge, and J. L. Margrave, *J. Am. Chem. Soc.* **102**, 7393 (1980).

(2) G. A. Ozin, D. F. McIntosh, and S. A. Mitchell, *J. Am. Chem. Soc.*, **103**, 1574 (1981).

(3) G. A. Ozin, J. M. Parnis, S. A. Mitchell, and J. Garcia-Prieto in H. Grünewald (Ed.): *Chemistry for the Future*, Pergamon Press, Oxford, 1984, pp 95-105.

(4) G. A. Ozin, D. F. McCaffrey, and J. M. Parnis, *Angew. Chem. Int. Ed. Engl.*, **25**, 1072 (1986).

(5) Z. H. Kafafi, R. H. Hauge, W. E. Billups, and J. L. Margrave, *J. Am. Chem. Soc.*, **109**, 4775 (1987).

(6) Z. H. Kafafi, R. H. Hauge, L. Fredin, W. E. Billups, and J. L. Margrave, *Chem. Commun.*, 1230 (1983).

(7) M. Park, R. H. Hauge, Z. H. Kafafi, and J. L. Margrave, *Chem. Commun.*, 1570 (1985).

# CYCLOPROPABENZENES AND ALKYLIDENECYCLOPROPABENZENES
# A SYNERGISTIC RELATION BETWEEN THEORY AND EXPERIMENT

Yitzhak Apeloig*, Miriam Karni and Dorit Arad
Department of Chemistry, Technion-Israel Institute
of Technology, Haifa 32000, Israel

**ABSTRACT.** Ab initio calculations, mostly with the 3-21G basis-set were carried out for cyclopropabenzene (1), the linear and angular cyclobutacyclopropabenzenes (2 and 3 respectively), the linear- and angular dicyclopropabenzenes 4 and 5 respectively, tricyclopropabenzene (6), alkylidenecyclopropabenzene (7) and two calecene homologues, i.e., 8 - containing an exocyclic cyclopropene moiety and 9 - containing an exocyclic cyclopentadiene moiety. The geometries and the total and relative energies of compounds 1-9 are reported. The calculated geometries of 1 and 2 are in excellent agreement with recent low-temperature X-ray structures. In all molecules, except 6, the fused bonds are shorter than the other aromatic bonds, in conflict with the Mills-Nixon "bond fixation" concept. The following strain energies were calculated for compounds 1-6 (in kcal/mol): 1 (70, also measured experimentally), 2 (102); 3 (103); 4 (133); 5 (140); 6 (217). The surprising theoretical prediction that the strain of the cyclopropane ring in 2 is slightly lower than that in 1, has been confirmed by recent experiments. Compounds 7-9 exhibit considerable polarization of the π-electrons, so that the cyclopropabenzene skeleton is positively charged in 7 and 9 and negatively charged in 8. Consequently, the direction of the dipole moment in 8 is opossite to that in 7 and 9.

## Results and Discussion.

Cycloproparenes have attracted considerable interest for almost 25 years.[1] Yet due to synthetic difficulties and the relatively high reactivities of these molecules the experimental knowledge on their fundamental properties is limited.[1] Encouraged by the recent experimental[2] confirmation of our earlier theoretical prediction[3] of the geometry of cyclopropabenzene (1), we have extended our theoretical study to the multiply-annelated systems 2-6, of which only 2 and 3 have been synthesized. We report here the results of ab initio calculations (mostly with the 3-21G basis set) for these compounds and also for the related alkylidenecyclopropabenzenes 7-9. The geometries (see Figure), strain, electronic structures, and the question of "bond fixation" in these compounds will be discussed briefly. Below we mention some of the major points of interest.

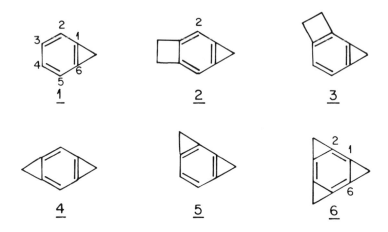

The high reliability of the calculated 3-21G geometries is nicely exemplified by the calculated 3-21G geometry of 2 (see Figure), which is in excellent agreement (including the acute angle of only 110° at $C_2$) with a recent low-temperature X-ray structure.[4] In all molecules, except 5 and 6, the fused $C_1$-$C_6$ bond is shorter than the other aromatic bonds. These geometries indicate (if relevant[3]) "bond fixation" in the **opposite direction to that predicted by Mills and Nixon.**[5] 6 is the only molecule which behaves in the "Mills-Nixon sense", the $C_1$-$C_6$ distance is 1.381A, much longer than the $C_1$-$C_2$ distance of 1.338A.

The angularly fused compounds are by several kcal mol$^{-1}$ less stable than the corresponding linearly fused isomers. The strain in 1 is ca. 70 kcal mol$^{-1}$ (experimentally[1] and theoretically[3]). We find that the effect on the total strain of the molecule of fusion of additional rings is approximately <u>additive</u>. Thus, the strain in 4 and 5 is roughly double and in 6 roughly triple than the strain in 1. Relatively small deviations from additivity are found. E.g., 5 is less stable than 4 by 7 kcal mol$^{-1}$ and the calculated total strain of 6 is 217 kcal mol$^{-1}$, 7 kcal mol$^{-1}$ higher than calculated by assuming strain additivity. The strain in the cyclopropyl ring of the multiply-fused compounds is slightly decreased by linear fusion of a second small ring (e.g. 2) and slightly increased by angular fusion (e.g. 5). This suggests that ring opening (e.g. by hydrogenation) of the cyclopropyl ring in compounds 2-6 is either, slightly more exothermic (e.g. by 3.5 kcal mol$^{-1}$ in 6), or slightly more endothermic (e.g., by 2.6 kcal mol$^{-1}$ in 2) than a similar reaction of 1. Recent experiments have indeed showed that the cleavage of the cyclopropyl ring in 1 occurs faster than in $2^6$ in full agreement with the theoretical prediction.

We will also report calculations for 3 prototype molecules (7-9) of a related novel group of compounds - the alkylidenecycloproparenes, some of which have been recently synthesized.[7]

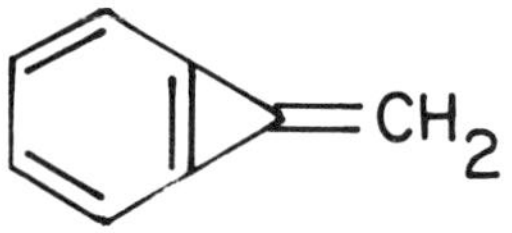

7

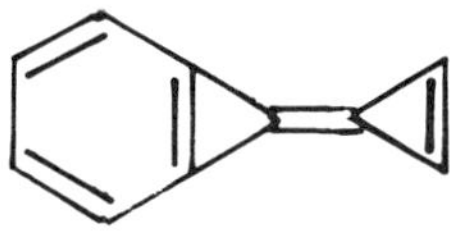

8

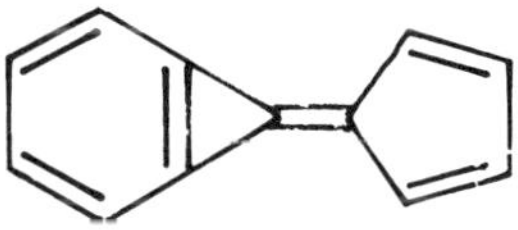

9

The calculations reveal considerable polarization of the π-electrons. In 7 and more strongly in 9 the cyclopropabenzene skeleton is postively charged, while in 8 it is negatively charged. Consequently, the dipole moments in 8 on one hand and in 7 and 9 on the other hand point in opposite directions.

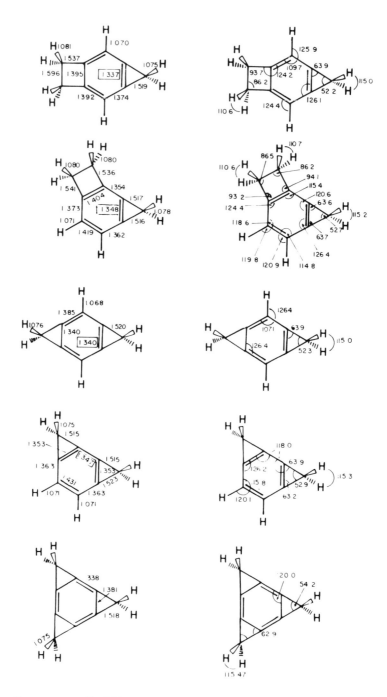

FIGURE. 3-21G CALCULATED GEOMETRIES OF 2 - 6.

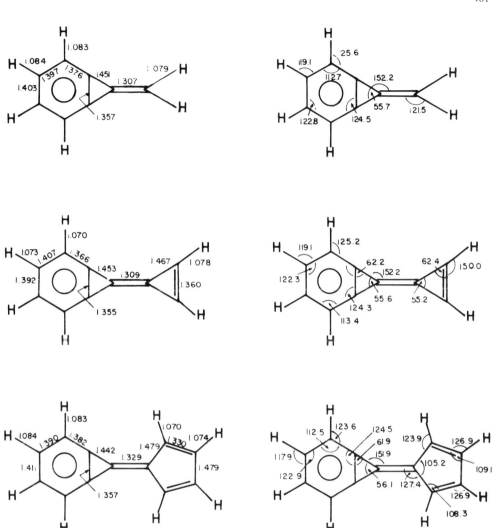

FIGURE. STO-3G CALCULATED GEOMETRIES OF COMPOUNDS 7 - 9.

## Acknowledgment

This research was supported by the Fund for the Promotion of Research at the Technion and the S. Faust Research Fund.

## References

1. (a) W.E. Billups, W.A. Rodin and M.M. Haley, Tetrahedron, **44**, 1305 (1988); (b) B. Halton, Chem. Rev. **73**, 113 (1973); (c) B. Halton, Ind. Eng. Chem. Prod. Res. Rev., **19**, 349 (1980).
2. R. Neidlein, D. Christen, V. Poignee, R. Boese, D. Blaser, A. Gieren, C. Ruiz-Pirez and T. Hubner, Angew. Chem. Int. Ed. Eng., **27**, 294 (1988).
3. Y. Apeloig and D. Arad, J. Amer. Chem. Soc., **108**, 3241 (1986).
4. R. Boese, personal communication and paper in press in Angew. Chemie (1988).
5. W.H. Mills and I.G. Nixon, J. Chem. Soc., 2150 (1930).
6. P. Vollhardt and A. Stanger, personal communication and paper in press in Angew. Chemie (1988).
7. B. Halton and P.J. Stang, Acc. Chem. Res., **20**, 443 (1987).

# HARNESSING STRAIN: FROM [1.1.1]PROPELLANES TO TINKERTOYS

Josef Michl*, Piotr Kaszynski, Andrienne C. Friedli,
Gudipati S. Murthy, Huey-Chin Yang, Randall E. Robinson,
Neil D. McMurdie, and Taisun Kim

*Center for Structure and Reactivity,
Department of Chemistry,
The University of Texas at Austin,
Austin, Texas 78712-1167*

ABSTRACT. We describe the initial results of an effort to develop a molecular-scale civil engineering construction set analogous to the children's "Tinkertoy" set. End-functionalized "staffs" were obtained by oligomerization of [1.1.1]propellane, promoted by strain relief. The first steps have been taken towards joining them with suitable connectors. The functionalized staffs have other interesting properties; their surface activity and liquid crystalline behavior are described briefly.

INTRODUCTION. SURFACE MOBILES AND DESIGNER SOLIDS

One of the joys of childhood is the construction of simple or complex mechanical structures from toothpicks stuck into horse chestnuts. An advanced version of this pursuit is offered by the commercially available "Tinkertoy" set[1] of wooden rods and connectors. The connectors are flat spools with round holes on the sides, into which the sticks can be inserted (Figure 1).

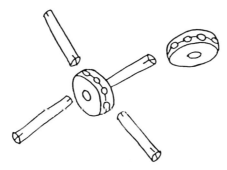

Figure 1. The construction elements of a "Tinkertoy" set.

Spurred on by my son's incessant demands for investment in the European version of toy construction sets, the "Lego",[2] and apparently not really having quite grown up myself, I have engaged my research group in an attempt to develop a "Tinkertoy" set of rods and connectors on a molecular scale, and can now report the initial results. Perhaps later, we shall be able to add more complex elements to the basic set of rods and connectors and thus arrive at a molecular "Lego" construction set as well.

In addition to the exciting challenge represented by tasks such as the total synthesis of a molecular swing or windmill and the decoration of a surface with a set of such constructs, possibly accompanied by statues of mounted knights with servants next to them, the project offers the intriguing possibility that interesting and useful properties could be designed into these "molecular mobiles." After all, the swing seats or the wing tips could carry charges or magnets and mediate communication between electromagnetic fields and a gas jet. Controlled motion could also be achieved by the use of reversible photochemical isomerizations. Perhaps Maxwell-demon-like properties could be designed into such devices. The construction of miniature machines has captivated man's imagination for some time[3,4] and there is no reason why it needs to remain in the realm of speculation. After all, progress in microelectronics is beginning to approach the atomic level of miniaturization.

Even static structures, most likely much easier to build, could be of great interest. Given a completely controlled and reasonably inert network or scaffolding constructed from suitably functionalized molecular staffs and connectors, the opportunity arises to produce "designer solids", containing groups of predetermined electric, magnetic, optical, mechanical, chemical, photochemical, and other properties at a predetermined set of points in space. The structure could be quite lofty, permitting the oriented inclusion and the passage of selected solutes. The selection could be by predetermined size, controllable in a static fashion by the choice of staff lengths and the pattern of their connection, as well as a dynamic fashion, e.g., by reversible cross-linking with sulfhydryl-bearing side chains. Dynamic selection could also be based on solute charge, based on the imposition of adjustable outside potentials onto various layers of the solid. This could also serve for forced selective transport, say, to an electrode. The ability to introduce controllable electrical potentials at predetermined points will clearly be important and calls for the control of electrical conductivity of selected staffs. Suitable functionalization along the length of a staff could indeed produce conductivity for electrons, holes, or ions.

Many other applications for such designer solids come to mind. After all, they too, could contain movable parts located throughout the body of a lofty three-dimensional structure, whose motion would be dictated by various fields applied from the outside. Reversible optical storage, nonlinear optics, ferroelectric, ferromagnetic, piezoelectric, and perhaps even "piezomagnetic" applications suggest themselves. Semiconductivity, photoconductivity, low-dimensional conductivity, quantum well properties, and with luck, even superconductivity could be attainable, and applications in microelectronics, optoelectronics, optics, laser technology and elsewhere would surely be found. Most exciting, in all likelihood, would be the properties and applications of which one has not yet thought.

With this type of motivation in mind, the most important requirements for the molecular staffs are:

(i) facile synthesis,

(ii) rigid and linear structure,

(iii) stability to heat, air, and common chemicals,

(iv) small and well defined length increments,

(v) choice of end groups for reversible or irreversible attachment to connectors,

(vi) transparency in the visible and UV regions,

(vii) feasibility of functionalization at pre-selected points along the staff length, and

(viii) availability in electrically insulating and electrically conducting forms.

The synthesis of molecular "Tinkertoy" staffs represents only a first step in a three-phase project, and it is the only one on which I can report a fair amount of progress today. The second phase, the production of suitable connectors, is subject to a similar set of strict demands. The connectors must be

(i) readily available, with selective affinity for various terminal groups on the staffs,

(ii) easily attachable at a variety of preselected rigid angles (90°, 120°, 180°, ...), reversibly or irreversibly,

(iii) stable to heat, air, and common chemicals,

(iv) transparent or absorbing at preselected wavelengths, non-luminescent or luminescent,

(v) magnetic or non-magnetic,

(vi) photostable or phototransformable,

(vii) electrically neutral or charged, insulating or conducting, and

(viii) inert or amenable to oxidation-reduction processes.

The oxidation-reduction and photochemical processes could be used to control the geometry of the connection or even its very existence.

In the third phase, which has not yet begun, we shall attempt the actual construction of free-floating or surface-anchored structures and materials, using self-assembly or the principle of Merrifield polypeptide synthesis. This should permit nearly perfect control on the molecular level, but only in one dimension. E.g., starting with a perfect crystal surface, we should be able to design arbitrary structure in the direction normal to the surface, but there will be unlimited periodicity in the other two dimensions, dictated by epitaxial growth. Its nature will presumably be determined by the arrangement of adsorption sites on the initial surface, over which a fair degree of control can be exercised using standard techniques of surface science. The procedures currently common in microelectronics could be used to introduce aperiodic structural design in the dimensions parallel to the surface as well, albeit only on a coarser scale at present.

## OLIGOMERS OF [1.1.1]PROPELLANE AS STAFFS FOR MOLECULAR TINKERTOYS

It has occurred to us that end-functionalized oligomers of [1.1.1]propellane (1), whose formation we have observed accidentally in the course of other work, represent fine candidates for the staffs needed in a molecular-size Tinkertoy construction set, and subsequent work has confirmed this expectation. One of their advantages is their simple and highly linear structure. While innumerable half-timbered houses in Northern Europe prove that it is possible to build with crooked beams, it must surely be a civil engineer's nightmare and is best avoided.

A literature search revealed that we were not the first to observe oligomer formation in the course of radical addition across the central bond of 1 - this occurred in the laboratory of Wiberg during an investigation of the addition of BrCN, where the formation of a dimer and a trimer was noted.[5]

**Parent and Functionalized [n]Staffanes - Synthesis**

We have proposed[6] the trivial name [n]staffanes for the hydrogen-terminated parent oligomeric hydrocarbons 2 (X - Y - H) and use this

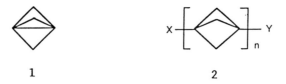

nomenclature hereafter. The numbering of carbons in these molecules is shown in formula 3.

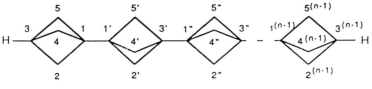

The easy availability of 1 is due to the breakthroughs in the laboratories of Wiberg[7] and Szeimies,[8] without which the progress I am reporting now would have been impossible. In the preparation of our starting material, 1, we rely on the procedure of Szeimies,[8] which we have scaled up,[9] working with 5-10 g batches of 1 at a time. So far, we have detected no obstacles to further scale-up. The process involves only two steps starting with the commercial methallyl dichloride 4:

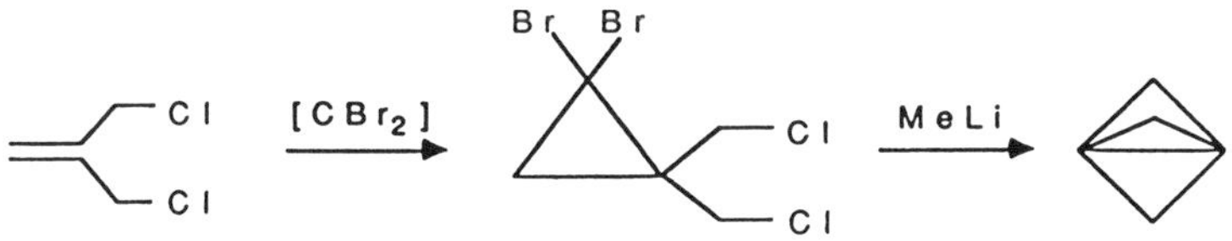

The first step relies on phase-transfer dibromocarbene addition[8] and proceeds in an about 30% yield. The second step is essentially quantitative when performed with MeLi in an ether-containing solution. At times, the presence of ether is detrimental in the subsequent step and since it is difficult to separate from 1, we avoid it altogether by using a n-BuLi-TMEDA combination.[6] Unfortunately, yields are then more erratic, although never lower than 30%; they increase when the reaction is carried out at higher dilution.

Radical-induced addition of many dozens of reagents to the central bond of 1 has been examined in our laboratory, typically with dibenzoyl peroxide as the initiator and under UV irradiation. Additions across C-H, C-X, X-X, S-S, S-H, S-X, P-H, and P-P bonds (X = halogen) have been observed and are believed to proceed by the standard chain mechanism. For example,

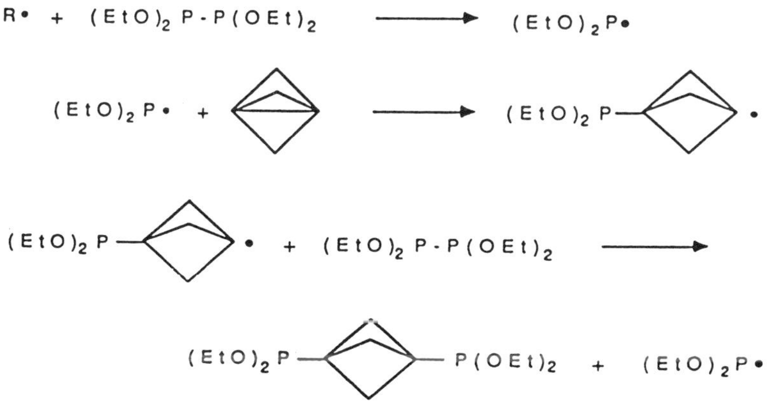

Also addition across the C-C bond of α-dicarbonyl compounds has been observed, and this is believed to involve a nucleophilic attack on a carbonyl group:

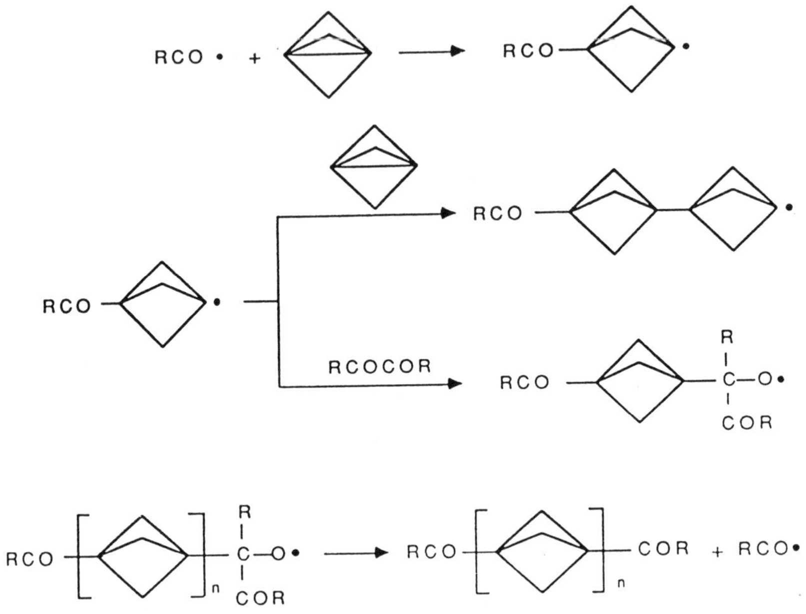

However, only a relatively small fraction of these reactions is useful for our purposes since most yield no oligomers in spite of the fac that a fair degree of control over the product distribution is available through a variation of concentrations of the reaction partners. For instance, the addition of biacetyl yields only the monomeric adduct,[9] while the addition of benzil yields a series of oligomeric diketones.

Under suitable conditions, insoluble "polymers" of very low molecula weight result. These are also accessible via anionic polymerization and we shall return to their properties below.

Additions that have yielded oligomeric products are listed in Table and are divided into those leading to [n]staffanes functionalized on a single end and those functionalized on both ends. The former group is of limited direct use for a Tinkertoy construction set, for terminal element only, but has other interesting properties.

Table I. End-Functionalized Oligomers 2

| Singly | | Doubly | |
|---|---|---|---|
| X | Y | X | Y |
| H | COOMe | AcS | SAc |
| H | COOH | MeS | SMe |
| H | CH(COOEt)$_2$ | BuS | SBu |
| H | C(COOEt)$_3$ | I | I |
| H | C(Ph)(COOEt)$_2$ | Br | Br |
| H | C(Me)(COOEt)$_2$ | Cl | SO$_2$Me |
| H | CH(COMe)COOEt | Cl | SO$_2$Ph |
| H | CH(CN)COOMe | PhCO | COPh |
| H | CH(CN)$_2$ | | |
| H | P(O)(OEt)$_2$ | | |

Further functionalization of singly functionalized [n]staffanes is possible. Thus, chlorocarbonylation with oxalyl chloride tends to occur preferentially at the terminal position:

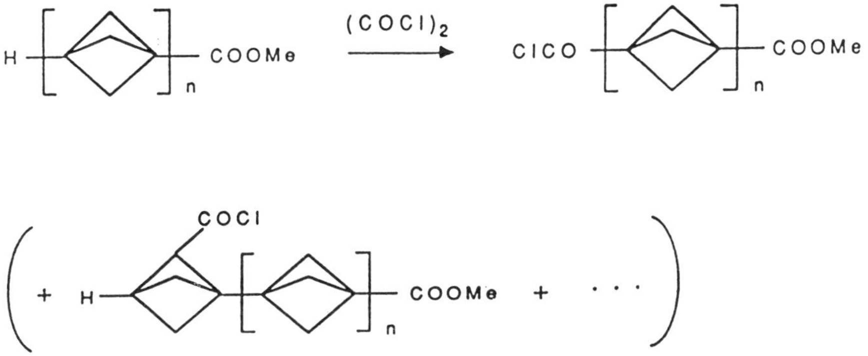

$$H \text{—[staff]}_n\text{—COOMe} \xrightarrow{(COCl)_2} ClCO\text{—[staff]}_n\text{—COOMe}$$

$$\left( + \quad H\text{—[staff]}\overset{COCl}{\text{—[staff]}_n}\text{—COOMe} + \cdots \right)$$

Functionalization of the CH$_2$ bridges is also important for our plans to attach conducting "wires" along the length of the staff. Attempts at free radical bromination have not been successful, but chlorination proceeds smoothly to yield geminal dichlorides. One of the two chlorine atoms can then be removed selectively with one equivalent of tri-n-butyltin hydride:

The chlorination has been run successfully with R = Br, COOH, COCl, and COOMe. It occurs also in the oligomers and permits a conversion of the intractable low-molecular weight "polymer" into a readily soluble material. We are now exploring further transformations of the bridge substituents.

Among oligomerization reactions that produce doubly end-functionalized staffanes, we have investigated the addition across the S-S bond of dialkyl and diacyl disulfides in most detail. Diacetyldisulfide is a particularly advantageous reaction partner since it gives a good distribution of oligomers, which have been isolated in pure state up to n = 5, and permits ready subsequent transformations of the end groups:

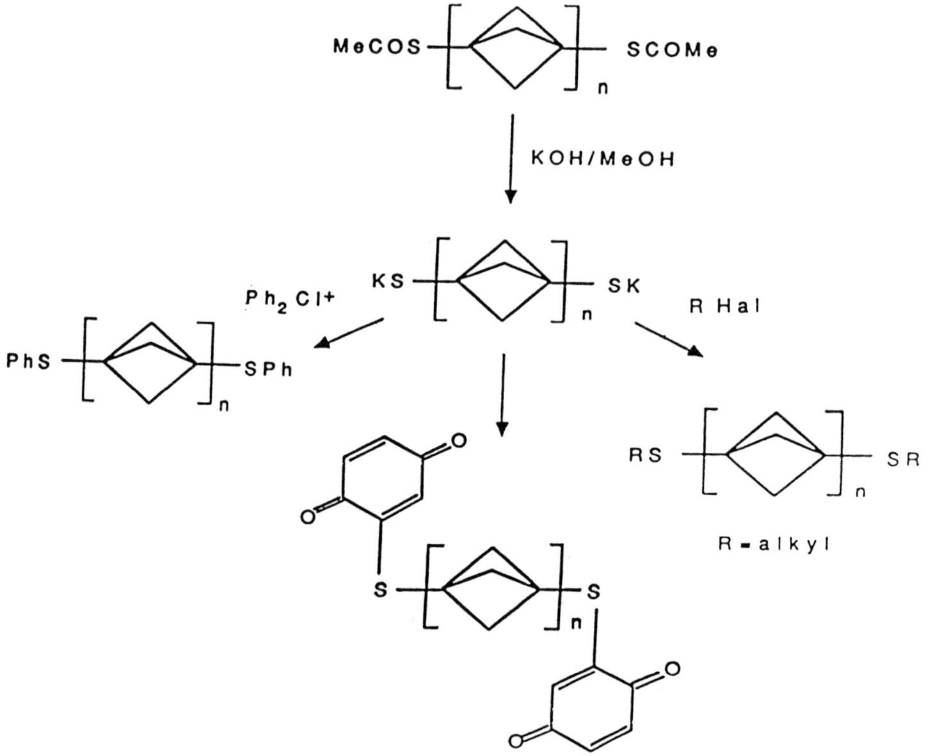

Partial hydrolysis of $3,3^{(n-1)}$-bisacetylthio[n]staffanes has provided access to staffs with differentiated ends.

The most straightforward route to the parent [n]staffanes is the oligomerizing addition of 1 across an H-H bond. We have accomplished this using a modification of a procedure developed in the laboratory of Mazur for additions to double bonds,[10] in which H atoms are produced by microwave-induced dissociation of $H_2$ in excess argon and allowed to react with a solution of the substrate at about -90°C. So far, we have isolated the first four [n]staffanes in pure state.

In summary, quite a few oligomers with n = 1 - 4 or 5 have already been isolated in pure state, from the parent hydrocarbons to singly and doubly end-functionalized derivatives, and further progress is likely to be rapid. Many promising avenues still remain to be explored, such as the use of anionic oligomerization.

## Parent and Functionalized [n]Staffanes - Properties

*Structure and Stability.* In spite of their high strain energy[11], [n]staffanes and their derivatives possess remarkable thermal stability and do not decompose until heated close to 300°C. Their melting points are high, much higher than those of the analogous functionalized alkanes with the same number of carbons.

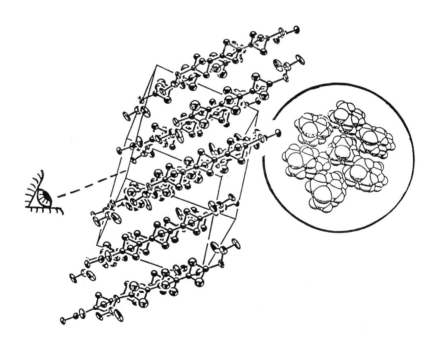

Figure 2. The unit cell in a crystal of 3,3"-bisacetylthio[3]staffane. In the end-on view of the packing, the acetylthio groups have been removed for clarity.

For instance, methyl [5]staffane-3-carboxylate melts above 300°C with decomposition, $C_{25}H_{51}COOCH_3$ at 60-62°C.[6] This is probably related to the very efficient crystal packing of the higher [n]staffane derivatives. An X-ray analysis of 3,3"-bisacetylthio[3]staffane showed that the staffs are arranged with their long axes parallel and only 4.6 Å apart, with the methylene groups of each staff meshed with the inter-cage spaces on the neighboring staffs (Figure 2). Each staff has six nearest neighbors.

The molecular structure, as revealed by X-ray analysis of five derivatives (Figure 3), contains few surprises. The molecules are linear and the inter-cage bond lengths are short, 1.47-1.48 Å, as predicted.[12]

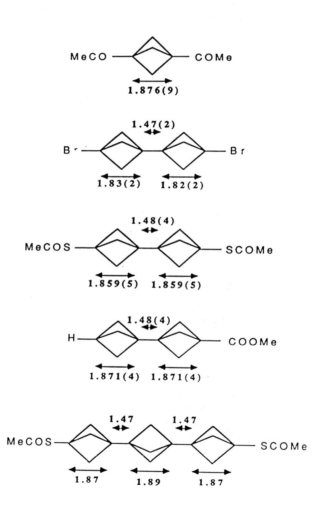

Figure 3. Bond lengths in [n]staffanes from X-ray analysis.

Very similar structures have been obtained recently for 3,3'-bismethylsulfonyl[2]staffane and for 3,3'-bismethylthio[2]staffane in the laboratory of Professor Szeimies.[13] The regularity of the X-ray structures permits one to estimate quite reliably the length increment from one staffane to the next at ~3.35 Å.

However, the intra-cage bridgehead-bridgehead separation is somewhat sensitive to the electronegativity of the bridgehead substituents. As could be anticipated,[14] the "normal" distance of 1.86-1.89 Å is shortened to 1.82-1.83 Å in 3,3'-dibromo[2]staffane, the bridgehead carbon using an orbital with higher p character in its bond to bromine.

The individual bicyclo[1.1.1]pentane cages in the string adopt a staggered conformation. The barrier to internal rotation is not known, but preliminary results from measurements of $^{13}C$ NMR relaxation suggest that it is very low.[15]

*Nuclear Magnetic Resonance.* The $^1H$ and $^{13}C$ chemical shifts observed for [n]staffanes and their derivatives follow very regular patterns and can be understood in terms of a few additive increments. Their use reproduces the observed $^1H$ shifts and the $^{13}C$ shifts with standard deviations of 0.002 and 0.08 ppm, respectively, for n ≥ 2.

The chemical shifts $\delta(n',X',n'',X'')$ of a $^1H$ nucleus in a bridge $CH_2$ group contained in a bicyclo[1.1.1]pentane cage separated by n' such cages from the terminal group X' and by n'' such cages from the other terminal group X'' is expressed as a sum of its value in an infinite chain, $\delta$ 1.335, of two corrections for chain end effects already present in the hydrocarbon, A(n') and A(n''), and of two corrections for effects of substituents other than hydrogen, S(n',X') and S(n'',X''), according to the simple equations given in Table II. In an analogous fashion, the chemical shift of the $^{13}C$ atom in such a bridge $CH_2$ group is given by end-effect corrections C(n') and C(n'') and substituent contributions U(n',X') and U(n'',X''). The chemical shift of a CH bridgehead hydrogen atom $\delta(n,X)$ is given by a contribution B(n) due to the other end, and a contribution of a substituent at that end, T(n,X), where n is the number of bicyclo[1.1.1]pentane cages separating the hydrogen from the other terminal group X. Finally, the chemical shift of a bridgehead carbon atom, C or CH, is given by end-effect contributions D(n') and D(n'') and substituent effects V(n',X') and V(n'',X''), where n' now is the number of bridgehead carbon atoms separating it from the terminal group X' and n'' is the number of bridgehead carbon atoms separating it from the other terminal group, X''. The equations for the chemical shifts and the magnitudes of the various increments for a series of substituents X are given in Table II.

We expect that this highly regular behavior will be of analytical utility in future work with these compounds. We hope that it will be possible to generalize the simple equations even to the cases of substitution in the $CH_2$ bridges.

The $^{13}C$ chemical shifts measured[6] on the low-molecular weight polymer, $\delta(CH_2)$ = 50.8 and $\delta(C)$ = 40.3, do not agree exactly with the expected values of 47.66 and 38.05, presumably due to environmental effects.

**Table II. Increments for NMR Chemical Shifts in [n]Staffanes and Their Derivatives**

$^1$H:
$$\delta(n',X',n'',X'') = 1.335 + A(n') + A(n'') + S(n',X') + S(n'',X'')$$
$$\delta(n,X) = 2.359 + B(n) + T(n,X)$$

|   |   |   | COOMe |   | SAc |   | PO(OEt)$_2$ |   |
|---|---|---|---|---|---|---|---|---|
| n | A | B | S | T | S | T | S | T |
| 0 | 0.245 | - | 0.251 | - | 0.377 | - | 0.269 | - |
| 1 | 0.024 | 0.091 | 0.038 | 0.051 | 0.045 | 0.403 | 0.021 | 0.191 |
| 2 | 0.007 | 0.001 | 0.005 | 0.032 | 0.003 |  | -0.002 | 0.020 |
| 3 | 0.002 | 0.006 | 0.000 | 0.004 | -0.001 |  | -0.007 | 0.003 |
| 4 | 0 | 0.002 | 0 | 0.000 | 0 |  | 0 | -0.006 |
| 5 | 0 | 0 | 0 | 0 | 0 | 0 | 0 | 0 |

$^{13}$C:
$$\delta(n',X',n'',X'') = 47.74 + C(n') + C(n'') + U(n',X') + U(n'',X'')$$
$$\delta(n',X',n'',X'') = 38.19 + D(n') + D(n'') + V(n',X') + V(n'',X'')$$

|   |   |   | COOMe |   | SAc |   | PO(OEt)$_2$ |   |
|---|---|---|---|---|---|---|---|---|
| n | C | D | U | V | U | V | U | V |
| 0 | 1.34 | -11.81 | 1.45 | 10.36 | 4.23 | 10.96 | 1.01 | 4.59 |
| 1 | 0.02 | 7.11 | 0.08 | -5.62 | 0.30 | -1.47 | -0.05 | -1.59 |
| 2 | 0.01 | 0.02 | 0.01 | -0.77 | 0.00 | -0.74 | -0.01 | -0.66 |
| 3 | 0.02 | -0.04 | 0.00 | 0.22 | 0.07 | 0.49 | 0.07 | 0.13 |
| 4 | 0 | -0.05 | 0 | -0.18 | 0 | -0.36 | 0 | -0.19 |
| 5 | 0 | -0.01 | 0 | 0.00 | 0 | 0.32 | 0 | 0.04 |
| 6 | 0 | -0.05 | 0 | -0.09 | 0 | -0.06 | 0 | 0.03 |
| 7 | 0 | -0.04 | 0 | -0.04 | 0 | 0.33 | 0 | 0.04 |
| 8 | 0 | 0 | 0 | 0 | 0 | 0 | 0 | 0 |

Like the bridgehead-bridgehead coupling constants in bicyclo[1.1.1]-pentane itself, those in the [2]staffanes are unusually large, presumably for the same reason:[16] overlap of the back lobes of the hybrid orbitals used by carbon atoms in forming bridgehead-bridgehead bonds. Thus, the coupling constant $^7J_{HP}$ in diethyl 3-[2]staffylphosphonate equals 1.7 Hz.[6]

*Vibrational Spectroscopy*. An example of an FTIR spectrum of a parent staffane is shown in Figure 4. A particularly interesting vibration, of likely importance for the understanding of low-frequency modes of the "designer solids" we hope to construct, is the doubly degenerate staff bending mode. Its appearance in [4]staffane is indicated in Figure 5.

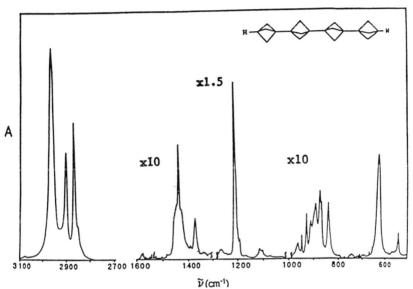

Figure 4. IR spectrum of [4]staffane obtained in $CCl_4$ and $CS_2$ solutions.

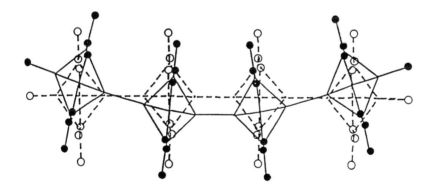

Figure 5. Calculated far IR skeletal bending mode of [4]staffane corresponding to the lowest frequency shown in Fig. 6. Equilibrium geometry is shown in broken lines (---) and displaced geometry in solid lines (—). The circles indicate positions of the hydrogen atoms.

As the staff length increases, the calculated bending frequency decreases dramatically (Figure 6). In addition to the lowest frequency staff bend, "overtone" staff bends begin to appear as normal modes starting with n = 3.

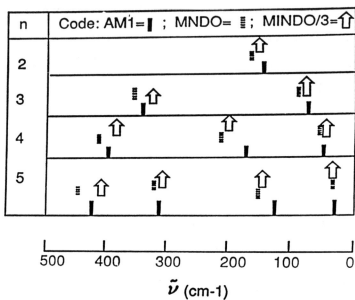

Figure 6. Far IR skeletal bending frequencies of [n]staffanes obtained from AM1, MNDO and MINDO/3 calculations.

The very low calculated frequency of these staff bending vibrations suggests that any scaffolding constructed from long staffs will be very easily deformable. This may be detrimental and it will perhaps be necessary to bundle the staffs for greater rigidity in construction. At the same time, this easy deformability may offer advantages as well, for instance in possible transduction of mechanical into electrical signals, and in making the new materials mechanically resilient.

## CONVENTIONAL APPLICATIONS OF [n]STAFFANE DERIVATIVES

In addition to the use of functionalized [n]staffanes as Tinkertoy construction elements, quite a few uses of more conventional nature suggest themselves. Thus, these compounds promise to be convenient spacers for investigations of the distance dependence of electron and excitation transfer, of substituent field effects, of magnetic coupling, etc. However, our experimental efforts in these directions are in their initial stages and have not yet borne fruit.

Furthermore, the rod-like shape of the staffanes suggests that they may exhibit liquid crystalline properties and that they may represent interesting surfactants when provided with a polar head group on one or both termini, and we have already obtained a preliminary confirmation of these expectations.

## Staffanes as Liquid Crystals

We find that disubstituted derivatives of [n]staffanes with n ≥ 3 have a tendency to form mesophases,[17] and some information on these is collected in Table III.

Table III. Mesogenic Properties of End-Functionalized [n]Staffanes[a]

| n | K | S | N | I |
|---|---|---|---|---|
| | 2[n] (X - Y - SCOCH$_3$) | | | |
| 2 | 99 | -- | -- | |
| 3 | 137[b] | 167[c] | 192 | |
| 4 | 107 | 211 | >280 dec. | |
| | 2[n] (X - Y - SC$_4$H$_9$) | | | |
| 2 | 8 | -- | -- | |
| 3 | 54.5 | 95[d] | -- | |
| 4 | 81 | 233 | -- | |

[a] Temperatures in °C; K - crystal, S - smectic, N - nematic, I - isotropic phase. [b] Both known crystalline modifications. [c] Smectic G. [d] Smectic B. These smectic phases were identified by comparison with microphotographs shown in G. W. Gray and J. W. G. Goodby, *Smectic Liquid Crystal Textures and Structures*, Leonard Hill, London, 1984; and D. Demus and L. Richter, *Textures of Liquid Crystals*, VEB, Leipzig, 1980.

These compounds have a strong tendency to form highly ordered smectic phases and are reminiscent of the closest related series, the "4,4'-dialkyl[2]rodanes" 5.[18]

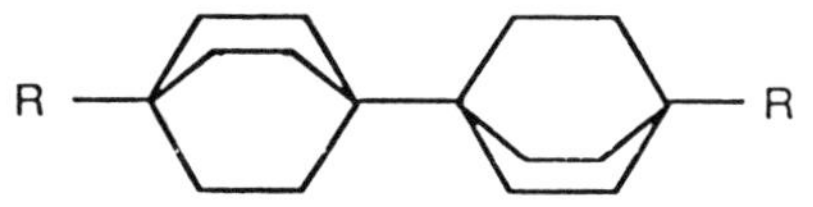

5

## Staffanes as Surfactants

The linear and relatively rigid nature of the staffane moiety suggests that its properties as a surfactant may be unusual. We have investigated the properties of the salts of [3]- and [4]staffane-3-carboxylic acids. Potassium [3]staffane-3-carboxylate lowers the surface tension of water (Figure 7) in a manner that corresponds to the quite unremarkable value of 107 Å$^2$ of the surface area per surfactant molecule.

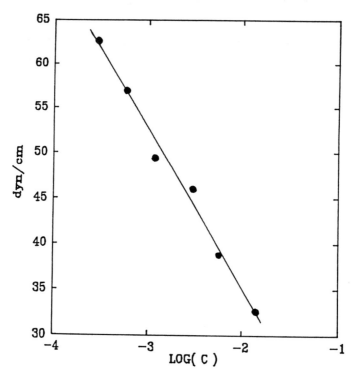

Figure 7. Surface tension $\gamma$ as a function of the concentration c of potassium [3]staffane-3-carboxylate, measured by the spinning drop method. The straight line is: $\gamma = -17.67 \log c - 0.229$.

The solubility of [4]staffane-3-carboxylate salts in water is sufficiently low that it has been possible to produce Langmuir-Blodgett monolayers from them in the usual fashion.[6,19] In Figure 8 we show the pressure-area isotherms for this acid for a few counterions, and for arachidic acid for comparison. The surface area is about 26 Å$^2$ per molecule of the [4]staffane-3-carboxylic acid. This can be compared with the 16.6 Å$^2$/molecule cross-section deduced from packing in a crystal. If the molecules in the monolayer were packed as tightly as they are in the crystal, one would have to conclude that they do not stand on end but are inclined at an angle of ~50° from the normal. However, our preliminary ellipsometric measurements suggest a thickness of close to 20 Å per monolayer. Taking into account the ~18-19 Å length of the salt molecule, this corresponds to no inclination of the long molecular axis from the normal. The matter is under further investigation.

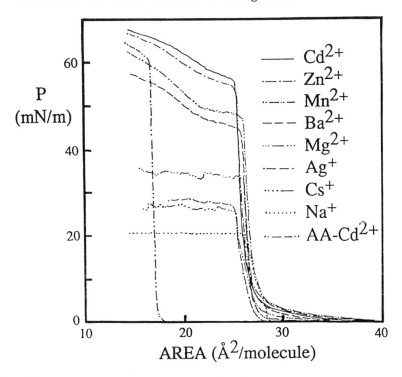

Figure 8. Pressure-area isotherms of the [4]staffane-3-carboxylic and arachidic (AA) acid with various counterions ($3 \times 10^{-4}$ M). pH was adjusted to 7.0 ± 0.1 using NaOH. Compression rate: 1 mm/sec.

Because of the high crystallinity of the staffanes, it is conceivable that these membranes and more complex Langmuir-Blodgett structures will be sturdier than those available from the more usual surfactant materials. This might be of considerable importance, as the limited stability of the conventional multilayer structures represents a severe hindrance to their practical use.

## "POLY[1.1.1]PROPELLANE"

Under appropriate conditions, both radical and anionic polymerization of 1 can be induced and produces a precipitate of a highly insoluble "polymer".[6] We have not been able to dissolve this product in any solvent at any temperature. It is highly polycrystalline, as judged by its X-ray diffraction pattern. Its CP MAS $^{13}$C NMR spectrum has already been mentioned and confirms the structure as that of poly[1.1.1]propellane. However, the presence of weak signals for the end group suggests that the molecular weight is very low. We have been able to obtain additional information on this by exhaustive chlorination of the insoluble material in $CCl_4$. This converts it into a quite soluble material whose molecular weight is so low that it cannot be determined by GPC analysis. Viscosity measurements[20] suggest that after chlorination the average number of monomer units in the molecule is no more than 10. However, we have not yet established that the chlorination process is not associated with a degradation of molecular weight. Still, it is quite probable that the increased solubility is at least in part due to the presence of substituents. The use of a substituted 1 as the starting monomer therefore is an obvious possibility to examine. However, a recent report from another laboratory[21] on the polymerization of a related [1.1.1]propellane, 6, is not particularly encouraging in this regard, since it, too, produced a highly insoluble low molecular weight "polymer."

The thermal stability of our poly[1.1.1]propellane is remarkable and quite similar to that of the lower oligomers. At about 300°C, violent decomposition occurs with loss of about 85% of weight and a dark residue is left behind.[6]

## MOLECULAR TINKERTOY CONNECTORS:  PRELIMINARY RESULTS

Our first attempts to connect several staffs together in a way that might be useful in the construction of actual two-dimensional and three-dimensional networks involved the carboxy group as the terminal group on the staff and the $Rh_2^{II}$ entity as the molecular "Tinkertoy" connector. The well-known structure of rhodium (II) carboxylates and their ability to form chains by ligating an additional bidentate group on each Rh atom[22] suggest strongly that this type of metal center will represent a relatively sturdy connector suitable for attachment of four carboxylate ends and two others, such as phosphines or amines.

Indeed, starting with rhodium (II) trifluoroacetate and 7, we have been able to prepare the cross-shaped structure 8 and characterize it spectrally.

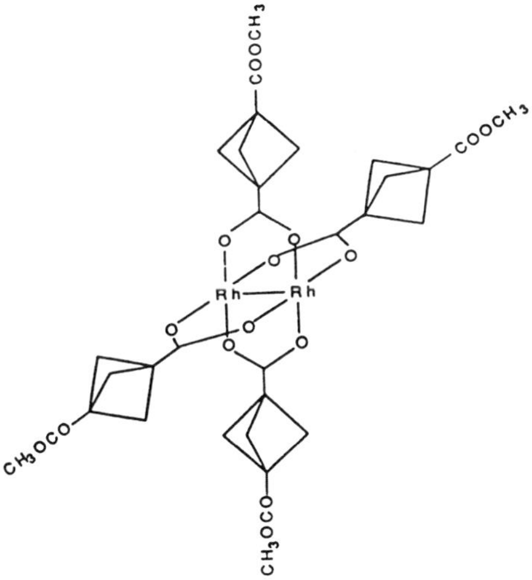

8

This represents merely a first tentative step in what we expect to become an extensive search for optimal molecular "Tinkertoy" connectors.

SUMMARY

Although the feasibility of a molecular-size Tinkertoy construction set has not yet been demonstrated in practice, we believe that the initial results are quite encouraging. There is no longer any doubt that suitably end-functionalized [n]staffanes of various lengths can be synthesized and possess most if not all of the desired properties. Attachment of conducting wires still needs to be demonstrated, and the rigidity of the staffs may be insufficient for some applications. It is, of course, quite possible that better structures for the purpose will be discovered in the future, but right now, the [n]staffane structures appear promising enough to pursue vigorously.

Acknowledgement

The initial work on the project was supported by the National Science Foundation. The more recent work has also benefitted from support by the Robert A. Welch Foundation and the Texas Advanced Research Program. Dr. Vincent M. Lynch determined the crystal structures.

1. Tinkertoy is a trade-mark of Playskool, Inc., Pawtucket, RI 02862, used to designate a children's toy construction set consisting of straight beams and connectors.
2. Lego is the trade-mark of LEGO System A/S, DK-7190 Billund, Denmark, and designates a children's toy construction set.
3. Feynman, R. P. *Saturday Review* 1960, *43*, April 2, 45.
4. Hameroff, S. R. *Ultimate Computing. Biomolecular Consciousness and NanoTechnology*; North Holland, Amsterdam, 1987; Chapter 10.
5. Wiberg, K. B.; Waddell, S. T.; Laidig, K. *Tetrahedron Lett.* 1986, *27*, 1553.
6. Kaszynski, P.; Michl, J. *J. Am. Chem. Soc.* 1988, *110*, 5225.
7. Wiberg, K. B.; Walker, F. M. *J. Am. Chem. Soc.* 1982, *104*, 5239. Wiberg, K. B. *Acc. Chem. Res.* 1984, *17*, 379.
8. Semmler, K.; Szeimies, G.; Belzner, J. *J. Am. Chem. Soc.* 1985, *107*, 6410.
9. Kaszynski, P.; Michl, J. *J. Org. Chem.*, in press.
10. Beeri, A.; Berman, E.; Vishkautsan, R.; Mazur, Y. *J. Am. Chem. Soc.* 1986, *108*, 6413.
11. The strain in bicyclo[1.1.1]pentane is 68 kcal/mol: Wiberg, K. B. *Angew. Chem., Int. Ed. Engl.* 1986, *25*, 312.
12. Ermer, O.; Lex, J. *Angew. Chem. Int. Ed. Engl.* 1987, *26*, 447.
13. Bunz, U.; Polborn, K.; Wagner, H.-U.; Szeimies, G. *Chem. Ber.*, preprint (1988).
14. Walsh, A. D. *Disc. Faraday Soc.* 1947, *2*, 18; Bent, H. A. *Chem. Rev.* 1961, *61*, 275.
15. Orendt, A.M.; Grant, D. M.; Friedli, A. C.; Michl, J., unpublished results.
16. Barfield, M.; Della, E. W.; Pigou, P. E. *J. Am. Chem. Soc.* 1984, *106*, 5051 and references therein.
17. Kaszynski, P.; Friedli, A. C.; Michl, J. *Mol. Cryst. Liq. Cryst. Lett.*, in press.
18. Reiffenrath, V.; Schneider, F. *Z. Naturforsch.* 1981, *36a*, 1006.
19. Gaines, G. L., Jr., in *Insoluble Monolayers at Liquid-Gas Interfaces*, Wiley, New York, 1966. Our monolayer work is done in collaboration with Prof. A.J. Bard.
20. Robinson, R. E.; Friedli, A. L.; Munk, P.; Du, Q.; Hattam, P.; Michl, J., unpublished results.
21. Schlüter, A.-D. *Angew. Chem. Int. Ed. Engl.* 1988, *27*, 296. Schlüter, A. D. *Macromolecules* 1988, *21*, 1208.
22. Cotton, F. A.; Felthouse, T. R. *Inorg. Chem.* 1981, *20*, 600.

# SYNTHESIS OF CYCLOPROPENES

W. E. Billups, M. M. Haley, G.-A. Lee
Department of Chemistry, Rice University
P.O. Box 1892, Houston, TX 77251, USA

ABSTRACT: The synthesis of cyclopropenes (in vacuo) using reagents adsorbed on inert surfaces to effect elimination reactions has been investigated. A superior route to very unstable cyclopropenes makes application of the ability of 2-functionalized cyclopropylsilanes to undergo facile ß-elimination in the presence of fluoride.
*Tetra-n*-butylammonium fluoride deposited on glass helices has been used to generate bicyclo[4.1.0]hept-(1,7)-ene, bicyclo[5.1.0]oct-(1,8)-ene, and several simple cyclopropenes, including cyclopropene itself, under conditions which allow spectroscopic characterization at low temperature.

# HOW TO GET STRUCTURES OF STRAINED COMPOUNDS: LOW TEMPERATURE STRUCTURES AND X-X ELECTRON DEFORMATION DENSITIES[1]

by Roland Boese* and Dieter Bläser
Institut für Anorganische Chemie der Universität–GH
Universitätsstr. 3–5, D–4300 Essen 1, FR Germany

Most of the highly strained molecules are either sensitive to heat or liquid under ambient conditions or both. This makes it difficult to get exact structural data, usually obtained by single crystal X–ray structure determination.

These problems may be overcome by two methods:

a) Crystallisation from solution at low temperatures, selecting and transferring the single crystal onto the diffractometer (with appropriate low temperature equipment) without interrupting the cooling chain.

b) Crystallisation by heating polycrystalline material, previously optained by cooling the liquid within a capillary below the melting point on the diffractometer.

For both methods we have developed special devices, combined with a computer controlled low temperature device and successfully employed for numerous structure determinations, including X–X electron density determinations.

A) Crystallisation in case (a) is performed in Schlenk tubes, surrounded with a cooling chamber and a vacuum isolated chamber. The coolant is temperature controlled for optimizing crystal nucleus growth and subsequent crystallisation.

The Schlenk tube is then coupled with a selection vessel, the crystals are washed into it together with the mother liquor and the supernatant solution is then decanted back.



*A. de Meijere and S. Blechert (eds.), Strain and Its Implications in Organic Chemistry, 485–488.*

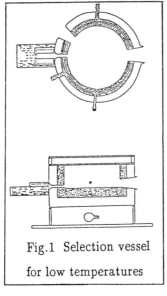

Fig.1 Selection vessel for low temperatures

The selection vessel (Fig. 1)[2] is surrounded with a cooling and a vaccum chamber, provided with an inlet for inert gas, which is cooled in the cooling chamber, prior to introduction into the inner chamber; it has an illumination from below and optical double wall glass on top. It has a detachable tiny cylindrical box, filled with powdered dry ice to pull the capillary containing the crystal, selected before, into this box. The capillary can be molten off and transferred together with the dry ice to the diffractometer. When the cooling gas on the diffractometer is turned on, the tiny dry ice box can be removed so that the cooling sequence is not even interrupted for a second.

B) Crystallisation in case (b) is performed with focussed infrared light[3]. The sample is first introduced into the capillary, sealed, and cooled below its melting point. After getting polycrystalline, an infrared beam is focussed onto the capillary with a parabolic mirror (Fig 2), until the sample melts within the focus. The

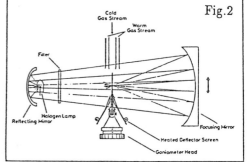

Fig.2

molten zone is now moved by a motion of the mirror until a single crystal grows.

This procedure has the advantage that the molten zone can be moved in any direction, twin crystals or satellites can be precisely molten again and the crystal is not obscured during the procedure.

A microprocessor control allows repetition of the procedure with a shift of the onset, varying of speed and lamp intensity to optimize unknown crystallisation parameters like crystallisation speed and temperature gradient.

Examples of structure determinations, performed with these procedures are shown in Figures 3–5.

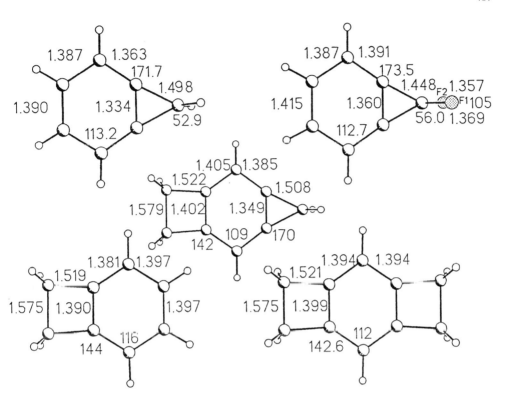

Fig.3 Benzoanneled cyclopropenes and cyclobutenes

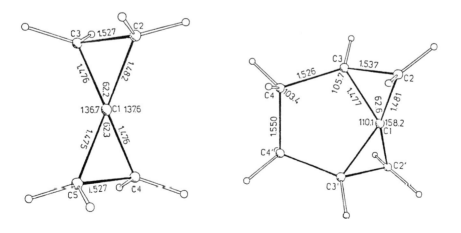

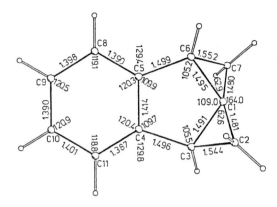

Fig. 4 Spiropentane as a tensile spring

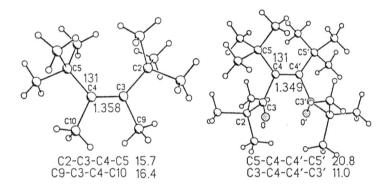

| C2-C3-C4-C5 | 15.7 |
| C9-C3-C4-C10 | 16.4 |

| C5-C4-C4'-C5' | 20.8 |
| C3-C4-C4'-C3' | 11.0 |

Fig. 5 Torsion of a double bond

References and Footnotes

1   Samples were provided by:
    U.Brinker, Bochum; G.Maier, Gießen; R.Neidlein, Heidelberg; W.R.Roth, Bochum; K.P.C.Vollhardt, Berkeley.

2   The authors gratefully acknowledge the elaborate work of the glasblowers Mr. D. Bonk and Mr. H. Klemz, who were able to build this delicate piece of equipment.

3   D.Brodalla, D.Mootz, R.Boese, W.Oßwald; *J. Appl. Cryst.* **18** (1985) 316

# NOVEL INTRAMOLECULAR 1,1-CYCLOADDITIONS OF DIAZOALLENES

Samuel Braverman and Ytzhak Duar
Department of Chemistry
Bar-Ilan University
Ramat-Gan 52100
Israel

ABSTRACT. Treatment of allenyl tosyl hydrazones 10a,b with BuLi in THF at 0°C, with subsequent reflux for 2 hrs, resulted in the exclusive formation of 4,4-dimethyl-6-alkylidene-1,2-diazabicyclo[3.1.0]hex-2-enes 12a,b, in high yields. Although rather unstable at room temperature, the formation of these products represent the first observation of a formal nitrene-type 1,1-cycloaddition of an allenyl substituted diazoalkane.

## BACKGROUND

The 1,3-addition of diazoalkanes to allenes is one of the first dipolar cycloadditions studied and is well documented.[1] The addition of diazomethane to allene was first reported in 1945.[2] Repeated later under carefully controlled condition[3] (sealed tube or teflon lined cylinder) it was shown to give the unstable 4-methylene-1-pyrazoline (1) described as sensitive to water, heat, light and oxygen. Photolysis of 1 at 185° gave trimethylenemethane (2) which is stable at liquid nitrogen temperatures, and thermolysis gave methylenecyclopropane (3).

1 2 3

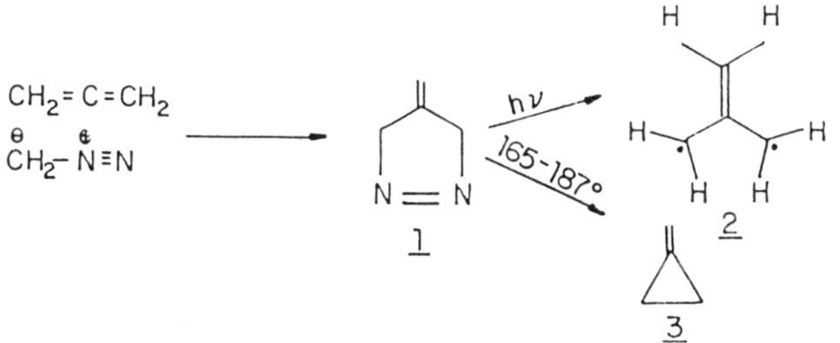

Subsequently, a variety of other intermolecular dipolar cycloadditions of diazoalkanes to allene and substituted



A. de Meijere and S. Blechert (eds.), Strain and Its Implications in Organic Chemistry, 489–492.

allenes has been performed, including the addition of 2-diazopropane to allene which gave a mixture of the relatively stable products 4-6.[4]

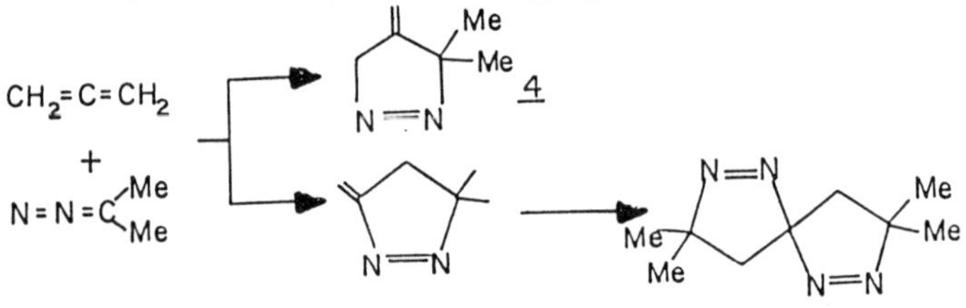

However, to the best of our knowledge no previous report on a similar intramolecular cycloaddition has been published. In view of our previous studies on the intramolecular cyclization of various diallenic systems such as 7 ⟶ 8,[5] we became interested in the thermal cyclization of the analogous diazoallenes 11a,b.

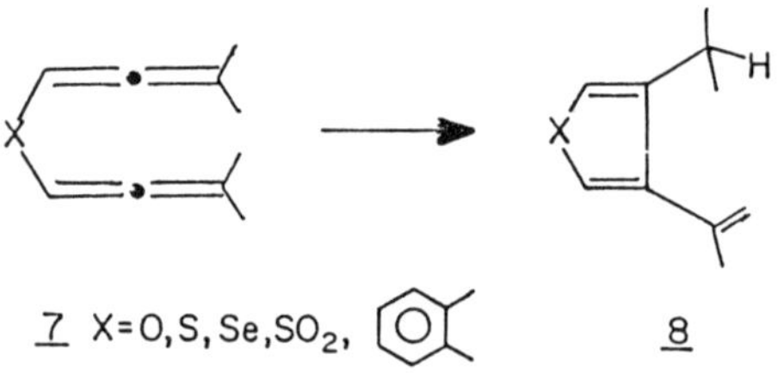

7 X=O, S, Se, SO$_2$, ⌬    8

## PREPARATION OF STARTING MATERIALS

Homoallenic aldehydes 9a,b were prepared by a literature procedure,[6] via the Claisen-Cope rearrangment of propargyl vinyl ethers derived from the p-toluenesulfonic acid catalyzed condensation of isobutyraldehyde and methyl-substituted propargyl alcohols. Conversion of aldehydes 9a,b to the corresponding tosyl hydrazones (10a,b) was readily achieved by treatment with tosyl hydrazine.

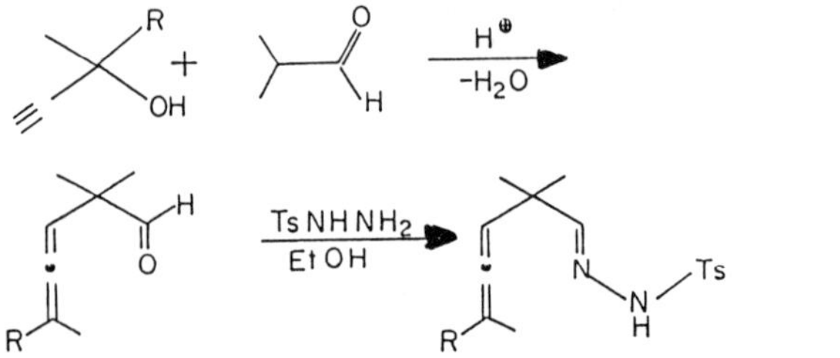

9 a R=CH$_3$
  b R=H

10 a R=CH$_3$ (Mp=112-113°C)
   b R=H (Mp=78-79°C)

## REACTION CONDITIONS AND PRODUCTS

Treatment of allenyl tosyl hydrazones 10a,b with BuLi in THF at 0°C, with subsequent reflux for 2 hrs resulted in the exclusive formation of 4,4-dimethyl-6-alkylidene-1,2-diazabicyclo[3.1.0]hex-2-enes 11a,b in high yields.

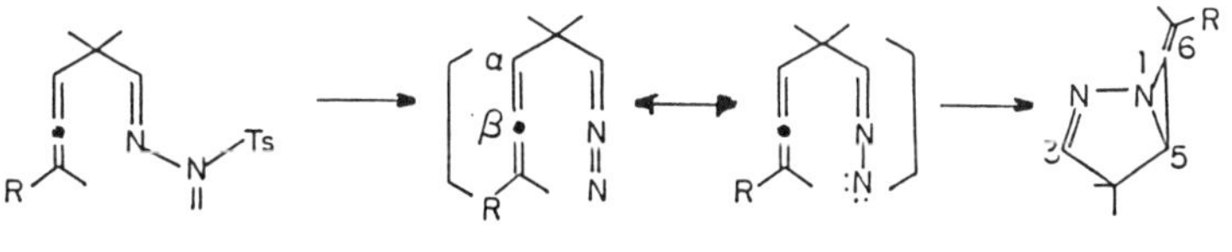

10 a R=CH$_3$
   b R=H

11 a R=CH$_3$
   b R=H

12 a R=CH$_3$
   b R=H

Although rather unstable at room temperature, the formation of these products represent the first observation of a formal <u>nitrene</u>-type 1,1-cycloaddition of an allenyl substituted diazoalkane. This experimental evidence clearly indicates that aziridine formation, which has been proved by Huisgen[7] not to occur during intermolecular cycloaddition reactions of diazomethanes and olefins, can take place intramolecularly.

## COMPARISON OF DIAZOLLENES VS. DIAZOALKENES

The results reported above are reminiscent of those previously reported on the intramolecular 1,1-cycloaddition of cyclic[8] and acyclic[9] allyl substituted diazomethanes.

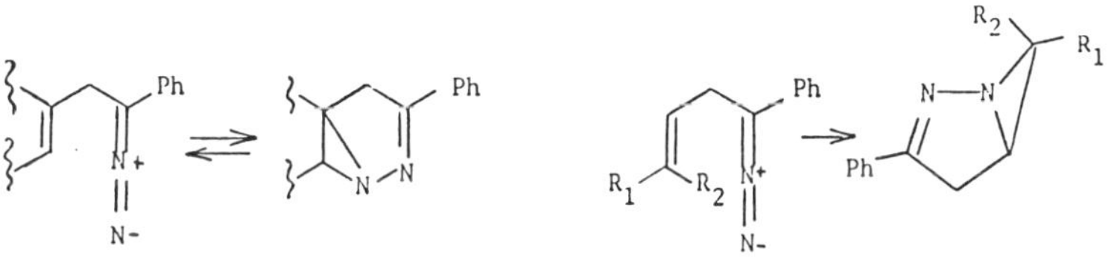

The authors[8,9] suggest two possible mechanisms. The first mechanism is a stereospecific addition of singlet nitrene to olefin, while the second mechanism is a two step mechanism involving formation of a cyclic six-membered 1,3-zwitterion or dipole in the first step, and ring closure of the dipole in the second step. The usual 1,3-dipolar cycloaddition is prevented in these particular cases due to the inability of the 1,3-dipole and dipolarophile to achieve the necessary orbital overlap of the π-system.

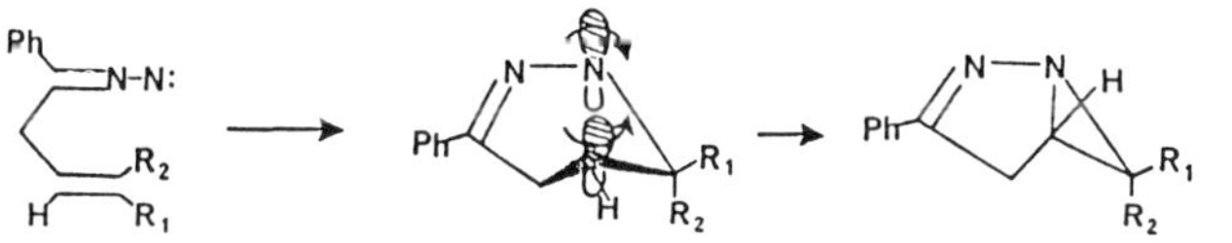

The ring strain likely to arise for both the intramolecular 1,3-dipolar cycloaddition and the formal nitrene addition to the β,τ-double bond of diazoallenes 11a,b, is responsible for the absence of these processes. The lack of stability of the observed products 12a,b may also be assigned to the excessive strain present in these structures.

REFERENCES

1. S.R. Landor, in The Chemistry of the Allenes, S.R. Landor, Ed., Academic Press, London, 1982, Vol. 2, Ch. 5.5.

2. I.A. D'Yakanov, Zhur. Obshch. Khim., 15, 473 (1945); Chem. Abstr., 40, 4718 (1946).

3. P. Dowd, J. Amer. Chem. Soc., 88, 2587 (1966); Acc. Chem. Res., 5, 242 (1972); R.J. Crawford and D.M. Cameron, J. Amer. Chem. Soc., 88, 2589 (1966); Canad. J. Chem., 52, 4033 (1974).

4. M. Schneider, O. Schuster and H. Rau, Chem. Ber., 110, 2180 (1977).

5. For an account see: S. Braverman, Phosphorus and Sulfur, 23, 297, (1985).

6. R.S. Bly and S.U. Koock, J. Am. Chem. Soc., 91, 3292 (1969).

7. R. Huisgen, R. Sustmann and K. Bunge, Chem. Ber., 105 1324 (1972).

8. Y. Nishizawa, T. Miyashi and T. Mukai, J. Am. Chem. Soc., 102, 1176, (1980).

9. A. Padwa and A. Rodriguez, Tetrahedron lett., 22, 187 (1981); A. Padwa and H.Ku, Tetrahedron lett., 21, 1009 (1980).

# EFFICIENT ROUTES TO OPTICALLY ACTIVE STRAINED COMPOUNDS

A. Fadel, B. Karkour, J. Ollivier and J. Salaün
Laboratoire des Carbocycles,
Université de Paris-Sud, 91405 ORSAY (France)

Optically active $\alpha$-alkylsuccinates, now readily available either from enzymatic resolution (1) or from the stereoselective alkylation of chiral imide enolates (2) underwent acyloin cyclization followed by stereoselective ring contraction to give 1-hydroxycyclopropanecarboxylic acids with high enantiomeric excesses (3,4).

(2R) > 95% ee (3R) (1S,2R) > 94% ee

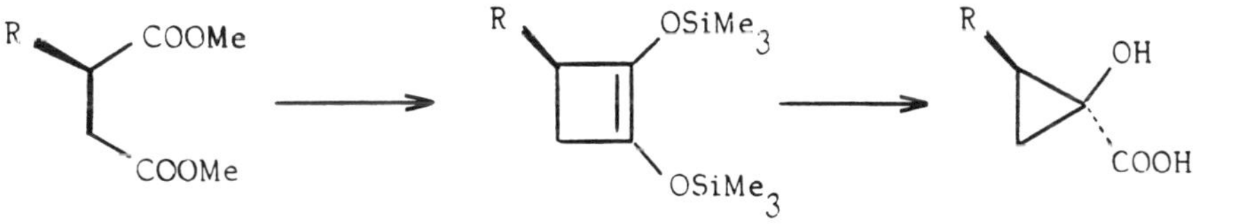

These cyclopropanols allowed the preparation of chiral cyclopropylvinylcarbinols which then, underwent regio- and stereospecific trifluoroborane-etherate induced C3 ⟶ C4 ring expansion into 2-vinylcyclobutanones, providing optically active building blocks for further useful ring expansions (4).

(1S,2R) (2R,3R)



*A. de Meijere and S. Blechert (eds.), Strain and Its Implications in Organic Chemistry, 493–494.*

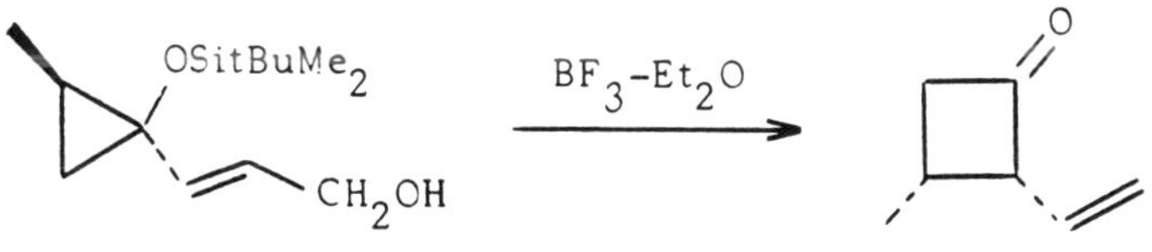

On the other hand, the corresponding chiral acetate underwent palladium induced C3→C4 ring expansion, from which a chirality transfer can be expected ? (5)

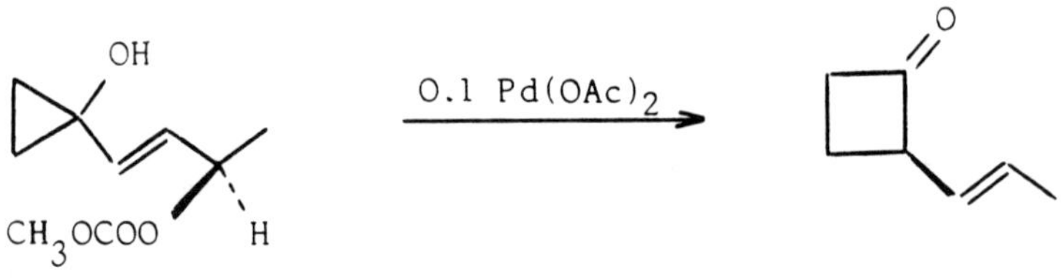

## REFERENCES

1) E. Guibé-Jampel, G. Rousseau and J. Salaün, J. Chem. Soc. Chem. Commun., 1987, 1080.

2) A. Fadel and J. Salaün, Tetrahedron Lett., submitted.

3) J. Salaün and B. Karkour, Tetrahedron Lett., 1987, 28, 4669 and 1988, 29, 1537.

4) J. Salaün, B. Karkour and J. Ollivier, Tetrahedron, 1988, in press.

5) J. Salaün and J. Ollivier, to be published.

# ANIONIC (3+2)-CYCLOREVERSION OF STRAINED CAGE NITRILES

Wolfram Grimme and Werner Neukam
Institut für Organische Chemie
der Universität zu Köln
Greinstraße 4, D-5000 Köln, West Germany

In the series of cage olefins **1, 2** and **3** only basketene (**1**) has been synthesized and its (4+2)-cycloreversion to syn-tricyclo[4.4.0.0$^{2,5}$]deca-3,7,9-triene has been demonstrated. The more strained compounds **2** and **3** are still unknown but their intermediacy has been postulated along (4+2)-cycloaddition pathways in hydrocarbon thermal rearrangements.

We have prepared the anionic two electron-one carbon counterparts of the above olefins by proton abstraction from the cage nitriles **4, 5** and **6** and report on their (3+2)-cycloreversion.

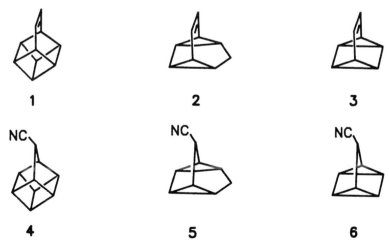

# REACTION OF AN ANGLE STRAINED CYCLOHEPTYNE WITH A STABILIZED STANNYLENE - SYNTHESIS AND X-RAY STRUCTURE OF A DISTANNACYCLOBUTENE SYSTEM

A.Krebs[a], A.Jacobsen-Bauer[a], E.Haupt[b], M.Veith[c], V.Huch[c]
Institute für Organische[a], Anorganische und Angewandte Chemie[b], Universität Hamburg, FRG
Institut für Anorganische Chemie, Universität des Saarlandes, Saarbrücken, FRG[c]

The first germirenes[1] were prepared by addition of germylenes to the angle strained cycloalkyne **1** which combines high reactivity of the triple bond with steric protection of the resulting 1:1-adduct.
However, the analogous reaction with the stabilized stannylene **2** did not yield the corresponding 1:1-adduct but the 1:2-adduct **3** in almost quantitative yield; **3** represents the first example of a distannacyclobutene system.

In solution an equilibrium exists between **3** and the educts **1** and **2** with the equilibrium shifted towards the educts on dilution. The structure **3**, proposed on the basis of $^1$H-, $^{13}$C- and $^{119}$Sn-nmr data, was proven by an X-ray structural analysis.
For the formation of **3** two routes may be discussed: A "direct" [2+2]-cycloaddition of a distannene (dimer of **2**) to **1** or a [1+2]-cycloaddition leading to a stannirene which is followed by an insertion of **2** into the strained carbon-tin bond.

---

1) A.Krebs, J.Berndt, Tetrahedron Lett. 24 (1983) 4083

# STRUCTURE AND REACTIVITY OF SMALL HETEROCYCLES CONTAINING GERMANIUM ATOMS

O.M. NEFEDOV, S.P. KOLESNIKOV, M.P. EGOROV
The Institute of Organic Chemistry USSR Academy of Sciences
117913 Moscow, Leninsky Prosp. 47, U.S.S.R.
In collaboration with A. KREBS (The Institute of Organic Chemistry, Hamburg University, F.R.G.) and Yu. T. STRUCHKOV (The Institute of Organoelement Compounds USSR Academy of Sciences, Moscow)

ABSTRACT. Some representatives of the relatively new classes of germacyclopropenes (germirenes) and 1,2-digermacyclobutenes have been synthesized, isolated and their structure and reactivity studied.

The last decade has brought a rapid development in the field of small-sized heterocycles. Two new classes of heterocycles - heterocyclopropenes (sila-, phospha-, boracyclopropenes) and 1,2-diheterocyclobutenes (1,2-disila-, 1,2-diphospha-, 1,2-dibora-, 1,2-dithiacyclobutenes) have been obtained in these years.
We report here the syntheses, structures, and reactivities of some germacyclopropenes and 1,2-digermacyclobutenes - a long sought class of unsaturated small heterocycles, which have been predicted in the early sixties /1/.
Germacyclopropene 1a, its silicon derivative, silirene 1b, and 1,2-digermacyclobutenes 2a-c were synthesized in nearly quantitative yields by the methods shown in Scheme 1 /2-10/.

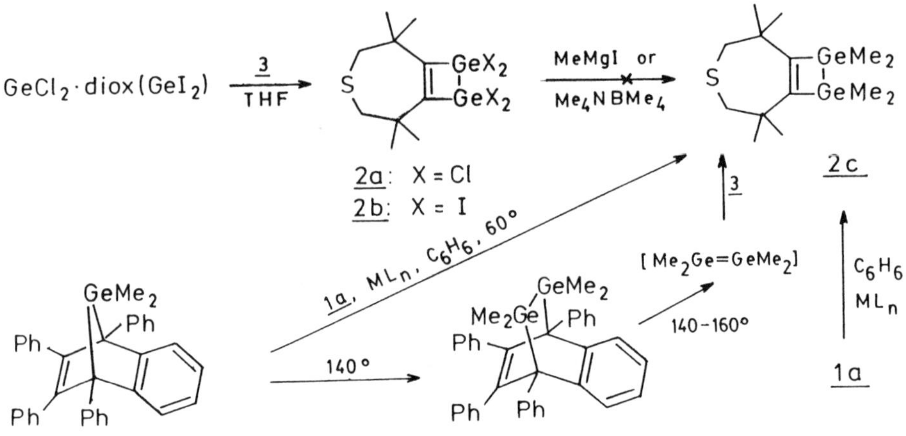

Scheme 1

It should be pointed out, that the reaction of tetramethyldigermene with acetylene 3 is the first example of [2+2]-cycloaddition of $Me_4Ge_2$ to a multiple bond.

Compounds 1a, b and 2a-c were isolated in pure form and characterized by $^1H$, $^{13}C$ NMR and mass- (electron impact and chemical ionization) spectral data. Germirene 1a and silirene 1b are thermally stable, low melting crystalline compounds (m.p. 40-42° and 28-30°C respectively), but very sensitive to air. Compounds 2a (m.p. 192-194°C), 2b (m.p. 230°C, dec.) and 2c (m.p. 80-82°C) are thermally stable and do not react with $O_2$ and MeOH.

The crystal structures of 1a,b and 2a, b have been determined by X-ray diffraction /4,5,8,10/. The structural parameters of germacyclopropene, silacyclopropene, and 1,2-digermacyclobutene fragments of 1a, b and 2a, b are presented in Fig. 1.

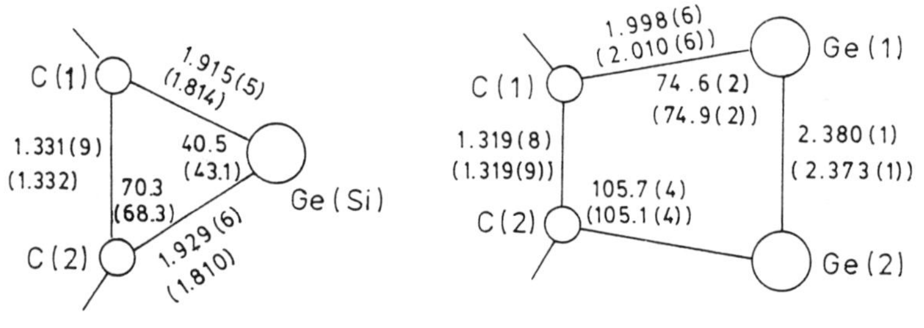

Fig. 1

An essential feature of the structures of 1a and 1b is that the bond lengths $E-C(sp^2)$ are noticeably shorter ( 4%, E=Ge; 3-3.5%, E=Si) than the standard $Ge-C(sp^2)$ (1.96-2.01 Å) and $Si-C(sp^2)$ (1.874 Å) bonds. The length of the double C (1)-C(2) bond is close to the standard C=C bond (1.337 Å), but considerably longer than in cyclopropene (1.296 Å). A

comparison of the structural data and properties of 1a, b (high thermal stability, low-field chemical shift of endocyclic carbon atoms in $^{13}C$ NMR spectra of 1a, b) with those of the aromatic borirene and non-aromatic phosphirene indicates that germirene 1a and silirene 1b fulfill the aromaticity criteria.
The digermacyclobutene ring of 2a, b is planar, the bond lengths Ge-X and Ge-C($sp^2$) are close to the standard values, but the C(1)-C(2) double bond is shorter than in cyclobutene (1.33 Å). It should be noted that the Ge-Ge bond length is shorter than the analogous bond in digermane (2.402 Å) and hexaphenyldigermane (2.437 Å).
The angle strain in germacyclopropenes causes an unusual reactivity of the endocyclic Ge-C($sp^2$) bond: all reactions which have been studied, result in its breaking and the formation of acyclic products or insertion products (scheme 2).

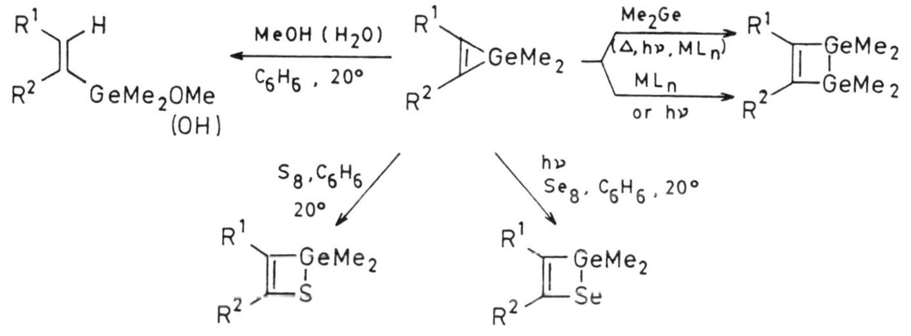


$R^1+R^2 = (H_3C)_2C-CH_2-S-CH_2-C(CH_3)_2$

Scheme 2

For the same reason, reactions of 2a and 2c all result in the cleavage of the Ge-Ge bond within in the strained four-membered ring (scheme 3).

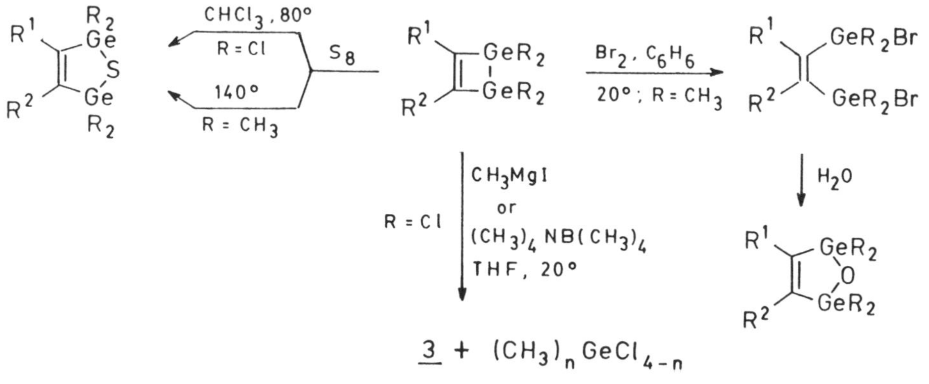


$R^1+R^2 = (H_3C)_2C-CH_2-S-CH_2-C(CH_3)_2$

Scheme 3

The Ge-Ge bond in 2a is more electrophilic than that in 2c, as judged, e.g., by the different reactivity of 2a and 2c towards sulphur (see scheme 3).

References

1. M.E. Vol'pin, Yu.D. Koreshkov, V.G. Dulova and D.N. Kursanov, *Tetrahedron* **18**, 107 (1962).
2. S.P. Kolesnikov, A. Krebs and O.M. Nefedov, *Izv. Akad. Nauk SSSR, Ser. Khim.*, **1983**, 2173.
3. A. Krebs and J. Berndt, *Tetrahedron Lett.* **24**, 4083 (1983).
4. A.A. Espenbetov, Yu.T. Struchkov, S.P. Kolesnikov and O.M. Nefedov *J. Organomet. Chem.* **275**, 33 (1984).
5. M.P. Egorov, S.P. Kolesnikov, Yu.T. Struchkov, M.Yu. Antipin, S.V. Sereda and O.M. Nefedov *ibid.*, **290**, C27 (1985).
6. O.M. Nefedov, S.P. Kolesnikov, M.P. Egorov, A.M. Galminas and A. Krebs, *Izv. Akad. Nauk SSSR, Ser. Khim* **1985**, 2834.
7. O.M. Nefedov, M.P. Egorov, A.M. Galminas, S.P. Kolesnikov, A. Krebs and J. Berndt, *J. Organomet. Chem.* **301**, C21 (1986).
8. S.P. Kolesnikov, M.P. Egorov, A.M. Galminas, A. Krebs and O.M. Nefedov, *Izv. Akad. Nauk SSSR, Ser. Khim.* **1985**, 2832.
9. S.P. Kolesnikov, M.P. Egorov, A.M. Galminas, Yu.T. Struchkov, A. Krebs and O.M. Nefedov, *ibid.* **1987**, 1835.
10. O.M. Nefedov, M.P. Egorov, S.P. Kolesnikov, A.M. Galminas, Yu.T. Struchkov, M.Yu. Antipin and S.V. Sereda, *ibid.* **1986**, 1693.

## RELIEF OF STRAIN ENERGY AS DRIVING FORCE IN RING-OPENING POLYMERIZATIONS OF A [1.1.1]PROPELLANE

Harald Bothe, Arnulf-Dieter Schlüter*
Max-Planck-Institut für Polymerforschung
Postfach 3148,
D-6500 Mainz, F.R.G.

Recently, the [1.1.1]propellane 1 was reported to be the first hydrocarbon monomer which cleanly polymerizes with breaking of a CC σ-bond.[1a] In a ring-opening polymerization involving the central bond of 1, homopolymer 2, which has a rigid rod-like structure, was obtained. We now observed spontaneous copolymerization of propellane 1 and a typical vinyl monomer, acrylonitrile 3, giving the alternating copolymer **4**.[1b]

1

2

4

5

6

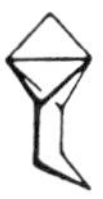

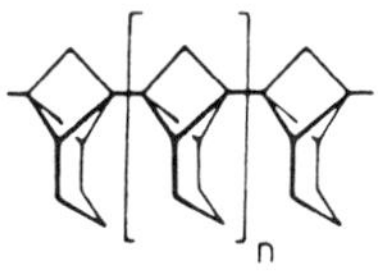

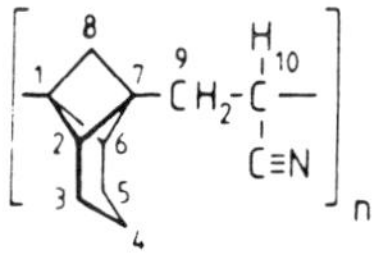

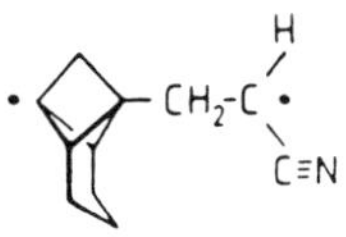

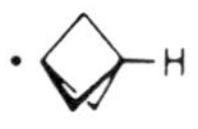

Monomers 1 and 3 react upon mixing under nitrogen in the temperature range from -25°C to 20°C, either neat or in heptane



*A. de Meijere and S. Blechert (eds.), Strain and Its Implications in Organic Chemistry, 503–505.*

solution. In several independent runs, utilizing both normal glassware and teflon vessels, the polymerization started spontaneously and reproducibly, even if light was excluded rigorously. Polymer **4** was obtained by precipitation with methanol from a chloroform solution in isolated yields ranging from 30 - 44% depending on the reaction conditions. A significant amount of oligomeric material of similar structure was also obtained giving a total mass balance of better than 90% based on **1**. Elemental analysis established that the product was a 1:1 copolymer, and detailed examination of its $^{13}$C-NMR spectrum lead to the conclusion that it was a perfectly alternating, head-to-tail, and atactic copolymer.[1b]

The experimental data confirm that only the central bond of propellane **1** was during the course of the polymerization. Breaking of this bond is associated with a loss of strain energy on the order of 30 kcal/mole. Presumably, this considerable relief of strain is responsible for the low activation energy of the polymerization as reflected by its spontaneity even at temperatures as low as -25°C. It is interesting to note that only the central bond of **1** breaks, even though propellane **1** consists of seven highly strained CC-bonds, the breaking of any of which would result in a comparable loss of strain energy.

We assume that the polymerization of **1** and **3** starts from the initially formed 1,5-diradical **5** and proceeds further by a radical mechanism.[2] Species **5** consists of two radical-bearing moieties, for each of which profound evidence exists in the literature. Especially important in this respect is the work of Maillard and Walton who proved that radical **6** was a surprisingly persistent radical.[3] Additional evidence can be derived from the structure of copolymer **4**. The high regularity of this structure rules out zwitterionic intermediates. Such species would have a high propensity to rearrange as observed for the

1-bicyclo[1.1.1]pentyl-cation.[4]

A determination of the exact molecular weight (m. w.) of **4** turned out to be difficult due to aggregation phenomena. We hope to overcome problems related to this with temperature dependent light scattering experiments. First results indicate a m. w. on the order of 100,000. From solutions of **4** in chloroform, transparent and flexible films were obtained, also revealing relatively high molecular weight.

Preliminary experiments show that propellane **1** not only undergoes spontanous copolymerization with acrylonitrile, but also with vinylidene chloride, styrene, and ethyl acrylate. Similar behavior of parent [1.1.1]propellane was also observed. These results will be published in due course.

Literature

1a) A.-D. Schlüter, <u>Angew. Chem</u>. 100, 283 (1988); A.-D. Schlüter, <u>Macromolecules</u> 21 (1988) in press; (b) H. Bothe, A.-D. Schlüter, submitted.

2) H. K. Hall, Jr., Angew. Chem. 95, 448 (1983).

3) B. Maillard, J. C. Walton, <u>J. Chem. Soc., Chem Commun</u>. 1983, 900;

4) K. B. Wiberg, V. Z. Williams, Jr., <u>J. Am. Chem. Soc</u>. 89, 3373 (1967).

# DIELS-ALDER REACTIONS OF SILOXYALLYLIDENECYCLOPROPANES: FACILE SYNTHESES OF SPIRO[2.5]OCTAN-3-ONES

Thies Thiemann, Stephan Kohlstruk, Gerhard Schwär and Armin de Meijere[*]
Institut für Organische Chemie der Universität Hamburg
Martin-Luther-King-Platz 6, D-2000 Hamburg 13, West-Germany

Although 1,1-disubstituted dienes are known to be quite unreactive in Diels-Alder-reactions, allylidenecyclopropanes due to the strain inherent in their methylenecyclopropane moiety have been found to react smoothly with activated enes.[1,2b] Moreover, it has been shown that the spirocyclopropane group in the cycloadducts can be reduced to a gem-dimethyl group.[1a] While such cycloadditions have been done with unsubstituted allylidenecyclopropane,[1] and recently also a 3-heterosubstituted allylidenecyclopropane has been synthesized,[2] a 2-heterosubstituted analog has not been reported to our knowledge. The 2-trialkylsiloxyallylidenecyclopropanes 2 can easily be prepared by silylenolization[3] from the readily available cyclopropylideneketones 1.[4]

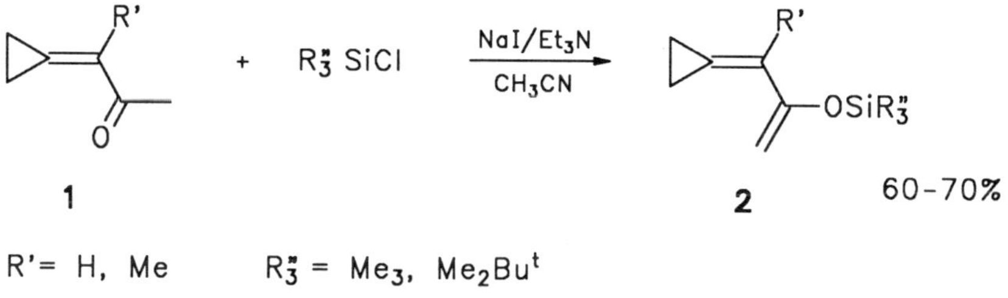

R' = H, Me      $R_3'' $ = Me$_3$, Me$_2$Bu$^t$

Diels-Alder-reactions of 2[5] with activated enes under thermal conditions give 3-siloxy-spiro[2.5]oct-2-enes of type 4, which can be hydrolysed to the corresponding spiro[2.5]octan-3-ones 5.

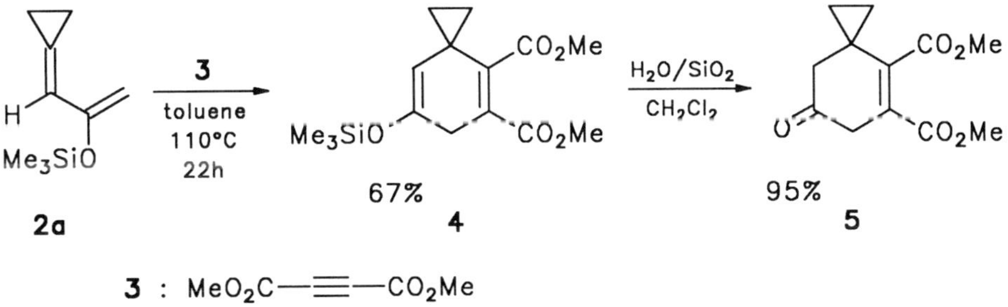

3 : MeO$_2$C—≡—CO$_2$Me

While doubly activated enes give moderate to good yields, less reactive enes such as cyclohexenone (6) fail to react at temperatures up to 150°C. Under Lewis-acid-catalysis,[6] however, cyclohexenone undergoes cycloaddition with 2 to give the diketone 7 in 30% yield. 3-acetoxy-2-methylcyclopentenone (8) did not cycloadd to 2a under these conditions. 4-acetyldispiro[2.1.2.3]decan-9-one (10) was isolated instead (56% yield); this apparently resulted from hydrolysis of 2a to the cyclopropylideneketone 1a and its cycloaddition to 2a.

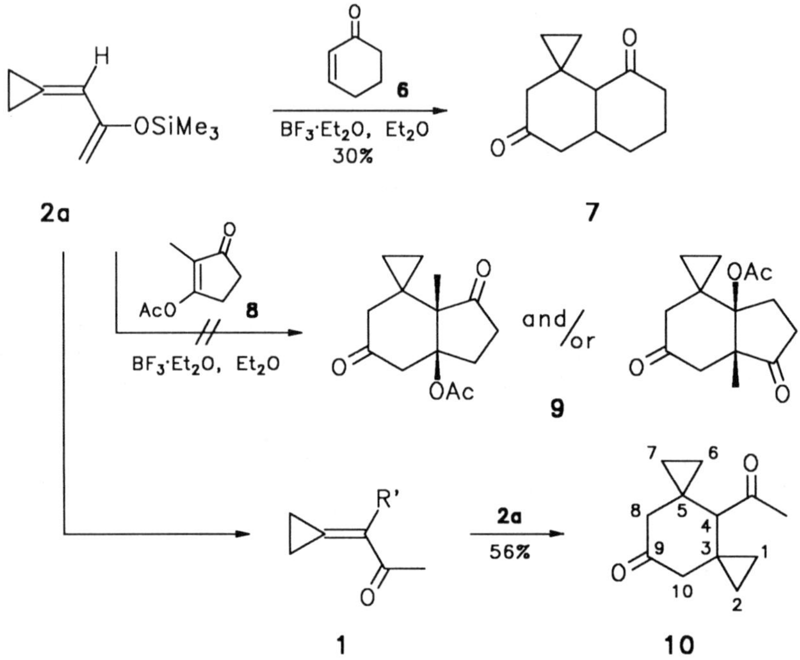

**References**

1.  1a) F. Zutterman, A. Krief, *J. Org. Chem.* **48** (1983) 1135. - 1b) T. Tsuji, R. Kikuchi, Sh. Nishida, *Bull. Chem. Soc. Jpn.* **58** (1985) 1603. - 1c) L.A. Paquette, G.J. Wells, G. Wickham, *J. Org. Chem.* **49** (1984) 3618.

2.  2a) J.A. Stafford, J.E. McMurry, *Tetrahedron Lett.* **29** (1988) 2531. - 2b) B.B. Snider, Y. Kulkarni, *Org. Prep. Proc. Int.* **18** (1986) 7.

3.  3a) P. Cazeau, F. Moulines, O. Laporte, F. Duboudin, *J. Organomet. Chem.* **201** (1980) C9. - 3b) P. Cazeau, F. Duboudin, F. Moulines, O. Babot, J. Dunogues, *Tetrahedron* **43** (1987) 2089.

4.  4a) D. Spitzner, H. Swoboda, *Tetrahedron Lett.* **27** (1986) 1281. - 4b) A. Lechevallier, F. Huet, J.M. Conia, *Tetrahedron* **39** (1983) 3307.

5.  2-Trialkylsiloxyallylidenecyclopropanes can only be stored for a limited length of time under argon at -40°C and should be used as soon as possible.

6.  J. Das, R. Kubela, G.A. MacAlpine, Z. Stojanac, Z. Valenta, *Can. J. Chem.* **57** (1979) 3308.

## CYCLOPROPYL GROUP CONTAINING AMINO ACIDS FROM α-CHLOROCYCLOPROPYLIDENACETATES

*Norbert Krass, Ludger Wessjohann, Dahai Yu, Armin de Meijere**

Institut für Organische Chemie der Universität Hamburg

Martin-Luther-King Platz 6, 2000 Hamburg 12, West-Germany

Cyclopropane amino acids have attracted the attention of synthetically and biologically oriented chemists throughout the last decade. Most efforts have been directed towards 1-aminocyclopropane carboxyclic acid (ACC), the natural precursor of the phytohormone ethylene,[1] and its derivatives. Other natural cyclopropane amino acids as well as artificial ones were used as enzyme inhibitors for receptor and metabolism studies.[2] We report here a new synthetic approach to a wide variety of cyclopropyl-substituted amino acids including the natural products cleonin (1) (1-hydroxycyclopropyl-glycin)[3] and 1-methylcyclopropyl-glycin (2).[4]

1

2

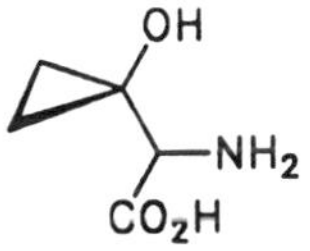

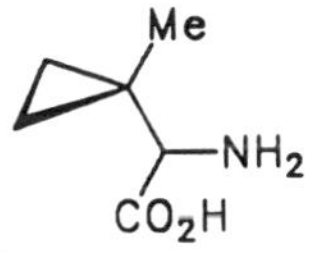

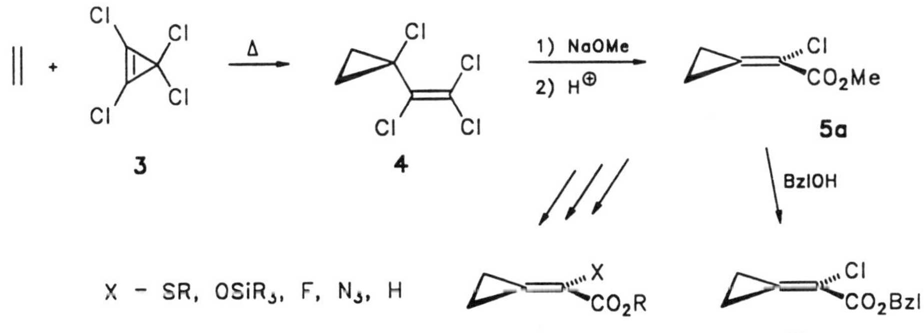

The universal starting material, leading to α- as well as β-amino acids, is methyl 2-chloro-cycloproylidenacetate (5a), which can be prepared in three steps from ethylene



*A. de Meijere and S. Blechert (eds.), Strain and Its Implications in Organic Chemistry, 509–512.*

and commercially available tetrachlorocyclopropene (3).[5] 5a combines high reactivity with multifunctionality; its α-chloro substituent can be replaced in a three step sequence (addition, substitution, elimination) to give a series of analogously α-substituted cyclopropylidenacetates 6.[6,7]

The 1,4-addition of nucleophiles (Michael-addition) onto 5 and 6, which is promoted by the strain release upon changing the $sp^2$ to a $sp^3$ carbon center in the cyclopropane moiety leads to 1'-substituted 2-chloro-2-cyclopropyl acetates 7-Nu in high yields[6,7] (scheme 1). With the use of suitable oxygen and nitrogen nucleophiles the resulting addition/substitution products can be deprotected to contain free hydroxy and amino groups respectively without destroying the cyclopropyl- or other functional groups.

Scheme 1.

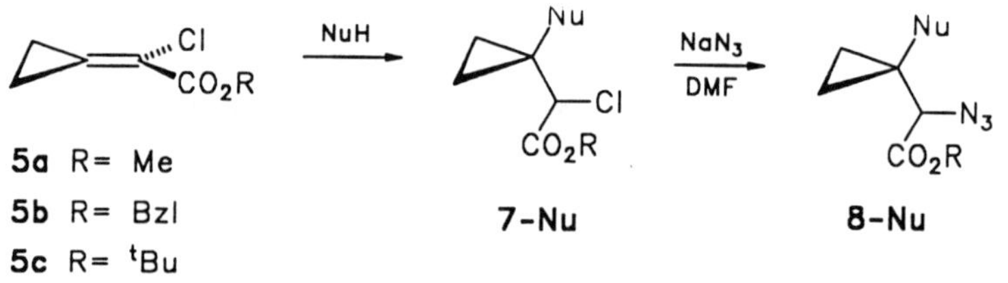

5a R= Me
5b R= Bzl
5c R= $^t$Bu

| 7-Nu | Reagent | Yield | 8-Nu Yield |
|---|---|---|---|
| OBzl | BzlONa/BzlOH | 50% | 65% |
| N(Bzl)$_2$ | HNBzl$_2$/THF | 97% | 90% |
| N(Boc)$_2$ | KNBoc$_2$/CH$_2$Cl$_2$ | 64% | – |
| N=CPh$_2$ | HN=CPh$_2$/MeOH | 98% | 80% |
| NPht | KNPht/DMF/H$_2$O | 18% | – |
| CH$_2$CO$_2^t$Bu | LiCH$_2$CO$_2^t$Bu/HMPT | 91% | 94% |
| Me | LiCuMe$_2$/Et$_2$O | 61% | 98% |
| SMe | NaSMe/THF | 78% | 96% |
| H | NaBH$_4$/CHCl$_3$/$^i$PrOH | 75% | 98% |

The benyloxy group in 7-OBzl, introduced by addition of benzyl alkoxide to 5b, can be deprotected by hydrogenation in the final product 8-OBzl. The nitrogen derivatives dibenzylamine, di-Boc-amine, benzophenonimine and potassium phthalimide in contrast to simple primary or secondary amines cleanly lead to 1,4-adducts 7-Nu without successive chlorine substitution or amide formation. The best yield was achieved with benzophenonimine to give 7-N=CPh$_2$. To our knowledge such a 1,4-addition of an imine to an acrylic ester derivative has not been reported before. The more classical phthalimide addition gave the worst yields.

Lithium t-butyl acetate enolate gave 7-CH$_2$CO$_2$Bu$^t$ in 91 % yield, whereas the yield of 7-Me with lithium dimethylcuprate did not exceed 61 %. Selective reduction of 5a to 7-H is achieved with sodiumborohydride.

The 1'-substituted 2-chloro-2-cyclopropylacetates 7-Nu can be further transformed by nucleophilic substitution of the chlorine; reductive dechlorination (Pd/C, H$_2$ or Zn/Cu, THF/H$_2$O) or α-halogenation to 2,2-disubstituted cyclopropylacetates has also been achieved. Treatment of 7-Nu with sodium azide in DMF most readily led to precursors of various cyclopropylglycin derivatives 8-Nu with yields usually exceeding 80 % (scheme 1). Catalytic hydrogenation over Pd/C was successful with most azidoesters 8-Nu and proceeded with high yields. In addition to an easy workup this method offered the advantage of deprotecting one or two of the other functional groups at the same time.

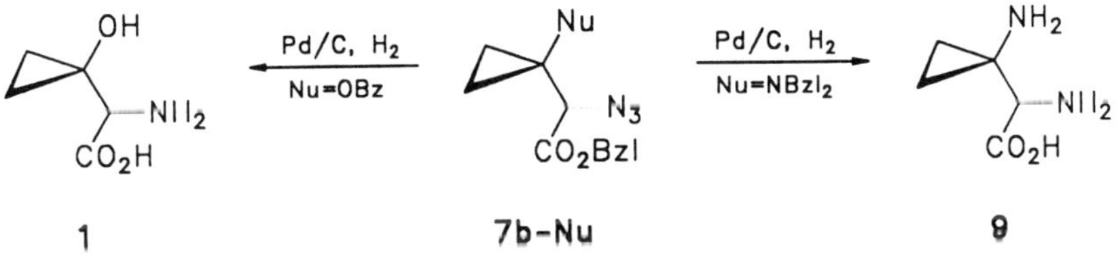

Especially the benzylester 7b-Nu, obtained either from the corresponding starting material 5b or by transesterification of 7a-Nu, directly leads to free amino acids 1 and 9.

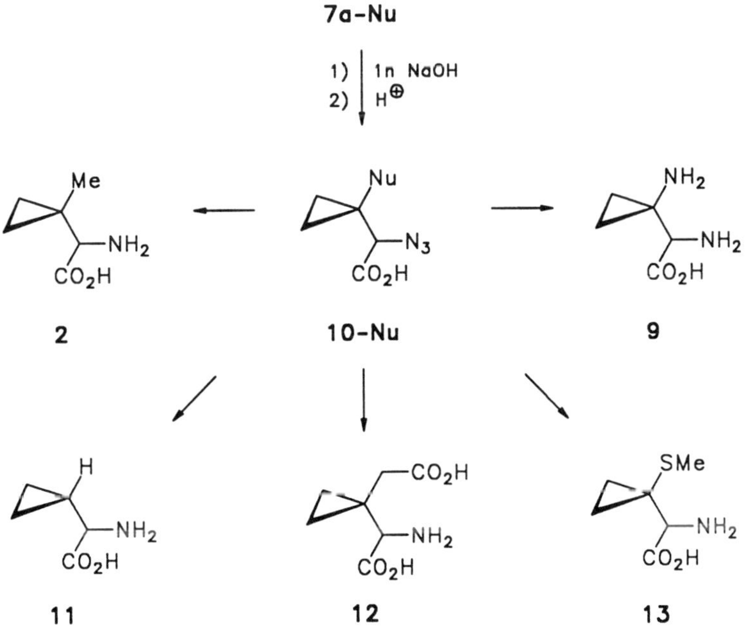

Alternatively, methylester 7a-Nu with functional groups Nu stable to aqueous base and acid, were hydrolyzed with 1N sodium hydroxide and the azidoacids 10-Nu converted to the free amino acids, e.g. 2, 9, 11, 12 and 13 by catalytic reduction with good to excellent yields.

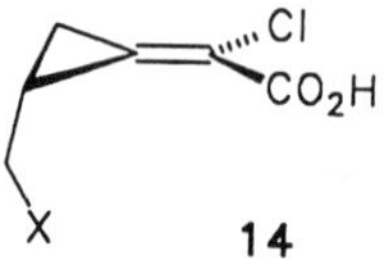

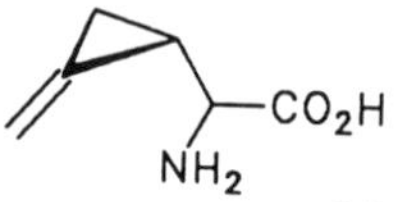

The same strategy can be applied to the synthesis of further interesting natural and non-natural amino acids. We are currently exploring the corresponding routes from appropriately substituted cyclopropylidenacetates 14 to (methylenecyclopropyl)glycin (15), 3,4-methanoproline 16 and other analogues like 17 and 18.

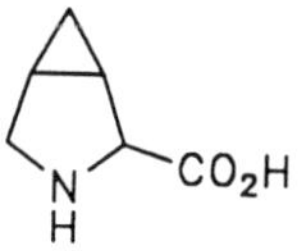

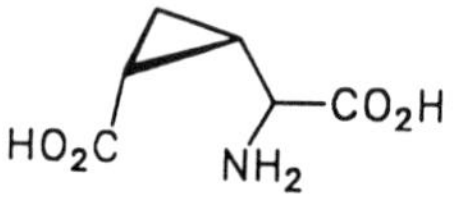

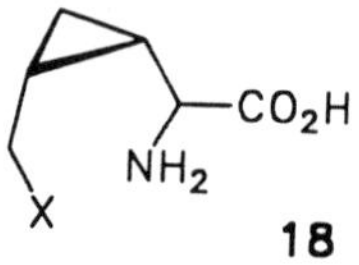

### References

1. Z. Procházka, *Chemiché Listy* **79** (1985) 1043.

2. C. J. Suckling, *Angew. Chem.* **100** (1988) 555; *Angew. Chem. Int. Ed. Engl.* **27** (1988) 537. - J. E. Baldwin, R. M. Adlington, B. Domayne-Hayman, G. Knight, H. H. Ting, *J. Chem. Soc., Chem. Comm.* (1987) 1661.

3. K. Kato, T. Takita, H. Umezawa, *Tetrahedron Lett.* **21** (1980) 4925.

4. J.-I. Shoji, R. Sakazaki, T. Kato, Y. Yoshimura, S. Matsuura, K. Tori, *J. Antibiotics* **34** (1981) 371.

5. Th. Liese, G. Splettstößer, A. de Meijere, *Angew. Chem.* **94** (1982) 799; *Angew. Chem. Int. Ed. Engl.* **21** (1982) 784. - Th. Liese, F. Seyed-Mahdavi, A. de Meijere, *Org. Synth.* (1988) in press.

6. F. Seyed-Mahdavi, S. Teichmann, A. de Meijere, *Tetrahedron Lett.* **27** (1986) 6185. F. Seyed-Mahdavi, Dissertation, Universität Hamburg 1986.

7. Th. Liese, S. Teichmann, A. de Meijere, *Synthesis* (1988)25. - S. Teichmann, Dissertation, Universität Hamburg 1988.

## ELIMINATIVE FISSION OF STRAINED CYCLOALKOXIDES - THE QUESTION OF ELECTROPHILIC CATALYSIS

*Susan Wyn Roberts and Charles J.M. Stirling*
Department of Chemistry, University of Wales,
Bangor, LL57 2UW, UK

Earlier work[1] on eliminatice fission of cycloalkoxides has established that for aqueous solutions, enforced catalysis of ring fission is operative:

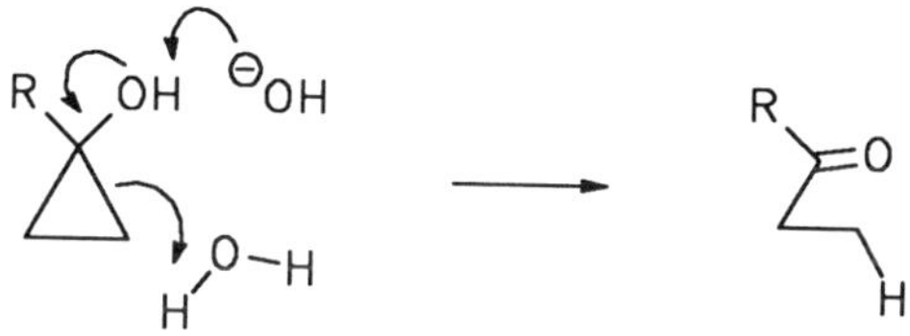


Reactions of this type have now been performed under strictly anhydrous conditions with variation of the counter cation and in the presence and absence of added electrophiles other than protons so as to address the questions:

(i) What is the effect of counter cation on the ease of cyclopropoxide cleavage?

(ii) What is the role, if any, of added electrophile (alkyl halide or aldehyde) on the ease of ring fission?

(iii) What products are obtained in the presence of added electrophiles and what is their origin?

[1] A. Thibblin and W.P. Jencks, *J. Am. Chem. Soc.*, 1979, **101**, 4963.



*A. de Meijere and S. Blechert (eds.), Strain and Its Implications in Organic Chemistry, 513.*

# SPECIAL REACTIVITY OF 8,11-DIHALO[5]METACYCLOPHANE

P.A. Krackman, W.H. de Wolf*, F. Bickelhaupt
Scheikundig Laboratorium, Vrije Universiteit Amsterdam, Postbus 7161
1007 MC Amsterdam, Netherlands

The accessibility on gram scale of 8,11-dihalo[5]metacyclophane **2** from tetrahalopropellanes **1** promised a considerable improvement of the synthesis of the parent [5]metacyclophane **2d**.

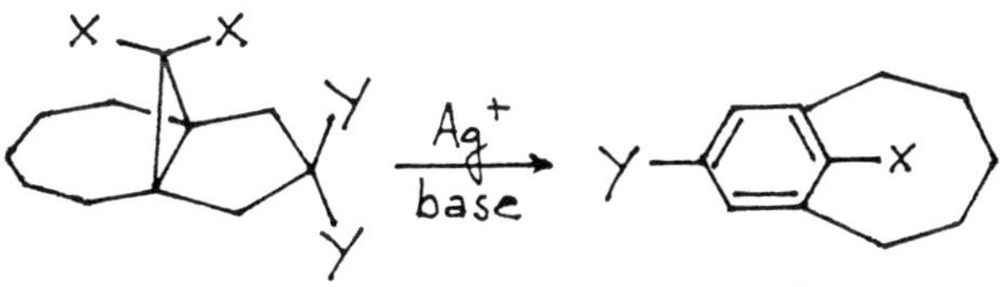

2a X=Y=Cl
2b X=Y=Br
2c X=Br ,Y=Cl
2d X=Y=H or D
2e X=H(D) ,Y=Cl
2f X=t-Bu ,Y=Cl

However, this reduction only succeeded in the case of **2b** in reaction with tert-butyllithium followed by quenching with $CH_3OH$ or $CH_3OD$. Under similar reaction conditions **2c** gave no **2e** but **3**.

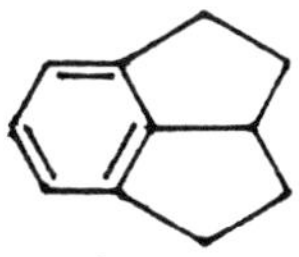

Compound **2a** followed again a different course of reaction with tert-butyllithium: the most hindered chlorine atom (at C 11) was substituted by the tert-butyl group, yielding **2f**.
In order to obtain a deeper understanding of the mechanism of this aromatic substitution several nucleophiles, like $PhS^-$ ,$CN^-$ ,$NH_2^-$ and $RO^-$ ,were used in stead of the tert-butyl anion.
Substitutions were observed with $RO^-$( R= $CH_3$ ,$C_2H_5$ ,i-$C_3H_7$ ,H), sometimes followed by rearrangements. In some cases addition takes place.

# SMALL [N]CYCLOPHANES. WHERE IS THE LIMIT?

G.B.M. Kostermans, W.H. de Wolf[*], F. Bickelhaupt
Scheikundig Laboratorium, Vrije Universiteit Amsterdam, Postbus 7161
1007 MC Amsterdam, Netherlands

Dewar benzenes **1** and **3** are appropriate precursors of small [n] cyclophanes( n<8): in fact the complete skeleton of the cyclophane is present and the high energy content of the Dewar benzene system is available to compensate for any strain and instability in **2** or **4** which might tend to jeopardize the transformation.

On purely thermodynamic grounds one would surmise that the limit for this approach might be reached for n= 5. For [4]paracyclophane as well as for [4]metacyclophane the corresponding Dewar benzene is calculated to be more stable.

Though, recently the intermediate formation of [4]paracyclophane could be demonstrated by interception with acid during photolysis of 1,4-tetramethylene Dewar benzene ( **1**, n=4 ).
[4]Metacyclophane was formed on thermolysis of 2,6-tetramethylene Dewar benzene ( **3**, n=4 ). It dimerized in a Diels-Alder reaction yielding several unexpected products, among which [4.4]paracyclophane and [4.4]metacyclophane; it also could be intercepted with dienophiles.
This results as well as attempts to synthesize [3]para- and [3]metacyclophane will be presented.

# INDEX

| | |
|---|---|
| Acyloin cyclization | 493 |
| Addition reactions | 317 |
| Agrochemicals | 333 |
| Alcohol dehydrogenase | 190 |
| 6-(1-Alkenyl)bicyclo[3.1.0]hex-2-enes, thermolysis of | 39, 46 |
| Alkenylidene carbene addition | 59, 70, 72 |
| Alkenylidenecyclopropanes | 60, 73 |
| 2-Alkylidene-1,3,4-thiadiazolines | 61 |
| Alkylidenecyclopropabenzenes | 459 |
| α-Alkylsuccinates | 493 |
| Alkynyl cyclobutanols | 16 |
| Alkynyl cyclopropanols | 15 |
| Allamandin synthesis | 17 |
| Allamcin synthesis | 17 |
| Allene episulfide, structure of | 69 |
| Allene episulfides | 59, 63 |
| Allene oxides | 59 |
| Allenyl sulfones | 7 |
| Allylidenecyclopropanes | 120 |
| Alternating copolymers | 503 |
| Amino acids | 509 |
| Aminocyclopropane carboxylic acid | 198 |
| 16π Aromatic hydrocarbon | 393 |
| Arylcyclopropanes | 169, 173, 175, 185, 333, 346 |
| Asymmetric alkylation | 235 |
| Atoms in molecules | 354 |
| Aziridines | 25 |
| Basketane | 405, 407 |
| Benzoanellated small rings | 483 |
| 4,5-Benzo-1,2,4-cyclohexatriene | 128 |
| Benzvalene | 94 |
| β-cis-Bergamotene | 214 |
| β-trans-Bergamotene | 214 |
| Bicalicene | 393 |
| Bicycloannulation | 8 |
| Bicyclo[n.1.0]alk-1-enes, calculated energies of | 350 |
| Bicyclo[1.1.0]butanes | 361 |
| Bicyclo[1.1.0]butene | 350 |
| Bicyclobutene derivatives as reactive intermediates | 362, 365 |
| Bicyclo[1.1.0]but-1-ene derivatives, Diels-Alder reactions of | 362, 364 |
| Bicyclo[3.2.0]heptan-5-one | 250 |
| Bicyclo[4.1.0]heptene | 350 |
| Bicyclo[4.1.0]hept-1(7)-ene | 483 |
| Bicyclo[3.1.0]hexene | 350 |
| Bicyclo[2.1.1]hex-2-ene | 95 |
| Bicyclo[3.2.1]octa-2,6-diene synthesis | 39, 45, 46 |
| Bicyclo[4.1.1]octane | 83 |
| Bicyclo[3.3.0]octanes | 290 |
| Bicyclo[4.1.1]octene | 84 |
| Bicyclo[3.2.1]oct-2-ene | 114 |
| Bicyclo[5.1.0]oct-1(8)-ene | 483 |
| Bicyclo[2.1.0]pentene | 350 |
| S-Bioallethrin | 333 |
| Bisalkylidenecyclopropanes | 60, 73 |
| 6,7-Bisalkylidene-1,2,3,4,5-pentathiepanes | 66 |
| Bishomocubanes | 405 |
| Bishomocubanol | 412 |
| 1,4-Bishomohexaprismane ("garudane") | 269, 270, 275, 277 |
| 1,7-Bistrimethylsilyltricyclo[4.1.0.0$^{2,7}$]heptane | 143, 165 |
| Bis(trimethylsilyl)thioketene | 61 |
| Bond angle distortion | 349 |
| Bond fixation | 457 |
| Bond path angles | 356 |
| Brendanone | 423 |
| Brendene | 425 |
| Brexene | 425 |
| Bridged aromatic molecules | 297 |
| Bridged cyclopropenes, calculated vibrational spectra of | 354 |
| Bridged [1.1.1]propellanes | 361 |
| Bridgehead alkenes | 135 |
| Bridgehead alkenes, epoxidation of | 136 |
| Bridgehead dienes | 135 |
| Bridgehead functional groups | 406 |
| Burgess reagent | 2, 89 |
| 1,2,3-Butatrienes | 368 |
| 1,2,3-Butatriene episulfide | 59, 66, 67, 68 |

| | | | |
|---|---|---|---|
| t-Butylthioketene | 61 | [2+1]Cycloaddition | 57 |
| Cage bridgehead alcohols | 406 | [2+2]Cycloaddition | 210, 235, 236, 274 |
| Cage closure reactions | 424 | | |
| Cage compounds, synthesis of | 405 | [2+2]Cycloaddition, assymmetric | 236, 240 |
| Cage compounds, two bond cleavage reactions | 421 | [2+2]Cycloaddition, intramolecular | 07, 209, 211, 212, 220, 230, 247, 250, 251 |
| Cage contraction | 406 | | |
| Cage directed stereochemistry | 410, 414, 424 | [2+2]Cycloaddition, intramolecular photochemical | 272, 275 |
| Cage expansion | 407 | | |
| Cage olefins | 495 | [2+2]Cycloaddition, photochemical | 290 |
| Cage opening reactions | 410, 418 | [2+2]Cycloaddition, strain energy controlled photochemical | 274 |
| Calculations, AM1 | 259 | | |
| Carbapenem | 215 | [4+2]Cycloaddition | 317, 319 |
| Carbapenem, C-analogue of | 218 | Cycloalkynes, angle strained | 497 |
| Carbenoid cyclization | 361, 374 | Cyclobutane relay effect | 78 |
| Carbon-carbon bond activation | 451 | Cyclobutane relay orbitals | 77, 97 |
| Carbon-hydrogen bond activation | 451 | Cyclobutane-cyclopentane interconversion | 263 |
| 2,9-Carbonylbrendene | 424 | | |
| Carbyne | 454 | Cyclobutaniminium salts | 244 |
| Cation radical- anion radical pair | 146, 158 | Cyclobutanone, angular annulated | 227 |
| Cation radicals | 146, 172, 187 | Cyclobutanone spiroannulation | 1, 10, 17 |
| Cationic rearrangements | 431, 433 | Cyclobutanones | 244, 255 |
| Chalcogran | 53 | Cyclobutanones, bicyclic | 212, 215, 246 |
| Charge densities | 349 | Cyclobutanones, homochiral | 235 |
| Chemical reactivity | 314 | Cyclobutanones, tricyclic | 207, 230 |
| Chiral cyclopropylvinylcarbinols | 493 | Cyclobutene electrocyclizations | 25 |
| Chirality transfer | 494 | Cyclobutene opening | 29 |
| 2-Chlorocyclopropylidenacetates | 509 | Cyclobutenes | 2, 3, 4 |
| Chrysanthemic acid | 333, 334 | Cyclobutenones, electrocyclic ring opening of | 221 |
| Chrysanthemic acid, enantioselective synthesis of | 334 | Cyclobutylcyclopropylcarbinyl cation rearrangement | 425 |
| Chrysanthenone | 214 | 1,2,3-Cycloheptatriene | 366 |
| Claisen rearrangement | 293 | Cycloheptatriene iron tricarbonyl | 264 |
| Cleonin, synthesis of | 511 | 1,2-Cyclohexadiene | 121, 122, 124, 130 |
| Composite pseudofunctional groups | 8, 16 | | |
| Compression | 283, 285 | Cyclohexadiene hexatriene isomerization | 30 |
| Conjugation | 77 | | |
| Conrotary opening | 30 | 1,2,4-Cyclohexatriene | 121, 129, 130 |
| Copaene | 215 | Cyclopentadienone ketals | 78 |
| Cope rearrangement | 39, 46, 48, 109, 114, 135 | Cyclopentanoic sesquiterpenes | 431 |
| | | Cyclopentene annulation | 110 |
| Cryogenic matrices | 451 | $[2_n]$Cyclophanes | 297, 298, 301, 316 |
| Crystallisation technique | 485 | | |
| Cubane | 406 | Cyclophanes, multibridged | 317 |
| 6-Cyanobicyclo[3.1.1]heptane | 144, 155 | [2.2]Cyclophanes, strain energies of | 307 |
| 1-Cyanonaphthalene, as photosensitizer | 143, 150, 156, 160 | Cyclopropabenzenes | 457 |
| Cyclization terminators | 15, 16 | Cyclopropanation | 326, 327, 338 |
| Cyclizations | 8, 13, 16 | Cyclopropanation, intramolecular | 346 |
| Cycloaddition, intramolecular arene-olefi | 283, 290 | Cyclopropane | 454 |
| Cycloaddition, intramolecular dipolar | 54 | Cyclopropane amino acids | 509 |
| 1,1-Cycloaddition | 489 | Cyclopropanethione | 59, 67 |
| Cycloaddition, $[_\sigma 2 + _\pi 2]$ | 169, 171 | Cyclopropanol, ring opening of | 416, 513 |

| | | | |
|---|---|---|---|
| Cyclopropanols | 493 | Diels-Alder additions | 297, 317, 320, 507 |
| Cyclopropanone hydrate | 107 | | |
| Cycloproparenes | 457 | Diels-Alder reactions, cascade | 515 |
| Cyclopropene ring opening | 118 | Diels-Alder reactions, intramolecular | 54, 135 |
| Cyclopropene vinylcarbene rearrangement | 117 | Diels-Alder trapping reactions | 365, 367 |
| | | Diene synthesis | 6 |
| Cyclopropenes | 117, 349, 483 | 1,2-Digermacyclobutene | 500, 501 |
| Cyclopropenium ions, heteroatom substituted | 383 | Dihalocyclopropane reduction | 44 |
| | | Dihydrofuran annulation | 111 |
| Cyclopropenium ions, stability of substituted | 385 | Dihydronicotinamid | 189 |
| | | Dihydropyran | 117 |
| Cyclopropyl aldehyds | 180, 191 | Dihydropyridines | 192 |
| Cyclopropyl amines | 180, 194 | 9,10-Dimethylenetricyclo[5.3.0.0$^{2,8}$]- deca-3,5-diene | 77, 79, 88 |
| Cyclopropyl carbinyl to cyclobutyl rearrangements | 1 | | |
| | | Dispirocompounds | 432 |
| Cyclopropyl methanols | 180, 187, 191 | Dispiroundecanes | 431, 432 |
| Cyclopropyl modified substrate | 179 | Displacement vector | 284 |
| Cyclopropyl peptides | 202 | Disrotary reactions | 30 |
| Cyclopropylbenzylamines | 183 | Distannacyclobutene | 497 |
| Cyclopropylcarbinyl cation to cyclobutyl cation rearrangement | 148, 162 | Distorted benzene rings | 517 |
| | | Distortions, angular | 284 |
| Cyclopropylcarbinyl radical | 182 | Dithiocyclopropenethione | 393 |
| Cyclopropylcarbinyl radical ring opening | 177, 179, 190 | Divinylcyclopropane rearrangement | 44, 45, 46, 48, 109, 114 |
| Cyclopropylglyoxalate | 193 | Dodecahedrane, attempted synthesis of | 449 |
| Cyclopropylidenacetates | 509 | | |
| Cyclopropylpropanols | 180 | Donor-acceptor-substituted cyclopropanes | 51, 56 |
| [3+2]Cycloreversion | 495 | | |
| [4+2]Cycloreversion | 495 | Dynamic NMR | 264 |
| Cytochrome P450 | 184, 186 | Electron acceptor photosensitizer | 149, 150, 151, 152, 154, 165 |
| DDQ, oxidation with | 169, 171 | | |
| 1,2-Dehalogenation | 117, 118 | Electron population | 355 |
| Dehydrogenase, NAD-dependant | 180, 189 | Electrophilic catalysis | 513 |
| Deltamethrin | 333, 341 | Eliminative ring fission | 513 |
| Designer solids | 463 | Ene reaction | 354 |
| Desulfurization | 68 | Enzymatic resolution | 493 |
| Deviation angles | 443 | Enzyme catalyzed reactions | 177 |
| Dewar benzenes, shortly bridged | 515 | Enzyme inhibitors | 509 |
| Diaminochlorcyclopropenium ion | 383, 384, 387 | Enzyme mechanisms | 177 |
| Diaminocyclopropenethione | 388 | Epoxidation | 317, 320 |
| Diaminocyclopropenone | 387 | Epoxides, intramolecular attack on | 444 |
| 1,2-Diazabicyclo[3.1.0]hex-2-enes | 491 | Epoxides, strain energies of | 139 |
| Diazoallenes, intramolecular cycloadditions of | 489 | Ethylene, biosynthesis | 180, 198, 199 |
| | | Ethylene iron carbene complex | 452 |
| 2-Diazocarbonylcyclobutanones | 255, 256 | Ethynyl iron hydride | 452 |
| Diazoketones | 255 | Exited state oxidants | 150, 165 |
| Diazomethane | 452, 453 | Exothermic reaction | 495 |
| 1,3-Dication synthons | 7 | [4.4.4.4]Fenestrane | 289 |
| Dichloroketene | 208 | [4.4.4.5]Fenestrane | 289 |
| Dichloroketene, cycloaddition of | 432 | [5.5.5.5]Fenestrane, hydroboration of | 294 |
| 9,10-Dicyanoanthracene, as photosensitizer | 144, 150, 152 | [5.5.5.5]Fenestrane, stereoisomers of | 289, 292 |
| | | Fenestranes | 285, 287 |
| Dicyclopropabenzenes | 458 | Ferracyclobutane | 452 |

| | |
|---|---|
| Fischer carbene complexes | 56 |
| Fluorescence quenching by strained hydrocarbons | 147 |
| Functionalized enones | 52 |
| Garudane | 270 |
| Germacyclopropene | 499 |
| Germirene | 499 |
| Germirene, reactivity of | 500 |
| 1-Halobicyclo[1.1.0]butanes | 362, 369 |
| Halogenative rearrangement | 1, 2 |
| Hammett leaving groups | 441 |
| Helvetane | 259 |
| Hexaprismane | 270 |
| Homo-[1,5]-sigmatropic hydrogen migrat | 41 |
| HOMO/ LUMO interaction | 213 |
| HOMO/ LUMO, of ethylene, cumulenes | 242 |
| Homoallylic rearrangement | 421 |
| Homoconjugative stabilization | 416 |
| Homocubane | 405, 409 |
| Homocubanol | 412 |
| Homocuneane | 409, 414 |
| Homoenolate | 411 |
| Homoketonization, regiochemistry, stereochemistry | 410, 412, 415, 418 |
| 1,4-Hydrogen shift | 62 |
| 1,5-Hydrogen shift | 292 |
| Hydroxylase | 180, 187 |
| Imimium salts | 235 |
| INDO calculations | 385 |
| Information storage | 402 |
| Information storage system | 464 |
| Insecticides | 333 |
| Intramolecular displacement | 440 |
| Intramolecular nucleophilic substitution | 439, 440 |
| 1-Iodobicyclo[1.1.0]butanes | 362 |
| Ipomeamarone | 109, 112 |
| Ipso-substitution | 317 |
| Iron atoms | 451 |
| Iron carbene complexes | 452 |
| Isobullvalene | 91 |
| Isocomene | 222, 228 |
| Isocomene | 431, 433, 436 |
| Isomerisation reactions | 306, 314, 315 |
| Israelane | 259 |
| Ketene cycloaddition | 207 |
| Ketenes | 207, 208, 209 |
| Keteniminium salts | 235, 237, 238, 246 |
| Keteniminium salts, derived from aminoacids | 248 |
| Keteniminium salts, diastereomeric | 251 |
| Ketenophile | 208, 209 |
| $\alpha$-Ketenylcyclobutanones | 256 |
| Kinetics | 264 |
| $\beta$-Lactams, C-analogues | 215 |
| $\gamma$-Lactones | 240 |
| $\gamma$-Lactones, optically active | 235 |
| Layered organic molecules, electronic interactions in | 297 |
| Lewis acid catalysis | 508 |
| Line shape analysis | 264, 266 |
| Lipoxygenase | 187 |
| Liquid crystals | 477 |
| Lithiocyclopropenes | 117 |
| Low temperature technique | 485 |
| Lumibullvalene | 91, 92 |
| MAO inhibitors | 194 |
| Matrix isolation spectroscopy | 452 |
| MCSCF calculations | 63 |
| Medium sized rings | 12, 13, 14 |
| Mercaptohexadiene | 61 |
| [4.4]Metacyclophane, strain and aromaticity | 515 |
| [3]-, [4]Metacyclophanes | 515 |
| [5]Metacyclophanes | 517 |
| Metallacycle | 453 |
| Methyl 2-siloxycyclopropanecarboxylate | 51 |
| Methyl vinyl iron | 452 |
| cis-1-Methyl-2-vinylcyclopropane | 34 |
| Methylcyclopropane | 187, 185 |
| Methylenecyclobutylidene | 354 |
| Methylenecyclopropanes | 25, 33, 59 |
| Methylenecyclopropanethione | 67 |
| Methylenecyclopropanones | 59, 67 |
| Methylenefuran | 117 |
| 1-Methyltricyclo[4.1.0.0$^{2,7}$]heptane | 143, 151 |
| Michael addition | 53, 510 |
| Mitsunobu reaction | 294 |
| MM2 calculations | 259, 271, 274, 433, 450 |
| MM2' calculations | 261 |
| MMP2 Calculations | 301 |
| MNDO calculations | 283, 287, 289, 443, 450 |
| Mobile, molecular | 464 |
| Modhephene | 222, 431, 433, 436 |
| Molecular energy storage | 383, 394 |
| Molecular orbital calculations | 457 |
| Molecular orbital calculations, ab initio | 25, 259, 271, 349, 354, 457 |
| Molecular rearrangement | 151, 161 |
| Molecular-size construction set | 463 |
| Monoaminoxigenase | 180, 194 |
| Monooxigenases | 180, 183, 187 |
| Moore's hydrocarbon | 149, 150, 152 |

| | | | |
|---|---|---|---|
| Moore´s hydrocarbon, dimerization of | 152, 154 | Plumericin synthesis | 17 |
| Moore´s hydrocarbon, rhodium promoted rearrangement of | 161 | Polyacene | 383, 402 |
| | | Polycyclic cage compounds | 405 |
| Mutually perpendicular pi-ribbons | 78 | Polycyclic cage compounds, strain energy of | 405 |
| Narcissistic coupling | 449 | | |
| Natural products | 53 | Polycyclic carbon compounds | 283 |
| Nef-reaction | 53 | Polycyclic compounds | 283 |
| Nitrile anions | 495 | Polyhedranes | 269 |
| Nitroalkanes | 53 | Polyquinanes | 447 |
| Nitrone | 54 | Poly[1.1.1]propellanes | 503 |
| Norbornadiene | 383, 397 | PQQ dependant enzymes | 180, 196 |
| Norbornadiene, donor-acceptor substitut | 397 | Prezizaene | 43 |
| Nucleophilic aromatic substitution | 517 | Prismanes | 259 |
| Olefinic strain | 349 | [n]-Prismanes | 269, 280 |
| One-electron carbon-carbon bond | 156, 149 | [7]-Prismane, its homo- and secologues | 278 |
| Optically active strained compounds | 493 | [8]Prismane | 280 |
| Organometallic rearrangements | 263 | Propellane formation | 361 |
| Orthogonal $\pi$-systems | 425 | [1.1.1]Propellane, reactions of | 466 |
| Overlap population | 385 | [1.1.1]Propellane synthesis | 361, 372, 374 |
| Oxa cage compounds | 424 | [1.1.1]Propellane telomers | 463, 466 |
| 1-Oxa-2,3-cyclohexadiene | 121, 125, 130 | [1.1.1]Propellane telomers, properties of | 471 |
| 1-Oxa-3,4-cyclohexadiene | 121, 124 | [1.1.1]Propellanes, functionalized | 467 |
| Oxidase | 177 | [1.1.1]Propellanes, polymerization of | 503 |
| Oxidation potentials | 158 | [1.1.1]Propellanes, thermal behavior of | 377 |
| Oxidation-reduction reactions | 177 | [2.1.1]Propellane, attempted synthesis of | 371 |
| Oxigen rebound mechanism | 183, 185 | [3.1.1]Propellane | 362, 364, 369, 370 |
| Oxiranes | 25 | | |
| Palladium(0)-catalyzed coupling | 44 | [3.3.3]Propellanes | 432 |
| [2.2]Paracyclophanediene, oxidation of | 325 | [4.1.1]Propellane | 362, 364, 370 |
| [2.2]Paracyclophanes, cyclopropanation of | 326, 327 | Prostaglandin | 251 |
| | | Pyrethrin I | 333, 343 |
| [2.2]Paracyclophanes, acid catalyzed isomerizations of | 314, 315 | Pyrethroids | 333, 346 |
| | | Pyrrolidine | 54 |
| [2.2]Paracyclophanes, Diels-Alder additio | 320 | Pyrroline annulation | 111, 113 |
| [2.2]Paracyclophanes, epoxidation of | 322 | Pyrrolizidine diols | 109, 113 |
| [2.2]Paracyclophanes, methylsubstituted | 302, 303 | Pyruvate dehydrooxidase | 193 |
| [2.2]Paracyclophanes, photochemistry of | 308 | Pyruvate oxidase | 193 |
| [2.2]Paracyclophanes, thermolysis of | 304, 305 | Quadricyclane | 143, 161, 383, 398 |
| [3]-, [4]Paracyclophane | 515 | | |
| [4.4]Paracyclophane | 515 | Quadricyclane to norbornadiene rearrangement | 143, 146 |
| Pauson-Khand annulation | 447, 449 | | |
| Penicillin, biozynthesis | 180, 201 | Quadricyclane, donor-acceptor substitute | 398, 399 |
| Pentalenene | 222, 228 | Quadricyclane, HOMO of | 144 |
| Perhydrotriquinacene-1,4,7-trione | 447 | Quadricyclane, oxidation of | 144, 185 |
| Pericyclic reactions | 267 | Quadrone | 46, 49 |
| Peterson olefination | 224 | Radical disproportionation | 149, 157, 158, 159, 163 |
| Phosphorus ylides | 338, 343 | | |
| Photochemical rearrangement | 257 | Radical intermediates, probe for | 179, 180, 185, 190 |
| Photodesilylation | 165 | | |
| Photoexcited iron atoms | 453 | Ramberg-Bäcklund ring contraction | 87 |
| $\beta$-Pinene | 214 | Reactive intermediates | 256 |
| Planar chirality | 298 | Rearrangements | 352 |
| Planarizing angular distortions | 283, 287, 289 | Rehm-Weller plot | 148, 161 |

| | | | |
|---|---|---|---|
| Relay conjugation | 99 | Spiro[2.5]octan-3-ones | 507 |
| Reporter groups | 178 | Spiro[4.4]nonatetraene | 77 |
| Retigeranic acid | 109, 111 | Spontaneous copolymerization | 503 |
| Retro-ene reactions | 26, 34 | Spread | 284 |
| RHF Closed-Shell SCF calculation | 69 | [n]Staffanes, properties of | 471 |
| Rigid-rod polymer | 463, 479, 503 | [n]Staffanes, synthesis of | 466 |
| Ring cleavage | 51, 349 | Stannylenes | 497 |
| Ring closure | 81, 439 | Stereoselective alkylation | 493 |
| C4 - C3 Ring contraction | 493 | Stereoselectivity | 56 |
| C3 - C4 Ring expansion | 493 | Strain | 349, 357, 457 |
| Ring-opening polymerization | 503 | Strain and aromaticity | 517 |
| 3-Ring | 440 | Strain controlled one bond | |
| 4-Ring | 440 | cleavage reactions | 410 |
| 5-Ring | 440 | Strain effects | 297, 307 |
| Secohexaprismane, synthesis of | 272, 274 | Strain energies | 299, 349, 358 |
| Seco-homocubanone | 423 | Strain energies | 383, 405 |
| Selenoacetals | 343 | Strain relief | 463, 495 |
| Selenoalkyllithiums | 343, 345 | Strain, harnessing of | 445, 463 |
| Selenobenzyllithiums | 345 | Strained aromatic molecules | 297 |
| Selenoxyalkyllithiums | 343 | Strained cage compounds | 405 |
| Semibullvalene | 79 | Strained compounds | 457 |
| Sesquiterpene synthesis | 40 | Strained cycloalkoxides | 513 |
| Sesquiterpenes | 431 | Strained hydrocarbons, | |
| Shapiro reaction | 447 | ionization potentials of | 144, 145 |
| Sigmahaptotropic rearrangements | 263, 267 | Strained hydrocarbons, | |
| [2,2]Sigmahaptotropic | | oxidation potentials of | 144, 145 |
| rearrangement, kinetic data | 266 | Strained molecules | 487 |
| [4,4]Sigmahaptotropic rearrangement | 263 | Strained organic compounds | 283 |
| [1,3]Sigmatropic migration | 90 | Strained polycyclic cage compounds | 259 |
| Sigmatropic rearrangement | 292, 293 | Strained rings, chemistry of | 383 |
| [1,5]Sigmatropic rearrangement | 92 | Strained systems, valence | |
| [3,3]Sigmatropic rearrangement | 92, 293 | isomerizations of | 394, 400, 402 |
| 2-Siloxyallylidenecyclopropanes, | | Structure analysis | 485 |
| Diels-Alder reactions of | 507 | Substituent directed one bond | |
| Silver trifluoroacetate | 143, 162 | cleavage reaction | 417 |
| Silyl group, $\alpha$-effect | 64 | Substrate binding | 178 |
| Single electron transfer | 143, 165, 170, 172 | Sulfur ylides | 336, 341, 342 |
| | | Surfactants | 478 |
| Single electron transfer, photoinduced | 146 | Symmetry deformation coordinates | 283 |
| Sinularene | 40 | Taxane skeleton, approach to | 109, 114 |
| Small ring formation, alignment in | 442 | TCNE | 169 |
| Small rings by ring closure | 439 | Tefluthrin | 333 |
| Small-ring propellanes, chemistry of | 361 | Tetrachlorocyclopropene | 118, 384, 509 |
| | | Tetracoordinate carbon | 283, 284 |
| Soil insecticides | 333 | Tetracyclic structures | 285 |
| Spin saturation transfer | 265 | Tetramethylallene episulfide | 62 |
| Spiro-activation | 173, 175, 176 | 2,2,5,5-Tetramethylcyclopentanethione | 71 |
| | | Thermal rearrangement | 256 |
| Spiroalkanes, bi-, tri- and tetracyclic | 286 | Thiamine pyrophosphate | 193 |
| Spiroconjugation | 77, 99 | Thietanone | 66 |
| Spirocyclopropanes | 169, 173, 175 | Thiiranoradialene | 66 |
| 5-Spirocyclopropyl-2-butenolides | 255, 256 | Thionation of methylenecyclopropanones | 67 |
| Spiro[m.n]alkanes | 283 | | |

| | | | |
|---|---|---|---|
| Thioxyallyl ion | 59, 65, 67, 70, 73 | Vinylcyclopropanes | 34, 55, 56, 109, 333, 345 |
| Through-bond interaction | 77, 78, 97 | Vinylcyclopropanols | 8, 10, 16 |
| Through-space interaction | 77 | Vinylketenes | 211, 212, 218 |
| Tinkertoys, molecular | 463 | Vinylketenes, by regioselective deprotonation | 214 |
| Torquoelectronics | 25 | Vinyloxirane rearrangement | 112 |
| Torsionally distorted double bonds | 133 | Vinyloxiranes | 109 |
| Torsionally distorted double bonds, strain energy | 134 | X-ray diffraction | 485 |
| Transannular reaction | 291 | X-ray structural analysis | 497 |
| Transesterification | 511 | X-ray structures | 284 |
| Transition structures, calculated | 441 | X-X electron deformation densities | 485 |
| Triadhelvetane | 260 | | |
| Triaminocyclopropenium ion | 384, 387 | | |
| Triaxane | 414 | | |
| Tricyclic structures | 285 | | |
| Tricyclopropabenzenes | 458 | | |
| Tricycloundecanes | 432 | | |
| Tricyclo[2.1.0.0$^{2,3}$]pentane | 287 | | |
| Tricyclo[3.1.0.0$^{2,6}$]hexane | 370 | | |
| Tricyclo[3.3.0.0$^{2,6}$]octa-3,7-diene | 79 | | |
| Tricyclo[4.1.0.0$^{2,7}$]hepta-3,4-diene | 121, 122 | | |
| Tricyclo[4.1.0.0$^{2,7}$]heptane | 149, 156, 163, 165 | | |
| Tricyclo[4.1.0.0$^{4,7}$]hept-1(7)-ene | 365 | | |
| Tricyclo[5.1.0.0$^{2,8}$]octane system | 363 | | |
| Tricyclo[5.3.0.0$^{2,8}$]deca-3,5,9-triene | 79, 86 | | |
| Tricyclo[5.5.0.0$^{2,8}$]dodecatetraene | 77, 79, 82, 85 | | |
| 1-Trimethylsilylmethylbicyclo-[3.2.0]hept-6-ene | 143, 161 | | |
| 1-Trimethylsilyltricyclo-[4.1.0.0$^{2,7}$]heptane | 144, 161, 165 | | |
| Triquinacene | 447 | | |
| Triquinane, angularly annulated | 207, 222, 225, 227, 431, 447 | | |
| Triquinane, linearly annulated | 207, 222 | | |
| Triquinanes | 431 | | |
| Trithiocyclopropenium ion | 383, 389 | | |
| Trithiocyclopropenium ions, heterocycles from | 390 | | |
| Twist | 283, 285 | | |
| Umpolung | 53 | | |
| Vicinal alkylation | 235 | | |
| Vinyl iron hydroxide | 452 | | |
| Vinylaziridine rearrangement | 111 | | |
| Vinylaziridines | 109 | | |
| Vinylcarbene | 117, 118 | | |
| Vinylcyclobutanes | 34, 79, 266 | | |
| Vinylcyclobutanones | 255 | | |
| 2-Vinylcyclobutanones | 493 | | |
| Vinylcyclopropane carboxylates | 333, 334 | | |
| Vinylcyclopropane rearrangement | 113 | | |
| Vinylcyclopropanes | 34, 55, 56, | | |